9780323912419

VOLUME ONE HUNDRED AND SIXTY TWO

CURRENT TOPICS IN DEVELOPMENTAL BIOLOGY

Fertilization and Activation of Development

CURRENT TOPICS IN DEVELOPMENTAL BIOLOGY

"A meeting-ground for critical review and discussion of developmental processes"
A.A. Moscona and Alberto Monroy (Volume 1, 1966)

VOLUME ONE HUNDRED AND SIXTY TWO

CURRENT TOPICS IN DEVELOPMENTAL BIOLOGY

Fertilization and Activation of Development

Edited by

PAUL M. WASSARMAN
Icahn School of Medicine at Mount Sinai, United States

EVELINE S. LITSCHER
Icahn School of Medicine at Mount Sinai, United States

ELSEVIER

ACADEMIC PRESS
An imprint of Elsevier

Academic Press is an imprint of Elsevier
125 London Wall, London, EC2Y 5AS, United Kingdom
50 Hampshire Street, 5th Floor, Cambridge, MA 02139, United States
525 B Street, Suite 1650, San Diego, CA 92101, United States

First edition 2025

ISBN: 978-0-323-91241-9
ISSN: 0070-2153

For information on all Academic Press publications
visit our website at https://www.elsevier.com/books-and-journals

Publisher: Zoe Kruze
Editorial Project Manager: Devwart Chauhan
Production Project Manager: Maria Shalini
Cover Designer: Arumugam Kothandan

Typeset by MPS Limited, India

Contents

Contributors

Mariano G. Buffone
Instituto de Biología y Medicina Experimental (IBYME), National Research Council of Argentina (CONICET), Buenos Aires, Argentina

Anne E. Carlson
Department of Biological Sciences, University of Pittsburgh, Pittsburgh, PA, United States

Julio C. Chávez
Departamento de Genética del Desarrollo y Fisiología Molecular, Instituto de Biotecnología, UNAM, Cuernavaca, Mexico

Alberto Darszon
Departamento de Genética del Desarrollo y Fisiología Molecular, Instituto de Biotecnología, UNAM, Cuernavaca, Mexico

Arturo Hernández-Cruz
Departamento de Neuropatología Molecular y Laboratorio Nacional de Canalopatías, Instituto de Fisiología Celular UNAM, Ciudad Universitaria, Ciudad de México, Mexico

Laurinda A. Jaffe
Department of Cell Biology, University of Connecticut Health Center, Farmington, CT, United States

Aimee Jaramillo-Lambert
Department of Biological Sciences, University of Delaware, Newark, DE, United States

Tomokazu Kawashima
Department of Plant and Soil Sciences, University of Kentucky, Plant Science Building, Lexington, KY, United States

Kayla M. Komondor
Department of Biological Sciences, University of Pittsburgh, Pittsburgh, PA, United States

Amber R. Krauchunas
Department of Biological Sciences, University of Delaware, Newark, DE, United States

Eveline S. Litscher
Department of Cell, Developmental, and Regenerative Biology, Icahn School of Medicine at Mount Sinai, New York, NY, United States

Guillermina M. Luque
Instituto de Biología y Medicina Experimental (IBYME), National Research Council of Argentina (CONICET), Buenos Aires, Argentina

Hitoshi Nishimura
Department of Life Science, Faculty of Science and Engineering, Setsunan University, Neyagawa, Osaka, Japan

Richard Oko
Department of Biomedical and Molecular Sciences, Queen's University, Kingston, ON, Canada

Enrique I. Oliver
Department of Physiology and Biophysics, Miller School of Medicine, University of Miami, Miami, FL, United States

Corie M. Owen
Department of Cell Biology, University of Connecticut Health Center, Farmington, CT, United States

Ana Romarowski
Instituto de Biología y Medicina Experimental (IBYME), National Research Council of Argentina (CONICET), Buenos Aires, Argentina

Takako Saito
Department of Applied Life Sciences, Faculty of Agriculture, Shizuoka University, Shizuoka, Japan

Hitoshi Sawada
Graduate School of Science, Nagoya University, Nagoya, Japan

Yoshihiro Shimada
Department of Life Science, Faculty of Science and Engineering, Setsunan University, Neyagawa, Osaka, Japan

Meng-Xiang Sun
State Key Laboratory of Hybrid Rice, College of Life Sciences, Wuhan University, Wuhan, Hubei, P.R. China

Peter Sutovsky
Division of Animal Sciences; Department of Obstetrics, Gynecology and Women's Health, University of Missouri, Columbia, MO, United States

Karl Swann
School of Biosciences, Cardiff University, Sir Martin Evans Building, Museum Avenue, Cardiff, Wales, United Kingdom

Claudia Sánchez-Cárdenas
Departamento de Genética del Desarrollo y Fisiología Molecular, Instituto de Biotecnología, UNAM, Cuernavaca, Mexico

Jonathon M. Thomalla
Department of Molecular Biology and Genetics, Cornell University; Department of Biomedical Sciences, Cornell University, College of Veterinary Medicine, Ithaca, NY, United States

Filip Tirpak
Division of Animal Sciences, University of Missouri, Columbia, MO, United States

Pablo E. Visconti
Department of Veterinary and Animal Sciences, University of Massachusetts, Amherst, MA, United States

Wei Wang
State Key Laboratory of Hybrid Rice, College of Life Sciences, Wuhan University, Wuhan, Hubei, P.R. China

Paul M. Wassarman
Department of Cell, Developmental, and Regenerative Biology, Icahn School of Medicine at Mount Sinai, New York, NY, United States

Mariana F. Wolfner
Department of Molecular Biology and Genetics, Cornell University, Ithaca, NY, United States

Hanxian Xiong
School of Life Science and Technology, Wuhan Polytechnic University, Wuhan, Hubei, P.R. China

Xuebing Yang
State Key Laboratory of Animal Biotech Breeding, College of Biological Sciences, China Agricultural University, Beijing, P.R. China

Hua Zhang
State Key Laboratory of Animal Biotech Breeding, College of Biological Sciences, China Agricultural University, Beijing, P.R. China

Yan Zhang
State Key Laboratory of Animal Biotech Breeding, College of Biological Sciences, China Agricultural University, Beijing, P.R. China

Yilin Zhang
Department of Plant and Soil Sciences, University of Kentucky, Plant Science Building, Lexington, KY, United States

Michal Zigo
Division of Animal Sciences, University of Missouri, Columbia, MO, United States

Preface

For many years after the discovery of sperm in males and eggs in females it was thought by "spermists" and "ovists" that either sperm alone or eggs alone, respectively, gave rise to offspring (Cobb, 2012; Farley, 1982). However, in the late 18th century Lazzaro Spallanzani (1729-1799) proposed that fertilization requires both male sperm and female ova and in the following century Eduard Strasburger (1844-1912) described fertilization in flowering plants, Oskar Hertwig (1849-1922) and Hermann Fol (1845-1892) observed penetration of sea urchin sperm into eggs, and Jacques Loeb (1859-1924) proposed that fertilization led to activation of eggs by altering the egg's ooplasm. These and other advances led to a burst of research in the 20th century on the biochemical/molecular basis of fertilization and egg activation in a variety of plants and animals.

Fertilization, the fusion of a single sperm (haploid) with an unfertilized egg (haploid), triggers *activation* of the egg and transforms it into a developing embryo (diploid) that in time gives rise to a new individual of the species. Today, several gamete components that participate in species-restricted sperm-egg recognition (e.g., zona pellucida proteins ZP2 and ZP3) and sperm-egg fusion (e.g., proteins Izumo1, Juno, CD9, and Spaca6) and several changes in the egg's ooplasm (e.g., changes in Ca^{2+}, Zn^{2+}, and phospholipase C) and the egg's proteome and cellular state (e.g., cortical reaction, zona reaction, inactivation MPF, activation APC, and changes in mRNA content, translation, and transcription) following gamete fusion are known (Elofsson, Han, Bianchi, Wright, & Jovine, 2024; Krauchunas & Wolfner, 2013; Machaty, 2024; Wassarman & Litscher, 2022). Collectively, the changes at fertilization and activation result in conversion of unfertilized eggs, paused in a late stage of meiosis, into mitotic zygotes (fertilized eggs) that are totipotent and capable of giving rise to new individuals that consist of tens-of-trillions of cells, hundreds of differentiated cell-types, and several tissue-types.

This volume of *Current Topics in Developmental Biology,* titled "Fertilization and Activation of Development", includes a dozen chapters that address many aspects of egg and sperm biology, fertilization, and egg activation in both plants and animals. Chapters 1 and 2 deal with flowering plants, Chapters 3 and 4 with ascidians and nematodes, Chapter 5 with amphibians, Chapter 6 with flies, sea urchins, etc., and Chapters 7–12 with mammals. We are grateful to all the contributors for their scholarly chapters submitted to the publisher on-time and thank Hua Zhang for providing a relevant image for the cover of the volume.

PAUL M. WASSARMAN AND EVELINE S. LITSCHER

References

Cobb, M. (2012). An amazing 10 years: The discovery of egg and sperm in the 17th century. *Reproduction in Domestic Animals, 47*(4), 2–6.

Elofsson, A., Han, L., Bianchi, E., Wright, G. J., & Jovine, L. (2024). Deep learning insights into the architecture of the mammalian egg-sperm fusion synapse. *Elife, 13*, RP93131.

Farley, J. (1982). *Gametes and Spores: Ideas about Sexual Reproduction, 1750–1914*. Baltimore, MD: Johns Hopkins University Press 312.

Krauchunas, A. R., & Wolfner, M. F. (2013). Molecular changes during egg activation. *Current Topics in Developmental Biology, 102*, 267–292.

Machaty, Z. (2024). The signal that stimulates mammalian embryo development. *Frontiers in Cell and Developmental Biology, 12*, 1474009.

Wassarman, P. M., & Litscher, E. S. (2022). Zona pellucida proteins as receptors for binding of mouse sperm to eggs. *Trends in Developmental Biology, 15*, 1–13.

CHAPTER ONE

Gamete activation for fertilization and seed development in flowering plants

Wei Wang[a], Hanxian Xiong[b], and Meng-Xiang Sun[a,*]
[a]State Key Laboratory of Hybrid Rice, College of Life Sciences, Wuhan University, Wuhan, Hubei, P.R. China
[b]School of Life Science and Technology, Wuhan Polytechnic University, Wuhan, Hubei, P.R. China
*Corresponding author. e-mail address: mxsun@whu.edu.cn

Contents

Abstract

Double fertilization is a defining feature of flowering plants, in which two male gametes (sperm cells) fuse with two female gametes (egg and central cell) to trigger embryogenesis and endosperm development. Gamete activation before fertilization is essential for the success of fertilization, while gamete activation after fertilization is the prerequisite for embryo and endosperm development. The two phases of activation are an associated and continuous process. In this review, we focus on current understanding of gamete activation both before and after fertilization in flowering plants, summarize and discuss the detailed cellular and molecular mechanisms underlying gamete activation for fertilization or initiation of embryogenesis and endosperm development.

1. Introduction

Sexual reproduction in angiosperms depends on the successful fusions of two pairs of male and female gametes during double fertilization. In this process,

Current Topics in Developmental Biology, Volume 162
ISSN 0070-2153, https://doi.org/10.1016/bs.ctdb.2024.10.009

one sperm cell fuses with the egg cell, producing a diploid embryo and the second sperm fertilizes the di-haploid central cell, giving rise to the triploid endosperm. The importance of these two fusion events is reflected by the fact that they establish the necessary basis for embryo and endosperm development in seeds, which not only passes on maternal and paternal genetic information to the next generation for population continuation of the species, but also provides majority of plant food for humans and acts as the main source for livestock feed both directly and indirectly.

Unlike animals, the sperm cells of flowering plants are immotile, and encapsulated in the male gametophyte (pollen grain or pollen tube) (Fig. 1A) (Berger & Twell, 2011; Hafidh & Honys, 2021). Transportation of the male gametes to the female gametes absolutely depends on the activity of pollen vegetative cell (Glöckle et al., 2018; Motomura et al., 2021; Zhang et al., 2017). In model plant *Arabidopsis thaliana*, pollen grains released from the anther land on and adhere to the papillae of a stigma, then hydrate and germinate there. Tip-growing pollen tube penetrates the stigma surface and navigates through tissues of pistils to reach the ovule

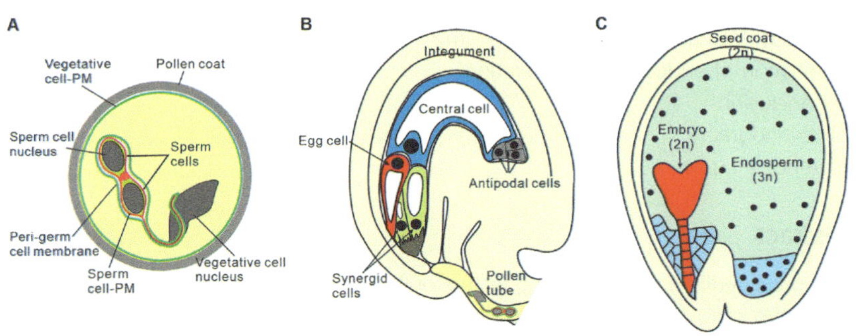

Fig. 1 Schematic representation of the mature pollen grain, ovule and seed organization in *Arabidopsis thaliana*. (A) The mature pollen grain, with a 'cell in cell' structure. Two sperm cells are surrounded by the peri-germ cell membrane (green) originating from the vegetative cell and a long cytoplasmic projection extending from one sperm cell tightly connects to the vegetative cell nucleus. Sperm cell membrane is in red, sperm cytoplasm in light green, sperm cell wall material in white (probably present), membrane of the vegetative cell in green, cytoplasm of the vegetative cell in light yellow, nuclei in gray. (B) The seven-celled female gametophyte in *Arabidopsis* ovule. The egg cell and two synergid cells are located close to the micropyle, which is the entrance for the pollen tube. The central cell with a large vacuole situated in the center and three antipodal cells are positioned on the opposing chalazal side. (C) Developing *Arabidopsis* seed after fertilization. Two fertilization products, the diploid embryo and the triploid endosperm, are wrapped by the maternal tissue seed coat.

where the female gametophyte (embryo sac) is located (Johnson, Harper, & Palanivelu, 2019). In most flowering plants, including *Arabidopsis*, the embryo sac is a seven-celled but eight-nucleate structure, which generates two dimorphic female gametes, a haploid egg cell and a homodiploid central cell, along with two additional haploid cell types, synergid and antipodal cells, at each pole (Fig. 1B) (Hater, Nakel, & Gross-Hardt, 2020; Skinner & Sundaresan, 2018). The synergid cells are close to the micropyle opening of an ovule, secrete chemical attractants and function in pollen tube guidance and reception (Higashiyama & Takeuchi, 2015; Higashiyama & Yang, 2017). The three antipodal cells are functionally unknown and situated on the chalazal side of the ovule.

The progamic phase, before the rupture of the pollen tube, is a period of specific interactions between pollen or pollen tube with maternal reproductive tissues, which is undertaken by the vegetative cell from the male side. During the process, involved mechanisms and key molecules have been well summarized (Dresselhaus, Sprunck, & Wessel, 2016; Hater et al., 2020; Higashiyama & Takeuchi, 2015; Higashiyama & Yang, 2017; Johnson et al., 2019; Zhong & Qu, 2019; Zhong et al., 2024; Zhou & Dresselhaus, 2019) and will not be discussed in this review. After sperm cells discharge from a pollen tube, live imaging of double fertilization in *Arabidopsis* showed that two gamete fusion events occurred within four to ten minutes (Hamamura et al., 2011), indicating that gamete activation, subsequent gamete adhesion and fusion must be highly regulated to be efficient enough to ensure the success of fertilization (Sprunck, 2020).

After the plasma membrane fusion of female and male gametes (plasmogamy), each sperm cell nucleus migrates towards its respective nucleus in the egg cell or central cell, and once contacted, gamete nuclear fusion occurs (Dresselhaus et al., 2016; Sprunck, 2020). Before gamete fusion, division of the egg cell and central cell is prevented, as it requires activation by paternal factors delivered from sperm cells. After fertilization, the egg cell is rapidly activated to establish the zygotic embryogenesis program to initiate development, which needs an associated process: maternal transcript clearance and zygotic genome activation (Dresselhaus & Jürgens, 2021; Zhao, Shi, Wang, & Sun, 2022). Whereas, after karyogamy, the central cell nucleus immediately undergoes mitosis, but not followed by cell division, leading to the syncytial endosperm developmental phase (Fig. 1C) (Li & Berger, 2012; Olsen, 2020).

In this review, we mainly focus on the knowledge of gamete activation before and after fertilization in flowering plants, including mutual gamete activation before plasmogamy and subsequent activation of the egg and central cell to initiate zygote development and primary endosperm cell division after karyogamy. This gained knowledge sheds new light on complex regulatory mechanisms required for the timely and successful fusion of two gametes during double fertilization, and also provides potential for applications of synthetic apomixis in crop breeding.

2. Sperm cell activation

During double fertilization, usually, just one pollen tube reaches the embryo sac, and the attraction of additional pollen tubes (polytubey) is prevented and occurs only in cases of failed fertilization (Beale, Leydon, & Johnson, 2012; Kasahara et al., 2012; Maruyama et al., 2013). Thus, the two female gametes in the ovule are confronted with only one pair of sperm and the fusion of one sperm cell with one of the female gametes must be guaranteed to avoid the risk of reproductive failure (Sprunck, 2020). During this process, a rapid fertilization time course has been monitored by live-cell imaging, in which released sperm cells immediately reach the boundary between the two female gametes (8.8 ± 5.5 s) after pollen tube discharge, and gamete fusion occurs after a period of immobility with an average time of 7.4 ± 3.3 min (Hamamura et al., 2011). Therefore, sperm cell activation events appear to occur in a rapid, efficient and coordinated manner to ensure the success of double fertilization, which includes some detailed events, e.g. peri-germ cell membrane release, cell wall degradation, locally assemblage or activation of the cell-adhesion and fusion machinery at sperm surface and subsequent sperm pair separation (Figs. 2 and 3).

2.1 Peri-germ cell membrane release and cell wall degradation

In flowering plants, the male gametophyte exhibits a unique topology, having cells within a cell, whereby the two sperm cells linked together are enclosed by the peri-germ cell membrane from the pollen vegetative cell and a tail-like protrusion extending from one of the sperm cells tightly connects to the outer membrane of the vegetative nuclear envelop (Dumas, Knox, & Gaude, 1985; McCue, Cresti, Feijó, & Slotkin, 2011; Russell & Cass, 1981; Sugi et al., 2024). The vegetative nucleus and the two sperm cells are connected in a stable association named male germ unit (MGU) (Dumas et al., 1985), which functions as a mechanistic element for delivery of the

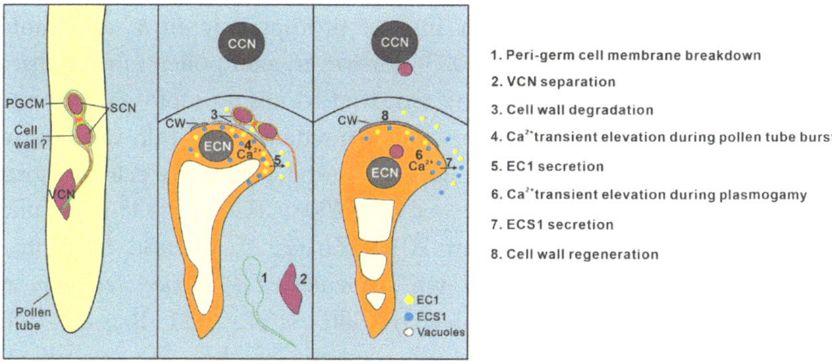

Fig. 2 Sperm and egg cell activation events during fertilization. See Sections 2.1–3.2 for details. Abbreviations: PGCM, peri-germ cell membrane; SCN, sperm cell nuclei; VCN, vegetative cell nucleus; CCN, central cell nucleus; ECN, egg cell nucleus; CW, cell wall.

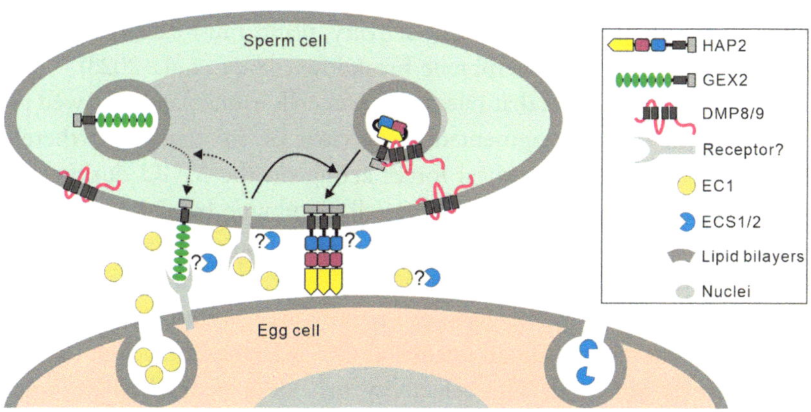

Fig. 3 Fertilization-related proteins acting at the sperm-egg cell interface. Upon sperm arrival, EC1 proteins are secreted by the egg cell, possibly via binding to a receptor on the sperm surface to activate the sperm cells. As part of sperm activation process, DMP8 and DMP9 respond to EC1 and translocate HAP2 from endomembrane compartments to the plasma membrane to facilitate gamete fusion. GEX2 is a sperm adhesion protein, possibly binding to a receptor on the egg cell. After successful plasmogamy, ECS1/2 are secreted to the extracellular space from the egg cell. Potential substrates for ECS1/2 are indicated by question marks. Proposed or identified molecular pathways are indicated by dotted or solid arrows, respectively. See Section 2.2 for details.

two sperm cells to the site of gamete interactions. During the growth phase, pollen tubes maintain an eyeglass-like peri-germ cell membrane morphology and it is assumed that the peri-germ cell membrane has a homeostasis system that resists continuously mechanical stresses induced by stretching, twisting, and moving back and forth in intense protoplasmic flow to maintain membrane integrity (Sugi et al., 2023). However, after pollen tube discharge, peri-germ cell membrane must be removed to ensure rapid gamete interactions (Sprunck, 2020). In earlier studies, in which sperm cells were isolated from mature pollen grains or pollen tubes, the peri-germ cell membrane easily ruptured in vitro (Borges et al., 2008; Russell, 1986; Sprunck, Hackenberg, Englhart, & Vogler, 2014; Zhang, Campenot, Mcgann, & Cass, 1992), and it also broke down in ovules of *Plumbago zeylanica*, and maize after sperm cell delivery in vivo (Gilles et al., 2021; Russell, 1983). Recently, the timing of peri-germ cell membrane breakdown was clarified in *Arabidopsis thaliana*, i.e., fragmented peri-germ cell membrane around sperm nuclei occurs 1 min after pollen tube discharge (Sugi et al., 2023). Considering that gamete mergers occur within 7.4 $\pm$ 3.3 min, even a slight delay of peri-germ cell membrane fragment perhaps reduces the success rate of fertilization. Since precocious peri-germ cell membrane disappearance from sperm cells is detrimental for sperm delivery, plants must exactly control the timing of peri-germ cell membrane breakdown (Sugi et al., 2023).

Based on the findings that released sperm cells immediately showed peri-germ cell membrane fragmentation in vitro or in vivo, it is therefore assumed that strong mechanical forces acting on the MGU during pollen tube discharge may trigger peri-germ cell membrane breakdown; whereas, osmotic stress, ROS and Ca^{2+} signal, arising simultaneously with pollen tube discharge could not be ruled out (Sprunck, 2020; Sugi et al., 2023). During the removal of peri-germ cell membrane, sperm plasma membrane closely apposed must maintain intact and thus high membrane selectivity for breakdown is necessary. This selectivity may be explained by the unique phospholipid composition of peri-germ cell membrane that phosphatidylinositol-4,5-bisphosphate is enriched, or the peri-germ cell membrane special location of molecules that are required for membrane elimination (Gilles et al., 2021; Sugi et al., 2023). Up to now, only four proteins, two small GTP-binding rho of plants (PsROP1 and AtROP9), LA and related protein 6c (AtLARP6c), and MATRILINEAL (MTL)/NOT LIKE DAD (NLD)/ZmPHOSPHOLIPASE-A (ZmPLA1) have been demonstrated to be peri-germ cell membrane localized (Billey et al., 2021; Gilles et al., 2021; Li et al., 2013; Lin, Wang, Zhu, & Yang, 1996). Among the limited number of proteins, NLD is the only one that specifically localizes to the peri-germ cell membrane in pollen, depending on lipid anchoring via S-acylation together with electrostatic interactions (Gilles et al., 2021). The absence of NLD induces haploid embryos with solely maternal genome

(Gilles et al., 2017; Kelliher et al., 2017; Liu et al., 2017). However, whether it functions in peri-germ cell membrane elimination remains unknown. Thus, identifying novel peri-germ cell membrane specific proteins and validating their function in selective peri-germ cell membrane breakdown, will undoubtedly lead to the progress in understanding the triggering signals and underlying mechanisms.

In addition to the peri-germ cell membrane, the extracellular matrix that surrounds sperm cells perhaps acts as another putative physical obstacle to prevent the lipid bilayers of two opposite sex into close apposition, and thus need to be removed during the process of gamete activation. This has been confirmed in the unicellular green alga, *Chlamydomonas reinhardtii*, in which activated gametes triggered by flagellar adhesion secrete a serine protease that converts an extracellularly-stored pro-metalloprotease to an active matrix-degrading enzyme for cell wall shedding (Kinoshita, Fukuzawa, Shimada, Saito, & Matsuda, 1992; Pan & Snell, 2000; Snell, Eskue, & Buchanan, 1989). In flowering plants, although the ultrastructural studies in *Papaver rhoeas* and *Plumbago zeylanica* reported that the generative cell and sperm cells possess a thin extracellular matrix, and the sperm-associated cell wall components was already removed from the sperm cell surface inside *Plumbago* ovules (McCue et al., 2011; Russell & Cass, 1981; Russell, 1983), it is not yet clear whether *Arabidopsis* sperm cells are coated by any cell wall material when delivered to the egg-central cell surface. How the sperm cell wall is removed during double fertilization in flowering plants is still a mystery.

All in all, rapid peri-germ cell membrane breakdown and cell wall degradation could be the initial step in sperm cell activation, which expose the sperm cell surface beforehand to facilitate adhesion, recognition, and fusion with the egg or central cells. Thus, analyzing the mechanisms to expose plasma membrane upon sperm cell activation after pollen tube discharge will become an important issue for understanding double fertilization in future research.

2.2 Gain of adhesion and fusion competence-four fertilization essential proteins

Fertilization depends on the fusion of two haploid gametes of opposite sex. Although the details of fertilization are highly variable for the different plant taxa, the common principles for cell–cell fusion in general also apply for flowering plants. During pre-fusion attachment, adhesion proteins bring the opposing plasma membranes into close proximity (a distance of ~10 nm), and the following physical merging of the membranes is mediated by fusogens, which have the ability to overcome energy barriers that impede fusion (Brukman, Uygur, Podbilewicz, & Chernomordik, 2019; Hernandez & Podbilewicz, 2017; Martens & McMahon, 2008).

In *Arabidopsis*, to date, four fertilization-essential proteins have been identified acting on the sperm surface (Fig. 3). The adhesion factor discovered in flowering plants is GENERATIVE EXPRESSED2 (GEX2), as evidenced by that *gex2* sperm cells frequently failed to attach to the female gamete in an in-situ adhesion assay (Mori, Igawa, Tamiya, Miyagishima, & Berger, 2014). The adhesion proteins, *Chlamydomonas* plus-gamete specific expressed FUS1 and mammalian sperm protein IZUMO1, share a common domain architecture with GEX2, consisting of an ectodomain constituted by Ig-like domains, a transmembrane domain, and a short cytoplasmic tail (Ferris, Woessner, & Goodenough, 1996; Inoue, Ikawa, Isotani, & Okabe, 2005; Misamore, Gupta, & Snell, 2003; Pinello, Liu, & Snell, 2021). This indicates that during evolution reproductive proteins often show rapid changes in their primary amino acid sequences, while the structural backbone may remain relatively unchanged (Pinello et al., 2021). Since disruption of a single copy GEX2 did not completely block double fertilization, as evidenced by 50% reduction in seed set in homozygous *gex2* mutants (Mori et al., 2014), other as-yet unidentified sperm membrane proteins may act alone or redundantly with GEX2. It is also conceivable that there is a partner protein of GEX2 on the surface of the egg and central cell, comparable to mammalian IZUMO1 binding to its counterpart egg receptor JUNO (a glycophosphatidylinositol-anchored folate receptor 4) (Bianchi, Doe, Goulding, & Wright, 2014; Ohto et al., 2016), and *Chlamydomonas* FUS1 forming a receptor pair with a newly identified, lineage-restricted protein Minus Adhesion Receptor 1 (MAR1) on minus gametes (Pinello et al., 2021).

The conserved membrane protein HAPLESS 2 (HAP2), also called GENERATIVE CELL SPECIFIC 1 (GCS1), turned out to be essential for gamete fusion in many unicellular organisms, in several animal taxa and in higher plants (Camacho-Nuez et al., 2017; Cole et al., 2014; Ebchuqin et al., 2014; Hirai et al., 2008; Kawai-Toyooka et al., 2014; Liu et al., 2008; Mori, Kuroiwa, Higashiyama, & Kuroiwa, 2006; Okamoto et al., 2016; von Besser, Frank, Johnson, & Preuss, 2006). The findings that *hap2* gametes are able to adhere but fail to fuse and the HAP2 extracellular domain sharing structural similarity with class II viral membrane fusion proteins fully demonstrate its function as a fusogen (Fedry et al., 2017; Mori et al., 2006; Mori et al., 2014; Pinello et al., 2017; Valansi et al., 2017). The structural studies along with in vivo evidence indicate that HAP2 undergoes transitions from a prefusion form into homotrimers during the fusion reaction (Fedry et al., 2017; Zhang et al., 2021). Although the core structure of HAP2 is conserved across eukaryotes, the membrane insertion motif to drive gamete fusion is evolutionarily diversified, reflecting accommodation to the diversity of membrane environments it has encountered during eukaryotic evolution (Fedry et al., 2018).

Beside the fusogen HAP2/GCS1, two Domain of Unknown Function 679 (DUF679) Membrane Proteins (DMP8 and DMP9), with redundant roles in gamete fusion, have been discovered in *Arabidopsis* sperm cells (Cyprys, Lindemeier, & Sprunck, 2019; Takahashi et al., 2018). *Arabidopsis* DMP8 and DMP9 are small tetra-spanning membrane proteins and localized to the sperm plasma membrane. Impairing gene function of *DMP8* and *DMP9* through either RNAi, Crispr-Cas9 or T-DNA insertion, could result in fertilization failure, as indicated by one to four unfused *dmp8 dmp9* sperm cells appearing in unfertilized ovules (Cyprys, Lindemeier, & Sprunck, 2019; Takahashi et al., 2018; Wang, Xiong et al., 2022; Xiong, Wang, & Sun, 2021). Importantly, the attachment assay using female gamete protoplasts showed that DMP8/9 function at the fusion step of fertilization rather than at gamete adhesion step. Moreover, single-fertilization events with the stronger tendency towards the central cell were observed in *dmp8 dmp9* mutants, leading to a conclusion that sperm-egg fusion is more dependent on DMP8/9 than sperm-central cell fusion (Cyprys et al., 2019). These results may imply tighter restrictions for sperm-egg cell fusion.

Since HAP2/GCS1 in *Arabidopsis* is stored on endomembrane compartments, *Arabidopsis* sperm are not competent for fusion when they first arrive in the vicinity of the egg cell. Thus, regulated exocytosis is required for of HAP2 delivery to the PM during the process of double fertilization, which is induced by five egg cell–secreted, small, cysteine-rich proteins, termed EC1.1 to EC1.5. EC1-GFP fusion proteins accumulate in spherical vesicle-like structures within the egg cell before fertilization, and were detected extracellularly only after sperm delivery, especially in the apical region of the degenerating synergid cell, where gamete fusion takes place (Sprunck et al., 2012). The five EC1 family members act redundantly and only in *4xec1* mutants and triple *ec1.1/4/5* knockout mutants heterozygous for the RNAi construct downregulating *EC1.2* and *EC1.3* (*ec1*–RNAi/+), seed development is affected (Rademacher & Sprunck, 2013; Sprunck et al., 2012; Wang et al., 2024). Importantly, EC1 peptides exogenously applied to sperm cells could stimulate the relocation of HAP2 from internal membranes to the surface of sperm cells, indicating that the egg cell secretes EC1 proteins to achieve rapid sperm activation and, thus, in the timely exposure of fusogenic membrane regions to acquire fusion competence. Recently, three different homozygous *5xec1* lines were generated and a severely reduced seed set as low as 13–14 % was observed, indicating a central role of EC1 proteins in double fertilization. Notably, the fertility is not completely abolished in the absence of sperm-activating EC1 proteins, which may either point to the presence of another as-yet unidentified mechanism for sperm activation or to the possibility that additional fusion machinery different from HAP2 exists in the central cell (Wang et al., 2024). Furthermore, a recent study showed that *Arabidopsis* thaliana DMP8/9 act downstream of EC1 and are required for EC1-induced redistribution of HAP2 to the sperm cell surface. DMP9 interacts with HAP2 and colocalizes with HAP2 on organelle-like structures in sperm cells. The formation of PM patches with HAP2 and DMP9 at sperm surface is indicative of an EC1-induced local formation of fusion competent sites (Wang, Xiong et al., 2022).

In addition to the delivery of HAP2 to sperm surface, the fusogenic conformational change of HAP2 is also necessary for its membrane insertion capacity. In *Chlamydomonas*, during the gamete membrane fusion reaction, HAP2 undergoes a fusogenic conformational change to become an SDS-resistant trimer to mediate gamete fusion, as evidenced by the findings that although the L310E and the L448E (hydrophobic residues at the trimer interface) minus gametes were fully capable of undergoing mating structure adhesion when mixed with WT plus gametes, the SDS-resistant trimeric form failed in both, and both mutant gametes were unable to fuse (Zhang et al., 2021). More importantly, HAP2/GCS1 reconfiguration is initiated during gamete adhesion when MAR1 binds to the adhesion protein FUS1, indicating that species-specific membrane attachment is the gateway to the activation of the fusion machinery (Pinello et al., 2021; Zhang et al., 2021). Although gamete adhesion can be functionally and biochemically separated from membrane fusion, they seem to be closely linked. Whether species-specific adhesion initiating the fusogenic reorganization of HAP2 also holds true for flowering plants need to be investigated in future study.

Sperm adhesion is also compromised in *ec1*-RNAi ovules (Cyprys et al., 2019), indicating that EC1 proteins serve a twofold function during the activation of male gametes, mediating the re-localization of HAP2 and recruiting sperm adhesion factors to sperm surface. Thus, it remains to be investigated whether GEX2 or unknown sperm adhesion factors act downstream of EC1. The findings that sperm pair remain connected in the absence of EC1 (Mori et al., 2014; Sprunck et al., 2012), and the fact that the separation of fusion-defective *gcs1*(*hap2*) sperm cell is frequently impaired (Mori et al., 2014), suggest that sperm pair separation depends on the initiation of membrane fusion. Since sperm cells can separate from each other in egg cell or central cell single fertilized ovules (Cyprys et al., 2019; Mori et al., 2014; Ron, Saez, Williams, Fletcher, & McCormick, 2010), it can be speculated that either fusion in double fertilization can induce the separation of the two sperm cells. It seems that egg and central cells have evolved their own signaling pathways to regulate precisely timed sperm separation, in order to avoid fusion of both sperm cells with one of the female gametes.

Recently, EC1 has been shown to be responsible for egg-cell dependent preferential fertilization (Wang et al., 2024). Traditionally, based on the concept of sperm dimorphism, preferential fertilization has been reported in *Plumbago* and *Zea mays*, in which two sperm cells have preferential fusion targets for fertilization (Chen, Hu, Guo, & Guan, 1995; Roman, 1948; Russell, 1985; Saito et al., 2002). However, most angiosperms generate two isomorphic sperm cells and the two isomorphic sperm cells are shown to be functionally equivalent in *Arabidopsis*, each of them is capable of fertilizing both female gametes (Hamamura et al., 2011; Ingouff et al., 2009; Kong, Lau, & Jurgens, 2015; Pagnussat, Yu, & Sundaresan, 2007). Thus, it was unclear whether there is priority between the egg cell and central cell fertilization. Using different mutants whose pollen yield a single sperm-like cell (SLC), it was found that the egg cell is preferentially fertilized in *Arabidopsis* double fertilization (Li, Hou et al., 2022; Wang et al., 2024). Although different types of SLC are functionally equivalent in their ability to fertilize the egg and central cell, each type of SLC clearly prefers fusion with the egg cell in the presence of both female gametes (Wang et al., 2024). This indicates that the observed preferential fertilization depends on the characteristics of the egg cell. Furthermore, a certain threshold level of secreted EC1 proteins has been shown to contribute to the preferential fertilization (Wang et al., 2024). Considering the observation that signal intensities of secreted EC1–GFP are highest at the egg cell membrane during gamete interaction and fusion, and that EC1 proteins act as promising signaling factors to regulate the local activity of the adhesion/fusion machinery at the sperm cell membrane (Sprunck et al., 2012), the difference in the speed or the ability of the gain of adhesion and fusion competence due to the spatial position of the sperm pair at the site of direct gamete interactions may explain why sperm–egg fertilization is prioritized.

Interestingly, four fertilization-essential proteins related to gamete activation have evolved across flowering plants and their functions in fertilization seem to be conserved. The seed set in *dmp8/9* mutants was fully restored by DMP8/9-like proteins from tobacco, soybean, and *Amborella trichopoda*, providing evidences for that DMP8/9-like proteins from flowering plant species can functionally replace each other during fertilization (Wang, Xiong et al., 2022). Moreover, egg-cell-expressed AmTr*EC1*-like from *A. trichopoda* could rescue seed set in the *5xec1* mutant (Wang et al., 2024). Since EC1 proteins are also expressed in egg cells of basal angiosperms and grasses, it is therefore likely that that EC1-induced sperm activation exists already in the most recent common ancestor of all extant flowering plants and this might be a general phenomenon in angiosperms (Chen et al., 2017; Flores-Tornero et al., 2019; Sprunck et al., 2014; Sprunck, 2020).

3. Egg cell activation

To date, our knowledge of egg cell activation in angiosperms is very restricted (Dresselhaus & Jurgens, 2021; Peng, Li, Ding, & Tian, 2018; Sukawa & Okamoto, 2018). In this section, we focus on the changes that occur in the egg cell from sperm release to fusion of the gametic nuclei, and also include post-fertilization activation of egg cells which initiates zygote development (Figs. 2 and 4A). While based mainly on investigations of in vivo fertilization, to provide a richness, studies of in vitro fertilization are also tapped into our discussion.

3.1 Egg cell activation before fertilization

Egg cell activation might begin with a detectable transient calcium elevation correlated with sperm cell release (Dresselhaus & Jurgens, 2021). In vivo studies in *Arabidopsis* have elucidated a transient Ca^{2+} spike before sperm–female gamete fusions, correlated with mechanic deformation of female gametes during pollen tube burst (Denninger et al., 2014; Hamamura et al., 2014). In the absence of pollen tube discharge, no Ca^{2+} spike was detected. When a gamete fusion mutant *gcs1/hap2* was used, both female gametes respond with a single Ca^{2+} wave to pollen tube burst, indicating that the first level of female gamete activation is achieved by the torrent of the explosively released pollen tube contents (Denninger et al., 2014; Hamamura et al., 2014). The function of this Ca^{2+} elevation remains

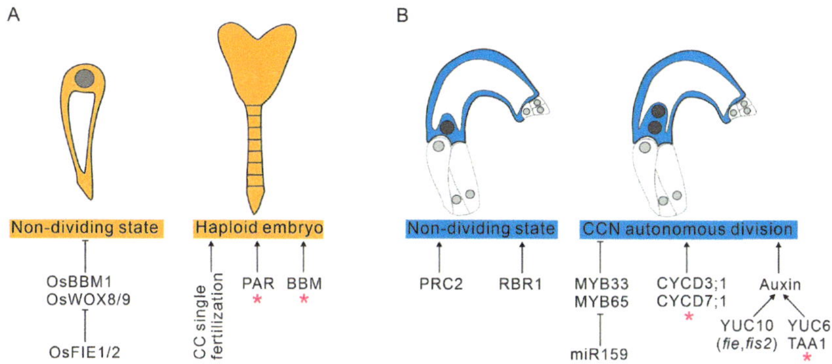

Fig. 4 The regulatory pathways of fertilization-independent embryogenesis and endosperm nucleus proliferation. (A) The regulation network inhibiting egg cell division before fertilization and signaling pathway promoting haploid embryogenesis without the fusion of a sperm cell with an egg cell. (B) The regulation of endosperm nucleus autonomous proliferation. Asterisks indicate the ectopic gene expression in the egg or central cell. See Sections 3.3.2 and 4 for details.

unclear, and it is likely to serve as a trigger for the secretion of fertilization-related protein from the egg cell, such as the sperm cell activator EC1 mentioned above (Dresselhaus et al., 2016).

The target area for sperm entry and fusion during fertilization is located at the chalazal region of the egg cell, and this area should show two features associated with fertilization: absence of a cell wall before gamete fusion, and formation of a new cell wall after the fusion. Whether the mature egg cell at the central cell-adjoining site is covered by any cell wall material seems to be variable across flowering plants. In *Arabidopsis*, this cell wall boundary is discontinuous, as indicated by the presence of thin furrows in between the thick cell wall structures (Tekleyohans, Mao, Kagi, Stierhof, & Gross-Hardt, 2017). Microtubule depletion domain 1 (MIDD1), which are necessary for the formation of pits in the xylem cell wall by eliminating cortical microtubules from the plasma membrane, displays polarized patches at the boundary between the central cell and the egg apparatus, suggesting the involvement of MIDD1 in regulating the nonuniform distribution of the cell wall materials (Oda & Fukuda, 2012; Tsutsui, Sato, Susaki, & Higashiyama, 2019). Therefore, further studies are necessary to elucidate whether pollen tube discharge, short Ca^{2+} spike in gametes, or gamete interaction triggers the local secretion of cell-degrading enzymes to remove residual cell wall material at the egg or central cell fertilization site as part of the gamete activation process.

3.2 Egg cell activation at fertilization

In animals, sperm entry induces Ca^{2+}- waves or repetitive oscillations in the oocyte/egg, which is generally considered to be the main signature of egg activation (Kashir, Deguchi, Jones, Coward, & Stricker, 2013; Satouh & Ikawa, 2018). In response to such Ca^{2+} elevations, the oocyte/egg becomes activated and progresses through the cell cycle while also undergoes such processes as cortical granule exocytosis, polyspermy prevention, and pronuclear development (Kashir et al., 2013).

Live imaging of Ca^{2+} dynamics in the cytosol of *Arabidopsis* egg cells have shown that a second Ca^{2+} transient occurs during successful plasmogamy, starting from the chalazal region of the egg cell and lasting only 3 to 5 min on average (Denninger et al., 2014; Hamamura et al., 2014). In vitro fertilization (IVF) experiments in maize also show Ca^{2+} influx induction after egg-sperm fusion, which spreads as a wave from the fusion site to the entire zygote (Antoine et al., 2000; Digonnet, Aldon, Leduc, Dumas, & Rougier, 1997). During IVF, cell wall material was detected after gamete fusion, and the establishment of a cellulosic cell wall can be achieved by Ca^{2+} ionophore induced Ca^{2+} influx without sperm cell involvement (Antoine et al., 2000; Kranz, Vonwiegen, & Lorz, 1995). Thus, it is very likely that that the second Ca^{2+} transient elevation in vivo mediates the renovation of the egg's extracellular matrix by inducing cell wall formation, thereby establishing a physical block to polyspermy, functionally comparable to the fertilization envelope in animals (Dresselhaus et al., 2016; Tekleyohans et al., 2017). In addition, the second Ca^{2+} elevation appear to be associated with karyogamy, as indicated by the findings that during IVF in rice an artificial increase in cytosolic Ca^{2+} can mimic additional sperm entry to accelerate the fusion of two egg nuclei (Ohnishi, Kokubu, Kinoshita, & Okamoto, 2019).

Upon fusion with a sperm cell, the egg cell also can aid in fortifying the polyspermy block by restricting the sperm number arriving at the female gametes as a sign of egg cell activation. Two aspartic proteases EGG CELL-SPECIFIC1 (ECS1) and ECS2 have been reported to be secreted from the egg cell to prevent polytubey by cleaving the pollen tube attractor LURE1 (Yu et al., 2021). The secretion of ECS1 and ECS2 is tightly regulated, as secretion of ECS1-GFP and ECS2-GFP by the egg cell is detectable only after fertilization (Susaki et al., 2021) and no secretion signals is visible when pollinated with *hap2/+* mutants whose sperm cells are able to adhere to but fail to fuse with the egg cell (Yu et al., 2021). Thus, plant egg cells have established a relatively fast post-fertilization block to polytubey by sensing successful fertilization and secreting proteases responsible for degradation of attraction factors. Synergids in embryo sacs are key to the attraction of pollen tubes (Higashiyama et al., 2001; Okuda et al., 2009; Takeuchi & Higashiyama, 2012). Thus, the disintegration of the persistent synergid is vital to the termination of pollen tube attraction. While fertilization of the central cell induces fusion between the persistent synergid and endosperm, the EIN3 signaling pathway, but not the ethylene pathway, is predominantly activated by the egg cell fertilization to trigger disintegration of the synergid nucleus (Heydlauff et al., 2022; Li, Lyu et al., 2022; Maruyama et al., 2015; Völz, Heydlauff, Ripper, von Lyncker, & Gross-Hardt, 2013).

The deactivation or degradation of gamete adhesion and fusion machinery after fertilization can also act as the membrane block to polyspermy. JUNO, the receptor of sperm cell surface protein IZUMO, disappears from the egg plasma membrane within 40 min after fertilization, suggesting its involvement in the establishment of the block to polyspermy (Bianchi et al., 2014). Similarly, in *Chlamydomonas* the adhesion pair FUS1/MAR1 and the fusogen HAP2 become nearly undetectable after gamete fusion, rendering the gamete resistant to further fusion events (Liu, Misamore, & Snell, 2010; Pinello et al., 2021). Compared with the central cell, the egg cell is more stringent to prevent polyspermy (Grossniklaus, 2017; Nagahara, Takeuchi, & Higashiyama, 2021; Scott, Armstrong, Doughty, & Spielman, 2008). Therefore, it will be necessary to confirm whether the similar mechanism for blocking polyspermy is present in flowering plants.

In brief, in comparison with studies in animals, the research about egg activation for fertilization in plants is, despite recent advances, still in its infancy. It is known that this process is tightly associated with Ca^{2+} elevations and secreted proteins induced by fertilization. Several fundamental questions await future investigation, such as how the egg cell senses the release of sperm cells, what is the function of Ca^{2+} oscillations, how cell wall material is removed and regenerated, and how EC1 and ECS1 are triggered to secrete from the egg cell, etc.

3.3 Egg cell activation after karyogamy

After fusion of the gametic nuclei, the egg cell becomes a zygote, which is the starting point for embryogenesis and initiates a regulated cell division program to develop into a mature embryo (Lau, Slane, Herud, Kong, & Jürgens, 2012). Although it is widely accepted that the combination of parental factors could trigger embryogenesis, how this happens in higher plants is poorly understood until recent great progress that has been made in zygotic genome activation (ZGA) and parthenogenesis.

3.3.1 *Zygotic genome activation (ZGA)*

The maternal-to-zygotic transition is a critical part of early post-fertilization development (Dresselhaus & Jurgens, 2021; Tadros & Lipshitz, 2009; Zhao & Sun, 2015). In animals, early embryo development is controlled by the maternal mRNAs and proteins stored in the oocyte until zygote genome activation that takes place after one to several cell divisions depending on the species (Jukam, Shariati, & Skotheim, 2017; Schier, 2007; Tadros & Lipshitz, 2009; Vastenhouw, Cao, & Lipshitz, 2019). In contrast to animals, the comparative transcriptome analysis of egg cells and zygotes in the eudicot *Arabidopsis* and monocots rice and maize, have revealed that ZGA in flowering plants is rapidly initiated and occurs shortly after fertilization (Anderson et al., 2017; Chen et al., 2017; Schulz & Harrison, 2019; Zhao et al., 2019). In *Arabidopsis*, ~13 % of genes in the genome began to be transcribed before the first zygotic division. Upregulated genes in early zygotes (14 h after pollination) are enriched in DNA transcription and RNA biosynthesis, while upregulated genes in elongated zygote (24 h after pollination) encode proteins involved in cell division related pathways, suggestive of two successive activation waves during ZGA (Zhao et al., 2019). In maize zygotes at 12 h after pollination, 3605 genes (9.1 % of annotated genes) were found to be upregulated compared with their expression in the egg and sperm (Chen et al., 2017). In rice, thousands of genes are differentially regulated after fertilization but only 181 genes are *de novo* expressed in zygotes (Anderson et al., 2017). In addition, treatment with drugs like a-amanitin blocking transcription of the zygotic genome revealed that ZGA is required for zygote elongation and division (Zhao et al., 2011; Zhao et al., 2019). Thus, ZGA in flowering plants occurs at the zygotic stage to generate novel transcripts, which are essential for zygote development and asymmetric zygote division.

Paternal genome activation after fertilization and its involvement in early embryogenesis is another critical question to be answered. In *Arabidopsis*, early reports indicate that at 2–4 celled and globular stage embryos the transcripts were maternally derived (Alaniz-Fabián, Orozco-Nieto, Abreu-Goodger, & Gillmor, 2022; Autran et al., 2011). There is also genetic evidence that 40 of 49 paternal alleles of EMBRYODEFECTIVE (EMB) genes could not fully complement the lack of maternal function until 3–5 days after pollination, which indicating delayed paternal genome activation in early embryogenesis (Del Toro-De León, García-Aguilar, & Gillmor, 2014). Whereas, several hybrid transcriptome analyses of *Arabidopsis* revealed paternal contributions in elongated zygote and early embryo (Nodine & Bartel, 2012; Zhao et al., 2019), which was further demonstrated in the transcriptomes of both apical and basal cell lineages of early proembryos (Zhao et al., 2020). In rice, allelic specific transcriptome analysis revealed that early zygotic transcripts were preferentially maternally derived, and this is supported by male genome methylation, which is predominantly remodeled to match maternal allelic levels upon fertilization and persists after the first zygotic division (Anderson et al., 2017; Liu et al., 2023). Although it is no doubt that paternal genome activation occurs shortly after gamete fusion in plants, to what extent paternal genome contributes to early embryogenesis remains to be elucidated.

3.3.2 Egg cell activation for fertilization-independent embryogenesis

In a process known as haploid induction, embryo development can also be activated without egg cell and sperm uniting. In recent decades, great attention has been paid to the formation of haploid embryos due to their important applications in crop breeding. Haploid breeding technology is a powerful technique to improve plant breeding efficiency through the production of doubled haploids to rapidly produce homozygous plants, in which the breeding time can be shorten to several generations (Jacquier et al., 2020; Shen, Qu, & Zhao, 2023).

Except the hypothesis that in vivo haploid embryo development occurs after successful double fertilization followed by uniparental chromosome elimination in the zygote, another potential mechanism to explain the haploid formation is a single fertilization-induced haploid, in which the egg cell develops into an embryo without fertilization, while the central cell is fertilized normally to form the endosperm (Fig. 4A) (Jacquier et al., 2020; Shen et al., 2023). ZmDMP, a paternal player in haploid embryogenesis, has been identified from maize. Although a knockout solely in *ZmDMP* induced a haploid induction rate at 0.1–0.3 %, similar to spontaneous haploid induction in maize, it can significantly increase haploid induction rates from 2 % to about 10 % in combination of *MATRILINEAL/NOT LIKE DAD/ZmPHOSPHOLIPASE-A1* mutation (Zhong et al., 2019). Depletion of the gamete-fusion-related DMP8/9 in *Arabidopsis* or orthologs in multiple dicotyledonous plants also leads to the production of maternal haploids, indicating the conserved role of DMP in egg cell activation and haploid induction in flowering plants (Li, Xiao et al., 2022; Long et al., 2024; Wang, Xia et al., 2022; Zhang, Yin et al., 2022; Zhang, Zhang et al., 2022; Zhao, Yuan et al., 2022; Zhong et al., 2020; Zhong, Chen et al., 2022; Zhong, Wang et al., 2022). Recently, *Arabidopsis KOKOPELLI* (*KPL*), known to lead to single fertilization events, have been shown to trigger in planta maternal haploid induction at a haploid induction rate of 0.34% (Jacquier et al., 2023). In *ecs1 ecs2* mutants, wild type sperm could fuse with egg cell but its nucleus failed in fusion with egg cell nucleus, which also results in egg cell activation and haploid offspring generation (Mao et al., 2023; Zhang et al., 2023). All together, these observations suggest that the mutants impaired in gamete interactions, especially resulting in single-fertilization with the central cell, might induce haploid embryos. However, detailed genetic and cytologic dissection of the egg cell activation in these cases are obviously required.

Egg cells can also bypass fertilization to be triggered for embryogenesis by ectopic expression of a transcriptional factor BABY BOOM (BBM). In naturally apomictic *Pennisetum squamulatum*, *PsASGR-BABY BOOM-like* (*PsASGR-BBML*) gene is expressed in the egg cells before fertilization and govern parthenogenesis (Conner, Mookkan, Huo, Chae, & Ozias-Akins, 2015). In addition to triggering somatic embryogenesis, ectopic expression of the *PsASGR-BBML* or *Setaria italica BBM* in the egg cell can trigger egg cells to initiate embryogenesis in diverse species, including tobacco, pearl millet, maize, and rice (Boutilier et al., 2002; Chahal, Conner, & Ozias-Akins, 2022; Conner et al., 2015; Conner, Podio, & Ozias-Akins, 2017; Zhang, Conner, Guo, & Ozias-Akins, 2020). It is further found that parthenogenesis is induced, when rice endogenous *OsBBM1* and *OsBBM4*, or maize endogenous *ZmBBM1* and *ZmBBM2* are ectopically expressed in the rice or maize egg cells (Khanday, Skinner, Yang, Mercier, & Sundaresan, 2019; Qi et al., 2023; Skinner, Mallari, Zafar, Cho, & Sundaresan, 2023; Vernet et al., 2022; Wei et al., 2023). Similarly, in dicots *Arabidopsis*, *Brassica napus* and *Solanum lycopersicon*, ectopic *BBML* expression in the egg cell is sufficient to bypass the fertilization requirement for embryogenesis, even if the haploid induction rate is low (Chen et al., 2022). All of these data suggest that BBM could be a trigger of embryogenesis in flowering plants. More importantly, de novo expression of *OsBBM1* in the rice zygote is initially expressed from the male gamete, indicating the existence of a mechanism that male-borne factor can activate expression of key ZGA factor genes for triggering embryogenesis (Khanday et al., 2019). Thus, the transcription factors, that exhibit paternal allele-specific *de novo* expression, might be the prime candidates for triggering the egg cell activation to initiate embryogenesis in flowering plants, especially in crops (Dresselhaus & Jurgens, 2021).

To maintain normal seed development, prevention of precocious egg cell activation for embryogenesis is necessary before fertilization. Heterologous expression of *PARTHENOGENESIS* (*PAR*) gene from apomictic dandelion in sexually reproductive lettuce, could induce haploid embryo-like structures in the absence of fertilization, and overexpression of *PAR* in foxtail millet could also create haploid plants (Huang et al., 2024; Underwood et al., 2022). PAR encodes a K2–2 zinc finger domain protein, with an ethylene-responsive element-binding factor-associated amphiphilic repression (EAR) motif, and may function as a transcriptional repressor. There are two *PAR* homologs in *Arabidopsis*, named *DAZ3* and *TREE1*, which are abundantly expressed in sperm cells, but not in egg cells. More strikingly, *DAZ3* messenger RNA is rapidly degraded in fertilized egg cell after delivery by male gametes, as indicated by that its transcripts could be detected in zygotes 14 h after fertilization but not 24 h after fertilization (Zhao et al., 2019). Thus, in sexual reproduction, paternal PAR protein may release a block on embryogenesis through suppressing inhibitors of the embryo developmental program. Whether PAR can act in parallel or synergistically with the transcription activator BBM to mediate fertilization-independent egg cell activation and parthenogenesis, needs to be further investigated.

Prevention of autonomous division of the egg cell before fertilization is also related to epigenetic modification. A recent study in rice has revealed a novel function of two Fertilization-Independent Seed-Polycomb Repressive Complex2 (FIS-PRC2) genes, *OsFIE1* and *OsFIE2* in suppressing cell division of the egg apparatus in the absence of fertilization (Wu et al., 2023). In single *Osfie2* or *Osfie* double mutants, ovules display asexual pre-embryo-like structures and transcriptomic analysis showed that male genome-expressed *OsBBM1* and *OsWOX8/9* were activated in the asexual embryos. This suggests that rice egg cell adopt PRC2 to maintain the repressive state before fertilization, possibly through silencing of the maternal alleles of male genome-expressed genes (Wu et al., 2023). Recently, Chromomethylases3b (CMT3b) was also proposed to prevent the precocious expression of zygotic genes in rice egg cells through DNA methylation, as indicated by those 83 zygotic genes upregulated in *cmt3b* egg cell (Li et al., 2024). It is tempting to investigate whether epigenetic modification may function as a universal mechanism in flowering plants to maintain the egg cell quiescent before fertilization to ensure reproductive integrity.

4. Initiation of endosperm proliferation

In sexual plants, the central cell resumes cell division in a fertilization-dependent manner. A body of knowledge that explains how the central cell is prevented from cell division prior to fertilization has been built up in *Arabidopsis* and rice (Fig. 4B). Notably, the pivotal role of preventing autonomous cell division is played by the PRC2 related genes, which catalyzes the trimethylation of histone H3 on lysine 27 (H3K27me3) and mainly represses the expression of maternally inherited genes (Derkacheva & Hennig, 2014; Hsieh et al., 2011). Disruptions in members of the FIS–PRC2 complex cause autonomous endosperm development in *Arabidopsis* and rice, potentially attributed to the activation of the auxin biosynthesis pathway in the central cell in the absence of fertilization (Chaudhury et al., 1997; Cheng et al., 2020; Figueiredo, Batista, Roszak, & Kohler, 2015; Grossniklaus, Vielle-Calzada, Hoeppner, & Gagliano, 1998; Guitton et al., 2004; Kohler et al., 2003; Li et al., 2014; Luo et al., 1999; Ohad et al., 1999; Sun et al., 2014; Tonosaki et al., 2021; Xu et al., 2021). Indeed, in *Arabidopsis* auxin is necessary for endosperm proliferation, and exogenous application of auxin 2,4-d or co-expressing auxin biosynthesis gene *TRYPTOPHAN AMINOTRANSFERASE OF ARABIDOPSIS 1* (*TAA1*) and *YUC6* in the central cell is sufficient to induce autonomous endosperm formation in unfertilized ovules (Figueiredo et al., 2015).

During seed development, gamete fusion triggers central cell division, and to date, two sperm factors have been reported to be delivered to the central cell to promote endosperm nuclear division. The first identified molecule is miR159 that is required for seed development by triggering endosperm nuclear division (Zhao et al., 2018). Loss of paternal miR159 causes defective initiation of endosperm nuclear divisions, leading to retention of maternal *MYB33* and *MYB65* in the fertilized central cell, which need to be eliminated by paternal miR159 after fertilization. Furthermore, ectopic expression of a miR159-resistant version of *MYB33* in the endosperm significantly inhibits initiation of endosperm nuclear division, mimicking arrested endosperm nuclear division in loss-of-paternal miR159 (Zhao et al., 2018). These data indicated that miR159 is a sperm-borne factor to trigger endosperm nuclear division by clearing the maternally inherited roadblock. During fertilization, the sperm cell also delivers a D-type cyclin to the central cell to resume the cell cycle. Before fertilization, the central cell is arrested in S phase by the activity of the RETINOBLASTOMA RELATED1 (RBR1) protein (Simonini, Bencivenga, & Grossniklaus, 2024). Upon fertilization, sperm cell cycle component CYCD7;1 is delivered to the central cell to promote RBR1 degradation and S phase completion, thus triggering central cell proliferation and ensuring endosperm formation. Embryo development was unaffected, when pollinated with *cycd3;1 cycd3;2 cycd3;3 cycd7;1* (*cyc4D*) pollen, indicating that paternal CYCD7;1-dependent RBR1 degradation is specific to the central cell (Simonini et al., 2024). Considering that the endosperm cell division is just delayed in *cyc4D* mutants, other fertilization dependent signals to reactivate the cell cycle must exist.

5. Conclusion

In past decades, tremendous progress has been made on understanding the process of gamete activation before and after fertilization, which mainly includes the discovery of Ca^{2+} transient and protein secretion during egg activation; the identification of cell surface proteins with key roles in gamete interaction; clarification in the understanding of ZGA in plants, as well as great breakthroughs in parthenogenesis and haploid induction. It is conceivable that the basic research related to cellular and molecular mechanisms for gamete activation and the initiation of embryogenesis and endosperm development will undoubtedly aid us in the identification of new factors that can be leveraged for the engineering of synthetic apomixis in crops, or optimizing haploid induction systems to achieve a highly efficient crop breeding. Despite recent advances, however, remarkable voids in our understanding of fundamental processes remain. The fact that double fertilization involves a sequence of rapidly occurring events taking place deeply within the ovule have contributed to the slow progress in understanding this process. In recent years, the advancement in isolating gametes and single-cell sequencing have accumulated considerable amount of useful data of different cell types including gametes, zygote and early endosperm. Careful analysis of these data may help to identify those key regulators in gamete activation through reverse genetics approaches. Improvement of resolution in living cell imaging will also enable to monitor the real dynamics of critical processes during fertilization, especially, for tracking the transient or compartmental behaviors of cells or proteins during the process of activation. We are convinced that the advances in technology will provide a considerable boost to ongoing research and many fascinating avenues in this field will be unraveled for deeply understanding of molecular mechanisms underlying gamete activation and fertilization in near future.

Acknowledgment

We thank the following funding sources for support: Nature Science Foundation of China (32130031; 31991201; 32200702).

We apologize to colleagues whose work was not mentioned in this brief review due to space limitations.

References

Alaniz-Fabián, J., Orozco-Nieto, A., Abreu-Goodger, C., & Gillmor, C. S. (2022). Hybridization alters maternal and paternal genome contributions to early plant embryogenesis. *Development (Cambridge, England), 149*, dev201025.

Anderson, S. N., Johnson, C. S., Chesnut, J., Jones, D. S., Khanday, I., Woodhouse, M., et al. (2017). The zygotic transition is initiated in unicellular plant zygotes with asymmetric activation of parental genomes. *Developmental Cell, 43*, 349–358.

Antoine, A. F., Faure, J. E., Cordeiro, S., Dumas, C., Rougier, M., & Feijo, J. A. (2000). A calcium influx is triggered and propagates in the zygote as a wavefront during in vitro fertilization of flowering plants. *Proceedings of the National Academy of Sciences of the United States of America, 97*, 10643–10648.

Autran, D., Baroux, C., Raissig, M. T., Lenormand, T., Wittig, M., Grob, S., et al. (2011). Maternal epigenetic pathways control parental contributions to *Arabidopsis* early embryogenesis. *Cell, 145*, 707–719.

Beale, K. M., Leydon, A. R., & Johnson, M. A. (2012). Gamete fusion is required to block multiple pollen tubes from entering an *Arabidopsis* ovule. *Current Biology, 22*, 1090–1094.

Berger, F., & Twell, D. (2011). Germline specification and function in plants. *Annual Review of Plant Biology, 62*, 461–484.

Bianchi, E., Doe, B., Goulding, D., & Wright, G. J. (2014). Juno is the egg Izumo receptor and is essential for mammalian fertilization. *Nature, 508*, 483–487.

Billey, E., Hafidh, S., Cruz-Gallardo, I., Litholdo, C. G., Jean, V., Carpentier, M. C., et al. (2021). LARP6C orchestrates posttranscriptional reprogramming of gene expression during hydration to promote pollen tube guidance. *The Plant Cell, 33*, 2637–2661.

Borges, F., Gomes, G., Gardner, R., Moreno, N., McCormick, S., Feijo, J. A., et al. (2008). Comparative transcriptomics of *Arabidopsis* sperm cells. *Plant Physiology, 148*, 1168–1181.

Boutilier, K., Offringa, R., Sharma, V. K., Kieft, H., Ouellet, T., Zhang, L. M., et al. (2002). Ectopic expression of BABY BOOM triggers a conversion from vegetative to embryonic growth. *The Plant Cell, 14*, 1737–1749.

Brukman, N. G., Uygur, B., Podbilewicz, B., & Chernomordik, L. V. (2019). How cells fuse. *Journal of Cell Biology, 218*, 1436–1451.

Camacho-Nuez, M., Hernandez-Silva, D. J., Castaneda-Ortiz, E. J., Paredes-Martinez, M. E., Rocha-Martinez, M. K., Alvarez-Sanchez, M. E., et al. (2017). *Hap2*, a novel gene in *Babesia bigemina* is expressed in tick stages, and specific antibodies block zygote formation. *Parasites & Vectors, 10*, 568.

Chahal, L. S., Conner, J. A., & Ozias-Akins, P. (2022). Phylogenetically distant *BABY BOOM* genes from *Setaria italica* induce parthenogenesis in rice. *Frontiers in Plant Science, 13*.

Chaudhury, A. M., Ming, L., Miller, C., Craig, S., Dennis, E. S., & Peacock, W. J. (1997). Fertilization-independent seed development in *Arabidopsis thaliana*. *Proceedings of the National Academy of Sciences of the United States of America, 94*, 4223–4228.

Chen, B. J., Maas, L., Figueiredo, D., Zhong, Y., Reis, R., Li, M. R., et al. (2022). BABY BOOM regulates early embryo and endosperm development. *Proceedings of the National Academy of Sciences of the United States of America, 119*, e2201761119.

Chen, G., Hu, Z., Guo, F., & Guan, X. (1995). Male gametophyte development in *Plumbago zeylanica*: Cytoplasm localization and cell determination in the early generative cell. *Protoplasma, 186*, 79–86.

Chen, J., Strieder, N., Krohn, N. G., Cyprys, P., Sprunck, S., Engelmann, J. C., et al. (2017). Zygotic genome activation occurs shortly after fertilization in maize. *The Plant Cell, 29*, 2106–2125.

Cheng, X. J., Pan, M. Y., E, Z. G., Zhou, Y., Niu, B. X., & Chen, C. (2020). Functional divergence of two duplicated *Fertilization Independent Endosperm* genes in rice with respect to seed development. *The Plant Journal, 104*, 124–137.

Cole, E. S., Cassidy-Hanley, D., Pinello, J. F., Zeng, H., Hsueh, M., Kolbin, D., et al. (2014). Function of the male-gamete-specific fusion protein HAP2 in a seven-sexed ciliate. *Current Biology, 24*, 2168–2173.

Conner, J., Podio, M., & Ozias-Akins, P. (2017). Haploid embryo production in rice and maize induced by *PsASGR-BBML* transgenes. *Plant Reproduction, 30*, 41–52.

Conner, J. A., Mookkan, M., Huo, H., Chae, K., & Ozias-Akins, P. (2015). A parthenogenesis gene of apomict origin elicits embryo formation from unfertilized eggs in a sexual plant. *Proceedings of the National Academy of Sciences of the United States of America, 112*, 11205–11210.

Cyprys, P., Lindemeier, M., & Sprunck, S. (2019). Gamete fusion is facilitated by two sperm cell-expressed DUF679 membrane proteins. *Nature Plants, 5*, 253–257.

Del Toro-De León, G., García-Aguilar, M., & Gillmor, C. S. (2014). Non-equivalent contributions of maternal and paternal genomes to early plant embryogenesis. *Nature, 514*, 624–627.

Denninger, P., Bleckmann, A., Lausser, A., Vogler, F., Ott, T., Ehrhardt, D. W., et al. (2014). Male-female communication triggers calcium signatures during fertilization in Arabidopsis. *Nature Communications, 5*, 4645.

Derkacheva, M., & Hennig, L. (2014). Variations on a theme: Polycomb group proteins in plants. *Journal of Experimental Botany, 65*, 2769–2784.

Digonnet, C., Aldon, D., Leduc, N., Dumas, C., & Rougier, M. (1997). First evidence of a calcium transient in flowering plants at fertilization. *Development (Cambridge, England), 124*, 2867–2874.

Dresselhaus, T., & Jürgens, G. (2021). Comparative embryogenesis in Angiosperms: Activation and patterning of embryonic cell lineages. *Annual Review of Plant Biology, 72*, 641–676.

Dresselhaus, T., Sprunck, S., & Wessel, G. M. (2016). Fertilization mechanisms in flowering plants. *Current Biology, 26*, 125–139.

Dumas, C., Knox, R. B., & Gaude, T. (1985). The spatial association of the sperm cells and vegetative nucleus in the pollen grain of *Brassica*. *Protoplasma, 124*, 168–174.

Ebchuqin, E., Yokota, N., Yamada, L., Yasuoka, Y., Akasaka, M., Arakawa, M., et al. (2014). Evidence for participation of GCS1 in fertilization of the starlet sea anemone *Nematostella vectensis*: Implication of a common mechanism of sperm-egg fusion in plants and animals. *Biochemical and Biophysical Research Communications, 451*, 522–528.

Fedry, J., Forcina, J., Legrand, P., Pehau-Arnaudet, G., Haouz, A., Johnson, M., et al. (2018). Evolutionary diversification of the HAP2 membrane insertion motifs to drive gamete fusion across eukaryotes. *PLoS Biology, 16*, e2006357.

Fedry, J., Liu, Y. J., Pehau-Arnaudet, G., Pei, J. M., Li, W. H., Tortorici, M. A., et al. (2017). The ancient gamete fusogen HAP2 is a eukaryotic class II fusion protein. *Cell, 168*, 904–915.

Ferris, P. J., Woessner, J. P., & Goodenough, U. W. (1996). A sex recognition glycoprotein is encoded by the plus mating-type gene *fus1* of *Chlamydomonas reinhardtii*. *Molecular Biology of the Cell, 7*, 1235–1248.

Figueiredo, D. D., Batista, R. A., Roszak, P. J., & Kohler, C. (2015). Auxin production couples endosperm development to fertilization. *Nature Plants, 1*, 15184.

Flores-Tornero, M., Proost, S., Mutwil, M., Scutt, C. P., Dresselhaus, T., & Sprunck, S. (2019). Transcriptomics of manually isolated *Amborella trichopoda* egg apparatus cells. *Plant Reproduction, 32*, 15–27.

Gilles, L. M., Calhau, A. R. M., La Padula, V., Jacquier, N. M. A., Lionnet, C., Martinant, J. P., et al. (2021). Lipid anchoring and electrostatic interactions target NOT-LIKE-DAD to pollen endo-plasma membrane. *Journal of Cell Biology, 220*, e202010077.

Gilles, L. M., Khaled, A., Laffaire, J. B., Chaignon, S., Gendrot, G., Laplaige, J., et al. (2017). Loss of pollen-specific phospholipase NOT LIKE DAD triggers gynogenesis in maize. *The EMBO Journal, 36*, 707–717.

Glöckle, B., Urban, W. J., Nagahara, S., Andersen, E. D., Higashiyama, T., Grini, P. E., et al. (2018). Pollen differentiation as well as pollen tube guidance and discharge are independent of the presence of gametes. *Development (Cambridge, England), 145*, dev152645.

Grossniklaus, U. (2017). Polyspermy produces tri-parental seeds in maize. *Current Biology, 27*, 1300–1302.

Grossniklaus, U., Vielle-Calzada, J. P., Hoeppner, M. A., & Gagliano, W. B. (1998). Maternal control of embryogenesis by MEDEA, a polycomb group gene in *Arabidopsis*. *Science (New York, N. Y.), 280*, 446–450.

Guitton, A. E., Page, D. R., Chambrier, P., Lionnet, C., Faure, J. E., Grossniklaus, U., et al. (2004). Identification of new members of Fertilisation Independent Seed Polycomb Group pathway involved in the control of seed development in *Arabidopsis thaliana*. *Development (Cambridge, England), 131*, 2971–2981.

Hafidh, S., & Honys, D. (2021). Reproduction multitasking: The male gametophyte. *Annual Review of Plant Biology, 72*, 581–614.

Hamamura, Y., Nishimaki, M., Takeuchi, H., Geitmann, A., Kurihara, D., & Higashiyama, T. (2014). Live imaging of calcium spikes during double fertilization in *Arabidopsis*. *Nature Communications, 5*, 4722.

Hamamura, Y., Saito, C., Awai, C., Kurihara, D., Miyawaki, A., Nakagawa, T., et al. (2011). Live-cell imaging reveals the dynamics of two sperm cells during double fertilization in *Arabidopsis thaliana*. *Current Biology, 21*, 497–502.

Hater, F., Nakel, T., & Gross-Hardt, R. (2020). Reproductive multitasking: The female gametophyte. *Annual Review of Plant Biology, 71*, 517–546.

Hernandez, J. M., & Podbilewicz, B. (2017). The hallmarks of cell-cell fusion. *Development (Cambridge, England), 144*, 4481–4495.

Heydlauff, J., Serbes, I. E., Vo, D., Mao, Y. B., Gieseking, S., Nakel, T., et al. (2022). Dual and opposing roles of EIN3 reveal a generation conflict during seed growth. *Molecular Plant, 15*, 363–371.

Higashiyama, T., & Takeuchi, H. (2015). The mechanism and key molecules involved in pollen tube guidance. *Annual Review of Plant Biology, 66*, 393–413.

Higashiyama, T., Yabe, S., Sasaki, N., Nishimura, Y., Miyagishima, S., Kuroiwa, H., et al. (2001). Pollen tube attraction by the synergid cell. *Science (New York, N. Y.), 293*, 1480–1483.

Higashiyama, T., & Yang, W. C. (2017). Gametophytic pollen tube guidance: Attractant peptides, gametic controls, and receptors. *Plant Physiology, 173*, 112–121.

Hirai, M., Arai, M., Mori, T., Miyagishima, S., Kawai, S., Kita, K., et al. (2008). Male fertility of malaria parasites is determined by GCS1, a plant-type reproduction factor. *Current Biology, 18*, 607–613.

Hsieh, T. F., Shin, J. Y., Uzawa, R., Silva, P., Cohen, S., Bauer, M. J., et al. (2011). Regulation of imprinted gene expression in *Arabidopsis* endosperm. *Proceedings of the National Academy of Sciences of the United States of America, 108*, 1755–1762.

Huang, Y., Liang, Y. Q., Xie, Y. Y., Rao, Y. C., Xiong, J., Liu, C. L., et al. (2024). Efficient haploid induction via egg cell expression of dandelion *PARTHENOGENESIS* in foxtail millet (*Setaria italica*). *Plant Biotechnology Journal, 22*, 1797–1799.

Ingouff, M., Sakata, T., Li, J., Sprunck, S., Dresselhaus, T., & Berger, F. (2009). The two male gametes share equal ability to fertilize the egg cell in *Arabidopsis thaliana*. *Current Biology, 19*, 19–20.

Inoue, N., Ikawa, M., Isotani, A., & Okabe, M. (2005). The immunoglobulin superfamily protein Izumo is required for sperm to fuse with eggs. *Nature, 434*, 234–238.

Jacquier, N. M. A., Calhau, A. R. M., Fierlej, Y., Martinant, J. P., Rogowsky, P. M., Gilles, L. M., et al. (2023). In planta haploid induction by *kokopelli* mutants. *Plant Physiology, 193*, 182–185.

Jacquier, N. M. A., Gilles, L. M., Pyott, D. E., Martinant, J. P., Rogowsky, P. M., & Widiez, T. (2020). Puzzling out plant reproduction by haploid induction for innovations in plant breeding. *Nature Plants, 6*, 610–619.

Johnson, M. A., Harper, J. F., & Palanivelu, R. (2019). A fruitful journey: Pollen tube navigation from germination to fertilization. *Annual Review of Plant Biology, 70*, 809–837.

Jukam, D., Shariati, S. A. M., & Skotheim, J. M. (2017). Zygotic genome activation in vertebrates. *Developmental Cell, 42*, 316–332.

Kasahara, R. D., Maruyama, D., Hamamura, Y., Sakakibara, T., Twell, D., & Higashiyama, T. (2012). Fertilization recovery after defective sperm cell release in *Arabidopsis*. *Current Biology, 22*, 1084–1089.

Kashir, J., Deguchi, R., Jones, C., Coward, K., & Stricker, S. A. (2013). Comparative biology of sperm factors and fertilization-induced calcium signals across the animal kingdom. *Molecular Reproduction and Development, 80*, 787–815.

Kawai-Toyooka, H., Mori, T., Hamaji, T., Suzuki, M., Olson, B. J. S. C., Uemura, T., et al. (2014). Sex-specific posttranslational regulation of the gamete fusogen GCS1 in the isogamous volvocine alga *Gonium pectorale*. *Eukaryotic Cell, 13*, 648–656.

Kelliher, T., Starr, D., Richbourg, L., Chintamanani, S., Delzer, B., Nuccio, M. L., et al. (2017). MATRILINEAL, a sperm-specific phospholipase, triggers maize haploid induction. *Nature, 542*, 105–109.

Khanday, I., Skinner, D., Yang, B., Mercier, R., & Sundaresan, V. (2019). A male-expressed rice embryogenic trigger redirected for asexual propagation through seeds. *Nature, 565*, 91–95.

Kinoshita, T., Fukuzawa, H., Shimada, T., Saito, T., & Matsuda, Y. (1992). Primary structure and expression of a gamete lytic enzyme in *Chlamydomonas reinhardtii*: Similarity of functional domains to matrix metalloproteases. *Proceedings of the National Academy of Sciences of the United States of America, 89*, 4693–4697.

Kohler, C., Hennig, L., Bouveret, R., Gheyselinck, J., Grossniklaus, U., & Gruissem, W. (2003). *Arabidopsis* MSI1 is a component of the MEA/FIE Polycomb group complex and required for seed development. *The EMBO Journal, 22*, 4804–4814.

Kong, J., Lau, S., & Jurgens, G. (2015). Twin plants from supernumerary egg cells in *Arabidopsis*. *Current Biology, 25*, 225–230.

Kranz, E., Vonwiegen, P., & Lorz, H. (1995). Early cytological events after induction of cell division in egg cells and zygote development following in vitro fertilization with angiosperm gametes. *The Plant Journal, 8*, 9–23.

Lau, S., Slane, D., Herud, O., Kong, J. X., & Jürgens, G. (2012). Early embryogenesis in flowering plants: setting up the basic body pattern. *Annual Review of Plant Biology, 63*, 483–506.

Li, J., & Berger, F. (2012). Endosperm: Food for humankind and fodder for scientific discoveries. *New Phytologist, 195*, 290–305.

Li, L., Hou, S., Xiang, W., Song, Z., Wang, Y., Zhang, L., et al. (2022). The egg cell is preferentially fertilized in *Arabidopsis* double fertilization. *Journal of Integrative Plant Biology, 64*, 2039–2046.

Li, W. H., Li, Q. Y., Lyu, M. H., Wang, Z. J., Song, Z. H., Zhong, S. W., et al. (2022). Lack of ethylene does not affect reproductive success and synergid cell death in *Arabidopsis*. *Molecular Plant, 15*, 354–362.

Li, Y. F., Li, D., Xiao, Q., Wang, H. D., Wen, J., Tu, J. X., et al. (2022). An *in planta* haploid induction system in *Brassica napus*. *Journal of Integrative Plant Biology, 64*, 1140–1144.

Li, S., Zhou, L. Z., Feng, Q. N., McCormick, S., & Zhang, Y. (2013). The C-terminal hypervariable domain targets *Arabidopsis* ROP9 to the invaginated pollen tube plasma membrane. *Molecular Plant, 6*, 1362–1364.

Li, S. S., Zhou, B., Peng, X. B., Kuang, Q., Huang, X. L., Yao, J. L., et al. (2014). *OsFIE2* plays an essential role in the regulation of rice vegetative and reproductive development. *New Phytologist, 201*, 66–79.

Li, X., Zhu, B., Lu, Y., Zhao, F., Liu, Q., Wang, J. H., et al. (2024). DNA methylation remodeling and the functional implication during male gametogenesis in rice. *Genome Biology, 25*, 84.

Lin, Y. K., Wang, Y. L., Zhu, J. K., & Yang, Z. B. (1996). Localization of a Rho GTPase implies a role in tip growth and movement of the generative cell in pollen tubes. *The Plant Cell, 8*, 293–303.

Liu, C. X., Li, X., Meng, D. X., Zhong, Y., Chen, C., Dong, X., et al. (2017). A 4-bp insertion at *ZmPLA1* encoding a putative phospholipase A generates haploid induction in maize. *Molecular Plant, 10*, 520–522.

Liu, Q., Ma, X., Li, X., Zhang, X. R., Zhou, S. L., Xiong, L. Z., et al. (2023). Paternal DNA methylation is remodeled to maternal levels in rice zygote. *Nature Communications, 14*, 6571.

Liu, Y. J., Misamore, M. J., & Snell, W. J. (2010). Membrane fusion triggers rapid degradation of two gamete-specific, fusion-essential proteins in a membrane block to polygamy in. *Development (Cambridge, England), 137*, 1473–1481.

Liu, Y. J., Tewari, R., Ning, J., Blagborough, A. M., Garbom, S., Pei, J. M., et al. (2008). The conserved plant sterility gene *HAP2* functions after attachment of fusogenic membranes in *Chlamydomonas* and *Plasmodium* gametes. *Genes & Development, 22*, 1051–1068.

Long, L., Feng, Y. M., Shang, S. Z., Zhao, J. R., Hu, G. Y., Xu, F. C., et al. (2024). In vivo maternal haploid induction system in cotton. *Plant Physiology, 194*, 1286–1289.

Luo, M., Bilodeau, P., Koltunow, A., Dennis, E. S., Peacock, W. J., & Chaudhury, A. M. (1999). Genes controlling fertilization-independent seed development in *Arabidopsis thaliana*. *Proceedings of the National Academy of Sciences of the United States of America, 96*, 296–301.

Mao, Y. B., Nakel, T., Serbes, I. E., Joshi, S., Tekleyohans, D. G., Baum, T., et al. (2023). ECS1 and ECS2 suppress polyspermy and the formation of haploid plants by promoting double fertilization. *Elife, 12*, e85832.

Martens, S., & McMahon, H. T. (2008). Mechanisms of membrane fusion: Disparate players and common principles. *Nature Reviews. Molecular Cell Biology, 9*, 543–556.

Maruyama, D., Hamamura, Y., Takeuchi, H., Susaki, D., Nishimaki, M., Kurihara, D., et al. (2013). Independent control by each female gamete prevents the attraction of multiple pollen tubes. *Developmental Cell, 25*, 317–323.

Maruyama, D., Völz, R., Takeuchi, H., Mori, T., Igawa, T., Kurihara, D., et al. (2015). Rapid elimination of the persistent synergid through a cell fusion mechanism. *Cell, 161*, 907–918.

McCue, A. D., Cresti, M., Feijó, J. A., & Slotkin, R. K. (2011). Cytoplasmic connection of sperm cells to the pollen vegetative cell nucleus: Potential roles of the male germ unit revisited. *Journal of Experimental Botany, 62*, 1621–1631.

Misamore, M. J., Gupta, S., & Snell, W. J. (2003). The *Chlamydomonas* Fus1 protein is present on the mating type plus fusion organelle and required for a critical membrane adhesion event during fusion with minus gametes. *Molecular Biology of the Cell, 14*, 2530–2542.

Mori, T., Igawa, T., Tamiya, G., Miyagishima, S. Y., & Berger, F. (2014). Gamete attachment requires GEX2 for successful fertilization in *Arabidopsis*. *Current Biology, 24*, 170–175.

Mori, T., Kuroiwa, H., Higashiyama, T., & Kuroiwa, T. (2006). Generative cell specific 1 is essential for angiosperm fertilization. *Nature Cell Biology, 8*, 64–71.

Motomura, K., Takeuchi, H., Notaguchi, M., Tsuchi, H., Takeda, A., Kinoshita, T., et al. (2021). Persistent directional growth capability in *Arabidopsis thaliana* pollen tubes after nuclear elimination from the apex. *Nature Communications, 12*, 2331.

Nagahara, S., Takeuchi, H., & Higashiyama, T. (2021). Polyspermy block in the central cell during double fertilization of *Arabidopsis thaliana*. *Frontiers in Plant Science, 11*, 588700.

Nodine, M. D., & Bartel, D. P. (2012). Maternal and paternal genomes contribute equally to the transcriptome of early plant embryos. *Nature, 482*, 94–97.

Oda, Y., & Fukuda, H. (2012). Initiation of cell wall pattern by a Rho- and microtubule-driven symmetry breaking. *Science (New York, N. Y.), 337*, 1333–1336.

Ohad, N., Yadegari, R., Margossian, L., Hannon, M., Michaeli, D., Harada, J. J., et al. (1999). Mutations in FIE, a WD polycomb group gene, allow endosperm development without fertilization. *The Plant Cell, 11*, 407–415.

Ohnishi, Y., Kokubu, I., Kinoshita, T., & Okamoto, T. (2019). Sperm entry into the egg cell induces the progression of karyogamy in rice zygotes. *Plant and Cell Physiology, 60*, 1656–1665.

Ohto, U., Ishida, H., Krayukhina, E., Uchiyama, S., Inoue, N., & Shimizu, T. (2016). Structure of IZUMO1-JUNO reveals sperm-oocyte recognition during mammalian fertilization. *Nature, 534*, 566–569.

Okamoto, M., Yamada, L., Fujisaki, Y., Bloomfield, G., Yoshida, K., Kuwayama, H., et al. (2016). Two HAP2-GCS1 homologs responsible for gamete interactions in the cellular slime mold with multiple mating types: Implication for common mechanisms of sexual reproduction shared by plants and protozoa and for male-female differentiation. *Developmental Biology, 415*, 6–13.

Okuda, S., Tsutsui, H., Shiina, K., Sprunck, S., Takeuchi, H., Yui, R., et al. (2009). Defensin-like polypeptide LUREs are pollen tube attractants secreted from synergid cells. *Nature, 458*, 357–361.

Olsen, O. A. (2020). The modular control of cereal endosperm development. *Trends in Plant Science, 25*, 279–290.

Pagnussat, G. C., Yu, H. J., & Sundaresan, V. (2007). Cell-fate switch of synergid to egg cell in *Arabidopsis eostre* mutant embryo sacs arises from misexpression of the BEL1-like homeodomain gene *BLH1*. *The Plant Cell, 19*, 3578–3592.

Pan, J. M., & Snell, W. J. (2000). Signal transduction during fertilization in the unicellular green alga. *Current Opinion in Microbiology, 3*, 596–602.

Peng, L., Li, Z. K., Ding, X. L., & Tian, H. Q. (2018). Advances in the study of egg activation of higher plants. *Zygote (Cambridge, England), 26*, 435–442.

Pinello, J. F., Lai, A. L., Millet, J. K., Cassidy-Hanley, D., Freed, J. H., & Clark, T. G. (2017). Structure-function studies link class II viral fusogens with the ancestral gamete fusion protein HAP2. *Current Biology, 27*, 651–660.

Pinello, J. F., Liu, Y. J., & Snell, W. J. (2021). MAR1 links membrane adhesion to membrane merger during cell-cell fusion in *Chlamydomonas*. *Developmental Cell, 56*, 3380–3392.

Qi, X. T., Gao, H. M., Lv, R. Y., Mao, W. B., Zhu, J. J., Liu, C. L., et al. (2023). CRISPR/dCas-mediated gene activation toolkit development and its application for parthenogenesis induction in maize. *Plant Communications, 4*, 100449.

Rademacher, S., & Sprunck, S. (2013). Downregulation of egg cell-secreted EC1 is accompanied with delayed gamete fusion and polytubey. *Plant Signaling & Behavior, 8*, e27377.

Roman, H. (1948). Directed fertilization in maize. *Proceedings of the National Academy of Sciences of the United States of America, 34*, 36–42.

Ron, M., Saez, M. A., Williams, L. E., Fletcher, J. C., & McCormick, S. (2010). Proper regulation of a sperm-specific *cis*-nat-siRNA is essential for double fertilization in *Arabidopsis*. *Genes & Development, 24*, 1010–1021.

Russell, S. D. (1983). Fertilization in *Plumbago zeylanica*: Gametic fusion and fate of the male cytoplasm. *American Journal of Botany, 70*, 416–434.

Russell, S. D. (1985). Preferential fertilization in *Plumbago*: Ultrastructural evidence for gamete-level recognition in an angiosperm. *Proceedings of the National Academy of Sciences of the United States of America, 82*, 6129–6132.

Russell, S. D. (1986). Isolation of sperm cells from the pollen of *Plumbago zeylanica*. *Plant Physiology, 81*, 317–319.

Russell, S. D., & Cass, D. D. (1981). Ultrastructure of the sperms of *Plumbago zeylanica*.1. Cytology and association with the vegetative nucleus. *Protoplasma, 107*, 85–107.

Saito, C., Nagata, N., Sakai, A., Mori, K., Kuroiwa, H., & Kuroiwa, T. (2002). Angiosperm species that produce sperm cell pairs or generative cells with polarized distribution of DNA-containing organelles. *Sexual Plant Reproduction, 15*, 167–178.

Satouh, Y., & Ikawa, M. (2018). New insights into the molecular events of mammalian fertilization. *Trends in Biochemical Sciences, 43*, 818–828.

Schier, A. F. (2007). The maternal-zygotic transition: Death and birth of RNAs. *Science (New York, N. Y.), 316*, 406–407.

Schulz, K. N., & Harrison, M. M. (2019). Mechanisms regulating zygotic genome activation. *Nature Reviews. Genetics, 20*, 221–234.

Scott, R. J., Armstrong, S. J., Doughty, J., & Spielman, M. (2008). Double fertilization in *Arabidopsis thaliana* involves a polyspermy block on the egg but not the central cell. *Molecular Plant, 1*, 611–619.

Shen, K., Qu, M. X., & Zhao, P. (2023). The roads to haploid embryogenesis. *Plants, 12*, 243.

Simonini, S., Bencivenga, S., & Grossniklaus, U. (2024). A paternal signal induces endosperm proliferation upon fertilization in *Arabidopsis*. *Science (New York, N. Y.), 383*, 646–653.

Skinner, D. J., Mallari, M. D., Zafar, K., Cho, M. J., & Sundaresan, V. (2023). Efficient parthenogenesis via egg cell expression of maize BABY BOOM 1: A step toward synthetic apomixis. *Plant Physiology, 193*, 2278–2281.

Skinner, D. J., & Sundaresan, V. (2018). Recent advances in understanding female gametophyte development. *F1000Research, 7*, 804.

Snell, W. J., Eskue, W. A., & Buchanan, M. J. (1989). Regulated secretion of a serine protease that activates an extracellular matrix-degrading metalloprotease during fertilization in *Chlamydomonas*. *Journal of Cell Biology, 109*, 1689–1694.

Sprunck, S. (2020). Twice the fun, double the trouble: Gamete interactions in flowering plants. *Current Opinion in Plant Biology, 53*, 106–116.

Sprunck, S., Hackenberg, T., Englhart, M., & Vogler, F. (2014). Same same but different: Sperm-activating EC1 and ECA1 gametogenesis-related family proteins. *Biochemical Society Transactions, 42*, 401–407.

Sprunck, S., Rademacher, S., Vogler, F., Gheyselinck, J., Grossniklaus, U., & Dresselhaus, T. (2012). Egg cell-secreted EC1 triggers sperm cell activation during double fertilization. *Science (New York, N. Y.), 338*, 1093–1097.

Sugi, N., Calhau, A. R. M., Jacquier, N. M. A., Millan-Blanquez, M., Becker, J. D., Begcy, K., et al. (2024). The peri-germ cell membrane: poorly characterized but key interface for plant reproduction. *Nature Plants*.

Sugi, N., Izumi, R., Tomomi, S., Susaki, D., Kinoshita, T., & Maruyama, D. (2023). Removal of the endoplasma membrane upon sperm cell activation after pollen tube discharge. *Frontiers in Plant Science, 14*, 1116289.

Sukawa, Y., & Okamoto, T. (2018). Cell cycle in egg cell and its progression during zygotic development in rice. *Plant Reproduction, 31*, 107–116.

Sun, X. C., Ling, S., Lu, Z. H., Ouyang, Y. D., Liu, S. S., & Yao, J. L. (2014). *OsNF-YB1*, a rice endosperm-specific gene, is essential for cell proliferation in endosperm development. *Gene, 551*, 214–221.

Susaki, D., Suzuki, T., Maruyama, D., Ueda, M., Higashiyama, T., & Kurihara, D. (2021). Dynamics of the cell fate specifications during female gametophyte development in Arabidopsis. *PLoS Biology, 19*, e3001123.

Tadros, W., & Lipshitz, H. D. (2009). The maternal-to-zygotic transition: A play in two acts. *Development (Cambridge, England), 136*, 3033–3042.

Takahashi, T., Mori, T., Ueda, K., Yamada, P., Nagahara, S., Higashiyama, T., et al. (2018). The male gamete membrane protein DMP9/DAU2 is required for double fertilization in flowering plants. *Development (Cambridge, England), 145*, dev170076.

Takeuchi, H., & Higashiyama, T. (2012). A species-specific cluster of defensin-like genes encodes diffusible pollen tube attractants in *Arabidopsis*. *PLoS Biology, 10*, e1001449.

Tekleyohans, D. G., Mao, Y. B., Kagi, C., Stierhof, Y. D., & Gross-Hardt, R. (2017). Polyspermy barriers: A plant perspective. *Current Opinion in Plant Biology, 35*, 131–137.

Tonosaki, K., Ono, A., Kunisada, M., Nishino, M., Nagata, H., Sakamoto, S., et al. (2021). Mutation of the imprinted gene *OsEMF2a* induces autonomous endosperm development and delayed cellularization in rice. *The Plant Cell, 33*, 85–103.

Tsutsui, H., Sato, Y., Susaki, D., & Higashiyama, T. (2019). Microtubule depletion domain 1 localizes at the boundary between female gametes in *Arabidopsis thaliana*. *Molecular Reproduction and Development, 86*, 925.

Underwood, C. J., Vijverberg, K., Rigola, D., Okamoto, S., Oplaat, C., Op den Camp, R. H. M., et al. (2022). A *PARTHENOGENESIS* allele from apomictic dandelion can induce egg cell division without fertilization in lettuce. *Nature Genetics, 54*, 84–93.

Valansi, C., Moi, D., Leikina, E., Matveev, E., Grana, M., Chernomordik, L. V., et al. (2017). *Arabidopsis* HAP2/GCS1 is a gamete fusion protein homologous to somatic and viral fusogens. *Journal of Cell Biology, 216*, 571–581.

Vastenhouw, N. L., Cao, W. X., & Lipshitz, H. D. (2019). The maternal-to-zygotic transition revisited. *Development (Cambridge, England), 146*, dev161471.

Vernet, A., Meynard, D., Lian, Q. C., Mieulet, D., Gibert, O., Bissah, M., et al. (2022). High-frequency synthetic apomixis in hybrid rice. *Nature Communications, 13*, 7963.

Völz, R., Heydlauff, J., Ripper, D., von Lyncker, L., & Gross-Hardt, R. (2013). Ethylene signaling is required for synergid degeneration and the establishment of a pollen tube block. *Developmental Cell, 25*, 310–316.

von Besser, K., Frank, A. C., Johnson, M. A., & Preuss, D. (2006). *Arabidopsis HAP2* (*GCS1*) is a sperm-specific gene required for pollen tube guidance and fertilization. *Development (Cambridge, England), 133*, 4761–4769.

Wang, N., Xia, X. Z., Jiang, T., Li, L. L., Zhang, P. C., Niu, L. F., et al. (2022). *In planta* haploid induction by genome editing of *DMP* in the model legume *Medicago truncatula*. *Plant Biotechnology Journal, 20*, 22–24.

Wang, W., Xiong, H., Cyprys, P., Malka, R., Flores-Tornero, M., Zhao, P., et al. (2022). DMP8 and 9 regulate HAP2/GCS1 trafficking for the timely acquisition of sperm fusion competence. *Proceedings of the National Academy of Sciences of the United States of America, 119*, e2207608119.

Wang, W., Malka, R., Lindemeier, M., Cyprys, P., Tiedemann, S., Sun, K. T., et al. (2024). EGG CELL 1 contributes to egg-cell-dependent preferential fertilization in. *Nature Plants, 10*, 268–282.

Wei, X., Liu, C. L., Chen, X., Lu, H. W., Wang, J., Yang, S. L., et al. (2023). Synthetic apomixis with normal hybrid rice seed production. *Molecular Plant, 16*, 489–492.

Wu, X. B., Xie, L. Q., Sun, X. Z., Wang, N. N., Finnegan, E. J., Helliwell, C., et al. (2023). Mutation in Polycomb repressive complex 2 gene *OsFIE2* promotes asexual embryo formation in rice. *Nature Plants, 9*, 1848–1861.

Xiong, H. X., Wang, W., & Sun, M. X. (2021). Endosperm development is an autonomously programmed process independent of embryogenesis. *The Plant Cell, 33*, 1151–1160.

Xu, X. Y., Zhiguo, E., Zhang, D. P., Yun, Q. B., Zhou, Y., Niu, B. X., et al. (2021). *OsYUC11*-mediated auxin biosynthesis is essential for endosperm development of rice. *Plant Physiology, 185*, 934–950.

Yu, X., Zhang, X., Zhao, P., Peng, X., Chen, H., Bleckmann, A., et al. (2021). Fertilized egg cells secrete endopeptidases to avoid polytubey. *Nature, 592*, 433–437.

Zhang, G. C., Campenot, M. K., Mcgann, L. E., & Cass, D. D. (1992). Flow cytometric characteristics of sperm cells isolated from pollen of *Zea mays* L. *Plant Physiology, 99*, 54–59.

Zhang, J., Huang, Q. P., Zhong, S., Bleckmann, A., Huang, J. Y., Guo, X. Y., et al. (2017). Sperm cells are passive cargo of the pollen tube in plant fertilization. *Nature Plants, 3*, 17079.

Zhang, J., Pinello, J. F., Fernandez, I., Baquero, E., Fedry, J., Rey, F. A., et al. (2021). Species-specific gamete recognition initiates fusion-driving trimer formation by conserved fusogen HAP2. *Nature Communications, 12*, 4380.

Zhang, J. Z., Yin, J., Luo, J. Y., Tang, D., Zhu, X. J., Wang, J., et al. (2022). Construction of homozygous diploid potato through maternal haploid induction. *Abiotech, 3*, 163–168.

Zhang, X. L., Zhang, L. L., Zhang, J. S., Jia, M. A., Cao, L. G., Yu, J., et al. (2022). Haploid induction in allotetraploid tobacco using DMPs mutation. *Planta, 255*, 98.

Zhang, X., Shi, C., Li, S., Zhang, B., Luo, P., Peng, X., et al. (2023). A female in vivo haploid-induction system via mutagenesis of egg cell-specific peptidases. *Molecular Plant, 16*, 471–480.

Zhang, Z. F., Conner, J., Guo, Y. P., & Ozias-Akins, P. (2020). Haploidy in tobacco induced by *PsASGR-BBML* transgenes via parthenogenesis. *Genes, 11*, 1072.

Zhao, J., Xin, H. P., Qu, L. H., Ning, J., Peng, X. B., Yan, T. T., et al. (2011). Dynamic changes of transcript profiles after fertilization are associated with *de novo* transcription and maternal elimination in tobacco zygote, and mark the onset of the maternal-to-zygotic transition. *Plant Journal, 65*, 131–145.

Zhao, P., Shi, C., Wang, L., & Sun, M. X. (2022). The parental contributions to early plant embryogenesis and the concept of maternal-to-zygotic transition in plants. *Current Opinion in Plant Biology, 65*, 102144.

Zhao, X. Y., Yuan, K. W., Liu, Y. X., Zhang, N., Yang, L. M., Zhang, Y. Y., et al. (2022). *In vivo* maternal haploid induction based on genome editing of *DMP* in *Brassica oleracea*. *Plant Biotechnology Journal, 20*, 2242–2244.

Zhao, P., & Sun, M. X. (2015). The maternal-to-zygotic transition in higher plants: Available approaches, critical limitations, and technical requirements. *Current Topics in Developmental Biology, 113*, 373–398.

Zhao, P., Zhou, X., Zheng, Y., Ren, Y., & Sun, M. X. (2020). Equal parental contribution to the transcriptome is not equal control of embryogenesis. *Nature Plants, 6*, 1354–1364.

Zhao, P., Zhou, X. M., Shen, K., Liu, Z. Z., Cheng, T. H., Liu, D. N., et al. (2019). Two-step maternal-to-zygotic transition with two-phase parental genome contributions. *Developmental Cell, 49*, 882–893.

Zhao, Y. S., Wang, S. Y., Wu, W. Y., Li, L., Jiang, T., & Zheng, B. L. (2018). Clearance of maternal barriers by paternal miR159 to initiate endosperm nuclear division in *Arabidopsis*. *Nature Communications, 9*, 5011.

Zhong, S., & Qu, L. J. (2019). Peptide/receptor-like kinase-mediated signaling involved in male-female interactions. *Current Opinion in Plant Biology, 51*, 7–14.
Zhong, S., Zhao, P., Peng, X. B., Li, H. J., Duan, Q. H., & Cheung, A. Y. (2024). From gametes to zygote: Mechanistic advances and emerging possibilities in plant reproduction. *Plant Physiology, 195*, 4–35.
Zhong, Y., Chen, B., Li, M., Wang, D., Jiao, Y., Qi, X., et al. (2020). A *DMP*-triggered in vivo maternal haploid induction system in the dicotyledonous *Arabidopsis*. *Nature Plants, 6*, 466–472.
Zhong, Y., Chen, B. J., Wang, D., Zhu, X. J., Li, M. R., Zhang, J. Z., et al. (2022). *In vivo* maternal haploid induction in tomato. *Plant Biotechnology Journal, 20*, 250–252.
Zhong, Y., Wang, Y., Chen, B., Liu, J., Wang, D., Li, M., et al. (2022). Establishment of a *dmp* based maternal haploid induction system for polyploid *Brassica napus* and *Nicotiana tabacum*. *Journal of Integrative Plant Biology, 64*, 1281–1294.
Zhong, Y., Liu, C., Qi, X., Jiao, Y., Wang, D., Wang, Y., et al. (2019). Mutation of *ZmDMP* enhances haploid induction in maize. *Nature Plants, 5*, 575–580.
Zhou, L. Z., & Dresselhaus, T. (2019). Friend or foe: Signaling mechanisms during double fertilization in flowering seed plants. *Plant Development and Evolution, 131*, 453–496.

CHAPTER TWO

Cytoskeletal dynamics of gamete nuclear migration in flowering plants, animals, and yeast

Yilin Zhang and Tomokazu Kawashima*
Department of Plant and Soil Sciences, University of Kentucky, Plant Science Building, Lexington, KY, United States
*Corresponding author. e-mail address: tomo.k@uky.edu

Contents

Abstract

Gamete nuclear migration is a critical process during fertilization in flowering plants, yet its molecular mechanisms remain poorly understood. Recent studies have highlighted the essential role of cytoskeletal elements, particularly F-actin, in directing sperm nuclear migration, which differ from the microtubule-driven migration in animals. We summarize the process of sperm nuclear migration in plants and the

Current Topics in Developmental Biology, Volume 162
ISSN 0070-2153, https://doi.org/10.1016/bs.ctdb.2024.10.010

involvement of Class XI myosin XI-G in *Arabidopsis*, along with the ROP8-SCAR2 pathway's ARP2/3-independent mechanism for F-actin nucleation. We also provide a comparative overview of examples from sea urchins, *C. elegans*, mice and yeast contrasting these mechanisms with those in plants. Finally, we outline possible future research directions related to sperm nuclear migration in plants. This review highlights the need for further exploration of pre- and post-fertilization processes, emphasizing their importance in plant cytoskeleton biology and the coordinated development of seeds.

1. Introduction

Each generation of sexually reproducing species relies on the fusion of maternal and paternal genomes through fertilization. For this fusion to occur, the nuclei of male and female gametes must undergo directed migration. In flowering plants, it is typically the male gamete nuclei that migrate towards the female gamete nuclei, whereas in animals and yeast, both male and female gamete nuclei migrate cooperatively toward the center of the cell. Beyond this distinction, there are several other mechanistic differences among plants, animals, and yeast. Notably, in animals and yeast, the centrosome transitions with the male nucleus during nuclear migration, and this migration is regulated by microtubules and spindle bodies organized by the centrosomes (Manandhar et al., 2005). On the other hand, flowering plants lack centrosomes, and nuclear migration is governed by filamentous actin (F-actin) rather than microtubules (Carvalho-Santos et al., 2011; Kawashima et al., 2014). This raises an important question: How do microtubules and F-actin control the migration of male and female nuclei in animals and flowering plants? Advances in microscopy have significantly enhanced our understanding of gamete nuclear migration (Shin, Yuan, & Kawashima, 2022). The use of cytoskeletal inhibitors has provided valuable insights into the role of cytoskeletal elements in this process, while molecular biology tools have enabled detailed investigations into the underlying mechanisms.

This review will focus on the recent discoveries in flowering plants and compare the distinct mechanisms that govern gamete nuclear migration in plants and animals. In animals, gamete nuclei are referred to as pronuclei during fertilization. By contrast, in plants, the term "nuclei" is used, as meiosis is completed in the plant gametes prior to fertilization. Since nuclear fusion does not occur immediately after gamete fusion, female gametes are still referred to as central cells or egg cells during the nuclear migration, rather than as zygotes.

2. Sperm nuclear migration in flowering plants

Double fertilization is a crucial process for the reproduction of flowering plants. When pollen contacts the stigma, it begins to hydrate and germinate, forming a pollen tube. The sperm cells travel along this tube to the micropylar end of the unfertilized ovule. Upon reaching the ovule, the tip of the pollen tube ruptures, releasing the sperm cells into the space between the central cell and the egg cell. The plasma membranes and cytoplasm of the two sperm cells then fuse with those of the egg cell and central cell, completing plasmogamy. Subsequently, the sperm nuclei enter the egg cell and central cell, where they fuse with the female nuclei, guided by actin cables to complete karyogamy (Kawashima et al., 2014). One sperm nucleus fuses with the egg cell nucleus, resulting in a zygote that develops into an embryo. The second sperm nucleus fuses with the diploid central cell nucleus to form triploid endosperm, which nourishes the developing embryo (Fig. 1) (Hamamura et al., 2012; Raghavan, 2003).

After plasmogamy, the sperm nuclei are captured by F-actin meshwork surrounding the female nucleus, which is anchored to the female gamete membrane (Kawashima et al., 2014). In the female gametes of *Arabidopsis thaliana* and the egg cell of rice, an F-actin meshwork structure radiates, while in the central cell of *Asparagaceae*, an F-actin mega-cable and parallel actin filaments have been observed (González-Gutiérrez, Gutiérrez-Mora, Verdín, & Rodríguez-Garay, 2021; Kawashima et al., 2014; Ohnishi, Hoshino, & Okamoto, 2014). The inward movement of the F-actin meshwork drives the male nucleus toward the female nucleus, ultimately

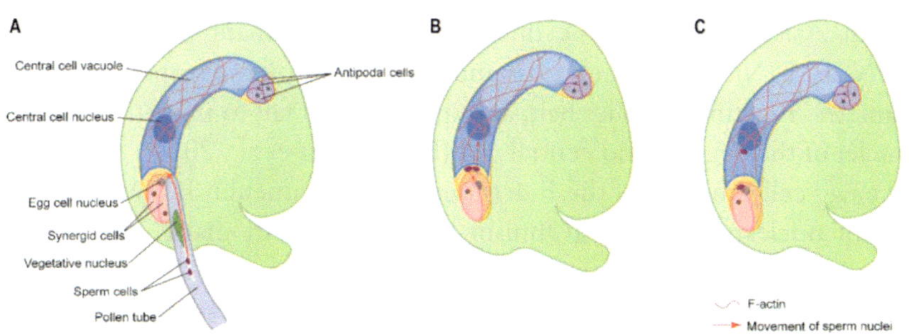

Fig. 1 Double fertilization and nuclear migration in *Arabidopsis*. (A) Sperm cells are released from the pollen tube and positioned between the egg cell and the central cell. (B) After plasmogamy, the two sperm nuclei are guided by F-actin towards the central cell and egg cell nuclei, respectively. (C) One sperm nucleus fuse with the central cell nucleus, while the other fuses with the egg cell nucleus.

leading to their fusion and the completion of karyogamy, which has been shown in *Arabidopsis*, rice, tobacco and maize (Kawashima et al., 2014; Ohnishi & Okamoto, 2017; Ohnishi et al., 2014; Peng et al., 2017). This process significantly differs from animals, where migration is primarily guided by microtubules. In flowering plants, formins and ARP2/3 are well-characterized actin nucleators that polymerize G-actin into F-actin. ROP and WASP/SCAR regulate F-actin dynamics. The LINC complex is known to bridge the nucleus and F-actin, and motor proteins like myosin also involves in F-actin dynamic movement. This section will cover these factors individually, with a primary focus on findings in *Arabidopsis* and rice. Fertilization mechanisms in algae have been reviewed in Fatema et al. (2019).

2.1 F-actin is essential for sperm nuclear migration in both the egg cell and central cell

In most animals and yeast, microtubules control pronuclear migration (see sections on Animals and Yeast). However, in flowering plants, sperm nuclear migration depends on F-actin dynamics. In the *Arabidopsis* unfertilized ovule, a constant inward movement of the F-actin meshwork surrounding the central cell nucleus is observed. Regardless of pollen tube arrival or sperm cell release, the central cell initiates and maintains this inward F-actin movement and subsequently stops and disassembles the actin cables once fertilization is complete. These observations suggest that F-actin dynamics are primed to facilitate sperm nuclear migration. After plasmogamy with the central cell, actin cables capture the sperm nucleus and continuously migrate it toward the central cell nucleus, driven by ongoing inward movement (Fig. 1) (Kawashima et al., 2014).

In *Arabidopsis* transgenic lines with semi-dominant negative *ACTIN* (*DN-ACTIN*), where actin assembly is specifically disrupted in female gametes, F-actin dynamics halt, and sperm nuclei fail to migrate toward the nuclei of the egg cell and central cell (Kawashima et al., 2014). Similarly, in rice egg cells, the use of the F-actin inhibitor latrunculin B (Lat B) disrupts sperm nuclear migration (Ohnishi et al., 2014). In maize egg cells and tobacco central cells, treatment with F-actin inhibitors such as cytochalasin B and latrunculin A (Lat A) also restricts sperm nuclear migration (Peng et al., 2017). These results collectively demonstrate that F-actin dynamics are crucial for sperm nuclear migration.

By contrast, microtubules are not essential for sperm nuclear migration in *Arabidopsis*, as indicated by the fact that fertilization occurs in microtubule-defective *PORCINO/+* (*por/+*) mutants (Kawashima et al., 2014).

Similarly, sperm nuclear migration is not effected in the egg cells of rice, maize, and tobacco when treated with microtubule inhibitors like oryzalin and colchicine (Ohnishi & Okamoto, 2017; Peng et al., 2017). These genetic and pharmacological findings collectively demonstrate that sperm nuclear migration is independent of microtubules in flowering plants.

2.2 Formins, but not ARP2/3, contribute to F-actin assembly in female gametes

In plants, formins and the ARP2/3 complex are well-characterized factors involved in actin nucleation (Basu et al., 2008; Kovar et al., 2006; MILLARD et al., 2004). The ARP2/3 complex, originally discovered in *Acanthamoeba*, consists of seven subunits, including two actin-related proteins, ARP2 and ARP3, and five additional subunits (Machesky et al., 1994). ARP2/3 complex functions as an actin nucleator for the formation of new filament branches (Blanchoin et al., 2000; Dayel et al., 2001; Fišerová et al., 2006). Mutants lacking ARP2/3 complex subunits exhibit defects in epidermal cell morphology and trichome distortion due to the inability to nucleate branched actin filaments (Le et al., 2003; Mathur et al., 2003; Xu et al., 2024; Yanagisawa et al., 2015; Zhang et al., 2013). Despite the presence of ARP2 transcripts in *Arabidopsis* central cells, inhibition of the ARP2/3 complex with CK-666 or mutation of ARP2/3 complex genes (*apr2* single and *arp2;arp3* double mutants) did not affect the inward F-actin meshwork movement or seed development, indicating that ARP2/3 is not involved in sperm nuclear migration (Ali et al., 2020; Peng et al., 2017).

Formins are another group of actin nucleators in flowering plants (Cvrčková et al., 2004; Deeks et al., 2002). Formins primarily produce linear actin filaments and do not require pre-existing filaments for polymerization (Valencia & Quinlan, 2021). Inhibition of formins using SMIFH2, a small molecule inhibitor, resulted in reduced F-actin movement in *Arabidopsis* central cell. Additionally, the accumulation of F-actin around the central cell nucleus was not observed, further supporting the role of formins in regulating F-actin dynamics through their actin nucleation function during sperm nuclear migration (Ali & Kawashima, 2021).

2.3 ROPs and WAVE/SCAR promote F-actin movement in an ARP2/3-independent manner

The ARP2/3 complex is activated initially by Rho-GTPases and Wiskott–Aldrich syndrome protein (WASP) family proteins, including the WAVE/SCAR family (Machesky & Insall, 1998; Machesky et al., 1999;

Weaver et al., 2003). The WASP family consists of five members: WASP, N-WASP, and the WAVE/SCAR proteins—WAVE/SCAR1, WAVE/SCAR2, and WAVE/SCAR3 (Stradal et al., 2004). WAVE/SCAR proteins are known to activate ARP2/3 and contribute to F-actin formation. The use of wiskostatin, a chemical inhibitor of the WASP domain, impairs the movement of the central cell F-actin meshwork in *Arabidopsis* (Ali et al., 2020) and has similarly affected egg cells in tobacco and maize during *in vitro* fertilization (Peng et al., 2017). Furthermore, the movement of the central cell F-actin meshwork is significantly slower in *scar2–1* and *scar2–1; scar4–1* mutants compared to wild-type plants, indicating that SCAR2 is the primary SCAR factor crucial for regulating F-actin movement during sperm nuclear migration (Ali et al., 2020).

Several Rho-GTPases in plants (ROPs) influence the organization of actin filaments in various contexts, including tip-growing cells and pollen tubes (Craddock et al., 2012; Ou & Yi, 2022). *ROP8*, expressed in the central cell of *Arabidopsis*, is essential for F-actin dynamics, as *DN-ROP8* dominant-negative mutant showed impaired F-actin assembly (Kawashima et al., 2014). Although extensive research has been conducted on other Rho-GTPase family proteins, studies specifically related to sperm nuclear migration are limited (Gu et al., 2003; Ou & Yi, 2022).

In summary, an ARP2/3-independent ROP-SCAR pathway exists in female gametes, controlling inward F-actin meshwork movement necessary for sperm nuclear migration in flowering plants. Recently, a new factor/pathway affecting actin nucleation in *Arabidopsis* epidermal cells has been proposed, where inhibition of both formins and ARP2/3 increased actin filament nucleation (Xu et al., 2024). It remains to be determined whether this new pathway also operates in gamete cells. Identifying and characterizing such factors will significantly enhance our understanding of both plant reproduction and cell biology.

2.4 Class XI myosin XI-G is involved in the sperm nuclear migration in *Arabidopsis* central cell

The forces driving F-actin dynamics are generated through the interaction between actin and myosin. Myosin serves as a key link between the F-actin meshwork and the nuclei. In *Arabidopsis* central cell, Class XI myosin *XI-G* is expressed, and the *xi-g* knockout mutant exhibits significantly slower F-actin meshwork movement compared to the wild-type (Ali et al., 2020). Additionally, after applying the 50 μM 2,3-butanedione monoxime (BDM) to *Arabidopsis* and tobacco central cells inhibits myosin activity and impairs

F-actin meshwork movement (Ali et al., 2020; Kawashima et al., 2014). This indicates that Class XI myosin is a crucial factor in regulating active F-actin movement in plant gamete cells. Further, treating rice egg cells with both 50 mM BDM and 20 μM N-ethylmaleimide (NEM), another myosin inhibitor, also inhibits F-actin meshwork movement and arrests sperm nuclear migration (Ohnishi & Okamoto, 2017).

In *Arabidopsis* somatic cells, myosin XI-I interacts with outer nuclear membrane proteins, tail-anchored protein 1 (WIT1) and WIT2, anchoring to the nuclear membrane (Meier et al., 2017). Defects in nuclear movement have been observed in *xi-i* mutants (Tamura et al., 2013). WIT1 proteins also interact with WPP domain-interacting proteins (WIP), which form part of the Linker of Nucleoskeleton and Cytoskeleton (LINC) complex along with Sad1/UNC-84 (SUN) proteins (Evans et al., 2014; Zhou et al., 2012). Members of the LINC complex have been recently identified in maize (Gumber et al., 2019). While the *xi-i* mutant does not show fertilization defects (Tamura et al., 2013), it remains unclear whether XI-G interacts with WIT1/2. Further investigation is needed to elucidate the specific role of the LINC complex in sperm nuclear migration, if any.

2.5 Calcium concentration levels are associated with the speed of actin meshwork movement

In *Arabidopsis*, calcium concentration increases in the fertilized egg cell following plasmogamy (Denninger et al., 2014). In maize, karyogamy fails when the Ca^{2+}-channel inhibitor gadolinium (Gd^{3+}) is applied (Antoine et al., 2001). In rice, the F-actin meshwork surrounds the entire egg cell and nucleus, facilitating sperm nucleus migration toward the egg nucleus within approximately 20 to 30 min after plasmogamy. Following sperm-egg cell fusion, the movement speed of actin filaments is 1.4 times faster compared to unfused cells, and this increase in speed appears to correlate with elevated Ca^{2+} levels in the fused gametes. Notably, the nuclear fusion of two egg nuclei is significantly accelerated with the exogenous application of Ca^{2+} (Ohnishi & Okamoto, 2017; Ohnishi et al., 2019).

In *Arabidopsis* and tobacco, tip-localized ROPs regulate both the dynamics of F-actin and the oscillation of the Ca^{2+} gradient during tip growth (Fu et al., 2001; Gu et al., 2003). However, whether calcium plays a role in ROP-mediated signaling during nuclear migration remains to be elucidated.

3. Pronuclear migration in animals

In flowering plants, the organization of sperm nuclear migration is directed by the dynamics of F-actin meshwork. By contrast, in most animals, microtubules guide pronuclear migration during fertilization. Sperm cell migrate with sperm motility dependent on the sperm flagella or cilia, which are composed of microtubules (Gibbons, 1981). Upon reaching the egg cell, the sperm cell membrane ruptures, allowing the sperm pronucleus, along with the centrosome (absent in the case of mice), to enter the egg cell, completing sperm-egg fusion. In the sperm fused egg cell, male and female pronuclei migrate to the cell center and undergo the first embryonic mitosis (Dunkley et al., 2022; Meaders & Burgess, 2020; Scheffler et al., 2021).

Research on early embryos has largely focused on the polarized distribution of anterior-posterior and asymmetric division, with relatively limited studies on pronuclear migration (Chaigne et al., 2017; Galli et al., 2008; Xiong et al., 2011). Additionally, pronuclear migration varies across different animal species, complicating the identification of conserved mechanisms. Model organisms such as sea urchins, *Caenorhabditis elegans* (*C. elegans*), and mice have been frequently used to explore the intricate details of sperm-egg fusion.

3.1 Microtubules drive the migration of male and female pronuclei

The sea urchin was the first animal in which pronuclear migration was observed (Chambers, 1939). In sea urchins, the sperm aster, formed by microtubules, plays a role in positioning the male and female pronuclei, and the pronuclear migration occurs in three distinct phases. In the first phase, the sperm pronucleus penetrates the egg and migrates slowly with no clear direction. In the second phase, the microtubule sperm aster captures the female pronucleus and directs it toward the egg center. In the third phase, the male and female pronuclei approach the egg center together at a slow speed (Fig. 2) (Chambers, 1939; Tanimoto et al., 2016). The migration of the male pronucleus is inhibited by the application of microtubule inhibitors such as nocodazole and the dynein inhibitor ciliobrevin D. However, pronuclear migration still occurs in the presence of the F-actin inhibitor Lat B (Tanimoto et al., 2016).

In *C. elegans*, the male and female pronuclei are initially positioned at opposite ends of the egg cell, with the female pronucleus at the anterior and

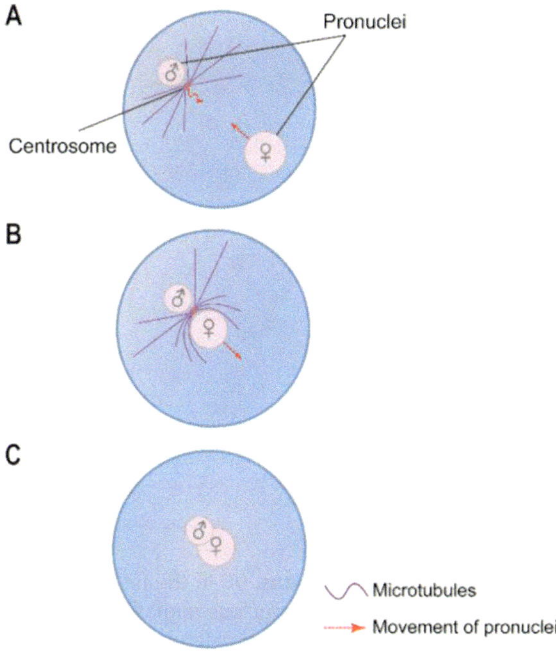

Fig. 2 Pronuclear migration in sea urchin egg cells. (A) The sperm pronucleus migrates in a slow speed with no direction in the first phase. (B) The microtubule sperm aster captures the female pronucleus and migrates to the cell center with male pronucleus in the second phase. (C) In the third phase, male and female pronuclei migrate to the cell center at slow speed, where nuclear fusion occurs.

the male pronucleus at the posterior. Establishing these poles is crucial for subsequent nuclear migration. Similar to sea urchins, there are three main phases of pronuclear migration in *C. elegans*. In the first phase, both the male pronucleus that associated with two centrosomes, and the female pronucleus migrate slowly toward the cell center. In the second phase, spindle microtubules generated by the centrosomes, capture and pull the female pronucleus toward the male pronucleus at an accelerated rate. The distance between the united pronuclei and the posterior pole at this stage is critical for ensuring proper nuclear migration and successful fertilization. In the third phase, the male and female pronuclei rotate 90° from the anterior-posterior axis to a vertical orientation while slowly migrating to the cell center, facilitating chromosome alignment and zygote formation (Fig. 3) (Kimura & Kimura, 2011; Shinar et al., 2011). Pronuclear migration defects have been observed in RNA-mediated interference (RNAi) experiments targeting gut on the exterior (*gex*), where the female pronucleus continued to migrate toward the

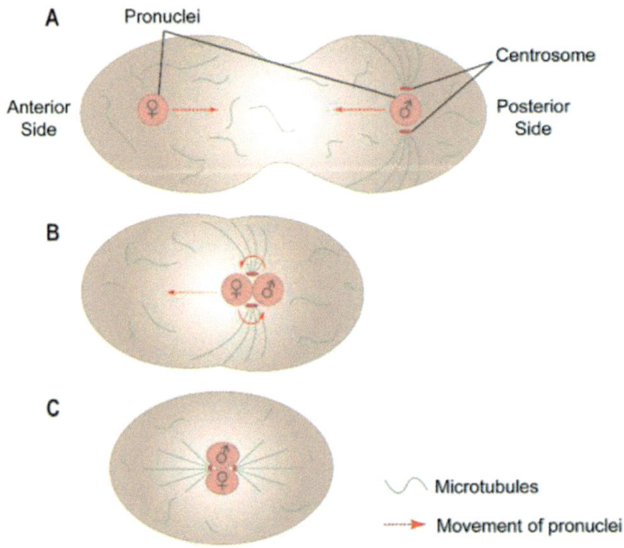

Fig. 3 Pronuclear migration in *C. elegans*. (A) In the first phase, female pronucleus and two centrosomes with the male pronucleus migrate towards the center of egg cell. (B) Microtubules capture and pull the female pronucleus toward the male pronucleus. The combination of the male and female pronuclei rotates 90° and migrates to the cell center. (C) Nuclear fusion occurs at the cell center.

posterior side after contacting the male pronucleus in the second phase (Xiong et al., 2011). Additional defects in nuclear migration were confirmed in RNAi mutants of microtubule-stabilizing factors such as transforming acidic coiled-coil protein (*tac-1*), dynein heavy chain-1 (*dhc-1*), and the human type I lissencephaly responsible gene *lis-1* (Cockell et al., 2004; Faulkner et al., 2000; Gönczy et al., 1999; Vazquez-Pianzola et al., 2022).

In conclusion, microtubules play a decisive role in pronuclear migration. Due to the lack of time-lapse live imaging data for other animals, the detailed mechanisms of male and female pronuclear migration remain elusive.

3.2 Both actin filaments and microtubules are involved in pronuclear migration in mice

In mice, both actin filaments and microtubules are essential for pronuclear migration, distinguishing this process from other animal species. Although the role of F-actin in controlling zygotic microtubule spindle positioning and asymmetric division in mice has been demonstrated (Chaigne et al. 2016, 2017), its specific function in pronuclear migration has been less explored. However, Scheffler et al. (2021) utilized time-lapse live-cell confocal and super-resolution microscopy, combined with *in vitro* fertilization techniques to observe that F-actin and microtubules are involved in distinct phases of pronuclear migration in mice.

During the first phase, a fertilization cone forms at the egg cell's surface upon sperm contact. After the sperm pronucleus entering the egg cell, it moves toward the cell center at a speed six times faster than the female pronucleus. This rapid movement is dependent on cortical actin filament nucleation factors Formin2 and Spire2 (Fig. 4A) (Montaville et al., 2014; Schuh, 2011). Formin2 and Spire2, enriched behind the male pronucleus, generate actin filaments that provide the force to push the male pronucleus towards the cell center. Overexpression of Spire2 enhances F-actin formation and accelerates the male pronucleus's movement. Conversely, expression of the dominant-negative-FH2 form of Formin2 impairs its interaction with Spire2, decreases cytoplasmic F-actin velocity, and slows

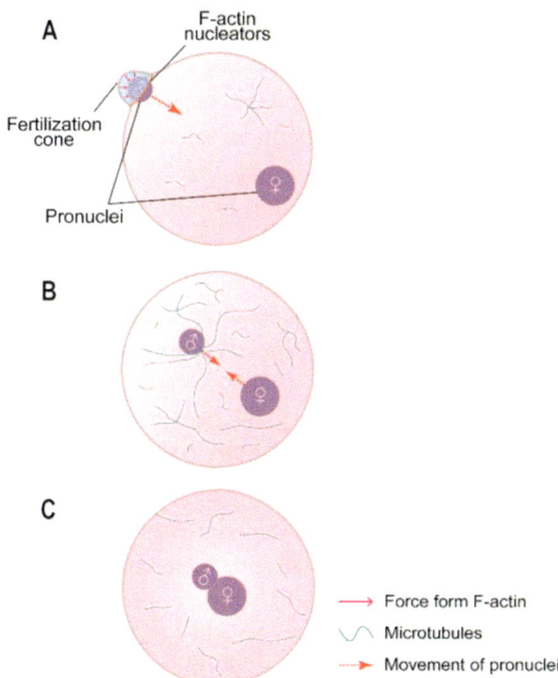

Fig. 4 Pronuclear migration in mouse egg cells. (A) In the first phase, male pronucleus is pushed by F-actin polymerized in the fertilization cone to the cell center. (B) In the second phase, both male and female pronuclei migrate towards egg cell center slowly. (C) Nuclear fusion occurs at the cell center.

the male pronucleus's movement, although it does not prevent the pronucleus from eventually reaching the cell center. This indicates that Formin2 and Spire2 primarily regulate F-actin on the cortex during the first phase. Additionally, the actin depolymerizing drug cytochalasin D inhibits pronuclear migration, suggesting that both cortical and cytoplasmic actin filaments are involved throughout this phase. On the other hand, the male pronucleus continues to migrate rapidly despite the application of the microtubule inhibitor nocodazole, indicating that microtubules are not crucial during the first phase (Scheffler et al., 2021). Moreover, the expression of the dominant-negative Myosin-Vb tail in fertilized mouse eggs results in defective nuclear migration, further implicating myosin playing important role in pronuclear migration (Chaigne et al., 2016).

In the second phase, once both pronuclei have moved near the cell center, they are guided together to complete their movement at a slower pace (Fig. 4B and C). Although nocodazole application does not affect the speed of male pronuclear migration in the first phase, it dramatically decreases the speed of pronuclear movement in the second phase. The pronuclei fail to fuse due to the excessive distance apart, indicating that microtubules are essential for this phase. Furthermore, the application of the dynein inhibitor p150-CC1 peptide reduces the movement of both male and female pronuclei, suggesting that the motor protein dynein together with microtubules, participates in pronuclear migration (Scheffler et al., 2021).

In summary, F-actin plays a continuous role throughout the entire pronuclear migration process in mice, while microtubules and dynein are crucial for guiding the pronuclei, especially during the second phase.

3.3 The interaction between SUN-domain and KASH-domain proteins, dependent on kinesin or dynein, is required for pronuclear migration

In animals, SUN-domain and KASH-domain protein families on the nuclear envelope of male and female pronuclei are crucial for linking the F-actin and microtubule networks during pronuclear migration. In *C. elegans*, SUN-1 interacts with the KASH protein ZYG-12. Disruption of ZYG-12 leads to the detachment of the centrosome from the male pronucleus, illustrating its role in pronuclear positioning (Malone et al., 2003; Minn et al., 2009). Additionally, another SUN-domain protein, UNC-84, interacts with the KASH-domain proteins UNC-83 and ANC-1 in *C. elegans* (Malone et al., 1999; Starr & Han, 2002; Starr et al., 2001).

The interaction between UNC-84 and UNC-83 is mediated by kinesin-1 and dynein (Fridolfsson et al., 2010; Meyerzon et al., 2009). Mutants lacking UNC-84 exhibit defects in pronuclear rotation and centration, highlighting the importance of these interactions in proper pronuclear migration (Xiong et al., 2011).

In summary, SUN-domain and KASH-domain proteins have been biologically demonstrated to play crucial roles in pronuclear migration in *C. elegans*. However, similar experimental evidence in plants is still lacking, and future research may provide insights into the roles of SUN and KASH proteins in plant cells.

4. Nuclear migration in yeast

In the sexual reproduction of yeast (*Saccharomyces cerevisiae*), two haploid yeast cells of different mating types (Mat a and Mat α) undergo cell-cell fusion. The haploid nuclei then engage in a process referred to as "nuclear mating" or "nuclear congression", where they move toward each other and complete karyogamy to form a diploid zygote (Herskowitz, 1988). Unlike the movement of the male nucleus towards the female nucleus in flowering plants and most animals, the nuclei in sexual cells Mat a and Mat α of yeast move towards the cell's center simultaneously, which was revealed by microscopy live-cell imaging (Fig. 5) (Gibeaux et al., 2013). This nuclear mating process depends on microtubules, which are polymerized and controlled by the spindle pole body (Rose et al., 1986). The nuclei, connected to the spindle pole bodies, move along the microtubules until the spindle pole bodies contact each other, leading to nuclear fusion.

4.1 Spc72 interacts with Kar1 and Kar3 to generate the microtubule-spindle pole body complex and facilitate movement along microtubules

Using the microtubule inhibitor nocodazole, significant blockage of nuclear mating was observed (Hašek et al., 1987). In mutants related to karyogamy, such as Kar1, Kar3, Kar4, and Kar9, which encode kinesin motors (Conde & Fink, 1976; Kurihara et al., 1994), as well as the β-tubulin gene Tub2 mutant (Huffaker et al., 1988), or the Bik1 mutant, which encodes a microtubule-associated protein (Berlin et al., 1990), all exhibit microtubule defects resulting in nuclear mating failure. These studies highlight that the spindle pole body acts as the microtubule

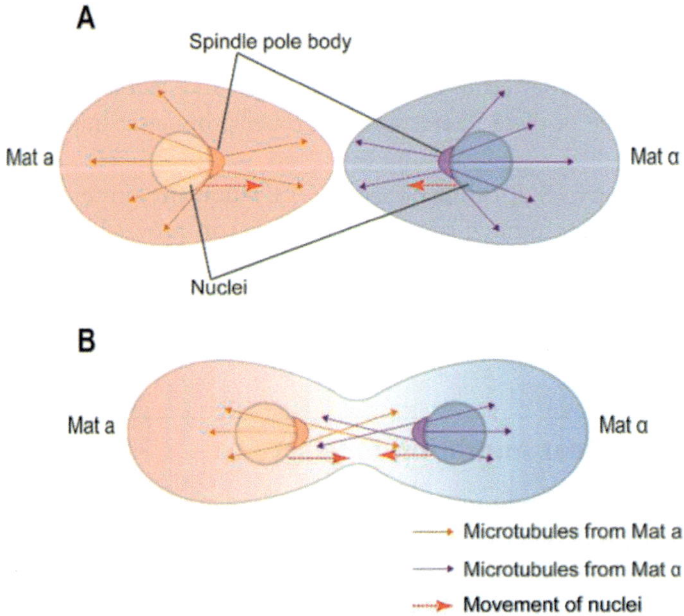

Fig. 5 Nuclear mating in yeast sexual cells. (A and B) Microtubules generating from spindle pole bodies on the nuclei of Mat a and Mat α guide nuclei migrate towards the cell's center simultaneously.

organizing center and guides the nuclei towards each other. Spc72 is a γ-tubulin complex-binding protein localized at the spindle pole body, while Kar1 is a kinesin motor protein also localized at the spindle pole body and microtubules. In Mat α, Kar1Δ15 mutants with defective Spc72 interaction domains, and wild-type Mat a, microtubules polymerized from Mat α disconnect from the opposite side of the spindle pole bodies during nuclear mating. This indicates that Spc72 interacts with Kar1 to form a microtubule-spindle pole body complex, guiding the two nuclei from opposite sides closer to the center (Fig. 5) (Gibeaux et al., 2013; Pereira et al., 1999). Spc72 also binds with Kar3 to make pulling forces on microtubules for nuclear mating (Gibeaux et al., 2013). In yeast, the centrosome embeds directly into the nuclear envelope, which differs from animals. The Sad1 protein, an inner nuclear membrane SUN-family protein, accumulates at the nuclear periphery and spindle pole body, playing a role in the structure of the spindle pole body by maintaining its functional interface with the nuclear membrane and providing an anchor for microtubule motor proteins (Hagan & Yanagida, 1995).

In yeast, even when one microtubule-spindle pole body is defective, the coordinated pulling forces making by both spindle poles enhance the overall robustness of nuclear congression, ensuring successful nuclear fusion and providing greater stability compared to mechanisms observed in plants and animals.

5. Conclusion

Recent advances in understanding gamete nuclear migration in flowering plants have provided valuable insights into this complex process, yet several questions remain unresolved. Key discoveries include the identification of Class XI myosin XI-G as crucial for nuclear migration and the characterization of the ROP8-SCAR2 pathway, which promotes F-actin nucleation through an ARP2/3-independent mechanism (Ali et al., 2020). This finding contrasts with the ARP2/3-dependent mechanisms observed in somatic cells (Xu et al., 2024). However, experimental evidence linking F-actin interactions with the KASH-SUN complex and motor proteins during nuclear migration in sexual reproduction is still lacking, despite observations of such interactions in stomatal cells (Moser et al., 2024). The specific genes regulating F-actin dynamics in *Arabidopsis* egg cells have not been identified. Although *in vitro* fertilization experiments have provided insights into sperm nuclear migration in species like rice, tobacco, and maize (Ali et al., 2020; Peng et al., 2017), detailed mechanisms in central cells remain unclear. The role of calcium signaling in modulating the rate of inward F-actin movement during sperm nuclear migration also need further investigation (Ohnishi & Okamoto, 2017).

We are only beginning to unravel the molecular intricacies of gamete nuclear migration in flowering plants. In flowering plants, only F-actin is employed, whereas in animals, either microtubules or both cytoskeletal elements are recruited. Moreover, F-actin is already polymerized and exhibits dynamic movement in unfertilized ovules, whereas in most animals, microtubule dynamics typically begin after gamete fusion. In flowering plants, the immotility of sperm cells necessitates reliance on F-actin of the egg and central cell to facilitate nuclear migration. Conversely, in animals, sperm cells are motile, utilizing microtubule-based flagella for movement toward the egg, with microtubules subsequently driving pronuclear migration after fusion. A key question is whether the immotility of plant sperm cells has driven the evolution of actin-dependent nuclear migration systems. This highlights the need to explore the evolutionary and functional reasons behind these distinct strategies in flowering plants and animals. Exploring plant-to-animal evolutionary process may offer insight into how these divergent mechanisms have evolved to suit the distinct reproductive strategies of the two kingdoms.

The coordination of processes from pollen migration and plasmogamy to sperm nuclear migration, karyogamy, and the initiation of embryo and endosperm development is critical for successful seed development. In flowering plants, a distinct mechanism has evolved where F-actin plays a central role instead of microtubules (reviewed in Davidson & Wood, 2016; Dresselhaus et al., 2016; Paez-Garcia et al., 2018). Further investigation into nuclear migration and related pre- and post-fertilization processes will not only deepen our understanding of plant reproduction but also provide valuable insights into plant cytoskeleton biology and the evolutionary pathways of plant reproduction.

References

Ali, M. F., Fatema, U., Peng, X., Hacker, S. W., Maruyama, D., Sun, M.-X., & Kawashima, T. (2020). ARP2/3-independent WAVE/SCAR pathway and class XI myosin control sperm nuclear migration in flowering plants. *Proceedings of the National Academy of Sciences, 117*, 32757–32763. https://doi.org/10.1073/pnas.2015550117.

Ali, M. F., & Kawashima, T. (2021). Formins control dynamics of F-actin in the central cell of Arabidopsis thaliana. *Plant Signaling & Behavior, 16*, 1920192. https://doi.org/10.1080/15592324.2021.1920192.

Antoine, A.-F., Faure, J.-E., Dumas, C., & Feijó, J. A. (2001). Differential contribution of cytoplasmic Ca2+ and Ca2+ influx to gamete fusion and egg activation in maize. *Nature Cell Biology, 3*, 1120–1123. https://doi.org/10.1038/ncb1201-1120.

Basu, D., Le, J., Zakharova, T., Mallery, E. L., & Szymanski, D. B. (2008). A SPIKE1 signaling complex controls actin-dependent cell morphogenesis through the heteromeric WAVE and ARP2/3 complexes. *Proceedings of the National Academy of Sciences, 105*, 4044–4049. https://doi.org/10.1073/pnas.0710294105.

Berlin, V., Styles, C. A., & Fink, G. R. (1990). BIK1, a protein required for microtubule function during mating and mitosis in Saccharomyces cerevisiae, colocalizes with tubulin. *The Journal of Cell Biology, 111*, 2573–2586. https://doi.org/10.1083/jcb.111.6.2573.

Blanchoin, L., Amann, K. J., Higgs, H. N., Marchand, J.-B., Kaiser, D. A., & Pollard, T. D. (2000). Direct observation of dendritic actin filament networks nucleated by Arp2/3 complex and WASP/Scar proteins. *Nature, 404*, 1007–1011. https://doi.org/10.1038/35010008.

Carvalho-Santos, Z., Azimzadeh, J., Pereira-Leal, JoséB., & Bettencourt-Dias, M. (2011). Tracing the origins of centrioles, cilia, and flagella. *Journal of Cell Biology, 194*, 165–175. https://doi.org/10.1083/jcb.201011152.

Chaigne, A., Campillo, C., Voituriez, R., Gov, N. S., Sykes, C., Verlhac, M.-H., & Terret, M.-E. (2016). F-actin mechanics control spindle centring in the mouse zygote. *Nature Communications, 7*, 10253. https://doi.org/10.1038/ncomms10253.

Chaigne, A., Terret, M.-E., & Verlhac, M.-H. (2017). Asymmetries and symmetries in the mouse oocyte and zygote. In J.-P. Tassan, & J. Z. Kubiak (Eds.). *Asymmetric cell division in development, differentiation and cancer* (pp. 285–299). Cham: Springer International Publishing. https://doi.org/10.1007/978-3-319-53150-2_13.

Chambers, E. L. (1939). The movement of the egg nucleus in relation to the sperm aster in the echinoderm egg. *Journal of Experimental Biology, 16*, 409–424. https://doi.org/10.1242/jeb.16.4.409.

Cockell, M. M., Baumer, K., & Gönczy, P. (2004). lis-1 is required for dynein-dependent cell division processes in *C. elegans* embryos. *Journal of Cell Science, 117*, 4571–4582. https://doi.org/10.1242/jcs.01344.

Conde, J., & Fink, G. R. (1976). A mutant of Saccharomyces cerevisiae defective for nuclear fusion. *Proceedings of the National Academy of Sciences of the United States of America, 73*, 3651–3655.

Craddock, C., Lavagi, I., & Yang, Z. (2012). New insights into Rho signaling from plant ROP/Rac GTPases. *Trends in Cell Biology, 22*, 492–501. https://doi.org/10.1016/j.tcb.2012.05.002.

Cvrčková, F., Novotný, M., Pícková, D., & Žárský, V. (2004). Formin homology 2 domains occur in multiple contexts in angiosperms. *BMC Genomics, 5*, 44. https://doi.org/10.1186/1471-2164-5-44.

Davidson, A. J., & Wood, W. (2016). Unravelling the actin cytoskeleton: A new competitive edge? *Trends in Cell Biology, 26*, 569–576. https://doi.org/10.1016/j.tcb.2016.04.001.

Dayel, M. J., Holleran, E. A., & Mullins, R. D. (2001). Arp2/3 complex requires hydrolyzable ATP for nucleation of new actin filaments. *Proceedings of the National Academy of Sciences, 98*, 14871–14876. https://doi.org/10.1073/pnas.261419298.

Deeks, M. J., Hussey, P. J., & Davies, B. (2002). Formins: Intermediates in signal-transduction cascades that affect cytoskeletal reorganization. *Trends in Plant Science, 7*, 492–498. https://doi.org/10.1016/S1360-1385(02)02341-5.

Denninger, P., Bleckmann, A., Lausser, A., Vogler, F., Ott, T., Ehrhardt, D. W., ... Grossmann, G. (2014). Male–female communication triggers calcium signatures during fertilization in Arabidopsis. *Nature Communications, 5*, 4645. https://doi.org/10.1038/ncomms5645.

Dresselhaus, T., Sprunck, S., & Wessel, G. M. (2016). Fertilization mechanisms in flowering plants. *Current Biology, 26*, R125–R139. https://doi.org/10.1016/j.cub.2015.12.032.

Dunkley, S., Scheffler, K., & Mogessie, B. (2022). Cytoskeletal form and function in mammalian oocytes and zygotes. *Current Opinion in Cell Biology, 75*, 102073. https://doi.org/10.1016/j.ceb.2022.02.007.

Evans, D. E., Pawar, V., Smith, S. J., & Graumann, K. (2014). Protein interactions at the higher plant nuclear envelope: Evidence for a linker of nucleoskeleton and cytoskeleton complex. *Frontiers in Plant Science, 5*. https://doi.org/10.3389/fpls.2014.00183.

Fatema, U., Ali, M. F., Hu, Z., Clark, A. J., & Kawashima, T. (2019). Gamete nuclear migration in animals and plants. *Frontiers in Plant Science, 10*. https://doi.org/10.3389/fpls.2019.00517.

Faulkner, N. E., Dujardin, D. L., Tai, C.-Y., Vaughan, K. T., O'Connell, C. B., Wang, Y., & Vallee, R. B. (2000). A role for the lissencephaly gene LIS1 in mitosis and cytoplasmic dynein function. *Nature Cell Biology, 2*, 784–791. https://doi.org/10.1038/35041020.

Fišerová, J., Schwarzerová, K., Petrášek, J., & Opatrný, Z. (2006). ARP2 and ARP3 are localized to sites of actin filament nucleation in tobacco BY-2 cells. *Protoplasma, 227*, 119–128. https://doi.org/10.1007/s00709-006-0146-6.

Fridolfsson, H. N., Ly, N., Meyerzon, M., & Starr, D. A. (2010). UNC-83 coordinates kinesin-1 and dynein activities at the nuclear envelope during nuclear migration. *Developmental Biology, 338*, 237–250. https://doi.org/10.1016/j.ydbio.2009.12.004.

Fu, Y., Wu, G., & Yang, Z. (2001). ROP gtpase-dependent dynamics of tip-localized F-actin controls tip growth in pollen tubes. *The Journal of Cell Biology, 152*, 1019–1032.

Galli, M., Heuvel, S., & van den (2008). Determination of the cleavage plane in early *C. elegans* embryos. *Annual Review of Genetics, 42*, 389–411. https://doi.org/10.1146/annurev.genet.40.110405.090523.

Gibbons, I. R. (1981). Cilia and flagella of eukaryotes. *Journal of Cell Biology, 91*, 107s–124s. https://doi.org/10.1083/jcb.91.3.107s.

Gibeaux, R., Politi, A. Z., Nédélec, F., Antony, C., & Knop, M. (2013). Spindle pole body-anchored Kar3 drives the nucleus along microtubules from another nucleus in preparation for nuclear fusion during yeast karyogamy. *Genes & Development, 27*, 335–349. https://doi.org/10.1101/gad.206318.112.

Gönczy, P., Pichler, S., Kirkham, M., & Hyman, A. A. (1999). Cytoplasmic dynein is required for distinct aspects of mtoc positioning, including centrosome separation, in the one cell stage Caenorhabditis elegans embryo. *The Journal of Cell Biology, 147*, 135–150.

González-Gutiérrez, A. G., Gutiérrez-Mora, A., Verdín, J., & Rodríguez-Garay, B. (2021). An F-Actin Mega-Cable Is Associated With the Migration of the Sperm Nucleus During the Fertilization of the Polarity-Inverted Central Cell of Agave inaequidens. *Frontiers of plant science, 12*. https://doi.org/10.3389/fpls.2021.774098.

Gu, Y., Vernoud, V., Fu, Y., & Yang, Z. (2003). ROP GTPase regulation of pollen tube growth through the dynamics of tip-localized F-actin. *Journal of Experimental Botany, 54*, 93–101. https://doi.org/10.1093/jxb/erg035.

Gumber, H. K., McKenna, J. F., Tolmie, A. F., Jalovec, A. M., Kartick, A. C., Graumann, K., & Bass, H. W. (2019). MLKS2 is an ARM domain and F-actin-associated KASH protein that functions in stomatal complex development and meiotic chromosome segregation. *Nucleus, 10*, 144–166. https://doi.org/10.1080/19491034.2019.1629795.

Hagan, I., & Yanagida, M. (1995). The product of the spindle formation gene sad1+ associates with the fission yeast spindle pole body and is essential for viability. *The Journal of Cell Biology, 129*, 1033–1047. https://doi.org/10.1083/jcb.129.4.1033.

Hamamura, Y., Nagahara, S., & Higashiyama, T. (2012). Double fertilization on the move. *Current Opinion in Plant Biology, Growth and Development, 15*, 70–77. https://doi.org/10.1016/j.pbi.2011.11.001.

Hašek, J., Rupeš, I., Svobodová, J., & Streiblová, E. (1987). Tubulin and actin topology during zygote formation of Saccharomyces cerevisiae. *Microbiology (Reading, England), 133*, 3355–3363. https://doi.org/10.1099/00221287-133-12-3355.

Herskowitz, I. (1988). Life cycle of the budding yeast Saccharomyces cerevisiae. *Microbiological Reviews, 52*, 536–553. https://doi.org/10.1128/mr.52.4.536-553.1988.

Huffaker, T. C., Thomas, J. H., & Botstein, D. (1988). Diverse effects of beta-tubulin mutations on microtubule formation and function. *The Journal of Cell Biology, 106*, 1997–2010. https://doi.org/10.1083/jcb.106.6.1997.

Kawashima, T., Maruyama, D., Shagirov, M., Li, J., Hamamura, Y., Yelagandula, R., ... Berger, F. (2014). Dynamic F-actin movement is essential for fertilization in Arabidopsis thaliana. *eLife, 3*, e04501. https://doi.org/10.7554/eLife.04501.

Kimura, K., & Kimura, A. (2011). Intracellular organelles mediate cytoplasmic pulling force for centrosome centration in the Caenorhabditis elegans early embryo. *Proceedings of the National Academy of Sciences, 108*, 137–142. https://doi.org/10.1073/pnas.1013275108.

Kovar, D. R., Harris, E. S., Mahaffy, R., Higgs, H. N., & Pollard, T. D. (2006). Control of the assembly of ATP- and ADP-actin by formins and profilin. *Cell, 124*, 423–435. https://doi.org/10.1016/j.cell.2005.11.038.

Kurihara, L. J., Beh, C. T., Latterich, M., Schekman, R., & Rose, M. D. (1994). Nuclear congression and membrane fusion: Two distinct events in the yeast karyogamy pathway. *Journal of Cell Biology, 126*, 911–923. https://doi.org/10.1083/jcb.126.4.911.

Le, J., El-Assal, S. E.-D., Basu, D., Saad, M. E., & Szymanski, D. B. (2003). Requirements for arabidopsis ATARP2 and ATARP3 during epidermal development. *Current Biology, 13*, 1341–1347. https://doi.org/10.1016/S0960-9822(03)00493-7.

Machesky, L. M., Atkinson, S. J., Ampe, C., Vandekerckhove, J., & Pollard, T. D. (1994). Purification of a cortical complex containing two unconventional actins from Acanthamoeba by affinity chromatography on profilin-agarose. *Journal of Cell Biology, 127*, 107–115. https://doi.org/10.1083/jcb.127.1.107.

Machesky, L. M., & Insall, R. H. (1998). Scar1 and the related Wiskott–Aldrich syndrome protein, WASP, regulate the actin cytoskeleton through the Arp2/3 complex. *Current Biology, 8*, 1347–1356. https://doi.org/10.1016/S0960-9822(98)00015-3.

Machesky, L. M., Mullins, R. D., Higgs, H. N., Kaiser, D. A., Blanchoin, L., May, R. C., ... Pollard, T. D. (1999). Scar, a WASp-related protein, activates nucleation of actin filaments by the Arp2/3 complex. *Proceedings of the National Academy of Sciences, 96*, 3739–3744. https://doi.org/10.1073/pnas.96.7.3739.

Malone, C. J., Fixsen, W. D., Horvitz, H. R., & Han, M. (1999). UNC-84 localizes to the nuclear envelope and is required for nuclear migration and anchoring during *C. elegans* development. *Development (Cambridge, England), 126*, 3171–3181. https://doi.org/10.1242/dev.126.14.3171.

Malone, C. J., Misner, L., Bot, N. L., Tsai, M.-C., Campbell, J. M., Ahringer, J., & White, J. G. (2003). The *C. elegans* hook protein, ZYG-12, mediates the essential attachment between the centrosome and nucleus. *Cell, 115*, 825–836. https://doi.org/10.1016/S0092-8674(03)00985-1.

Manandhar, G., Schatten, H., & Sutovsky, P. (2005). Centrosome reduction during gametogenesis and its significance1. *bire, 72*, 2–13. https://doi.org/10.1095/biolreprod.104.031245.

Mathur, J., Mathur, N., Kirik, V., Kernebeck, B., Srinivas, B. P., & Hülskamp, M. (2003). Arabidopsis CROOKED encodes for the smallest subunit of the ARP2/3 complex and controls cell shape by region specific fine F-actin formation. *Development (Cambridge, England), 130*, 3137–3146. https://doi.org/10.1242/dev.00549.

Meaders, J. L., & Burgess, D. R. (2020). Microtubule-based mechanisms of pronuclear positioning. *Cells, 9*, 505. https://doi.org/10.3390/cells9020505.

Meier, I., Richards, E. J., & Evans, D. E. (2017). Cell biology of the plant nucleus. *Annual Review of Plant Biology, 68*, 139–172. https://doi.org/10.1146/annurev-arplant-042916-041115.

Meyerzon, M., Fridolfsson, H. N., Ly, N., McNally, F. J., & Starr, D. A. (2009). UNC-83 is a nuclear-specific cargo adaptor for kinesin-1-mediated nuclear migration. *Development (Cambridge, England), 136*, 2725–2733. https://doi.org/10.1242/dev.038596.

Millard, T. H., Sharp, S. J., & Machesky, L. M. (2004). Signalling to actin assembly via the WASP (Wiskott-Aldrich syndrome protein)-family proteins and the Arp2/3 complex. *Biochemical Journal, 380*, 1–17. https://doi.org/10.1042/bj20040176.

Minn, I., Rolls, M. M., Hanna-Rose, W., & Malone, C. J. (2009). SUN-1 and ZYG-12, mediators of centrosome–nucleus attachment, are a functional SUN/KASH pair in Caenorhabditis elegans. *MBoC, 20*, 4586–4595. https://doi.org/10.1091/mbc.e08-10-1034.

Montaville, P., Jégou, A., Pernier, J., Compper, C., Guichard, B., Mogessie, B., ... Carlier, M.-F. (2014). Spire and formin 2 synergize and antagonize in regulating actin assembly in meiosis by a ping-pong mechanism. *PLoS Biology, 12*, e1001795. https://doi.org/10.1371/journal.pbio.1001795.

Moser, M., Groves, N. R., & Meier, I. (2024). Plant KASH proteins SINE1 and SINE2 have synergistic and antagonistic interactions with actin-branching and actin-bundling factors. *Journal of Experimental Botany, 75*, 73–87. https://doi.org/10.1093/jxb/erad400.

Ohnishi, Y., Hoshino, R., & Okamoto, T. (2014). Dynamics of male and female chromatin during karyogamy in rice zygotes. *Plant Physiology, 165*, 1533–1543. https://doi.org/10.1104/pp.114.236059.

Ohnishi, Y., Kokubu, I., Kinoshita, T., & Okamoto, T. (2019). Sperm entry into the egg cell induces the progression of karyogamy in rice zygotes. *Plant and Cell Physiology, 60*, 1656–1665. https://doi.org/10.1093/pcp/pcz077.

Ohnishi, Y., & Okamoto, T. (2017). Nuclear migration during karyogamy in rice zygotes is mediated by continuous convergence of actin meshwork toward the egg nucleus. *Journal of Plant Research, 130*, 339–348. https://doi.org/10.1007/s10265-016-0892-2.

Ou, H., & Yi, P. (2022). ROP GTPase-dependent polarity establishment during tip growth in plants. *New Phytologist, 236*, 49–57. https://doi.org/10.1111/nph.18373.

Paez-Garcia, A., Sparks, J. A., de Bang, L., & Blancaflor, E. B. (2018). Plant actin cytoskeleton: New functions from old scaffold. In V. P. Sahi, & F. Baluška (Eds.). *Concepts in cell biology – History and evolution* (pp. 103–137). Cham: Springer International Publishing. https://doi.org/10.1007/978-3-319-69944-8_6.

Peng, X., Yan, T., & Sun, M. (2017). The WASP-Arp2/3 complex signal cascade is involved in actin-dependent sperm nuclei migration during double fertilization in tobacco and maize. *Scientific Reports, 7*, 43161. https://doi.org/10.1038/srep43161.

Pereira, G., Grueneberg, U., Knop, M., & Schiebel, E. (1999). Interaction of the yeast γ-tubulin complex-binding protein Spc72p with Kar1p is essential for microtubule function during karyogamy. *The EMBO Journal, 18*, 4180–4195. https://doi.org/10.1093/emboj/18.15.4180.

Raghavan, V. (2003). Some reflections on double fertilization, from its discovery to the present. *New Phytologist, 159*, 565–583. https://doi.org/10.1046/j.1469-8137.2003.00846.x.

Rose, M. D., Price, B. R., & Fink, G. R. (1986). *Saccharomyces cerevisiae* nuclear fusion requires prior activation by alpha factor. *Molecular and Cellular Biology, 6*, 3490–3497. https://doi.org/10.1128/mcb.6.10.3490-3497.1986.

Scheffler, K., Uraji, J., Jentoft, I., Cavazza, T., Mönnich, E., Mogessie, B., & Schuh, M. (2021). Two mechanisms drive pronuclear migration in mouse zygotes. *Nature Communications, 12*, 841. https://doi.org/10.1038/s41467-021-21020-x.

Schuh, M. (2011). An actin-dependent mechanism for long-range vesicle transport. *Nature Cell Biology, 13*, 1431–1436. https://doi.org/10.1038/ncb2353.

Shin, J. M., Yuan, L., & Kawashima, T. (2022). Live-cell imaging reveals the cellular dynamics in seed development. *Plant Science, 325*, 111485. https://doi.org/10.1016/j.plantsci.2022.111485.

Shinar, T., Mana, M., Piano, F., & Shelley, M. J. (2011). A model of cytoplasmically driven microtubule-based motion in the single-celled Caenorhabditis elegans embryo. *Proceedings of the National Academy of Sciences, 108*, 10508–10513. https://doi.org/10.1073/pnas.1017369108.

Starr, D. A., & Han, M. (2002). Role of ANC-1 in tethering nuclei to the actin cytoskeleton. *Science (New York, N. Y.), 298*, 406–409. https://doi.org/10.1126/science.1075119.

Starr, D. A., Hermann, G. J., Malone, C. J., Fixsen, W., Priess, J. R., Horvitz, H. R., & Han, M. (2001). unc-83 encodes a novel component of the nuclear envelope and is essential for proper nuclear migration. *Development (Cambridge, England), 128*, 5039–5050. https://doi.org/10.1242/dev.128.24.5039.

Stradal, T. E. B., Rottner, K., Disanza, A., Confalonieri, S., Innocenti, M., & Scita, G. (2004). Regulation of actin dynamics by WASP and WAVE family proteins. *Trends in Cell Biology, 14*, 303–311. https://doi.org/10.1016/j.tcb.2004.04.007.

Tamura, K., Iwabuchi, K., Fukao, Y., Kondo, M., Okamoto, K., Ueda, H., ... Hara-Nishimura, I. (2013). Myosin XI-i links the nuclear membrane to the cytoskeleton to control nuclear movement and shape in arabidopsis. *Current Biology, 23*, 1776–1781. https://doi.org/10.1016/j.cub.2013.07.035.

Tanimoto, H., Kimura, A., & Minc, N. (2016). Shape–motion relationships of centering microtubule asters. *The Journal of Cell Biology, 212*, 777–787. https://doi.org/10.1083/jcb.201510064.

Valencia, D. A., & Quinlan, M. E. (2021). Formins. *Current Biology, 31*, R517–R522. https://doi.org/10.1016/j.cub.2021.02.047.

Vazquez-Pianzola, P., Beuchle, D., Saro, G., Hernández, G., Maldonado, G., Brunßen, D., ... Suter, B. (2022). Female meiosis II and pronuclear fusion require the microtubule transport factor Bicaudal D. *Development (Cambridge, England), 149*, dev199944. https://doi.org/10.1242/dev.199944.

Weaver, A. M., Young, M. E., Lee, W.-L., & Cooper, J. A. (2003). Integration of signals to the Arp2/3 complex. *Current Opinion in Cell Biology, 15*, 23–30. https://doi.org/10.1016/S0955-0674(02)00015-7.

Xiong, H., Mohler, W. A., & Soto, M. C. (2011). The branched actin nucleator Arp2/3 promotes nuclear migrations and cell polarity in the *C. elegans* zygote. *Developmental Biology, 357*, 356–369. https://doi.org/10.1016/j.ydbio.2011.07.008.

Xu, L., Cao, L., Li, J., & Staiger, C. J. (2024). Cooperative actin filament nucleation by the Arp2/3 complex and formins maintains the homeostatic cortical array in Arabidopsis epidermal cells. *The Plant Cell, 36*, 764–789. https://doi.org/10.1093/plcell/koad301.

Yanagisawa, M., Desyatova, A. S., Belteton, S. A., Mallery, E. L., Turner, J. A., & Szymanski, D. B. (2015). Patterning mechanisms of cytoskeletal and cell wall systems during leaf trichome morphogenesis. *Nature Plants, 1*, 1–8. https://doi.org/10.1038/nplants.2015.14.

Zhang, C., Mallery, E. L., & Szymanski, D. (2013). ARP2/3 localization in Arabidopsis leaf pavement cells: a diversity of intracellular pools and cytoskeletal interactions. *Frontiers in Plant Science, 4*. https://doi.org/10.3389/fpls.2013.00238.

Zhou, X., Graumann, K., Evans, D. E., & Meier, I. (2012). Novel plant SUN–KASH bridges are involved in RanGAP anchoring and nuclear shape determination. *The Journal of Cell Biology, 196*, 203–211. https://doi.org/10.1083/jcb.201108098.

CHAPTER THREE

Fertilization mechanisms in hermaphroditic ascidians and nematodes: Common mechanisms with mammals and plants

Hitoshi Sawada[a,*], Takako Saito[b,*], Yoshihiro Shimada[c], and Hitoshi Nishimura[c,*]
[a]Graduate School of Science, Nagoya University, Nagoya, Japan
[b]Department of Applied Life Sciences, Faculty of Agriculture, Shizuoka University, Shizuoka, Japan
[c]Department of Life Science, Faculty of Science and Engineering, Setsunan University, Neyagawa, Osaka, Japan
*Corresponding authors. e-mail address: sawada.hitoshi.r8@a.mail.nagoya-u.ac.jp; saito.takako@shizuoka.ac.jp; nishimura@lif.setsunan.ac.jp

Contents

Abstract

Most animals have male and female, whereas flowering plants are hermaphrodites. Exceptionally, a small population of invertebrates, including ascidians and nematodes, has hermaphrodite in reproductive strategies. Several ascidians exhibit strict self-sterility (or self-incompatibility), similar to flowering plants. Such a self-incompatibility mechanism in ascidian has been revealed to be very similar to those of flowering plants. Here, we describe the mechanisms of ascidian fertilization shared with invertebrates and mammals, as well as with plants. In the nematode *Caenorhabditis elegans,* having self-fertile hermaphrodite and male, several genes responsible for

Current Topics in Developmental Biology, Volume 162
ISSN 0070-2153, https://doi.org/10.1016/bs.ctdb.2025.01.010

fertilization are homologous to those of mammals. Thus, novel proteins responsible for fertilization will be easily disclosed by the analyses of sterile mutants. In this review, we focus on the same or similar reproductive strategies by shedding lights on the common mechanisms of fertilization, particularly in hermaphrodites.

1. Introduction

Fertilization is an essential process for sexual reproduction to create the next generation. Generally, living organisms produce haploid germ cells, usually sperm and egg, whose conspecific cell fusion triggers the activation of embryonic development (Barresi & Gilbert, 2020; Gallo & Costantini, 2012; Gert & Pauli, 2020; Litscher & Wassarman, 2018; Saito & Sawada, 2022; Sawada & Saito, 2022; Sawada, 2002; Vacquier & Swanson, 2011). Although spermatozoa have no extracellular matrix, the eggs of most species are covered with the proteinaceous egg coat called the vitelline coat (VC) or vitelline envelope (VE) in marine invertebrates, and with the zona pellucida (ZP) in mammals (Barresi & Gilbert, 2020; Sawada, 2002). Moreover, eggs carrying the proteinaceous coat are surrounded by cellular coat designated follicle cells (FCs) in ascidians, and by the cumulus cells (CCs) in mammals (Barresi & Gilbert, 2020; Sawada & Saito, 2022). Exceptionally, the eggs of nematodes or cnidarians presumably have no typical proteinaceous egg coat (Nishimura & L'Hernault, 2010; Yamashita, 1988), although at least jellyfish eggs are surrounded by the jelly layer (Yamashita, 1988).

Generally, animal fertilization comprises the following six steps (see Fig. 1):

(Step 1) Sperm chemotaxis
(Step 2) Sperm binding to the egg coat
(Step 3) Sperm acrosome reaction or sperm reaction
(Step 4) Sperm penetration of the egg coat
(Step 5) Gamete fusion
(Step 6) Polyspermy block

Steps 1 and 2 are critical in marine invertebrates, because many marine invertebrates, having the same spawning season, inhabit contiguously at the same rocky seashore. In Step 2, the issues of species-specificity and self-incompatibility (SI) or self/nonself-recognition are also included. Because most terrestrial or marine hermaphroditic animals, including sea hares, snails, and worms, perform sexual behavior and copulation, self/nonself-recognition and species-recognition in gamete

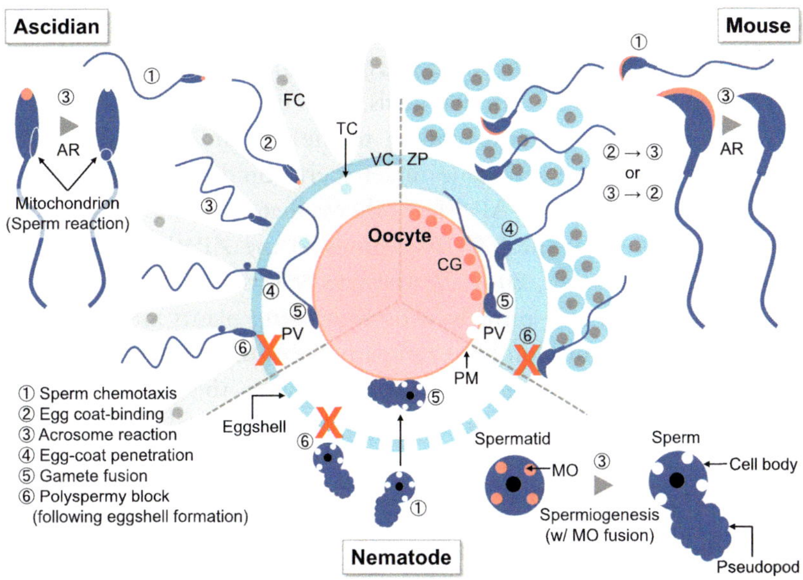

Fig. 1 Fertilization steps in ascidians, mice and nematodes. Left. In the ascidian (*Ciona intestinalis*), sperm are attracted to the eggs with the gradient of SAAF, a sperm chemoattractant (Step 1). After binding to the VC (Step 2), sperm undergoes sperm reaction, including vigorous sperm movement, mitochondrial swelling, sliding and shedding (Step 3). Simultaneously, self/nonself-recognition process also takes place. Whether sperm undergoes AR on the VC or after penetration is debated. After sperm rection, sperm penetrate the VC with the aid of lysin (Step 4), and finally fuse with the egg membrane (Step 5). After gamete fusion, polyspermy is blocked on the VC (Step 6). **Right.** In the mouse (*Mus musculus*), spermatozoa are attracted by chemoattractants, including cholesterol (Step 1). Then, there are two pathways: one is acrosome-intact sperm binding to the ZP (primary binding), which elicits the AR (Step 2 to Step 3). Alternatively, sperm undergoes AR during the passage of CC layer, and then bind to the ZP (Step 3 to Step 2). The acrosome-reacted sperm binding to ZP is designated secondary binding. Then, sperm penetrate the ZP and get into the perivitelline space (PV) (Step 4). Then, sperm bind to and fuse with the oocyte plasma membrane (PM) *via* the equatorial segment of the sperm head (Step 5) followed by the block of polyspermy (Step 6). **Bottom.** In the nematode (*Caenorhabditis elegans*), the production from spermatid into sperm, called spermiogenesis, takes place with an exocytosis of MO (membranous organelle) (Step 3). The resulting sperm with motility and fertility is guided to an oocyte in response to F-series prostaglandins (PGFs) that were secreted from the oocyte (Step 1). The oocyte has neither surrounding layers nor thick egg coats, so sperm presumably binds to and fuses with the oocyte PM directly (Step 5). Unlike in mice, the subcellular region where the sperm binds to and fuses with the oocyte PM is unclear. After gamete fusion, polyspermy is blocked by forming an eggshell (Step 6).

interactions are not necessarily required. However, such recognitions are important for hermaphroditic marine animals that release gametes to the surrounding seawater. In mammals, it is currently believed that there are two ways for the sperm binding to the ZP and acrosome reaction (AR). One is that the acrosome-intact sperm binds to the ZP (primary binding), which elicits the AR. The other is that acrosome-reacted sperm during the passage of the CC layer binds to the ZP (Fig. 1; Barresi & Gilbert, 2020; Jin et al., 2011; Wassarman & Litscher, 2022).

In contrast to most animals, most flowering plants are hermaphrodites. Sexual reproduction between gametes of different individuals has the advantage of creating genetic diversity in the next generation, which would be adaptable to possible environmental alterations. On the other hand, self-fertilization in hermaphrodites has significant benefits for maintaining respective species by a single individual. For example, some nematode species belonging to the genus *Caenorhabditis* have male and hermaphrodite, the latter being self-fertile. In angiosperms, approximately 60 %, but not all, of nearly 250,000 angiosperms are reported to be self-incompatible (de Nettancourt, 1997; Sawada et al., 2014). In the nematode *Caenorhabditis elegans*, hermaphrodites undergo the production of spermatids during the fourth larval (L4) stage, activation of the spermatids into motile self-sperm with fertility in the spermatheca, and production of eggs from the adult stage, enabling to reproduce siblings without copulation. On the other hand, hermaphrodites can create genetically diverged siblings after copulation with males. This system appears to be the best in reproductive strategies since efficiency to produce the next generation and acquisition of genetic divergence can be achieved at appropriate balances. Similar reproductive strategy is reported in the ascidian *Ciona savignyi*, which prefers nonself-fertilization but is self-fertile in the absence of nonself-sperm (Jiang & Smith, 2005).

Here, we focus on the fertilization mechanisms of ascidians and nematodes from the viewpoint of common mechanisms shared with mammals or plants. For the detailed mechanism of ascidian and vertebrate fertilization, see the following reviews: (Gert & Pauli, 2020; Lu & Ikawa, 2022; Saito & Sawada, 2022; Satoh, 1994; Sawada & Saito, 2022; Sawada, 2002). Regarding the fertilization mechanism in nematodes, see also the following reviews: (Nishimura & L'Hernault, 2010; Nishimura & L'Hernault, 2016).

2. Fertilization mechanisms in ascidians: Comparison with echinoderms, mammals and plants

2.1 Sperm chemotaxis

2.1.1 Ascidian sperm chemotaxis

The initial event of ascidian fertilization is the step of sperm attraction to conspecific eggs, which was first reported in *Ciona intestinalis* (Miller, 1975). Sperm chemoattractant in ascidians is called SAAF (sperm-activating and -attracting factor), which is released from eggs (Yoshida & Yoshida, 2011; Yoshida et al., 2002). *Ciona intestinalis* SAAF (Ci-SAAF) is genus-specific, *i.e.*, *Ciona savignyi* SAAF (Cs-SAAF) is identical to Ci-SAAF, and the chemical structure is a sulfated steroid, 3,4,7,26-tetrahydroxycholestane-3,26-disulfate (Yoshida et al., 2002; Yoshida et al., 2013). SAAFs of other species are also sulfated steroids (Yoshida & Yoshida, 2011). SAAF receptor on the sperm plasma membrane (PM) was identified as Ca^{2+}-ATPase (PMCA), which mediates the ATP-driven efflux of Ca^{2+} (Yoshida et al., 2018). SAAF can bind to PMCA, resulting in the activation of PMCA (Yoshida et al., 2018). This idea is supported by the fact that PMCA inhibitors blocked the chemotactic activity of SAAF (Yoshida et al., 2018). Store-operated Ca^{2+} channel is also involved in this chemotaxis (Yoshida et al., 2003). Under down-gradient of SAAF, the Ca^{2+} efflux activity of PMCA decreases during sperm circulating behavior, resulting in a transient increase in the intracellular Ca^{2+} concentration $[Ca^{2+}]_i$. "*The Ca^{2+} bursts induced around a local minimal SAAF concentration during circulation trigger a sequence of flagellar responses comprising quick turning followed by straight swimming to direct spermatozoa efficiently toward eggs*" (Shiba et al., 2008). The increased intracellular Ca^{2+} binds to sperm calaxin, a Ca^{2+} sensor suppressing the microtube sliding of outer-arm dynein, resulting in the generation and propagation of Ca^{2+}-induced asymmetric flagellar bending (Mizuno et al., 2012). A recent study using genetically modified *C. intestinalis* revealed that the ascidian Ca^{2+} channel CatSper3 is involved in sperm flagellar beating and motility (Kijima et al., 2023). Furthermore, soluble adenylyl cyclase and cAMP are also involved in ascidian sperm flagellar motility and chemotaxis (Shiba & Inaba, 2023). This is also common in mammals, sea urchins and corals (Speer et al., 2021; Vacquier et al., 2014).

2.1.2 Echinoderm sperm chemotaxis

In the sea urchin *Arbacia punctulata,* a tetradeca-peptide designated "Resact" was first identified as a chemoattractant (Ward et al., 1985). Resact belonging to the sperm-activating peptide (SAP) family, is contained in the egg jelly, and stimulates sperm motility (Barresi & Gilbert, 2020; Vacquier & Swanson, 2011). This peptide elicits rapid, transient increase of $[cGMP]_i$, hyperpolarization by K^+-channel-mediated K^+ influx, alkalinization *via* NHE, and eventual Ca^{2+} influx mediated by CatSper (Kaupp et al., 2003; Seifert et al., 2015; Ward et al., 1985). In the starfish *Asterias amurensis,* "Asterosap", a member of the SAP family, is reported to be a chemoattractant (Matsumoto et al., 2003; Nishigaki et al., 1996).

2.1.3 Mammalian sperm chemotaxis

Candidate chemoattractants in mammals have been proposed to be peptides, proteins, odorant (chemoattractant reacting to bourgeonal- or lyral-receptor) or progesterone (see a review by Yoshida & Yoshida, 2011). In particular, it is worth to note that the olfactory receptor hOR17-4 is expressed in human sperm and that bourgeonal has the chemoattracting activity to human sperm. However, any naturally occurring chemoattractants to the olfactory receptor have not been identified. Intriguingly, the human sperm-specific Ca^{2+} channel CTSR (CatSper) is directly activated by not only progesterone but also prostagrandins and odorants (Brenker et al., 2012; Rahban & Nef, 2020; Williams et al., 2015). Progesterone receptor in human sperm was identified as an orphan enzyme α/β hydrolase domain-containing 2 (ABHD2), which enables depletion of endogenous CatSper inhibitor (2-arachidonoylglycerol) from plasma membrane by progesterone binding (Miller et al., 2016). ABHD2 is not present in mouse sperm, which consists with the results that mouse sperm CatSper showed no response to progesterone (Lishko et al., 2011; Strünker et al., 2011). Since PLCδ4KO mouse sperm reduced the response to progesterone (Fukami et al., 2003), another signaling pathway may be involved in progesterone-mediated Ca^{2+} increase in mouse sperm.

Allurin, a 21-kDa CRISP (cysteine-rich secretory protein) family protein, was identified as a chemoattractant in amphibians (*Xenopus laevis* and *Xenopus tropicalis*). Notably, allurin has a chemoattracting activity toward frog and mouse sperm, implying the possible participation of allurin-like proteins in mammalian sperm chemotaxis (Burnett et al., 2012; Olson et al., 2001).

Taken together, peptides or steroids presumably function as sperm chemoattractants in many animals. The mechanisms of CatSper-mediated sperm chemotaxis seem to be similar among echinoderms, ascidians and mammals, although birds or amphibians lack *CatSper* genes.

2.1.4 Flowering plant pollen tube guidance

In flowering plants, attractants responsible for pollen tube guidance were identified as defensin-like cysteine-rich proteins, called LUREs in Torenia and Arabidopsis, and ZmEA1 in maize, which are expressed and secreted by synergid cells (see Fig. 2) (Dresselhaus et al., 2016; Higashiyama & Takeuchi, 2015; Higashiyama & Yang, 2017; Higashiyama et al., 2001; Higashiyama, 2010; Johnson et al., 2019). Among six AtLUREs, the AtLURE 1.2 receptor was identified as a PM protein kinase complex

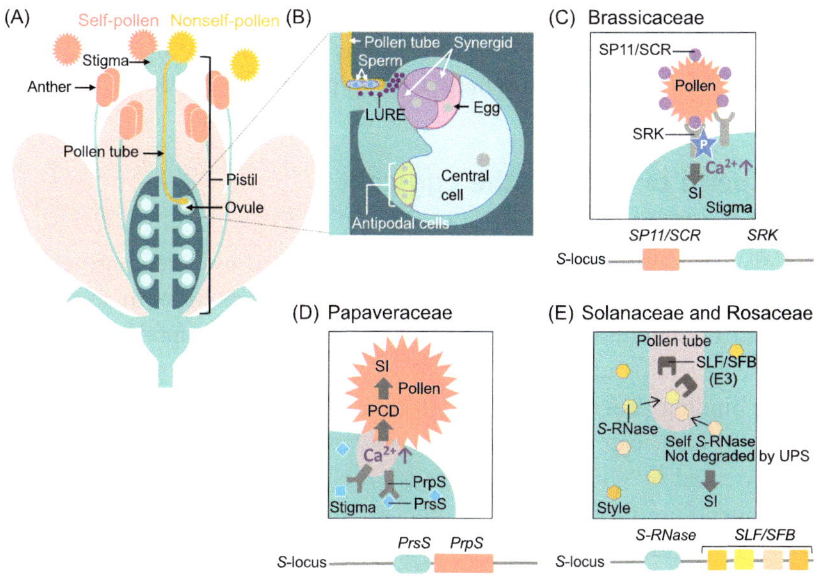

Fig. 2 Pollen tube guidance and self-incompatibility (SI) in flowering plants. There are two systems for the SI in angiosperms: The self-recognition system (Brassicaceae and Papaveraceae) and nonself-recognition system (Solanaceae and Rosaceae). Roughly 60 % of angiosperms are known to be self-incompatible. **(A, B)** In *Arabidopsis thaliana* (self-compatible species), pollen tube guidance was induced by LURE, which is secreted by two synergids. **(C)** In Brassicaceae, pollen *SP11/SCR* and pistil *SRK* genes are tightly linked and highly polymorphic. If pistil SRK recognizes the pollen SP11 as the same haplotype, pollen germination and pollen tube extension are inhibited, which results in SI. **(D)** In Papaveraceae, pollen *PrpS* and pistil *PrsS* genes are identified as SI genes, which are tightly linked in the same chromosome. When PrpS recognizes PrsS on the stigma as nonself, Ca^{2+} influx occurs, resulting in pollen programmed cell death with the aid of caspase-like protease. **(E)** In Solanaceae and Rosaceae, stigmatic *S*-RNases penetrate the membrane of elongating pollen tube and are ubiquitinated by SLF/SFB (E3), recognizing nonself *S*-RNase. After ubiquitination, nonself-*S*-RNases are degraded by UPS. As a result, only self *S*-RNase, which is not degraded by UPS, can degrade the self RNA in pollen tube, resulting in SI. For detail, see text.

consisting of MDIS1 (male discoverer 1), MIK (male-discoverer-interacting kinase) and PRK6 (pollen receptor kinase 6), which elicit the increase of $[Ca^{2+}]_i$ (Higashiyama & Yang, 2017). Although sperm cells are not directly attracted by the eggs in flowering plants, the mechanisms of Ca^{2+}-mediated sperm attraction in animals appear to be similar to the mechanisms of sperm-enclosed pollen tube guidance to ovule in flowering plants. Recently, a novel ovule-derived attracting factor, named AMOR (arabinogalactan polysaccharide; its reactive region is 4-Me-GlcA-β-(1→6)-Gal), was found to be necessary for pollen tube competence to respond to LURE in *Torenia fournieri* (Mizukami et al., 2016). In the case of peptides or proteins, it must be easy to inactivate the pollen tube guiding activity by the proteases released from synergid cells after gamete fusion to reject extra tubes (Yu et al., 2021).

2.2 Sperm binding to the egg coat

2.2.1 Ascidian sperm binding to the VC

Ascidian sperm binding to the egg coat is mediated by several proteins and carbohydrates. Such a conspecific sperm-VC interaction in ascidians is coupled to the steps of species-recognition and self/nonself-recognition, because several, but not all, ascidians exhibit self-sterility (or SI) in the sperm binding process to the VC (Satoh, 1994).

An ascidian egg is covered by a rigid proteinaceous egg coat that was initially called as "chorion", but later as "vitelline coat (VC)", because the VC components are produced by the oocyte rather than the FCs (Cotelli et al., 1981). An ascidian egg possesses the test cells (TCs) in the perivitelline space (PV) and a monolayer of FCs on the VC. The role of TCs in fertilization is not known. On the other hand, FCs are responsible for egg floatation (Lambert & Lambert, 1978), acquisition of self-sterility during oogenesis, gamete interaction and polyspermy block (Satoh, 1994; Sawada & Saito, 2022). In *C. intestinalis,* FCs are not necessarily required for fertilization, although the removal of FCs sometimes delayed fertilization and lowered the fertilization efficiency (Kawamura et al., 1988). In contrast, the attachment of FCs to the VC is indispensable for fertilization in *Halocynthia roretzi* (Fuke, 1983) and *Phallusia mammillata* (Honegger & Koyanagi, 2008).

H. roretzi sperm possess two trypsin-like proteases, designated ascidian acrosin and spermosin (Saito & Sawada, 2022; Sawada & Saito, 2022). The C-terminal CUB domain of proacrosin and the light chain of spermosin are involved in sperm binding to the C-terminal region of vitellogenin on the VC. Furthermore, sperm α-L-fucosidase is also involved in sperm binding to the VC, probably to the carbohydrate moiety of HrVC70 (Matsumoto et al., 2002; Sawada & Saito, 2022).

The sperm-VC interaction in *C. intestinalis* is also mediated by several gamete proteins and carbohydrates. In this species, sperm α-L-fucosidase is involved in the sperm binding to the VC (Hoshi, 1986; Hoshi et al., 1983, 1985, 1994; Mengerink & Vacquier, 2001). The binding partner of sperm fucosidase is thought to be 140-kDa, 125-kDa and 78-kDa fucosylated glycoproteins present on the fibrillar tuffs emerging on the outer surface of the VC (De Santis et al., 1980, 1983; Pinto et al., 1981). A 70-kDa sperm protein is also a candidate binding protein to the apolipoprotein B-like proteins on the VC, suggested by the data on Far Western blotting (Yamada et al., 2009). Sperm CiUrabin, a CRISP family protein, is involved in the binding to CiVC57, a homolog of HrVC70 (Yamaguchi et al., 2011). However, these proteins or carbohydrates are unlikely to be responsible for SI.

2.2.2 Ascidian SI

In *H. roretzi,* immature oocytes in the ovary are self-fertile while mature oocytes (eggs) are strictly self-sterile (Fuke & Numakunai, 1996; Fuke, 1983). Furthermore, the eggs treated with acidic seawater (at pH 2–3 for 1 min) become self-fertile. Since the VC-free eggs are self-fertile, Sawada and his colleagues isolated the VCs from immature and mature oocytes and compared the constituents of each VC. Intriguingly, the amount of a 70-kDa VC protein (HrVC70) in mature oocytes was considerably higher than that in immature oocytes. Furthermore, the number of nonself-sperm bound to HrVC70-immobilized beads was higher than that of self-sperm, and the pretreatment of sperm with nonself-HrVC70 more strongly inhibited fertilization than that with self-HrVC70. From these results, HrVC70 is perhaps a promising candidate for self-sterility-responsible sperm receptor on the VC. The cDNA cloning and *in situ* hybridization of HrVC70 (Harada & Sawada, 2007; Sawada et al., 2002, 2004) revealed that this protein is expressed as a 120-kDa precursor (HrVC120) in the oocyte, consisting of the N-terminal signal peptide, 12 epidermal growth factor (EGF)-like domains, one incomplete EGF-like domain, ZP domain, and a single transmembrane (TM) domain. Mature HrVC70 appears to be produced after processing by the FC trypsin-like protease HrOvochymase at Arg-668 (Mino & Sawada, 2016; Sawada & Saito, 2022). These results coincided with the observations that the acquisition of self-sterility during oocyte maturation was inhibited by trypsin inhibitors and was induced by endogenous and exogenous protease (Sakairi & Shirai, 1991; Fuke & Numakunai, 1999). HrVC70 shows polymorphisms at the regions between the third and fourth Cys residues and between EGF-like domains (Harada & Sawada, 2007; Sawada et al., 2002, 2004). Although the diversity is not so prominent, the polymorphism is not random. Two HrVC70-interacting proteins have been so far identified. One is HrUrabin, a CRISP family protein present in the detergent-insoluble membrane fractions (lipid rafts) of sperm, as revealed by Far Western blotting (Urayama et al., 2008). The other is a type-II TM protein with serine protease domain (HrTTSP-1), as revealed by yeast two-hybrid screening and pull-down assay (Harada & Sawada, 2007). Although HrUrabin and HrTTSP-1, localized on the sperm surface, mediate the sperm binding to HrVC70, they cannot distinguish self- and nonself-HrVC70. Therefore, further studies are necessary to explore sperm self/nonself-recognition proteins.

Historically, the SI system in *C. intestinalis* was first studied by Thomas Hunt Morgan since 1910 (Morgan, 1910: see also reviews by Saito & Sawada, 2022; Sawada & Saito, 2022). Morgan noticed that the SI is genetically controlled (Morgan, 1942, 1944). Intriguingly, he observed one-way cross-sterile combination, in addition to reciprocally cross-sterile combination, between selfed F1 siblings. To explain these phenomena, he proposed the "haploid sperm hypothesis", where the SI gene specificity is determined by haploid expression in sperm and by diploid expression in eggs. Given that the SI genes of sperm-donating individual ("male") is heterozygote (S1/S2) and that the egg-donating individual ("female") is homozygote (S1/S1 or S2/S2), either S1-sperm or S2-sperm recognizes S1/S1-eggs and S2/S2-eggs as nonself, which results in cross-fertile combination. But it's not *vice versa*, because both S1-sperm (from S1/S1 homozygote) and S2-sperm (from S2/S2 homozygote) recognize S1/S2-eggs (from S1/S2 heterozygote) as self, resulting in cross-sterile combination. Therefore, if this one-way cross-sterility was observed, the "male" individual's SI genes must be heterozygous and "female" individual's SI genes must be homozygous (Fig. 3A and B). These experiments were repeated by Murabe and Hoshi (2002) and Harada et al. (2008). Based on these criteria, SI-responsible genes were explored by using 50 marker genes, and three pairs of candidate SI genes, s/v-Themis-A, s/v-Themis-B, and s/v-Themis-B2, have been identified (Fig. 3C and D) (Harada et al., 2008; Sawada et al., 2020).

The sperm SI gene in the locus A (on chromosome 2q), encoding a protein that has a hyper-variable region, REJ (receptor for egg jelly) domain, GPS (G-protein coupled receptor (GPCR) proteolysis site)

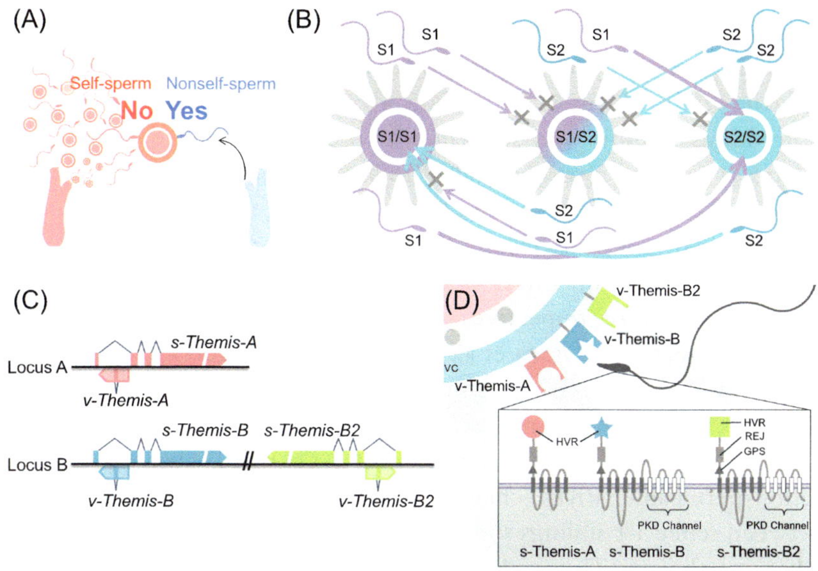

Fig. 3 ***C. intestinalis*** **self-incompatibility (SI) system and the responsible genes. involved in SI. (A)** Ascidians release sperm and eggs nearly simultaneously to the surrounding sea in the spawning season, but several species, including *Ciona intestinalis* or *Halocynthia roretzi,* show strict SI. Only nonself-sperm tightly bind to the VC and fertilize the eggs. **(B)** When self-fertilized F1 siblings are subjected to mutual cross-fertilization experiments, one-way cross-sterile combination is observed, which is hardly observed in natural crossing. To explain these phenomena, Morgan proposed a "haploid sperm hypothesis". In this hypothesis, the SI genes of "male" (sperm-donating individuals) and "female" (egg-donating individuals) are predicted to be heterozygous (S1/S2) and homozygous (S1/S1 or S2/S2), respectively. In addition, it is thought that sperm SI genes have haploid expression and that the egg SI genes have diploid expression. In this case, either S1-sperm or S2-sperm of heterologous male (S1/S2) can recognize homologous female's eggs (S1/S1-eggs or S2/S2-eggs) as nonself eggs, resulting in cross-fertilization as an efficiency of 50 % fertilization. On the other hand, homozygous sperm (S1-sperm or S1-sperm) recognize the heterozygous eggs (S1/S2-eggs) as self-eggs, which results in cross-sterile combination. Thus, if this one-way cross-sterility is observed, it is thought that the "male" individual's SI genes must be heterozygous and "female" individual's SI genes must be homozygous. From these criteria, SI genes were explored by examining 50 gene markers. **(C)** Three gene pairs were identified as a determinant of SI: (*s-Themis-A* and *v-Themis-A), (s-Themis-B* and *v-Themis-B),* and *(s-Themis-B2 and v-Themis-B2)* genes. In every case, *v-Themis* gene resides in the first intron of *s-Themis* gene, but transcribed in the opposite direction. **(D)** Given that all the three pairs of *s-Themis* and *v-Themis* genes are matched, fertilization never takes place whereas even a single pair is mismatched, fertilization is allowed. For details, see text.

domain, LH2 (lipoxygenase homology 2) domain, and five-pass TM domains, was designated *s(sperm)-Themis-A*. On the other hand, the egg SI gene in the locus A, encoding a protein containing a fibrinogen C-terminus homologous sequence with an N-terminal hyper-variable region, was called *v(vitelline coat)-Themis A* (Harada et al., 2008; Saito & Sawada, 2022; Sawada & Saito, 2022). Similarly, two SI gene pairs called *s/v-Themis-B* and *s/v-Themis-B2* are also located in the locus B (on chromosome 7q). Interestingly, three *v-Themis* genes are located in the first intron of respective *s-Themis* genes, but transcribed in the opposite direction (Harada et al., 2008). Basic structures of s/v-Themis-A and s/v-Themis-B/B2 are similar except for that s-Themis-B and -B2 possess 11-pass TM domains, instead of the 5-pass TM domain in s-Themis-A. The C-terminal regions of s-Themis-B and -B2 contain a cation-permeable PKD channel. The cross-fertilization experiment between selfed F1 siblings showed that sperm recognizes the VC as self, when all the *s/v-Themis-A*, *-B*, and *-B2* allele pairs (haplotypes) are matched. Even the one haplotype is mismatched, sperm recognizes the VC as nonself, albeit with some exceptional cases in s/v-Themis-A (Sawada et al., 2020). This rule is very strict in *s/v-Themis-B* or *-B2* haplotypes, but not so strict in *s/v-Themis-A*. It was also revealed that both *s-Themis-A* and *s-Themis-B/B2* genes were essential for SI, because individuals where these genes were targeted by TALEN showed self-fertility although the achievement of self-fertilization is slightly delayed in *s-Themis-A*-targeted individual's gametes. From these results, Sawada and his colleagues concluded that three SI pairs, *s/v-Themis-A*, *s/v-Themis-B*, and *s/v-Themis-B2*, must be necessary for the achievement of SI. Further studies are necessary to clarify the allele-specific protein-protein interaction between s-Themis and v-Themis on the structural basis. In mice, it is reported that PKDREJ, a homolog of *C. intestinalis* s-Themis, may be involved in ZP-evoked AR or sperm movement during the passage of female reproductive tract (Sutton et al., 2008). Intriguingly, PKDREJ is reported to evolve under positive selection and shows polymorphisms among individuals in humans and primates (Hamm et al., 2007). These features led us to speculate that mammalian sperm PKDREJ may be involved in fertilization efficiency between gametes with genetic proximity.

After sperm recognizes the VC as self, drastic sperm Ca^{2+} influx takes place, resulting in vigorous sperm movement on the VC followed by eventual termination of sperm motility or sperm detachment from the VC of self-eggs (Saito et al., 2012). These results led to investigate whether prevention of Ca^{2+} influx allows self-fertilization or not. It was found that the reduction of external Ca^{2+} concentration in ASW to one-tenth (~1 mM) makes the eggs self-fertile (Hashimoto et al., 2022).

Several other candidate factors responsible for SI in *C. intestinalis* have also been proposed. First, the proteasome is responsible for proteolysis of autoantigen within FC, and the generated peptides may be trapped by Ci-hsp70 on the surface of FC. This must be transferred to some protein on the VC presenting the autoantigen. This model was led by the results that two anti-hsp70 antibodies inhibited the acquisition of self-sterility during oogenesis and that immunocytochemistry using those antibodies showed the localization of hsp70 on the surface of FC (Marino et al., 1998; Marino et al., 1999). This idea was speculated from that hsp70 gene, which belongs to the MHC class-III gene, is thought to be an ancestral gene of MHC class-I and -II. However, the putative peptides associated with Ci-hsp70 and its sperm-recognition proteins have not been identified.

Kawamura and his collaborators attempted to identify a putative allorecognition factor from a different angle (Kawamura et al., 1987, 1991). They succeeded in identifying two factors: one is a non-allorecognizing glucose-enriched glycan that inhibits sperm binding to the VC, and the other is a Glu/Gln-enriched peptide modulator. These two factors may play roles as cofactors in the allorecognition of sperm receptors. However, the molecular mechanism remains elusive.

Using *C. intestinalis* type B, Khalturin and his collaborators cloned gonadal cDNAs that were expressed from genes with high polymorphism among individuals, using the method of suppression subtraction hybridization. They revealed the occurrence of highly polymorphic ZP proteins and sushi domain-containing protein vCRL1. They proposed that these proteins might be involved in fertilization and allorecognition in *C. intestinalis* type B (Kürn et al., 2007a; Kürn et al., 2007b). However, their genetic analysis and gene-knockdown experiments of vCRL1 showed that vCRL is not involved in self/nonself recognition during fertilization but is likely to be involved in the establishment and maintenance of the blood system (Khalturin et al., 2005; Sommer et al., 2012).

Although the above-mentioned candidate proteins were presumed to be involved in SI, convincing evidence on the basis of protein-protein interaction has not been obtained. Further studies are needed to show the structural and kinetic data clarifying the difference between self and nonself protein interaction.

2.2.3 Flowering plant SI

The s/v-Themis-mediated SI system in *C. intestinalis* is very similar to those of angiosperms (Fig. 2). Generally, the SI system is governed by a multiallelic gene pair, called haplotype, which contiguously resides in the SI locus of the same chromosome. The male and female SI genes are tightly linked and highly polymorphic. There are two systems in the angiosperm SI system: self-recognition system and nonself-recognition system. The former SI system is observed in Brassicaceae and Papaveraceae, and the latter is seen in Solanaceae and Rosaceae. In Brassicaceae, pollen SP11/SCR (small cysteine-rich protein) and pistil SRK (S-locus receptor kinase), whose genes are tightly linked, are highly polymorphic. Given that pistil SRK conceives pollen SP11 as the same haplotype, pollen germination and pollen tube extension are inhibited, which is mediated by Ca^{2+} increase and phosphorylation in stigmatic papilla cells. In Papaveraceae, pollen *PrpS* and pistil *PrsS* genes are identified as SI genes, which are contiguously reside in the same chromosome. When PrpS recognizes PrsS on the stigma as nonself, Ca^{2+} influx takes place resulting in apoptosis, which is mediated by caspase-like protease (Iwano & Takayama, 2012; Sawada et al., 2014). In Solanaceae and Rosaceae, *S*-RNase of stigma penetrate the membrane of pollen tube. A variety of E3 ligase designated SLF/SFB (*S*-locus F-box) in the pollen tube are capable of recognizing nonself *S*-RNase resulting in degradation by the proteasome. As a result, only self *S*-RNase survives and degrades the self RNA in pollen tube. Taken together, the SI system in Papaveraceae is very similar to the SI of *C. intestinalis* (Fig. 3).

2.2.4 Echinoderm sperm binding to the VE

In sea urchins, species-specific sperm binding to the VC is mediated by sperm Bindin, a 30-kDa acrosomal protein, and the EBR1 on the VE, consisting of an ADAMTS-like N-terminal domain followed by 19 repeats of CUB domains and thrombospondin type-1 domains (Kamei & Glabe, 2003). The C-terminal region of EBR1 is responsible for species-specificity. During the AR, Bindin is exposed to the surface during the acrosomal process and involved in the binding to EBR1 on the VE. Bindin is known to be essential for sperm binding to the VE and *Bindin* knockout sperm were incapable of fertilizing the sea urchin eggs (Wessel et al., 2021a, 2021b).

2.2.5 Mammalian sperm binding to the ZP

In mice, acrosome-intact spermatozoa first bind to ZP3 designated as "primary binding", which induces the AR. Then, acrosome-reacted sperm binds to

ZP2, which is called "secondary binding", followed by sperm penetration through the ZP (Wassarman & Litscher, 2001, 2022). Historically, ZP3 was believed to be a sole sperm receptor on the ZP in mice. As mouse ZP1, ZP2 and ZP3 were separated from solubilized ZP by SDS-PAGE, extracted from SDS-gels, and renatured, only ZP3 exhibited the potential inhibitory activity for sperm to bind to the ZP (Wassarman & Litscher, 2001). Later, it was found that the *O*-linked sugar chain attached to the C-terminal region of ZP3 is responsible for the sperm-ZP3 binding (Wassarman & Litscher, 2022).

Several candidate sperm proteins involved in the sperm binding were explored. One of the promising candidates is 1,4-galactosyltransferase (GalTase). GalTase located on the sperm surface was identified as a sperm primary binding receptor for ZP3 (Lopez et al., 1985; Miller et al., 1992; Shur & Hall, 1982; Shur et al., 1998). GalTase basically catalyzes the transfer reaction of galactose (Gal) from UDP-Gal to the terminal GlcNAc residues. However, since extracellular UDP-Gal is not present under fertilization environments, sperm-surface GalTase binds to terminal GlcNAc residue, one of the substrates (acceptor). Treatment of sperm with GalTase inhibitors, UDP-Gal, or GlcNAcase potently inhibited the sperm binding to ZP3. All of these results support the hypothesis that GalTase participates in the sperm binding to the ZP. Since the treatment of sperm with anti-GalTase IgG, but not with the Fab fragment of the same IgG, triggers the AR due to cross-linking of sperm GalTase with divalent antibody, GalTase may also be involved in the AR. Although many lines of evidence indicated the participation of GalTase in sperm binding to the ZP, GalTase-null mutant sperm was found to be fertile and was capable of binding to the ZP (Shur et al., 1998).

Then, Shur and his colleagues disclosed a new sperm binding protein to the ZP, called SED1, which is made up of two EGF-like domains and two discoidin domains (Ensslin & Shur, 2003; Shur et al., 2004). SED1 is localized on the sperm PM overlying the acrosome and is capable of specifically binding to the ZP of unfertilized eggs but not to that of fertilized eggs. Recombinant SED1 as well as anti-SED1 antibodies competitively inhibited sperm binding to the ZP (Ensslin & Shur, 2003). SED1 can bind to ZP3 strongly and also to ZP2 weakly (Ensslin & Shur, 2003). Furthermore, SED1-null males are subfertile, and their sperm diminished their binding to the ZP *in vitro*. These results indicate the importance of sperm SED1 for the sperm binding to ZP3 and also to ZP2.

Wassarman, Bleil, and their colleagues identified a sperm ZP3 receptor (ZP3R or sp56), containing sushi domains, as a candidate for sperm binding partner for ZP3 (Bleil & Wassarman, 1990; Cheng et al., 1994; Suzuki-Toyota et al., 1995). "*sp56 is a homomultimeric peripheral membrane protein overlying the sperm acrosome. Purified sp56, which binds to the ZP of unfertilized mouse eggs but not to fertilized embryos, blocks sperm-egg recognition*" (Bookbinder et al., 1995), which explains the polyspermy block mechanisms. However, sp56 was found to be nonessential for fertilization, since *Zp3r* gene knockout (KO) mice were fertile (Muro et al., 2012).

In pig, zonadhesin was identified as a species-specific binding protein to the ZP (Hardy & Garbers, 1994, 1995). The zonadhesin precursor consists of a meprin/A5 antigen/mu receptor tyrosine phosphatase (MAM) domain, a mucin-like domain and five von Willebrand factor-D (D0-D4) domains with a molecular mass of 26.7 kDa (Bi et al., 2003). After processing, zonadhesin consists of 45-kDa (p45, D1) and 105 kDa proteins (p105, D2-D3), and is localized on the outer acrosomal membrane, suggesting that this protein may be involved in the sperm binding to the ZP during the AR (Bi et al., 2003). The purified zonadhesin can bind to ZP (Bi et al., 2003), and also anti-zonadhesin inhibits the sperm binding to ZP (Tardif et al., 2010). However, *Zan* gene-null mutant male mice are fertile, and the efficiency of KO mouse sperm binding to the other mammalian egg ZP was enhanced (Tardif et al., 2010). Therefore, zonadhesin is involved in species-specific gamete binding but is not essential for fertilization.

Mouse ADAM3 (cyritestin) is also a candidate sperm binding protein to the ZP; *in vitro* pull-down assay using solubilized, biotinylated ZP implies that ADAM3 may directly bind to the ZP3 (Kim et al., 2005). Moreover, mutant mouse lines where either of 19 genes and two gene clusters are aberrant, such as *Clgn* (Yamaguchi et al., 2006), *Calr3* (Ikawa et al., 2011), *Pdilt* (Tokuhiro et al., 2012), *Adam1a* (Nishimura et al., 2004), *Adam2* (Nishimura et al., 2001), *Adam6* (Voronina et al., 2018), *Prss37* (Shen et al., 2013), *Prss55* (Shang et al., 2018), *Tryx5* (also known as *Prss59*: Zhang et al., 2020), *Tex101* (Fujihara et al., 2013), *Ly6k* (Fujihara et al., 2014), *t-Ace* (Yamaguchi et al., 2006), *Tpst2* (Marcello et al., 2011), *Rnase10* (Krutskikh et al., 2012), *Pmis2* (Yamaguchi et al., 2012), *Lypd4* (Fujihara et al., 2019), *Cmtm2a/2b* (Fujihara et al., 2018), *Nell2* (Kiyozumi et al., 2020), and *Ovch2* (Kiyozumi et al., 2020), and *Cst* (Fujihara et al., 2019) and *Pate* (Fujihara et al., 2019) clusters, produce sperm with loss of or mislocalization of cell-surface ADAM3, which are all incapable of binding to the ZP. Intriguingly, many of these mutant sperm also fail to migrate from the uterus into the oviduct.

Sperm arylsulfatase A is also likely to be involved in the sperm binding to the ZP (Xu et al., 2012). Arylsulfatase A is able to bind to ZP3, and also to ZP2 to a small extent. Even three residues responsible for the active site pocket of arylsulfatase are mutated, binding ability was maintained. This suggests that the enzymatic activity of arylsulfatase A is not necessary for the sperm binding ability to the ZP (Xu et al., 2012).

One of the candidate sperm proteins responsible for the binding to ZP2 is proacrosin (Howes & Jones, 2002; Howes et al., 2001; Jones, 1991; Urch & Patel, 1991). Proacrosin/acrosin appears to be involved in the binding to sulfated ZP protein, since ZP2 protein can bind to *Acr*-null sperm less effectively than to wild-type sperm (Howes et al., 2001). The binding is non-enzymatic, and the interaction is mediated by basic residues of (pro) acrosin and the ZP. ZP2 is capable of binding to acrosome-reacted sperm but not to acrosome-intact sperm. These results suggest the participation of (pro)acrosin in the secondary binding of acrosome-reacted sperm to ZP2. However, these results are at variance with the results that *Acr*-null male is fertile (Baba et al., 1994). In connection with this, it is intriguing to note that the ascidian proacrosin C-terminal pro-domain is involved in sperm binding to the VC, which is similar to the role of mammalian sperm proacrosin.

It has long been believed that sperm undergoes AR after sperm binding to the ZP. However, most but not all fertilizing sperm undergo the AR during the sperm passage through the CC layer (Jin et al., 2011). These findings gave us a paradigm shift; some acrosome-intact sperm binds to ZP3, which elicits the AR followed by sperm penetration through the ZP. However, another sperm undergoes the AR during the penetration through the CC layer, directly binds to ZP2, and penetrates the ZP (Wassarman & Litscher, 2022). In this case, sperm binding partner for ZP3 is not necessary for fertilization. This may explain why null-mutant male mice that produce sperm lacking either of the candidate ZP3-binding partners are fertile. While (pro)acrosin is not essential for fertilization, the other proteins such as SED1 or arylsulphatase A may compensate for the function of (pro)acrosin as a secondary binding protein.

2.3 Sperm AR or sperm reaction

2.3.1 Ascidian sperm reaction

Ascidian sperm does not possess a typical acrosome, but a tiny vesicle, at the tip of sperm head. However, it is still unclear whether the so-called "acrosome" undergoes the exocytosis upon sperm binding to the VC or not (De Santis et al., 1980; Fukumoto, 1988). Although De Santis and her colleagues observed a vesiculation of the acrosomal vesicle upon sperm binding to the VC (De Santis et al., 1980), Fukumoto observed the acrosome-intact sperm penetrating the VC (Fukumoto, 1988). To solve this apparent discrepancy, further detailed live-imaging studies must be necessary by using a transgenic ascidian expressing a fluorescent acrosomal protein.

Instead of the AR, ascidian sperm undergoes "sperm reaction", which is characterized by vigorous movement on the VC, and mitochondrial swelling and sliding through the tail, and its eventual shedding (Lambert & Epel, 1979; Lambert & Koch, 1988). During fertilization, the sperm PM covering the mitochondrion appears to interact with the VC, and sperm head and tail are thought to be inserted into the PV with the aid of the driving force of the mitochondrial sliding through the tail. However, according to this model, the sperm PM covering the mitochondrion or a VC-binding membrane protein connected with a mitochondrion must migrate together with the mitochondrion. Further detailed studies are necessary to prove this hypothesis.

2.3.2 Echinoderm sperm AR

In sea urchin, sperm undergoes the AR, exocytosis of the acrosome, followed by extrusion of the acrosomal process, which is induced by polymerization of filamentous actin. The AR is induced by fucose sulfate polymer (FSP) in the egg jelly (Alves et al., 1997; Barresi & Gilbert, 2020; Vacquier & Swanson, 2011). Fucose linkages and sulfated positions in FSP are different among species, which may cause species-specificity at least in part. Sperm receptor for FSP is called REJ (receptor for egg jelly). In particular, suREJ1 on the sperm surface is responsible for the AR. Asialoglycoprotein and SAP also participate in the AR at lower pH (Hirohashi & Vacquier, 2002). In sea urchins and other animals, Ca^{2+} influx is necessary for the AR.

It is reported that metalloprotease may be involved during the AR in sea urchins, since metal chelators or the metalloendopeptidase substrate Suc-Ala-Ala-Phe-MCA inhibited the AR (Farach et al, 1987). However, since these inhibitors are not specific to metalloprotease, participation of metalloprotease in AR remains unclear. Yokota and Sawada found that the proteasome inhibitor MG132 (Z-Leu-Leu-Leu-H) inhibited the egg jelly-induced AR but not the Ca^{2+}-ionophore A23187-induced AR (Yokota & Sawada, 2007). Therefore, Farach's findings might be explained by the participation of the proteasome instead of metalloendopeptidase during the AR, since Suc-Ala-Ala-Phe-MCA is also a substrate for the proteasome (Glas et al., 1998).

In starfish, Hoshi, Matsumoto and their co-workers studied the mechanism of AR. The AR was induced by three egg jelly factors, AR inducing substances (ARIS, a high-molecular-weight glycoprotein), Co-ARIS (cofactor of ARIS, a sulfated steroid), and asterosap (peptide belonging to the SAP family) (Matsumoto et al., 2008). It is proposed that asterosap first interacts with a sperm membrane receptor, guanylate cyclase, in the sperm tail. This results in the production of cGMP and increases of $[Ca^{2+}]_i$ and $[H^+]_i$; cGMP induces the increase of $[Ca^{2+}]_i$, which is mediated by a K^+-dependent Na^+/Ca^{2+} exchanger. ARIS alone is sufficient to induce the AR at high $[Ca^{2+}]$ SW or high pH SW (pH ~9.5) but not in regular SW, in which Co-ARIS or asterosap is required. It is also known that ARIS and asterosap can induce the sustained increase in $[Ca^{2+}]_i$ and the elevation of $[cAMP]_i$ (Hoshi et al., 2012; Matsumoto et al., 2008).

2.3.3 Mammalian sperm AR

Mammalian sperm AR is induced by sperm binding to ZP3 (Florman et al., 2008). Exocytosis of acrosomal vesicle takes place, resulting in formation of small vesicle mixed with the outer acrosomal membrane and the PM overlying the acrosome, which is called vesiculation. But, the acrosomal process is not extended. "*Sperm contact with the zona pellucida results in ZP3 activation of TRPC channels, leading to both Ca^{2+} entry and to activation of PI3 kinase. The resulting phosphorylation of PIP2 (phosphatidylinositol-4,5-bisphosphate) leads to the local accumulation of PIP3 (phosphatidylinositol-3,4–5-triphosphate). PIP3 provides docking sites for proteins with appropriate lipid binding domains and results in the activation of the serine/theonine protein kinases, Akt (Protein Kinase B) and PKCζ. These protein kinases mediate the downstream stages of sperm exocytosis*" (Florman et al., 2008). PLCδ is also involved in the AR in mice (Fukami et al., 2003). Ca^{2+} influx mechanism in mammalian sperm seems similar to the mechanisms of AR in echinoderms (Barresi & Gilbert, 2020; Matsumoto et al., 2008) and also similar to the sperm reaction in ascidians (Lambert & Epel, 1979; Lambert & Koch, 1988). In addition to Ca^{2+}, many ions and ion channels are involved in the induction of AR (Chávez et al., 2024; Florman et al., 2008). For detailed mechanisms, see the following reviews: Chávez et al., 2024; Florman et al., 2008. In mice, trypsin-like protease *Testp5* is also involved in AR, since *Test5*-null mice sperm barely underwent the ZP-induced AR albeit fertile

(Yamashita et al., 2008). As described in the previous section, GalTase is also involved in AR triggered by the binding of GalTase to ZP3 (Shur et al., 1998). In mammals (Sutovsky, 2011) and sea urchins (Yokota & Sawada, 2007), the sperm proteasome plays a key role in AR, although the natural substrate for the proteasome is not known.

As described above, it is notable that most, but not all, fertilizing sperm undergoes the AR before attachment to ZP (Jin et al., 2011). These findings led us to consider that the mechanism of the ZP-independent AR induction may be different from that of the ZP-dependent AR induction.

2.4 Sperm penetration of the egg coat

2.4.1 Ascidian sperm lysin

Among 15 protease inhibitors, two chymotrypsin inhibitors (chymostatin and potato proteinase inhibitor I) and three trypsin inhibitors (leupeptin, antipain and soybean trypsin inhibitor) potently inhibited the fertilization of *H. roretzi* at around 100 μM (Hoshi et al., 1981; Hoshi, 1985). Leupeptin and chymostatin were selected as representatives of trypsin inhibitors and chymotrypsin inhibitors, respectively, and their effects on fertilization of VC-free and -intact eggs were investigated. Chymostatin and leupeptin showed little and no inhibition toward the fertilization of VC-free eggs, respectively. These results, together with that proteinaceous inhibitors (soybean trypsin inhibitor and potato proteinase inhibitor I) blocked the fertilization, led to the conclusion that sperm trypsin-like and chymotrypsin-like proteases participate in the sperm penetration of the VC during fertilization of *H. roretzi,* most probably as VC-lysins.

By using Boc-Val-Pro-Arg-MCA, which is the strongest inhibitor toward fertilization among fluorogenic peptidyl-Arg (or Lys)-MCAs (trypsin substrates) tested (Sawada et al., 1982), two trypsin-like proteases (ascidian acrosin and spermosin) were purified from *H. roretzi* sperm (Sawada et al., 1984a). While ascidian acrosin had enzymatic properties similar to mammalian acrosin, spermosin possessed a narrow substrate specificity toward Boc-Val-Pro-Arg-MCA. Both enzymes were considered to participate in fertilization by comparing the effects of various peptidyl-argininals on fertilization and purified enzymes (Sawada & Someno, 1996; Sawada et al., 1984b). However, these proteases have little or no digesting activity towards the VC.

Therefore, chymotrypsin-like protease involved in fertilization was explored. By using Suc-Leu-Leu-Val-Tyr-MCA, which is the strongest inhibitor against fertilization among four substrates for chymotrypsin-like proteases, the 26S-like high-molecular-weight proteasome was purified from *H. roretzi* sperm, which is capable of degrading HrVC70 in a ubiquitin (Ub)/ATP-dependent manner (Sawada et al., 2002). After sperm addition, the VC appears to be ubiquitinated followed by proteasomal degradation (Sawada et al., 2002). By gamete interaction, the proteasome, ATP and ubiquitin are released (Sakai et al., 2003). In addition, when the supernatant after sperm reaction (sperm exudate) was subjected to gel filtration, the proteasomal antigen was detected at fractions around 1000 kDa, where the VC-digesting activity was detected (Saitoh et al., 1993). These results indicate that sperm proteasome is released, at least in part, to the surrounding seawater upon sperm reaction. Furthermore, HrVC70-ubiquitinating enzyme with an apparent molecular mass of 700 kDa was purified from the sperm exudate (Sakai et al., 2003). This enzyme was capable of ubiquitinating HrVC70 but not lysozyme, a widely used substrate for ubiquitination, and required 10 mM Ca^{2+} for its activity, which is the concentration of Ca^{2+} in seawater but not in the cytosol. Lys-234 of HrVC70 was found to be ubiquitinated with sperm exudate as revealed by site-directed mutants of recombinant HrVC70 fragments (Sawada & Saito, 2022; Sawada et al., 2002). Depletion of extracellular ATP by apyrase strongly inhibited the fertilization (Sakai et al., 2003). All of these results indicate that sperm ubiquitin-proteasome system (UPS) is partially released or exposed upon sperm reaction and is involved in the sperm penetration of the VC, probably as VC-lysin. Although the UPS plays a key role in intracellular protein breakdown, little is known about the extracellular roles of the proteasome. Therefore, these findings may be the first reports indicating the importance of the extracellular role of the proteasome.

The sperm proteasome is also involved in fertilization of the ascidian *C. intestinalis* type A (another name: *Ciona robusta*) (Brunetti et al., 2015; Sawada et al., 1998). In order to identify the proteasome subunits on the sperm surface, we carried out proteomic analysis by LC-MS/MS (Nakazawa et al., 2015). However, probably because of limited amounts of the proteasome subunits, we failed to detect the proteasomal subunit. Instead, several fragments of astacin-like metalloproteases with thrombospondin type-1 repeats (Tast) were identified. Five Tast proteases with a single TM domain in the N-terminal region (CiTast1, CiTast2a, CiTast2b, CiTast2c, CiTast2d) were found to be expressed in the sperm surface. Then, in order to investigate whether metalloproteases are involved in fertilization, the effects of metalloprotease inhibitors (hydroxamate derivatives) on fertilization were examined: GM6001 showed a strong inhibition toward fertilization but a negative control derivative (GM6001NC), lacking a reactive hydroxamate group, showed no inhibition. Furthermore, such a strong inhibition by GM6001 was not observed against the fertilization of VC-free eggs. The sperm binding to the VC was not affected by GM6001. From these results, it was concluded that four Tast metalloproteases are involved in sperm penetration of the VC during *C. intestinalis* fertilization.

Similarly, Tasts are also involved in fertilization of the ascidian *H. roretzi,* since metalloprotease inhibitors, GM6001, TAPI-0, and TAPI-1, strongly inhibited the fertilization. Four genes of Tasts *(HrTast1, HrTast2b, HrTast 2c,* and *HrTast3b)* are transcribed in the testis. All Tast proteins have no signal peptide but a single TM domain in the N-terminal region. These proteins were predicted to be a type II TM protein, locating the C-terminal region to the extracellular milieu. Since anti-HrTast2c antibody inhibited the fertilization, at least HrTast2c is considered to be involved in the fertilization. It was suggested that the proteasome and spermosin function at the early stage and Tasts function at the middle or late stage and acrosin functions at the very late stage during *H. roretzi* fertilization (Sawada & Saito, 2022; Sawada et al., 2024).

2.4.2 Echinoderm sperm lysin

In sea urchins, the effects of a variety of protease inhibitors on the fertilization of intact eggs and VE-impaired eggs were examined (Hoshi et al., 1979). Among the inhibitors tested, two chymotrypsin inhibitors, chymostatin and TPCK, inhibited the fertilization. Chymostatin inhibited the fertilization of intact eggs, but only weakly inhibited the VE-impaired eggs. From these results, chymotrypsin-like protease was thought to be a VE-lysin in sea urchin (Hoshi et al., 1979). Then, Yokota and Sawada reevaluated the effects of protease inhibitors on fertilization using the sea urchin *Pseudocentrotus deppressus* (Yokota & Sawada, 2007). It was found that proteasome inhibitors MG132 most potently inhibited the fertilization at 100 μM and proteasome inhibitor MG115 or lactacystin showed weak inhibition. MG132 inhibited the egg jelly-induced AR but not the ionophore A23187-induced AR, suggesting the participation of the proteasome in the AR, particularly in the process before Ca^{2+} influx. MG132 inhibited the fertilization with the egg jelly-induced reacted sperm. On the other hand, chymostatin or leupeptin showed no appreciable inhibition at 100 μM. MG132 did not inhibit the sperm binding to the VE of the eggs. Furthermore, proteasome antigen was confirmed in the supernatant of the AR by Western blotting. Immunocytochemistry revealed that the unfertilized eggs are ubiquitinated. Depletion of extracellular ATP by apyrase inhibited the fertilization. All of these data indicate the participation of the UPS in fertilization as a VE lysin.

2.4.3 Mammalian sperm lysin

After the report by Sawada et al. (2002), Sutovsky and his colleagues discovered that the sperm proteasome localized in the inner acrosomal membrane is involved in the sperm penetration through the ZP in mice and pigs, as revealed by the results of the effects of proteasome inhibitors and antibodies on fertilization of ZP-intact eggs and -free eggs and also by immunocytochemistry and immuno-electron microscopy (Sutovsky et al., 2004; Sutovsky, 2011). Localization of the proteasome in the acrosome was also clarified by using the transgenic pig line expressing eGFP-fused α6-subunit in the proteasome (Miles et al., 2013). They reported that the proteasome plays a key role not only in the sperm penetration through the ZP, but also in capacitation and AR (Kerns et al., 2016; Sutovsky, 2011). They also showed that ZP proteins are degraded by the sperm proteasome in mammals (Zimmerman et al., 2011). Similarly, sperm proteasome is reported to function as lysin in the quail (Sasanami et al., 2012).

Although it has long been believed that the acrosomal trypsin-like protease acrosin is a ZP-lysin, it was revealed that acrosin is not essential for fertilization in the mouse, using an *Acr* gene-null mice (Baba et al., 1994). Mouse acrosin is involved in the dispersal of acrosomal proteins during fertilization (Yamagata et al., 1998b). Instead of acrosin, it is thought that *p*-aminobenzamidine-sensitive acrosomal protease(s) participates in the sperm penetration of the ZP in mice (Yamagata et al., 1998a). Several other proteases may be cooperatively involved in sperm penetration of the ZP in mice (Honda et al., 2002; Kawano et al., 2010). On the other hand, acrosin was found to be essential for fertilization in hamsters (Hirose et al., 2020). These results suggest that the participation of acrosin depends on species. In hamsters, sperm acrosin must function as ZP-lysin, while in mice (Sutovsky et al., 2004), pigs (Sutovsky et al., 2004; Miles et al., 2013), quails (Sasanami et al., 2012), ascidians (Sawada et al., 1998; Sawada et al., 2002), and sea urchins (Yokota & Sawada, 2007), sperm proteasome seems to function as egg-coat lysin.

2.5 Gamete fusion

2.5.1 Ascidian gamete fusion

The mechanisms of gamete fusion in ascidians are not well known. It is reported that sperm metalloprotease may be involved in gamete fusion, since

Ca^{2+} chelators and peptide substrates for metalloendopeptidase inhibited the gamete fusion (De Santis et al., 1992). However, Ca^{2+} chelator or the peptide substrate Z-Gly-Phe-NH_2 are not specific to metalloproteases. Furthermore, GM6001, a hydroxamate-type metalloprotease inhibitor, did not inhibit the cleavage of VC-free eggs in *C. intestinalis* (Nakazawa et al., 2019) or *H. roretzi* (Sawada et al., 2024). From these results, metalloprotease is unlikely to be involved in gamete fusion, although a possibility that GM6001-insensitive metalloprotease is involved in gamete fusion will not be excluded. Since chymostatin weakly inhibited the fertilization of VC-free eggs, chymostatin-sensitive protease (possibly chymotrypsin-like protease) may play a role in gamete fusion. Further studies are necessary to explore a factor involved in ascidian gamete fusion.

2.5.2 Echinoderm gamete fusion

The mechanism of gamete fusion in sea urchins is poorly understood. Metalloprotease is presumed to be involved in sea urchin gamete fusion (Roe et al., 1988). However, as described above, convincing evidence about the participation of metalloprotease has not been reported.

2.5.3 Mammalian gamete fusion

In mice, the sperm inner acrosomal membrane-bound IZUMO1 (an immunoglobulin (Ig)-like TM protein) (Inoue et al., 2005), SPACA6 (sperm acrosome associated 6) (Lorenzetti et al., 2014), TMEM95 (transmembrane protein 95) (Lamas-Toranzo et al., 2020; Noda et al., 2020; Pausch et al., 2014), SOF1 (sperm-oocyte fusion required 1) (Noda et al., 2020), FIMP (fertilization influencing membrane protein) (Fujihara et al., 2020), and DCST1 and DCST2 (dendrocyte expressed seven transmembrane protein domain-containing) (Inoue et al., 2021; Noda et al., 2022) are sperm proteins which are essential for sperm-egg fusion. Mouse IZUMO1, SPACAS6, DCST1, and DCST2 orthologs are also present in fishes (Binner et al, 2022; Deneke et al., 2024; Noda et al.,2022), implying the common mechanism of gamete fusion.

Regarding egg proteins involved in gamete fusion, the GPI-anchored PM protein JUNO, an IZUMO1 receptor (Bianchi et al., 2014), and tetraspanin CD9 (Miyado et al., 2000) are known to be essential for sperm-egg fusion. Tetraspanin CD81 is also involved in sperm-egg fusion and AR (Tanigawa et al., 2008). IZUMO1 and JUNO are *trans*-partners with mutually interactable recognition (Aydin et al., 2016; Ohto et al., 2016), although these two proteins alone are insufficient for gamete fusion. In this connection, it is interesting to note that rat copper transport protein 2 (CTR2) expressed in sperm head and mid-tail and also on the oocyte surface are capable of directly interacting with IZUMO1 and JUNO, and play a role in sperm-oocyte fusion (Su et al., 2025). On the other hand, CD9 plays a crucial role in the microvilli biogenesis and molecular organization, such as JUNO localization, of the oocyte surface (Inoue et al., 2020; Umeda et al., 2020). Furthermore, CD9 is reported to be transferred to the sperm membrane with the aid of egg-derived esoxomes (Miyado et al., 2008). CD9-transferred sperm can fuse with CD9-deficient eggs (Miyado et al., 2008). Such a discovery gave us a paradigm shift and informed us the importance to investigate the mechanisms of cell fusion from different angles. Moreover, it is notable that MAIA (Fc receptor like 3), which interacts with the IZUMO1-JUNO complex, permits gamete fusion in humans (Vondrakova et al., 2022).

Although the tertiary structure of IZUMO1 is similar to those of SPACA6 and TMEM95, recombinant human SPACA6 and TMEM95 cannot interact with human JUNO. Recently, a novel factor TMEM81 with structural homology to IZUMO1 was identified by *in silico* screening of the interacting protein with IZUMO1 among ~1400 secreted/TM proteins expressed in zebrafish testis by using AlphaFold-Multimer (Deneke et al., 2024). By the analyses of a high interface-predicted template modeling score, SPACA6 and TMEM81 were predicted to interact with IZUMO1 (Deneke et al., 2024; Elofsson et al., 2024). TMEM81 is essential for gamete fusion in the mouse and zebrafish. They predicted that IZUMO1, SPACA6 and TMEM81 make a trimer. The N-terminal four helix bundles of IZUMO1 and SPACA6 make an interface with fish Bouncer, which is essential for species-specific binding of sperm to oolemma in fish (Deneke et al., 2024; Elofsson et al., 2024; Herberg et al., 2018). In mice, it was revealed that IZUMO1, SPACA6 and TMEM81 can make a trimer, which is predicted to interact with JUNO, albeit *via* the different interacting site. Such studies on the prediction of 3D structures and their interaction will be necessary for elucidation of another essential proteins for gamete interaction.

Although homologous proteins to IZUMO1 have not been identified in ascidians, it is notable that one Ig-like protein (XP_00427079.1) is exposed to the *C. intestinalis* sperm surface (Nakazawa et al., 2015). Such proteins might be involved in ascidian gamete interaction or fusion.

After gamete fusion, mouse astacin-like metalloprotease (ASTL, also called ovastacin) is released from cortical granules of the egg and hydrolyzes the N-terminal region of ZP2, inactivating the sperm receptor (Avella et al., 2013).

2.5.4 Flowering plant gamete fusion

In flowering plants, GCS1 (generative cell specific 1)/HAP2 (hapless 2) is known to be a male gamete-specific protein involved in gamete fusion, although whether GCS1 directly plays a pivotal role in cell fusion is still debated (Mori et al., 2006, 2015). This gene has been identified in not only flowering plants but also cnidarians, and phylogenetically lower animals and plasmodium (Hirai et al., 2008; Mori et al., 2015). This implies the common mechanism of gamete fusion in animals and plants. However, homologous genes have not been identified in ascidians or vertebrates. Notably, *GCS1* expression is always restricted to the male gamete, even in isogamy. Interestingly, *Chlamydomonas* GCS1/HAP2 is structurally similar to the class II viral membrane fusion proteins, including TBEV (E of tick-borne encephalitis virus) and EFF1 (*C. elegans* developmental fusogen), which make a trimer, consisting of three extracellular domains, a single TM domain, and a cytosolic tail (Fédry et al., 2017). Although it is predicted that trimeric GCS1 may be directly involved in membrane fusion without requiring female binding partner(Fédry et al., 2017), further studies are necessary to elucidate the detailed mechanisms. In any cate, there may be common molecules and mechanisms shared by plant- and animal- and also somatic-cell-fusion.

2.6 Polyspermy block

2.6.1 Ascidian egg polyspermy block

There are several reports about the polyspermy block in ascidians. *N*-acetylglucosamindase (GlcNAcase) purified from *H. roretzi* eggs participates in the polyspermy block (Matsuura et al., 1993). Such a glycosidase has little or no hydrolytic activity at around pH 8.0, but exhibits the binding activity to the VC *via* substrate-enzyme complex at seawater pH. A 70 kDa specific inhibitor or binding protein toward the egg GlcNAcase has been isolated from the VC (Matsuura et al.,1995). This protein may be HrVC70 itself, since 70-kDa HrVC70 is a sperm binding protein on the VC (Sawada et al., 2002, 2004), which possesses two *O*-linked sugar chains with fucose (Fuc) and *N*-acetylhexosamine (HexNAc) residues attached to Thr and Ser residues of HrVC70 (Sawada & Saito, 2022). It is interesting to note that sperm α-L-fucosidase-mediated sperm binding to HrVC70 would be inhibited by the binding of GlcNAcase, which is released from fertilized eggs to the carbohydrate moiety, containing both Fuc and HexNAc residues, in HrVC70.

2.6.2 Echinoderm egg polyspermy block

Two mechanisms to block polyspermy are reported in sea urchin eggs. One is the fast block to polyspermy by Na^+-influx-induced rapid and transient depolarization in membrane potential (Jaffe, 1976, 1980) and the other is the late block to polyspermy, in which protease(s) released by exocytosis of cortical granules degrade the sperm receptor on the VE (Barresi & Gilbert, 2020). However, the mechanism of the fast block to polyspermy is currently debated, because low-Na^+ SW-treated eggs undergo abnormal cleavage which is not due to polyspermy but to alterations in the cortical actin filament dynamics (Limatola et al., 2019; Dale, 2014).

2.6.3 Mammalian egg polyspermy block

After gamete fusion, mouse ovastacin is released from cortical granules of the egg and hydrolyzes the N-terminal region of ZP2, which inactivates the sperm receptor (Avella et al., 2013). This is the main polyspermy block mechanism in mammals. Dean and his colleagues clearly demonstrated that capacitated human sperm is capable of binding to the ZP of transgenic mice expressing human ZP2 but not mouse ZP2, which was not observed in the case of ZP1, ZP3, and ZP4. They also showed that mouse sperm can bind to the ZP of two-cell embryos, if ZP2 is not cleaved by ovastacin because of ZP2 mutant insusceptible to ovastacin or *Astl*-null mutant (see review: Avella et al., 2013). These results clearly indicated that ovastacin in oocyte cortical granule is released upon sperm fusion and digests the N-terminal region of ZP2, which is sufficient to block polyspermy. Recently, it was also found that post-fertilization cleavage of ZP2 stimulates the oligomerization of ZP2, allowing the cross-linking of ZP filaments and hardening of the ZP, which ensures the permanent block to polyspermy (Nishio et al., 2024). These molecular biological data indicate that ZP2, rather than ZP3, is a sperm receptor. However, these data appear to be inconsistent with many Wassarman's or Shur's biochemical data.

3. Fertilization mechanisms in nematodes: Comparison with mammals

A spermatozoon and an oocyte are specialized, haploid cell types, so *in vitro* experimental techniques to examine gene functions during fertilization are poorly available. Therefore, analyzing fertilization-defective mutants is essential to assess the mechanism of this phenomenon. The nematode *Caenorhabditis elegans* has long been used for genetics-based studies to elucidate the mechanisms of various biological phenomena, including fertilization (Brenner, 1974). The advantages of using *C. elegans* are that this animal can produce progeny *via* self-fertilization without outcrossing and that abundant genetic tools are available, leading to easy isolation and maintenance of sterile mutants. Since *C. elegans* is evolutionarily distant from mammals by approximately one billion years (Wheeler & Brändli, 2009), it is intriguing to ask how similar the molecular basis of *C. elegans* fertilization is to those of mammalian fertilization. This section reviews the findings regarding *C. elegans* fertilization and compares nematode and mouse genes involved during fertilization.

As shown in Fig. 1, *C. elegans* spermatids undergo spermiogenesis, by which the spermatids transform into sperm with motility and fertility (Step 3). The fertilization-competent sperm are guided to an oocyte by F-series prostaglandins (PGFs) that are secreted from the oocyte (Step 1, also described later). Since a *C. elegans* oocyte lacks a typical, thick egg coat like the ZP, the sperm guided by PGFs seems to bind to and fuse with the oocyte PM directly (Step 5). This feature enables to examine the mechanism of gamete fusion more straightforwardly. Unlike in ascidian and mouse sperm, a subcellular region(s) in *C. elegans* spermatozoa that interacts with the oocyte PM remains to be clarified.

C. elegans has a hermaphrodite instead of an actual female (Fig. 4, top). Hermaphrodites undergo spermatogenesis in the U-shaped gonad during the fourth larval (L4) stage. As L4 hermaphrodites develop into young adults, gametogenesis is wholly switched to oogenesis. Thus, adult hermaphrodites are somatically females containing self-spermatozoa. The adult hermaphrodite gonad carries female germ cells that are aligned by their developmental stages, immature cells are relatively distal. The fertilization-ready oocyte resides at the most proximal region of the gonad (Fig. 4A). As the first oocyte is ovulated into the spermatheca, ~300 spermatids that were produced during the L4 stage also move to the spermatheca with the oocyte (Fig. 4A). Those spermatids become motile, fertile self-sperm and then fertilize the ovulated oocyte (Fig. 4B). For successful embryogenesis, the fertilized oocyte moves into the uterus together with most of the remaining self-sperm that were not consumed by the last fertilization (Fig. 4C). The self-sperm in the uterus subsequently crawl back into the spermatheca, according to attraction mediated by PGFs, secreted from the oocyte residing in the spermatheca. This phenomenon is called "sperm guidance", functionally equivalent to sperm chemotaxis. By repeated

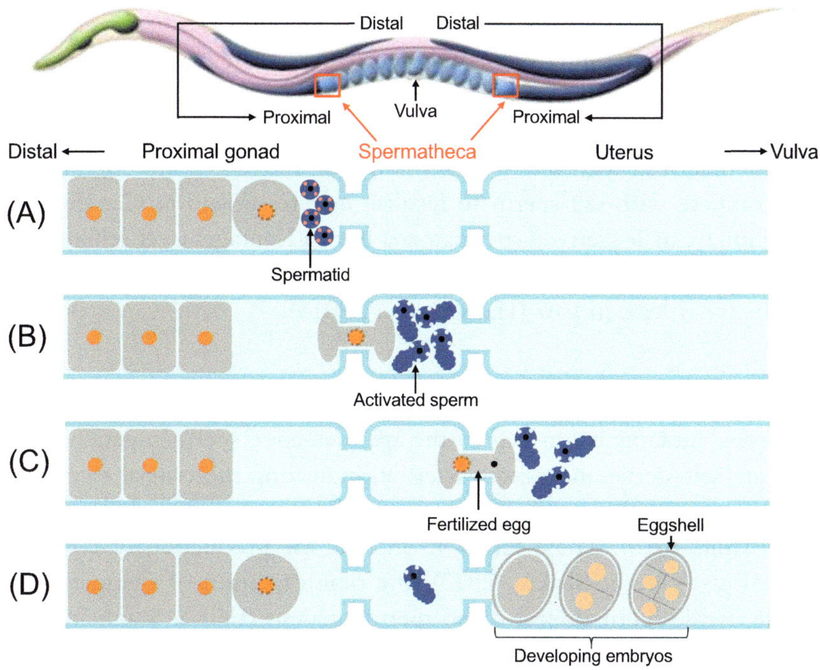

Fig. 4 An ovulation-fertilization cycle in the *C. elegans* hermaphroditic gonad. Top. A *C. elegans* hermaphrodite has a U-shaped gonad, which is symmetric around the single vulva. The adult gonad contains female germ cells that are aligned by their developmental stages; immature cells are relatively distal. The location of each spermatheca is outlined by red squares. **(A)** The fertilization-ready oocyte at the most proximal region of the gonad is ovulated into the spermatheca. Spermatids produced during the fourth larval (L4) stage also move into the spermatheca with the oocyte and then transform into motile self-sperm capable of fertilizing oocytes. **(B)** After fertilization, the fertilized oocyte moves into the uterus, accompanied by self-sperm that were not used for fertilization. **(C)** The self-sperm in the uterus crawl back into the spermatheca, according to sperm guidance mediated by oocyte-secreted PGFs (see also Step 1 in the nematode, Fig. 1). **(D)** After repeated ovulation-fertilization cycles, ~300 self-sperm are mostly consumed to produce ~300 fertilized oocytes, which will develop into the gastrula stage in the uterus and then be laid from the vulva. An image of an adult hermaphrodite at the top of this figure is found on the website of WormAtlas (http://www.wormatlas.org: Altun et al., 2002-2024), and is used here with permission.

ovulation-fertilization cycles, ~300 self-sperm are consumed by self-fertilization, and an adult hermaphrodite eventually produces ~300 self-progeny throughout its life, indicating that fertilization occurs quite efficiently in the spermatheca (Fig. 4D).

As males sire outcross-progeny by mating with hermaphrodites, the seminal fluids containing spermatids and spermatid-activating factor (SAF), presumably the serine protease TRY-5 (Smith & Stanfield, 2011), are ejaculated into the uterus of hermaphrodites. The spermatids are activated into spermatozoa with the SAF in the uterus, move into the spermatheca, and compete with self-sperm to fertilize oocytes (Ward & Carrel, 1979). Intriguingly, male-derived spermatozoa have advantages over self-sperm in fertilization due to *comp-1*, encoding a protein kinase with unknown specificity, at least in part (Hansen et al., 2015).

3.1 *C. elegans* genes belonging to the *spe* category

C. elegans mutants belonging to the *spe* (defective spermatogenesis) category are self-sterile and outcross-fertile, indicating the crucial roles of *spe* genes during male germline functions (L'Hernault, 2009; Nishimura & L'Hernault, 2010). These functions include the production of spermatids *via* meiosis (Chu & Shakes, 2013), the transformation of spermatids into motile, fertile spermatozoa, called spermiogenesis (Ellis & Stanfield, 2014; Ma et al., 2012; Shimada et al., 2023), and fertilization (Maniates & Singson, 2023; Marcello et al., 2013; Mei & Singson, 2021). Among *spe* mutants, those in the *spe-9* class produce otherwise normal spermatozoa incapable of fertilizing oocytes. Therefore, understanding the molecular basis for *C. elegans* fertilization is greatly aided by *spe-9* class mutants. Table 1 contains *C. elegans* proteins that participate during fertilization, including the SPE-*9* class (SPE-*9*, SPE-*36*, SPE-*38*, SPE-*41*/TRP-*3*, SPE-*42*, SPE-*49*, SPE-*45*, SPE-*51*).

3.1.1 spe-9 gene

SPE-9 is a TM protein with multiple EGF-like domains (Putiri et al., 2004; Singson et al., 1998). The gross structure of SPE-9 protein is similar to those of Delta-like proteins (DLLs), some of which act as ligands for the Notch family members. Thus, SPE-9 might be a ligand for a putative sperm receptor(s) on the oocyte PM. However, SPE-9 does not contain the Delta/Serrate/Lag-2 (DSL) domain, characterizing the DLL family (Cordle et al., 2008), suggesting that how SPE-9 acts as a ligand differs significantly from the DLL family. Time-lapse analysis on the consumption of self-spermatozoa during fertilization in the spermatheca suggests that SPE-9 functions at a later step(s) after gamete binding during fertilization (Nishimura et al., 2015).

Table 1 Summary of female and male factors involved in gamete interaction in animals and plants.

Steps and species (family *etc.*)	Female factor	Male factor	References
Sperm chemotaxis or pollen tube guidance			
Ciona intestinalis (ascidian)	SAAF (sulphated steroid)	PMCA (Ca^{2+}-ATPase)	Yoshida et al. (2002); Yoshida & Yoshida (2011)
Arbasia. panctulata (sea urchin)	Resact (14-aa P)	CatSper (Ca^{2+} channel)	Seifert et al. (2015); Ward et al. (1985)
Asterias amurensis (starfish)	Asterosap	ND	Nishigaki et al. (1996)
Xenopus laevis (frog)	Allurin (CRISP)	ND	Olson et al. (2001); Burnett et al. (2012)
Mus musculus (mouse)	Odorant	Olfactory receptor	Yoshida & Yoshida (2011)
Homo sapiens (human)	Odorant	Olfactory receptor	*ibid.*
	Choresterol, progesterone	CatSper	Brenker et al. (2012); Williams et al. (2015)
Arabidopsis thaliana (angiosperm)	AtLURE1/2 (defensin-like P) (from synergid cell)	LURE receptor (kinase)	Higashiyama et al. (2001); Higashiyama & Yang (2017); Higashiyama & Takeuchi (2015)
Torenia fournieri (angiosperm)	TfLURE1/2	LURE receptor	*ibid.*
Caenorhabditis elegans (nematode)	PGFs (F-series prostaglandins)	ND	Hoang et al. (2013)

(continued)

Table 1 Summary of female and male factors involved in gamete interaction in animals and plants. (*cont'd*)

Steps and species (family *etc.*)	Female factor	Male factor	References
Sperm binding to the egg coat and species-specificity in gamete interaction			
Ciona intestinalis (ascidian)	Fucosyl glycoproteins (140-kDa, 125-kDa, 78-kDa VCPs)	α-L-fucosidase	De Santis et al. (1983); Hoshi et al. (1983, 1985); Matsumoto et al. (2002); Pinto et al. (1981)
	CiVC57 (EGF-like repeats)	CiUrabin (CRISP)	Yamaguchi et al. (2011)
	Apolipoprotein B-like	70-kDa VC protein	Yamada et al. (2009)
Halocynthia roretzi (ascidian)	HrVC70 (12 EGF-like)	HrUrabin (CRISP), HrTTSP−1	Harada & Sawada (2007); Urayama et al. (2008)
	Fucosyl P (HrVC70?)	α-L-fucosidase	Matsumoto et al. (2002); Sawada & Saito (2022)
	30-kDa C-terminal of VG	HrProacrosn (CUB domain)	Kodama et al. (2001)
	25-kDa vWF-D	HrSpermosin L-chain	Kodama et al. (2002)
	ND	Hr20S-proteasome	Takizawa et al. (1993)
Phallussia mammillata (ascidian)	Glycoproteins	GlcNAcase	Godknecht & Honegger (1991); Hoshi et al. (1985)
Strongylocentrotus purpuratus (sea urchin)	EBR1 (CUB/ TSP-repeats)	Bindin	Glabe & Lennarz, 1979; Kamei & Glabe (2003); Wessel et al. (2021a, 2021b)

Mus musculus (mouse)	ZP3 (O-linked CHO)	GalTase	Miller et al. (1992); Shur & Hall (1982); Shur et al. (1998)
	ZP3, ZP2	SED1 (2 EGF, 2 discoidin domains)	Shur et al. (2006)
	ZP3	ZP3R (SP56: Sushi domains)	Bleil & Wassarman (1990); Cheng et al. (1994); Bookbinder et al. (1995)
	ZP2	proacrosin	Jones (1991); Urch & Patel (1991)
	ND	ADAM3 (cyritestin)	Kim et al. (2005)
	ZP2, ZP3	Arylsulfatase	Xu et al. (2012)
Sus sucrofa (pig)	ND	zonadhesin	Hardy & Garbers (1994, 1995)
Self-incompatibility			
Ciona intestinalis (ascidian)	v-Themis-A (fibrinogen C-term)	s-Themis-A (PKDREJ-like)	Harada et al. (2008); Harada & Sawada (2008)
	v-Themis-B (fibrinogen C-term)	s-Themis-B (PKD channel)	*ibid.*
	v-Themis-B2 (fibrinogen C-term)	s-Themis-B2 (PKD channel)	Sawada et al. (2020)
	Glx-enriched VCP	ND	Kawamura et al. (1987, 1988, 1991)
	Cihsp70	ND	Marino et al. (1998, 1999)

(continued)

Table 1 Summary of female and male factors involved in gamete interaction in animals and plants. (*cont'd*)

Steps and species (family *etc.*)	Female factor	Male factor	References
Halocynthia roretzi (ascidian)	HrVC70	ND	Sawada et al. (2004)
	v-Themis ?	s-Themis ?	Unpublished
Brassicaceae	SRK (kinase)	SP11/SCR (Cys-rich)	Sawada et al. (2014); Iwano & Takayama (2012); Murase et al. (2024)
Solanaceae/Rosaceae	S-RNase	SLF/SFB (E3 ligase)	*ibid.*
Papaveraceae	PrsS	PrpS	*ibid.*
Sperm penetration of the egg coat			
Ciona intestinalis (ascidian)	ND	UPS	Sawada et al. (1998)
	CiVC57 (substrate for CiTast)	CiTast (astacin-like)	Nakazawa et al. (2019)
	ND	Chymotrypsin-like protease	Pinto et al. (1990); Marino et al. (1992)
Halocynthia roretzi (ascidian)	HrVC70 (substrate for the UPS)	UPS	Sawada et al. (2002); Saitoh et al. (1993)
	ND	HrAcrosin	Sawada et al. (2014); Sawada et al. (1984a); Kodama et al. (2001)
	ND	HrSpermosin	Sawada et al. (2014); Sawada et al. (1984a); Sawada & Someno (1996); Kodama et al. (2002)
	ND	HrTast	Sawada et al. (1984b)

Mesocricetus auratus (golden hamster)	ZP	Acrosin	Hirose et al. (2020)
Mus musculus (mouse)	ZP	UPS	Sutovsky (2011)
Sus sucrofa (pig)	ZP	UPS	*ibid.*
Homo sapiens (human)	ZP	UPS	*ibid.*
			Zimmerman et al. (2011)
	ZP (ZPs)	Trypsin-like P (Acrosin?)	Saldívar-Hernández et al. (2015)
	ZP (ZP4)	Proteasome	*ibid.*
	ZP (ZP2)	Metalloprotease (MMT)	*ibid.*
Cell fusion or attachment between gametes			
Ciona intestinalis (ascidian)	ND	Metalloprotease	De Santis et al. (1992)
Mus musculus (mouse)	JUNO (GPI-anchored, folate receptor 4)	IZUMO (Ig-like)	Inoue et al. (2005); Bianchi et al. (2014); Ohto et al. (2016); Aydin et al. (2016)
	CD9 (tetraspanin, transferred to sperm membrane *via* exosome)		Miyado et al. (2000); Miyado et al. (2008)

(continued)

Table 1 Summary of female and male factors involved in gamete interaction in animals and plants. (*cont'd*)

Steps and species (family *etc.*)	Female factor	Male factor	References
	ND	SPACA6 (IZUMO1 BP)	Lorenzetti et al. (2014)
	ND	TMEM95	Lamas-Toranzo et al. (2020); Noda et al. (2020)
	ND	TMEM81 (IZUMO1 BP)	Deneke et al. (2024)
	ND	DCST1 (DCST1 BP)	Inoue et al. (2021); Deneke et al. (2024)
	ND	DCST2 (DCST1 BP)	Inoue et al. (2021); Deneke et al. (2024)
	ND	SOF1	Noda et al. (2020)
	ND	FIMP	Fujihara et al. (2020)
Dania rerio (zebrafish)	Bouncer (GPI, species-specific)	ND	Herberg et al. (2018)
Caenorhabditis elegans (nematode)	ND	SPE-9 (EGF-like)	Singson et al. (1998)
	ND	SPE-36 (EGF-like)	Krauchunas et al. (2023)
	ND	SPE-38 (SPE-41/TRP-3 BP)	Chatterjee et al. (2005)
	ND	SPE-41/TRP-3 (TRP channel)	Xu & Sternberg (2003); Takayama & Onami (2016)

	ND	SPE-42 (DCST2)	Kroft et al. (2005)
	ND	SPE-45 (Ig-like)	Nishimura et al. (2015); Singaravelu et al. (2015)
	ND	SPE-49 (DCST1)	Wilson et al. (2018)
	ND	SPE-51 (Ig-like)	Mei et al. (2023)
	EGG-1/EGG-2 (LDL-receptor-like)	ND	Kadandale et al. (2005)
Arabidopsis thaliana (angiosperm)	ND	GCS1/HAP2	Mori et al. (2006)
	ND	GEX2 (Ig-like)	Mori et al. (2014)
Chlamydomonas reinhardtii (green alga)	FUS1 in *plus* gamete	MAR1 (HAP2 BP) in *minus* gamete	Pinello et al. (2021)
Polyspermy block			
Ciona intestinalis (ascidian)	α-L-Fucosidase (Fucose BP on VC)	ND	De Santis et al. (1983)
Halocynthia roretzi (ascidian)	GlcNAcase (egg of FC) to GlcNAcase inhibitor (HrVC70 ?) on VC	ND	Matsuura et al. (1993, 1995)

(continued)

Table 1 Summary of female and male factors involved in gamete interaction in animals and plants. (*cont'd*)

Steps and species (family *etc.*)	Female factor	Male factor	References
Mus musculus (mouse)	Ovastacin (cortical granule) to ZP2	ND	Avella et al. (2013)
	GalTase to ZP3 (O-linked sugar)	ND	Miller et al. (1993)
Arabidopsis thaliana (angiosperm)	Asp endopeptidase (synergic cells)	pollen tube	Yu et al. (2021)
Caenorhabditis elegans (nematode)	CBD-1 (chitin BP)	ND	Johnston et al. (2010)
	CHS-1 (chitin synthase)	ND	*ibid.*
	EGG-1/EGG-2	ND	*ibid.*
	EGG-3 (tyrosine phosphatase-like)	ND	*ibid.*
	MARC-3 (ubiquitin ligase)	ND	Kawasaki et al. (2024)
	ND	SPE-11 (paternal factor)	Johnston et al. (2010)

Abbreviations: BP, binding protein; CRISP, cysteine-rich secretary protein; CHO, carbohydrate; CT, C-terminal; EGF, epidermal growth factor; Fucase, fucosidase; GalTase, 1,4-galactosyltransferase; GlcNAcase, N-acetylglucosaminidase; Ig, immunoglobulin; LDL, low-density lipoprotein; ND, not determined; P, protein; SRK, *S*-locus receptor kinase; SCR, *S*-locus cysteine-rich protein; TPS, thrombospondin type 1 repeat; VC, vitelline coat; VCP, VC protein; VG, vitellogenin; vWF-D, von Willebrand factor-D; UPS, ubiquitin-proteasome system,

3.1.2 *spe-36* and *spe-51* genes

SPE-36 (Krauchunas et al., 2023) and SPE-51 (Mei et al., 2023) contain an EGF-like and an Ig-like domains, respectively, and they are both secreted proteins. Moreover, these two proteins localize in membranous organelles (MOs) of spermatids and the cell body and the pseudopod of spermatozoa (Fig. 5), implying that SPE-36 and SPE-51 might be included in the same or separate protein complexes that play an essential role(s) during fertilization.

3.1.3 *spe-38* and *spe-41/trp-3* genes

SPE-38 is a four-pass TM protein (Chatterjee et al., 2005), whereas SPE-41/TRP-3 is an eight or nine-pass TM protein that mediates Ca^{2+} influx during fertilization as a transient receptor potential (TRP) channel (Takayama & Onami, 2016; Xu & Sternberg, 2003). The Ca^{2+} influx might be a crucial signal for successful gamete interactions, since some members of the TRP channel family are involved in lipid-processing events facilitating membrane-membrane fusion (Beech et al., 2009). Moreover, SPE-41/TRP-3 might form a homo- or hetero-tetramer like other members of the TRP channel family (Schindl & Romanin, 2007).

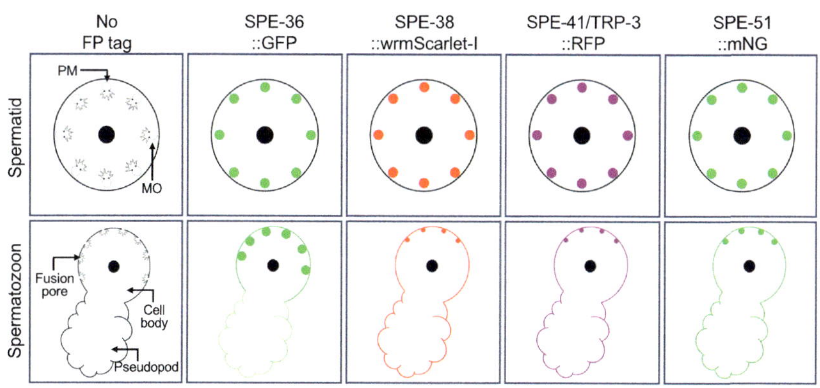

Fig. 5 Translocation of *C. elegans* SPE-9 class proteins during spermiogenesis. This figure is based on the live imaging data during spermiogenesis, which were obtained by monitoring the dynamics of SPE-9 class proteins tagged with fluorescent proteins (FPs), such as SPE-36::GFP (this FP's signal is shown by green) (Krauchunas et al., 2023), SPE-38::wrmScarlet-I (shown by red) (Zuo et al., 2023), SPE-41/TRP-3::TagRFP (shown by purple) (Takayama & Onami, 2016), and SPE-51:mNG (shown by green) (Mei et al., 2023). A *C. elegans* spermatid intracellularly contains numerous MOs, and SPE-9 class proteins shown here seem to be present on or within MOs. During spermiogenesis, those proteins translocate onto the entire sperm surface. Note that the SPE-36::GFP signal on the pseudopod of sperm is weaker than that on the cell body of the same sperm cell.

The localization profiles of SPE-38 and SPE-41/TRP-3 are similar in spermatids and spermatozoa (Fig. 5), they reside on MOs within spermatids and on the entire surface of spermatozoa (Takayama & Onami, 2016; Xu & Sternberg, 2003; Zuo, Mei, & Singson, 2023). Since the SPE-41/TRP-3 translocation is disturbed in *spe-38* mutant worms (Singaravelu et al., 2012), SPE-38 is presumably a regulatory protein *via* forming a protein complex with SPE-41/TRP-3.

3.1.4 spe-42 and spe-49 genes

SPE-42 (Kroft et al., 2005) and SPE-49 (Wilson et al., 2018) are six-pass and four-pass TM proteins, respectively, and both contain a dendritic cell-specific TM protein (DC-STAMP)-like and a C4C4-type RING-finger domains. Mammalian DC-STAMPs act in the cellular fusion of osteoclasts and foreign body giant cells (Miyamoto, 2006), implying that the DC-STAMP domains in SPE-42 and SPE-49 might play a key role during gamete fusion. On the other hand, SPE-42 and SPE-49 possibly associate with other sperm proteins *via* their RING finger domains (Borden, 2000). Since human CNOT4, containing a C4C4 RING finger, has the E3 ubiquitin-protein ligase activity (Albert et al., 2002; Hanzawa et al., 2001), it is also possible that SPE-42 and SPE-49 catalyze the ubiquitination of a sperm protein(s) that acts during gamete interactions. It is worth noting that human DC-STAMP domain-containing protein 1 (DCST1) was previously characterized as an E3 ubiquitin-protein ligase (Nair et al., 2016).

spe-42 and *spe-49* genes have orthologs in diverged organisms, including mammals, fish, and fruit flies, *spe-42* and *spe-49* correspond to *dcst2* and *dcst1* genes, respectively (Inoue et al., 2021; Noda et al., 2022) (Table 1). In mice, DCST1 and DCST2 proteins seem to associate with each other (Noda et al., 2022). Spermatozoa lacking both or either of *Dcst1* and *Dcst2* can bind to but fail to proceed to fuse with the oocyte PM (Inoue et al., 2021; Noda et al., 2022). Moreover, those mutant spermatozoa, as well as spermatozoa lacking izumo sperm-egg fusion protein 1 gene (*Izumo1*), barely contain sperm acrosome membrane-associated protein 6 (SPACA6) (Inoue et al., 2021). Since IZUMO1 is not contained in the DCST1/2 complex (Noda et al., 2022), DCST1/2 and IZUMO1 separately regulate the localization and/or function of SPACA6 (Deneke et al., 2024). As for zebrafish (Noda et al., 2022), mutant males deficient for *dcst1* and/or *dcst2* are largely or entirely sterile. A loss of either DCST1 or DCST2 results in lacking both DCST proteins in spermatozoa, suggesting that zebrafish DCST1/2 form a protein complex like mouse DCST1/2. Moreover, *dcst2*-disrupted spermatozoa can contact the oocyte PM but have a defect in stable binding to the oolemma. Intriguingly, *C. elegans spe-42* (*dcst2*) mutant spermatozoa are also incapable of stabilizing gamete interactions (Nishimura et al., 2015). Fruit flies are invertebrates like *C. elegans*, and the *Drosophila* gene sneaky (*snky*), an ortholog of *spe-49* (*dcst1*), is involved in the spermatozoon plasma membrane breakdown after fly spermatozoa enter eggs (Wilson et al., 2006). Given that *C. elegans* shares a particular mechanism for fertilization with other species, SPE-42 and SPE-49 might be included in the mechanism as a protein complex containing these two DCST proteins.

Fig. 6 exhibits the tertiary structure models of nematode, mouse, zebrafish, and fruit fly DCST1/2 predicted by AlphaFold2. All of the genes encoding these proteins are known to be indispensable for fertilization. To evaluate how those DCST1/2 orthologs and human DCST1/2 are similar in their topologies, template modeling scores (TM-scores) were compared

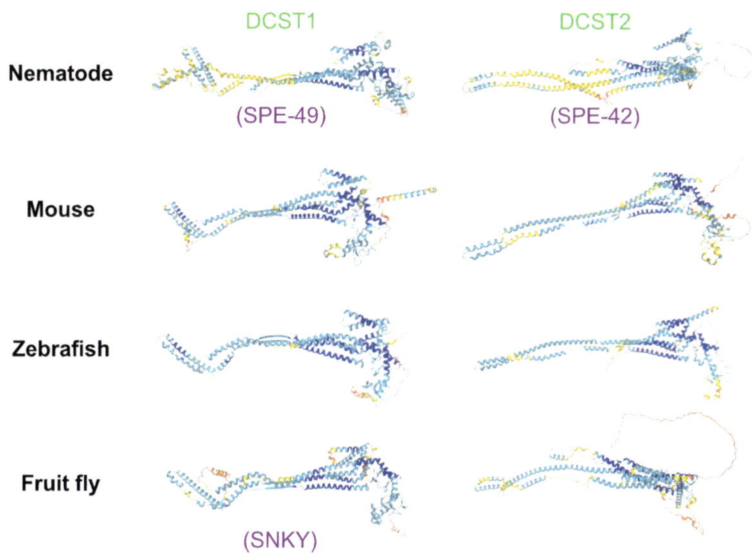

Fig. 6 Predicted tertiary structures of DCST1 and DCST2 orthologs from various species. The topologies of nematode, mouse, zebrafish, and fruit fly DCST1/2 proteins, whose encoded genes have been shown to participate in the fertility of sperm, are predicted by AlphaFold2. UniProt IDs used for the prediction are as follows: nematode (*Caenorhabditis elegans*) SPE-49 (DCST1), A0A2C9C337; nematode (*Caenorhabditis elegans*) SPE-42 (DCST2), Q17461; mouse (*Mus musculus*) DCST1, Q059Y8; mouse (*Mus musculus*) DCST2, A0A140LIJ0; zebrafish (*Danio rerio*) DCST1, A0A8M6YVA5; zebrafish (*Danio rerio*) DCST2, A0A8M6Z1B9; fruit fly (*Drosophila melanogaster*) DCST1 (SNKY), Q9VU52; fruit fly (*Zeugodacus cucurbitae*) DCST2, A0A0A1X0Q5.

(Fig. 7). The minimal TM-score is 0.44 (*C. elegans* SPE-42 *versus* *C. elegans* SPE-49). Owing to that TM-scores higher than 0.5 can judge that two compared proteins have the same fold (Xu & Zhang, 2010), the tertiary structures of DCST orthologs listed in Fig. 7 are likely to be similar or identical, implying that those DCSTs from a variety of species might possess similar missions during fertilization.

3.1.5 *spe-45* gene

This gene was independently identified by forward (Singaravelu et al., 2015) and reverse (Nishimura et al., 2015) genetic approaches. SPE-45 protein is a single-pass TM protein with an Ig-like domain, and this domain architecture is shared by mouse IZUMO1 and SPACA6 (Fig. 8),

	Ne SPE-49	Ne SPE-42	Hu DCST1	Hu DCST2	Mo DCST1	Mo DCST2	Ze DCST1	Ze DCST2	FrFl SNKY	FrFl DCST2
Ne SPE-49		0.44	0.50	0.48	0.46	0.45	0.49	0.53	0.51	0.46
Ne SPE-42			0.53	0.50	0.49	0.48	0.52	0.51	0.49	0.49
Hu DCST1				0.63	0.76	0.63	0.81	0.58	0.75	0.62
Hu DCST2					0.53	0.69	0.57	0.54	0.50	0.60
Mo DCST1						0.58	0.77	0.53	0.62	0.55
Mo DCST2							0.61	0.67	0.55	0.70
Ze DCST1								0.59	0.71	0.60
Ze DCST2									0.60	0.69
FrFl SNKY										0.54
FrFl DCST2										

Fig. 7 Comparison of the predicted topologies of DCST1 and DCST2 orthologs. Each number in this figure represents TM-score, obtained from the RCSB PDB database (https://www.rcsb.org/). This score is a metric to evaluate the topological similarity between two protein structures. A score of 1.0 indicates a perfect match between two structures, while scores below 0.17 correspond to randomly chosen unrelated proteins (Xu & Zhang, 2005). If scores higher than 0.5 can generally assume that two proteins have the same fold (Xu & Zhang, 2005). Except for human (*Homo sapiens*) DCST1 (UniProt ID, Q5T197) and DCST2 (Q5T1A1), predicted tertiary structures are the same as those shown in Fig. 4. Light blue, TM-score = 0.50–0.59, light green, TM-score = 0.60–0.69, light orange, TM-score = 0.70–0.79, light pink, TM = 0.80–0.89. Abbreviations used are: Ne, nematode, Hu, human, Mo, mouse, Ze, zebrafish, FrFl, fruit fly.

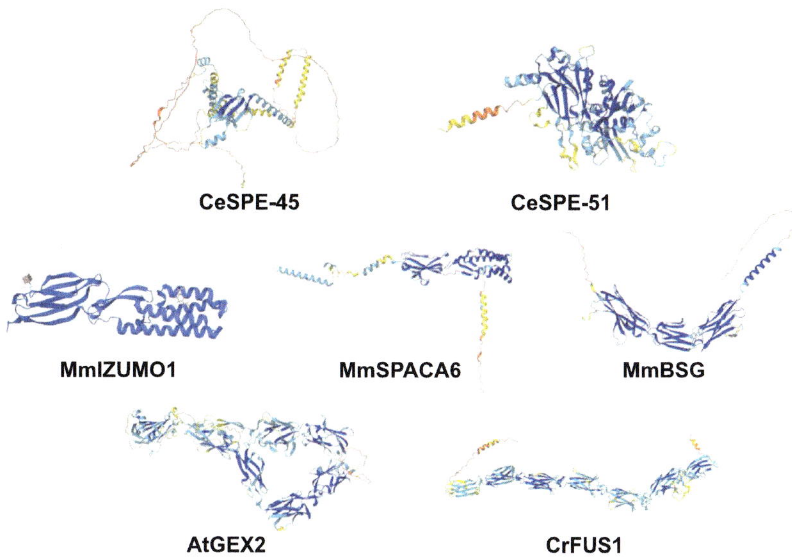

Fig. 8 Proteins containing Ig-like domains that play crucial roles during fertilization are conserved in various organisms. *Caenorhabditis elegans* SPE-45 (CeSPE-45), CeSPE-51, *Mus musculus* IZUMO1 (MmIZUMO1), MmSPACA6, MmBSG, *Arabidopsis thaliana* GEX2 (AtGEX2), and *Chlamydomonas reinhardtii* FUS1 (CrFUS1) are all Ig-like domain-containing proteins that play key roles during gamete interactions. Like in Fig. 6, the tertiary structures of these proteins are predicted by AlphaFold2. UniProt IDs used for the prediction are as follows: CeSPE-45, Q19879; CeSPE-51, O44564; MmIZUMO1, Q9D9J7; MmSPACA6, E9Q8Q8; MmBSG, P18572; AtGEX2, F4K4R6; CrFUS1, Q39594.

both of which are essential spermatozoon proteins for gamete fusion (Barbaux et al., 2020; Inoue et al., 2005). *spe-45* seems to act in the later phase of gamete interactions like *spe-9* (Nishimura et al., 2015; Takayama et al., 2021). Intriguingly, the Ig-like domain of SPE-45 is interchangeable with that of mouse IZUMO1 (Nishimura et al., 2015), suggesting that these two proteins are functionally equivalent to each other.

Besides nematode SPE-45 and SPE-51 and mouse IZUMO1 and SPACA6, Ig-like domain-containing proteins with indispensable roles during fertilization have been identified in various species (Fig. 8). An additional mouse example is Basigin (BSG), a single-pass TM protein containing two Ig-like domains. Both male and female mice lacking *Bsg* gene are sterile, and spermatogenesis is arrested in the mutant males (Igakura et al., 1998). Moreover, BSG protein seems to participate during spermatozoon interactions with the cumulus cells and the zona pellucida

(Saxena et al., 2002). *Arabidopsis* GEX2 (Mori & Igawa, 2014; Mori et al., 2014) and *Chlamydomonas* FUS1 (Ferris et al., 1996; Misamore et al., 2003) are both single-pass TM proteins with Ig-like filamin repeat domains. These two proteins are involved in gamete fusion or attachment. It still remains unclear how Ig-like domains found in the mouse, plant, and algal proteins are involved during gamete interactions. One hypothesis is that the Ig-like proteins shown in Fig. 8 associate with the same and/or other proteins containing Ig-like domains to form *cis*- and/or *trans*-protein complexes that are important for fertilization (Nishimura & L'Hernault, 2016), like *Hydractinia* allorecognition 1 (Alr1) and allorecognition 2 (Alr2) proteins. These Alr proteins are TM proteins with multiple Ig-like domains that play critical roles in recognition of self- *versus* non-self (allo-recognition) through *trans*-homotypic interactions of Alr proteins (Karadge et al., 2015). Recently, it was found that FUS1 and GEX2 are homologous proteins containing seven Ig-like domains (Fig. 8). Furthermore, FUS1 expressed in *plus* gametes binds to MAR1 (minus adhesion receptor 1) in *minus* gametes, which is associated with GCS1/HAP2 in *Chlamydomonas* (Pinello et al., 2021). In other words, FUS1 and MAR1 are gamete binding partners and are involved in gamete fusion mediated by GCS1/HAP2 (Pinello et al., 2021). However, the functions of Ig-like domains are remained to be solved.

3.2 *C. elegans* genes related to female germline functions

3.2.1 *C. elegans fertilization*

egg-1 and *egg-2* genes are paralogous to each other and encode type II TM proteins with eight low-density lipoprotein (LDL) receptor-like domains in their extracellular regions (Kadandale et al., 2005). Hermaphrodites lacking either *egg-1* or *egg-2* are sub-fertile, while those lacking both genes are entirely sterile (Kadandale et al., 2005). Moreover, wild-type spermatozoa can bind to but fail to fuse with the PM of *egg-1/2*-double knockout oocytes (Kadandale et al., 2005).

3.2.2 *C. elegans polyspermy block*

In *C. elegans*, eggshell chitin plays a critical role in polyspermy block as a barrier (step 6 in the nematode, Fig. 1), and its synthesis is initiated at fertilization. Hermaphrodites lacking the chitin synthase CHS-1, the CHS-1 substrate UDP-N-acetylglucosamine, the CHS-1-interacting protein EGG-3, or the sperm-supplied protein SPE-11 result in aberrant deposit of chitin (Johnston et al., 2010). The polyspermy barrier is also compromised by the loss of the chitin-binding protein CBD-1 or EGG-1/EGG-2.

Recently, MARC-3, a membrane-associated ubiquitin ligase that localizes on the PM and cortical puncta in oocytes, is involved in polyspermy block (Kawasaki et al., 2024). MARC-3 also functions in the selective degradation of maternal PM proteins and the transient accumulation of endosomal lysine 63-linked polyubiquitin after fertilization. The RING-finger domain of MARC-3 is required for polyspermy block, suggesting that a ubiquitination-mediated mechanism regulates polyspermy block and maternal membrane protein degradation during the oocyte-to-embryo transition.

In general, fertilization-related proteins evolve rapidly (Haerty et al., 2007; Swanson & Vacquier, 2002; Wyckoff et al., 2007), presumably in order to prevent the production of cross-species hybrids. Even if there is a conserved mechanism(s) that are required for fertilization among evolutionary distant species, it might be challenging to identify it by homology-based searches. With the development of experimental technologies, such as genome editing, the creation of fertilization-defective mutants has become more accessible in various organisms, resulting in the progress of understanding the fertilization mechanisms in each species. A particular mechanism(s) for fertilization that is conserved over the evolution will be elucidated in the near future.

4. Conclusions and perspectives

Sperm and egg proteins involved in fertilization in animals and plants were summarized in Table 1. As shown in the table, basic strategies involved in sexual reproduction seem to be much more similar than we previously thought. From this viewpoint, unexpected discovery obtained by biochemical or genetic studies in invertebrate reproduction may be applicable to the studies in mammalian fertilization. In addition, GCS1, a male-specific protein responsible for gamete fusion in plants, is commonly utilized in animals and parasites. However, the mechanisms of gamete fusion mediated by GCS1 is poorly understood. Further structural analyses using AlphaFold-multimer may solve the issues remained to be solved from different angle.

Acknowledgments

We thank Manabu Yoshida for his valuable comments. We also thank the National BioResource Project (AMED, Japan) for providing the *C. intestinalis* type A and *C. elegans* strains. This research was supported in part by JSPS KAKENHI (Japan) to HS (JP20K06621, JP17H03672, JP25291024, JP21112002) and TS (JP23H04738, JP23K05779) and HN (JP19K06448).

References

Albert, T. K., Hanzawa, H., Legtenberg, Y. I. A., de Ruwe, M. J., van den Heuvel, F. A. J., et al. (2002). Identification of a ubiquitin-protein ligase subunit within the CCR4-NOT transcription repressor complex. *The EMBO Journal, 21*, 355–364.

Altun, Z. F., Herndon, L. A., Wolkow, C. A., Crocker, C., Lints, R., et al. (2002–2024). (http://www.wormatlas.org).

Alves, A. P., Mulloy, B., Diniz, J. A., & Mourão, P. A. (1997). Sulfated polysaccharides from the egg jelly layer are species-specific inducers of acrosomal reaction in sperms of sea urchins. *The Journal of Biological Chemistry, 272*, 6965–6971.

Avella, M. A., Xiong, B., & Dean, J. (2013). The molecular basis of gamete recognition in mice and humans. *Molecular Human Reproduction, 19*, 279–289.

Aydin, H., Sultana, A., Li, S., Thavalingam, A., & Lee, J. E. (2016). Molecular architecture of the human sperm IZUMO1 and egg JUNO fertilization complex. *Nature, 534*, 562–565.

Baba, T., Azuma, S., Kashiwabara, S., & Toyoda, Y. (1994). Sperm from mice carrying a targeted mutation of the acrosin gene can penetrate the oocyte zona pellucida and effect fertilization. *The Journal of Biological Chemistry, 269*, 31845–31849.

Barbaux, S., Ialy-Radio, C., Chalbi, M., Dybal, E., Homps-Legrand, M., et al. (2020). Sperm SPACA6 protein is required for mammalian sperm-egg adhesion/fusion. *Scientific Reports, 10*, 5335.

Barresi, M. J. F., & Gilbert, S. F. (2020). Fertilization beginning a new organism. *Developmental Biology*, 216–247.

Beech, D. J., Bahnasi, Y. M., Dedman, A. M., & Al-Shawaf, E. (2009). TRPC channel lipid specificity and mechanisms of lipid regulation. *Cell Calcium, 45*, 583–588.

Bi, M., Hickox, J. R., Winfrey, V. P., Olson, G. E., & Hardy, D. M. (2003). Processing, localization and binding activity of zonadhesin suggest a function in sperm adhesion to the zona pellucida during exocytosis of the acrosome. *The Biochemical Journal, 375*, 477–488.

Bianchi, E., Doe, B., Goulding, D., & Wright, G. J. (2014). *Juno* is the egg *Izumo* receptor and is essential for mammalian fertilization. *Nature, 508*, 483–487.

Binner, M. I., Kogan, A., Panser, K., Schleiffer, A., Deneke, V. E., et al. (2022). The sperm protein Spaca6 is essential for fertilization in zebra- fish. *Frontiers in Cell Devlopmental Biology, 9*, 806982.

Bleil, J. D., & Wassarman, P. M. (1990). Identification of a ZP3-binding protein on acrosome-intact mouse sperm by photoaffinity crosslinking. *Proceedings of the National Academy of Sciences of the United States of America, 87*, 5563–5567.

Bookbinder, L. H., Cheng, A., & Bleil, J. D. (1995). Tissue- and species-specific expression of sp56, a mouse sperm fertilization protein. *Science (New York, N. Y.), 269*, 86–89.

Borden, K. L. (2000). RING domains: Master builders of molecular scaffolds? *Journal of Molecular Biology, 295*, 1103–1112.

Brenker, C., Goodwin, N., Weyand, I., Kashikar, N. D., Naruse, M., et al. (2012). The CatSper channel: A polymodal chemosensor in human sperm. *The EMBO Journal, 31*, 1654–1665.

Brenner, S. (1974). The genetics of *Caenorhabditis elegans*. *Genetics, 77*, 71–94.

Brunetti, R., Gissi, C., Pennati, R., Caicci, F., Gasparini, F., et al. (2015). Morphological evidence that the molecularly determined *Ciona intestinalis* type A and type B are different species: *Ciona robusta* and *Ciona intestinalis*. *Journal of Zoological and Systematics and Evolutionary Research, 53*, 186–193.

Burnett, L. A., Washburn, C. A., Sugiyama, H., Xiang, X., Olson, J. H., et al. (2012). Allurin, an amphibian sperm chemoattractant having implications for mammalian sperm physiology. *International Review of Cell and Molecular Biology, 295*, 1–61.

Chatterjee, I., Richmond, A., Putiri, E., Shakes, D. C., & Singson, A. (2005). The *Caenorhabditis elegans spe-38* gene encodes a novel four-pass integral membrane protein required for sperm function at fertilization. *Development (Cambridge, England), 132*, 2795–2808.

Chávez, J. C., Carrasquel-Martínez, G., Hernández-Garduño, S., Matamoros Volante, A., Treviño, C. L., et al. (2024). Cytosolic and acrosomal pH regulation in mammalian sperm. *Cells, 13*, 865.

Cheng, A., Le, T., Palacios, M., Bookbinder, L. H., Wassarman, P. M., et al. (1994). Sperm-egg recognition in the mouse: characterization of sp56, a sperm protein having specific affinity for ZP3. *The Journal of Cell Biology, 125*, 867–878.

Chu, D. S., & Shakes, D. C. (2013). Spermatogenesis. *Advances in Experimental Medicine and Biology, 757*, 171–203.

Cordle, J., Johnson, S., Tay, J. Z. Y., Roversi, P., Wilkin, M. B., et al. (2008). A conserved face of the Jagged/Serrate DSL domain is involved in Notch trans-activation and cis-inhibition. *Nature Structural & Molecular Biology, 15*, 849–857.

Cotelli, F., Andronico, F., De Santis, R., Monroy, A., & Rosati, F. (1981). Differentiation of the vitelline coat in the ascidian *Ciona intestinalis*: an ultrastructural analysis. *Wilhelm Roux's Archives of Developmental Biology, 190*, 252–258.

Dale, B. (2014). Is the idea of a fast block to polyspermy based on artifact? *Biochemical and Biophysical Research Communications, 450*, 1159–1165.

de Nettancourt, D. (1997). Incompatibility in angiosperms. *Sexual Plant Reproduction, 10*, 185–199.

De Santis, R., Jamunno, G., & Rosati, F. (1980). A study of the chorion and the follicle cells in relation to the sperm-egg interaction in the ascidian, *Ciona intestinalis*. *Developmental Biology, 74*, 490–499.

De Santis, R., Pinto, M. R., Cotelli, F., Rosati, F., Monroy, A., et al. (1983). A fucosyl glycoprotein component with sperm receptor and sperm-activating activities from the vitelline coat of *Ciona intestinalis* eggs. *Experimental Cell Research, 148*, 508–513.

De Santis, R., Shirakawa, H., Nakada, K., Miyazaki, S., Hoshi, M., et al. (1992). Evidence that metalloendoproteases are involved in gamete fusion of *Ciona intestinalis*, ascidia. *Developmental Biology, 153*, 165–171.

Deneke, V. E., Blaha, A., Lu, Y., Suwita, J. P., Draper, J. M., et al. (2024). A conserved fertilization complex bridges sperm and egg in vertebrates. *Cell, 187*, 7066–7078.

Dresselhaus, T., Sprunk, S., & Wessel, G. M. (2016). Fertilization mechanisms in flowering plants. *Current Biology, 26*, R125–R139.

Ellis, R. E., & Stanfield, G. M. (2014). The regulation of spermatogenesis and sperm function in nematodes. *Seminars in Cell & Developmental Biology, 29*, 17–30.

Elofsson, A., Han, L., Bianchi, E., Wright, G. J., & Jovine, L. (2024). Deep learning insights into the architecture of the mammalian egg-sperm fusion synapse. *eLife, 13*, RP93131.

Ensslin, M. A., & Shur, B. D. (2003). Identification of mouse sperm SED1, a bimotif EGF repeat and discoidin-domain protein involved in sperm-egg binding. *Cell, 114*, 405–417.

Farach, H. A. Jr, Mundy, D. I., Strittmatter, W. J., & Lennarz, W. J. (1987). Evidence for the involvement of metalloendopeptidases in the acrosome reaction in sea urchin sperm. *The Journal of Biological Chemistry, 262*, 5483–5487.

Fédry, J., Liu, Y., Péhau-Arnaudet, G., Pei, J., Li, W., et al. (2017). The ancient gamete fusogen HAP2 is a eukaryotic class II fusion protein. *Cell, 168*, 904–915.

Ferris, P. J., Woessner, J. P., & Goodenough, U. W. (1996). A sex recognition glycoprotein is encoded by the plus mating-type gene *fus1* of *Chlamydomonas reinhardtii*. *Molecular Biology of the Cell, 7*, 1235–1248.

Florman, H. M., Jungnickel, M. K., & Sutton, K. A. (2008). Regulating the acrosome reaction. *The International Journal of Developmental Biology, 52*, 503–510.

Fujihara, Y., Lu, Y., Noda, T., Oji, A., Larasati, T., et al. (2020). Spermatozoa lacking fertilization influencing membrane protein (FIMP) fail to fuse with oocytes in mice. *Proceedings of the National Academy of Sciences of the United States of America, 117*, 9393–9400.

Fujihara, Y., Noda, T., Kobayashi, K., Oji, A., Kobayashi, S., et al. (2019). Identification of multiple male reproductive tract-specific proteins that regulate sperm migration through the oviduct in mice. *Proceedings of the National Academy of Sciences of the United States of America, 116*, 18498–18506.

Fujihara, Y., Oji, A., Kojima-Kita, K., Larasati, T., & Ikawa, M. (2018). Co-expression of sperm membrane proteins CMTM2A and CMTM2B is essential for ADAM3 localization and male fertility in mice. *Journal of Cell Science, 131*, jcs221481.

Fujihara, Y., Okabe, M., & Ikawa, M. (2014). GPI-anchored protein complex, LY6K/TEX101, is required for sperm migration into the oviduct and male fertility in mice. *Biology of Reproduction, 90*, 60.

Fujihara, Y., Tokuhiro, K., Muro, Y., Kondoh, G., Araki, Y., et al. (2013). Expression of TEX101, regulated by ACE, is essential for the production of fertile mouse spermatozoa. *Proceedings of the National Academy of Sciences of the United States of America, 110*, 8111–8116.

Fukami, K., Yoshida, M., Inoue, T., Kurokawa, M., Fissore, R. A., et al. (2003). Phospholipase Cδ4 is required for Ca^{2+} mobilization essential for acrosome reaction in sperm. *Journal of Cell Biology, 161*, 79–88.

Fuke, M. T. (1983). Self and non-self recognition between gametes of the ascidian, *Halocynthia roretzi. Wilehm Roux's Archives of Developmental Biology, 192*, 347–352.

Fuke, M., & Numakunai, T. (1996). Establishment of self-sterility of eggs in the ovary of the solitary ascidian, *Halocynthia roretzi. Roux's Archives of Developmental Biology, 205*, 391–400.

Fuke, M., & Numakunai, T. (1999). Self sterility of eggs induced by exogenous and endogenous protease in the solitary ascidian, *Halocynthia roretzi. Molecular Reproduction and Development, 52*, 99–106.

Fukumoto, M. (1988). Fertilization in ascidians: Apical processes and gamete fusion in *Ciona intestinalis* spermatozoa. *Journal of Cell Science, 89*, 189–196.

Gallo, A., & Costantini, M. (2012). Glycobiology of reproductive processes in marine animals: The state of the art. *Marine Drugs, 10*, 2861–2892.

Gert, K. R., & Pauli, A. (2020). Species-specific mechanisms during fertilization. *Current Topics in Developmental Biology, 140*, 121–144.

Glabe, C. G., & Lennarz, W. J. (1979). Species-specific sperm adhesion in sea urchins. A quantitative investigation of bindin-mediated egg agglutination. *The Journal of Cell Biology, 83*, 595–604.

Glas, R., Bogyo, M., McMaster, J. S., Gaczynska, M., & Ploegh, H. L. (1998). A proteolytic system that compensates for loss of proteasome function. *Nature, 392*, 618–622.

Godknecht, A., & Honegger, T. G. (1991). Isolation, characterization, and localization of a sperm-bound N-acetylglucosamindase that is indispensable for fertilization in the ascidian, *Phallusia mammillata. Developmental Biology, 143*, 398–407.

Haerty, W., Jagadeeshan, S., Kulathinal, R. J., Wong, A., Ravi Ram, K., et al. (2007). Evolution in the fast lane: Rapidly evolving sex-related genes in *Drosophila. Genetics, 177*, 1321–1335.

Hamm, D., Mautz, B. S., Wolfner, M. F., Aquadro, C. F., & Swanson, W. J. (2007). Evidence of amino acid diversity-enhancing selection within humans and among primates at the candidate sperm-receptor gene PKDREJ. *American Journal of Human Genetics, 81*, 44–52.

Hansen, J. M., Chavez, D. R., & Stanfield, G. M. (2015). COMP-1 promotes competitive advantage of nematode sperm. *eLife, 4*, e05423.

Hanzawa, H., de Ruwe, M. J., Albert, T. K., van Der Vliet, P. C., Timmers, H. T., et al. (2001). The structure of the C4C4 ring finger of human NOT4 reveals features distinct from those of C3HC4 RING fingers. *Journal of Biological Chemistry, 276*, 10185–10190.

Harada, Y., & Sawada, H. (2007). Proteins interacting with the ascidian vitelline-coat sperm receptor HrVC70 as revealed by yeast two-hybrid screening. *Molecular Reproduction and Development, 74*, 1178–1187.

Harada, Y., & Sawada, H. (2008). Allorecognition mechanisms during ascidian fertilization. *The International Journal of Developmental Biology, 52*, 637–645.

Harada, Y., Takagaki, Y., Sugnagawa, M., Saito, T., Yamada, L., et al. (2008). Mechanism of self-sterility in a hermaphroditic chordate. *Science (New York, N. Y.), 320*, 548–550.

Hardy, D. M., & Garbers, D. L. (1994). Species-specific binding of sperm proteins to the extracellular matrix (zona pellucida) of the egg. *The Journal of Biological Chemistry, 269*, 19000–19004.

Hardy, D. M., & Garbers, D. L. (1995). A sperm membrane protein that binds in a species-specific manner to the egg extracellular matrix is homologous to von Willebrand factor. *The Journal of Biological Chemistry, 270*, 26025–26028.

Hashimoto, S., Kinjo, K., Saito, T., & Sawada, H. (2022). Removal of the block to self-fertilization by low-calcium artificial seawater in the ascidian *Ciona intestinalis. Zygote (Cambridge, England), 30*, 738–742.

Herberg, S., Bert, K. R., Schleiffer, A., & Pauli, A. (2018). The Ly6/uPAR protein Bouncer is necessary and sufficient for species-specific fertilization. *Science (New York, N. Y.), 361*, 1029–1033.

Higashiyama, T. (2010). Peptide signaling in pollen-pistil interactions. *Plant & Cell Physiology, 51*, 177–189.

Higashiyama, T., & Takeuchi, H. (2015). The mechanism and key molecules involved in pollen tube guidance. *Annual Review Plant Biology, 66*, 393–413.

Higashiyama, T., Yabe, S., Sasaki, N., Nishimura, Y., Miyagishima, S., et al. (2001). Pollen tube attraction by the synergid cell. *Science (New York, N. Y.), 293*, 1480–1483.

Higashiyama, T., & Yang, W.-C. (2017). Gametophytic pollen tube guidance: Attractant peptides, gametic controls, and receptors. *Plant Physiology, 173*, 112–121.

Hirai, M., Arai, M., Mori, T., Miyagishima, S., Kawai, S., et al. (2008). Male fertility of malaria parasites is determined by GCS1, a plant-type reproduction factor. *Current Biology, 18*, 607–613.

Hirohashi, N., & Vacquier, V. D. (2002). Egg fucose sulfate polymer, sialoglycan, and speract all trigger the sea urchin sperm acrosome reaction. *Biochemical and Biophysical Research Communications, 296*, 833–839.

Hirose, M., Honda, A., Fulka, H., Tamura-Nakano, M., Matoba, S., et al. (2020). Acrosin is essential for sperm penetration the zona pellucida in hamsters. *Proceedings of the National Academy of Sciences of the United States of America, 117*, 2513–2518.

Hoang, H. D., Prasain, J. K., Dorand, D., & Miller, M. A. (2013). A heterogeneous mixture of F-series prostaglandins promotes sperm guidance in the *Caenorhabditis elegans* reproductive tract. *PLoS Genetiics, 9* e1003271.

Honda, A., Siruntawineti, J., & Baba, T. (2002). Role of acrosomal matrix proteases in sperm-zona pellucida interactions. *Human Reproduction Update, 8*, 405–412.

Honegger, T. G., & Koyanagi, R. (2008). The ascidian egg envelope in fertilization: Structural and molecular features. *The International Journal of Developmantal. Biology, 52*, 527–533.

Hoshi, M. (1985). Lysins. *Biology of Fertilization, 2*, 431–462.

Hoshi, M. (1986). Sperm glycosidase as a plausible mediator of sperm binding to the vitelline envelope in ascidians. *Advances in Experimental Medicine and Biology, 207*, 251–260.

Hoshi, M., De Santis, R., Pinto, M. R., Cotelli, F., & Rosati, F. (1983). Is sperm α-L-fucosidase responsible for sperm-egg binding in *Ciona intestinalis*? *The Sperm Cell,* 107–110.

Hoshi, M., De Santis, R., Pinto, M. R., Cotelli, F., & Rosati, F. (1985). Sperm glycosidases as mediators of sperm-egg binding in the ascidians. *Zoological Science, 2*, 65–69.

Hoshi, M., Moriya, M., Aoyagi, T., Umezawa, H., Mohri, H., et al. (1979). Effects of hydrolase inhibitors on fertilization of sea urchins: I. Protease inhibitors. *Gamete Research, 2*, 107–119.

Hoshi, M., Moriyama, H., & Matsumoto, M. (2012). Structure of acrosome-reaction inducing substance in the jelly coat of starfish eggs. *Biochemical and Biophysical Research Communications, 425*, 595–598.

Hoshi, M., Numakunai, T., & Sawada, H. (1981). Evidence for participation of sperm proteinases in fertilization of the solitary ascidian, *Halocynthia roretzi*: Effects of protease inhibitors. *Developmental Biology, 86*, 117–121.

Hoshi, M., Takizawa, S., & Hirohashi, N. (1994). Glycosidases, proteases and ascidian fertilization. *Seminars in Developmental Biology, 5*, 201–208.

Howes, L., & Jones, R. (2002). Interactions between zona pellucida glycoproteins and sperm proacrosin/acrosin during fertilization. *Journal of Reproductive Immunology, 53*, 181–192.

Howes, E., Pascall, J. C., Engel, W., & Jones, R. (2001). Interactions between mouse ZP2 glycoprotein and proacrosin; a mechanism for secondary binding of sperm to the zona pellucida during fertilization. *Journal of Cell Science, 114*, 4127–4136.

Igakura, T., Kadomatsu, K., Kaname, T., Muramatsu, H., Fan, Q. W., et al. (1998). A null mutation in basigin, an immunoglobulin superfamily member, indicates its important roles in peri-implantation development and spermatogenesis. *Developmental Biology, 194*, 152–165.

Ikawa, M., Tokuhiro, K., Yamaguchi, R., Benham, A. M., Tamura, T., et al. (2011). Calsperin is a testis-specific chaperone required for sperm fertility. *The Journal of Biological Chemistry, 286*, 5639–5646.

Inoue, N., Hagihara, Y., & Wada, I. (2021). Evolutionarily conserved sperm factors, DCST1 and DCST2, are required for gamete fusion. *eLife, 10*, e66313.

Inoue, N., Ikawa, M., Isotani, A., & Okabe, M. (2005). The immunoglobulin superfamily protein Izumo is required for sperm to fuse with eggs. *Nature, 434*, 234–238.

Inoue, N., Saito, T., & Wada, I. (2020). Unveiling a novel function of CD9 in surface compartmentalization of oocytes. *Development (Cambridge, England), 147* dev189985.

Iwano, M., & Takayama, S. (2012). Self/non-self discrimination in angiosperm self-incompatibility. *Current Opinion in Plant Biology, 15*, 78–83.

Jaffe, L. A. (1976). Fast block to polyspermy in sea urchin eggs is electrically mediated. *Nature, 261*, 68–71.

Jaffe, L. A. (1980). Electric polyspermy block in sea urchins: Nicotine and low sodium experiments. *Development, Growth & Differentiation, 22*, 503–507.

Jiang, D., & Smith, W. C. (2005). Self- and cross-fertilization in the solitary ascidian *Ciona savignyi*. *The Biological Bulletin, 209*, 107–112.

Jin, M., Fujiwara, E., Kakiuchi, Y., Okabe, M., Satouh, Y., et al. (2011). Most fertilizing mouse spermatozoa begin their acrosome reaction before contact with the zona pellucida during in vitro fertilization. *Proceedings of the National Academy of Sciences of the United States of America, 108*, 4892–4896.

Johnson, M. A., Harper, J. F., & Palanivelu, R. (2019). A fruitful journey: Pollen tube navigation from germination to fertilization. *Annual Review of Plant Biology, 70*, 809–837.

Johnston, W. L., Krizus, A., & Dennis, J. W. (2010). Eggshell chitin and chitin-interacting proteins prevent polyspermy in *C. elegans*. *Current Biology, 20*, 1932–1937.

Jones, R. (1991). Interaction of zona pellucida glycoproteins, sulphated carbohydrates and synthetic polymers with proacrosin, the putative egg-binding protein from mammalian spermatozoa. *Development (Cambridge, England), 111*, 1155–1163.

Kadandale, P., Stewart-Michaelis, A., Gordon, S., Rubin, J., Klancer, R., et al. (2005). The egg surface LDL receptor repeat-containing proteins EGG-1 and EGG-2 are required for fertilization in *Caenorhabditis elegans*. *Current Biology, 15*, 2222–2229.

Kamei, N., & Glabe, C. C. (2003). The species specific egg receptor for sea urchin sperm adhesion is EBR1, a novel ADAMTS protein. *Genes & Development, 17*, 2502–2507.

Karadge, U. B., Gosto, M., & Nicotra, M. L. (2015). Allorecognition proteins in an invertebrate exhibit homophilic interactions. *Current Biology, 25*, 2845–2850.

Kaupp, U. B., Solzin, J., Hildebrand, E., Brown, J. E., Helbig, A., et al. (2003). The signal flow and motor response controling chemotaxis of sea urchin sperm. *Nature Cell Biology, 5*, 109–117.

Kawamura, K., Fujita, H., & Nakauchi, M. (1987). Cytological characterization of self-incompatibility in gametes of the ascidian, *Ciona intestinalis*. *Development, Growth & Differentiation, 29*, 627–642.

Kawamura, K., Fujita, H., & Nakauchi, M. (1988). Helper function of follicle cells in sperm-egg interactions of the ascidian, *Ciona intestinalis*. *Development, Growth & Differentiation, 30*, 693–703.

Kawamura, K., Nomura, M., Kameda, T., Shimamoto, H., & Nakauchi, M. (1991). Self-nonself recognition activity extracted from self-sterile eggs of the ascidian, *Ciona intestinalis*. *Development, Growth & Differentiation, 33*, 139–148.

Kawano, N., Kang, W., Yamashita, M., Koga, Y., Yamazaki, T., et al. (2010). Mice lacking two sperm serine proteases, ACR and PRSS21, are subfertile, but the mutant sperm are infertile in vitro. *Biology of Reproduction, 83*, 359–369.

Kawasaki, I., Sugiura, K., Sasaki, T., Matsuda, N., Sato, M., et al. (2024). MARC-3, a membrane-associated ubiquitin ligase, is required for fast polyspermy block in *Caenorhabditis elegans*. *Nature Communications, 15*, 792.

Kerns, K., Morales, P., & Sutovsky, P. (2016). Regulation of sperm capacitation by the 26S proteasome: An emerging new paradigm in spermatology. *Biology of Reproduction, 94*, 117.

Khalturin, K., Kürn, U., Pinnow, N., & Bosch, T. C. (2005). Towards a molecular code for individuality in the absence of MHC: screening for individually variable genes in the urochordate *Ciona intestinalis*. *Developmental and Comparative Immunology, 29*, 759–773.

Kijima, T., Kurokawa, D., Sasakura, Y., Ogasawara, M., Aratake, S., et al. (2023). CatSper mediates not only chemotactic behavior but also the motility of ascidian sperm. *Frontiers in Cell and Developmental Biology, 2*, 1136537.

Kim, E., Baba, D., Kimura, M., & Baba, T. (2005). Identification of a hyaluronidase, Hyal5, involved in penetration of mouse sperm through cumulus mass. *Proceedings of the National Academy of Sciences of the United States of America, 102*, 18028–18033.

Kiyozumi, D., Noda, T., Yamaguchi, R., Tobita, T., Matsumura, T., et al. (2020). NELL2-mediated lumicrine signaling through OVCH2 is required for male fertility. *Science (New York, N. Y.), 368*, 1132–1135.

Kodama, E., Baba, T., Kohno, N., Satoh, S., Yokosawa, H., et al. (2002). Spermosin, a trypsin-like protease from ascidian sperm: cDNA cloning, protein structures and functional analysis. *European Journal of Biochemistry, 269*, 657–663.

Kodama, E., Baba, T., Yokosawa, H., & Sawada, H. (2001). cDNA cloning and functional analysis of ascidian sperm proacrosin. *The Journal of Biological Chemistry, 276*, 24594–24600.

Krauchunas, A. R., Marcello, M. R., Looper, A., Mei, X., Putiri, E., et al. (2023). The EGF-motif-containing protein SPE-36 is a secreted sperm protein required for fertilization in *C. elegans*. *Current Biology, 33*, 3056–3064.

Kroft, T. L., Gleason, E. J., & L'Hernault, S. W. (2005). The *spe-42* gene is required for sperm-egg interactions during *C. elegans* fertilization and encodes a sperm-specific transmembrane protein. *Developmental Biology, 286*, 169–181.

Krutskikh, A., Poliandri, A., Cabrera-Sharp, V., Dacheux, J. L., Poutanen, M., et al. (2012). Epididymal protein Rnase10 is required for post-testicular sperm maturation and male fertility. *Federation of American Societies for Experimental Biology Journal, 26*, 4198–4209.

Kürn, U., Sommer, F., Bosch, T. C. G., & Khalturin, K. (2007a). In the urochordate *Ciona intestinalis* zona pellucida domain proteins vary among individuals. *Developmental and Comparative Immunology, 31*, 1242–1254.

Kürn, U., Sommer, F., Hemmrich, G., Bosch, T. C. G., & Khalturin, K. (2007b). Allorecognition in urochordates: Identification of a highly variable complement receptor-like protein expressed in follicle cells of *Ciona. Developmental and Comparative Immunology, 31*, 360–371.

Lamas-Toranzo, I., Hamze, J. G., Bianchi, E., Fernández-Fuertes, B., Pérez-Cerezales, S., et al. (2020). TMEM95 is a sperm membrane protein essential for mammalian fertilization. *eLife, 9*, e53913.

Lambert, C. C., & Epel, D. (1979). Calcium-mediated mitochondrial movement in ascidian sperm during fertilization. *Developmental Biology, 69*, 296–304.

Lambert, C. C., & Koch, R. A. (1988). Sperm binding and penetration during ascidian fertilization. *Development, Growth & Differentiation, 30*, 325–336.

Lambert, C. C., & Lambert, G. (1978). Tunicate eggs utilize ammonium ions for floatation. *Science, 200*, 64–65.

L'Hernault, S. W. (2009). The genetics and cell biology of spermatogenesis in the nematode *C. elegans. Molecular and Cellular Endocrinology, 306*, 59–65.

Limatola, N., Vasilev, F., Chun, J. T., & Santella, L. (2019). Sodium-mediated fast electrical depolarization does not prevent polyspermic fertilization in *Paracentrotus lividus* eggs. *Zygote (Cambridge, England), 27*, 241–249.

Lishko, P. V., Botchkina, I. L., & Kirichok, Y. (2011). Progesterone activates the principal Ca^{2+} channel of human sperm. *Nature, 471*, 387–391.

Litscher, E. S., & Wassarman, P. M. (2018). *Extracellular Matrix and Egg Coats, 130.*

Lopez, L. C., Bayna, E. M., Litoff, D., Shaper, N. L., Shaper, J. H., et al. (1985). Receptor function of mouse sperm surface galactosyltransferase during fertilization. *The Journal of Cell Biology, 101*, 1501–1510.

Lorenzetti, D., Poirier, C., Zhao, M., Overbeek, P. A., Harrison, W., et al. (2014). A transgenic insertion on mouse chromosome 17 inactivates a novel immunoglobulin superfamily gene potentially involved in sperm-egg fusion. *Mammalian Genome, 25*, 141–148.

Lu, Y., & Ikawa, M. (2022). Eukaryotic fertilization and gamete fusion at a glance. *Journal of Cell Science, 135* Jcs260296.

Ma, X., Zhao, Y., Sun, W., Shimabukuro, K., & Miao, L. (2012). Transformation: How do nematode sperm become activated and crawl? *Protein & Cell, 3*, 755–761.

Maniates, K. A., & Singson, A. (2023). Where are all the egg genes? *Frontiers in Cell and Developmental Biology, 11*, 1107312.

Marcello, M. R., Jia, W., Leary, J. A., Moore, K. L., & Evans, J. P. (2011). Lack of tyrosylprotein sulfotransferase-2 activity results in altered sperm-egg interactions and loss of ADAM3 and ADAM6 in epididymal sperm. *Journal of Biological Chemistry, 286*, 13060–13070.

Marcello, M. R., Singaravelu, G., & Singson, A. (2013). Fertilization. *Advances in Experimental Medicine and Biology, 757*, 321–350.

Marino, R., De Santis, R., Giuliano, P., & Pinto, M. R. (1999). Follicle cell proteasome activity and acid extract from the egg vitelline coat prompt the onset of self-sterility in *Ciona intestinalis* oocytes. *Proceedings of the National Academy of Sciences of the United States of America, 96*, 9633–9636.

Marino, R., DeSantis, R., Hirohashi, N., Hoshi, M., Pinto, M. R., et al. (1992). Purification and characterization of a vitelline coat lysin from *Ciona intestinalis* spermatozoa. *Molecular Reproduction and Development, 32*, 383–388.

Marino, R., Pinto, M. R., Cotelli, F., Lamia, C. L., & De Santis, R. (1998). The hsp70 protein is involved in the acquisition of gamete self-sterility in the ascidian *Ciona intestinalis*. *Development (Cambridge, England), 125*, 899–907.

Matsumoto, M., Hirata, J., Hirohashi, N., & Hoshi, M. (2002). Sperm-egg binding mediated by sperm alpha-L-fucosidase in the ascidian, *Halocynthia roretzi*. *Zoological Science, 19*, 43–48.

Matsumoto, M., Kawase, O., Islam, M. S., Naruse, M., Watanabe, S., et al. (2008). Regulation of the starfish sperm acrosome reaction by cGMP, pH, cAMP and Ca^{2+}. *The International Journal of Developmental Biology, 52*, 523–526.

Matsumoto, M., Solzin, J., Helbig, A., Hagen, V., Ueno, S., et al. (2003). A sperm-activating peptide controls a cGMP-signaling pathway in starfish sperm. *Developmental Biology, 260*, 314–324.

Matsuura, K., Sawada, H., & Yokosawa, H. (1993). Purification and properties of N-acetylglucosaminidase from eggs of the ascidian, *Halocynthia roretzi*. *European Journal of Biochemistry, 218*, 535–541.

Matsuura, K., Sawada, H., & Yokosawa, H. (1995). N-acetylglucosaminidase inhibitor isolated from the vitelline coat of ascidian eggs is a candidate sperm receptor. *Biochemical and Biophysical Research Communications, 213*, 311–316.

Mei, X., Maniates, K. A., Looper, A., Krauchunas, A. R., Druzhinina, M., et al. (2023). SPE-51, a sperm-secreted protein with an immunoglobulin-like domain, is required for fertilization in C. elegans. *Current Biology, 33*, 3048–3055.

Mei, X., & Singson, A. W. (2021). The molecular underpinnings of fertility: Genetic approaches in *Caenorhabditis elegans*. *Advanced Genetics, 2*, e10034.

Mengerink, K. J., & Vacquier, V. D. (2001). Glycobiology of sperm-egg interactions in deuterostomes. *Glycobiology, 11*, 37R–43R.

Miles, E. L., O'Gorman, C., Zhao, J., Samuel, M., Walters, E., et al. (2013). Transgenic pig carrying green fluorescent proteasomes. *Proceedings of the National Academy of Sciences of the United States of America, 110*, 6334–6339.

Miller, R. L. (1975). Chemotaxis of the spermatozoa of *Ciona intestinalis*. *Nature, 254*, 244–245.

Miller, D. J., Gong, X., Decker, G., & Shur, B. D. (1993). Egg cortical granule N-acetylglucosaminidase is required for the mouse zona block to polyspermy. *The Journal of Cell Biology, 123*, 1431–1440.

Miller, D. J., Macek, M. B., & Shur, B. D. (1992). Complementarity between sperm surface b- 1,4-galactosyltransferase and egg-coat ZP3 mediates sperm-egg binding. *Nature, 357*, 589–593.

Miller, M. R., Mannowetz, N., Iavarone, A. T., Safavi, R., Gracheva, E. O., et al. (2016). Unconventional endocannabinoid signaling governs sperm activation via the sex hormone progesterone. *Science, 352*, 555–559.

Mino, M., & Sawada, H. (2016). Follicle cell trypsin-like protease HrOvochymase: its cDNA cloning, localization, and involvement in the late stage of oogenesis in the ascidian *Halocynthia roretzi*. *Molecular Reproduction and Development, 83*, 347–358.

Misamore, M. J., Gupta, S., & Snell, W. J. (2003). The *Chlamydomonas* Fus1 protein is present on the mating type *plus* fusion organelle and required for a critical membrane adhesion event during fusion with *minus* gametes. *Molecular Biology of the Cell, 14*, 2530–2542.

Miyado, K., Yamada, G., Yamada, S., Hasuwa, H., Nakamura, Y., et al. (2000). Requirement of CD9 on the egg plasma membrane for fertilization. *Science (New York, N. Y.), 287*, 321–324.

Miyado, K., Yoshida, K., Yamagata, K., Sakakibara, K., Okabe, M., et al. (2008). The fusing ability of sperm is bestowed by CD9-containing vesicles released from eggs in mice. *Proceedings of the National Academy of Sciences of the United States of America, 105*, 12921–12926.

Miyamoto, T. (2006). The dendritic cell-specific transmembrane protein DC-STAMP is essential for osteoclast fusion and osteoclast bone-resorbing activity. *Modern Rheumatology, 16*, 341–342.

Mizukami, A. G., Inatsugi, R., Jiao, J., Kotake, T., Kuwata, K., et al. (2016). The AMOR arabinogalactan sugar chain induces pollen-tube competency to respond to ovular guidance. *Current Biology, 26*, 1091–1097.

Mizuno, K., Shiba, K., Okai, M., Takahashi, Y., Shitaka, Y., et al. (2012). Calaxin drives sperm chemotaxis by Ca^{2+}-mediated direct modulation of a dynein motor. *Proceedings of the National Academy of Sciences of the United States of America, 109*, 20497–20502.

Morgan, T. H. (1910). Cross- and self-fertilization in *Ciona intestinalis*. *Archiv für Entwicklungsmechanik der Organismen, 30*, 206–235.

Morgan, T. H. (1942). The genetic and the physiological problems of self-sterility in *Ciona*. V. The genetic problem. *Journal of Experimental Zoology, 90*, 199–228.

Morgan, T. H. (1944). The genetic and the physiological problems of self-sterility in *Ciona*. VI. Theoretical discussion of genetic data. *Journal of Experimental Zoology, 95*, 37–59.

Mori, T., & Igawa, T. (2014). Gamete attachment process revealed in flowering plant fertilization. *Plant Signaling & Behavior, 9*, e977715.

Mori, T., Igawa, T., Tamiya, G., Miyagishima, S., & Berger, F. (2014). Gamete attachment requires GEX2 for successful fertilization in *Arabidopsis*. *Current Biology, 24*, 170–175.

Mori, T., Kawai-Toyooka, H., Igawa, T., & Nozaki, H. (2015). Gamete dialogs in green lineages. *Molecular Plant, 8*, 1442–1454.

Mori, T., Kuroiwa, H., Higashiyama, T., & Kuroiwa, T. (2006). GENERATIVE CELL SPECIFIC 1 is essential for angiosperm fertilization. *Nature Cell Biology, 8*, 64–71.

Murabe, N., & Hoshi, M. (2002). Re-examination of sibling cross-sterility in the ascidian, *Ciona intestinalis*: Genetic background of the self-sterility. *Zoological Science, 19*, 527–538.

Murase, K., Takayama, S., & Isogai, A. (2024). Molecular mechanisms of self-incompatibility in Brassicaceae and Solanaceae. *Proceedings of the Japan Academy, Series B, 100*, 264–280.

Muro, Y., Buffone, M. G., Okabe, M., & Gerton, G. L. (2012). Function of the acrosomal matrix: Zona pellucida 3 receptor (ZP3R/sp56) is not essential for mouse fertilization. *Biology of Reproduction, 86*, 1–6.

Nair, S., Bist, P., Dikshit, N., & Krishnan, M. N. (2016). Global functional profiling of human ubiquitome identifies E3 ubiquitin ligase DCST1 as a novel negative regulator of Type-I interferon signaling. *Scientific Reports, 6*, 36179.

Nakazawa, S., Shirae-Kurabayashi, M., & Sawada, H. (2019). The role of metalloproteases in fertilization in the ascidian *Ciona robusta*. *Scientific Reports, 9*, 1009.

Nakazawa, S., Shirae-Kurayamashi, M., Otsuka, K., & Sawada, H. (2015). Proteomics of ionomycin-induced ascidian sperm reaction: Released and exposed sperm proteins in the ascidian *Ciona insetinalis*. *Proteomics, 15*, 4064–4079.

Nishigaki, T., Chiba, K., Miki, W., & Hoshi, M. (1996). Structure and function of asterosaps, sperm-activating peptides from the jelly coat of starfish eggs. *Zygote (Cambridge, England), 4*, 237–245.

Nishimura, H., Cho, C., Branciforte, D. R., Myles, D. G., & Primakoff, P. (2001). Analysis of loss of adhesive function in sperm lacking cyritestin or fertilin β. *Developmental Biology, 233*, 204–213.

Nishimura, H., Kim, E., Nakanishi, T., & Baba, T. (2004). Possible function of the ADAM1a/ADAM2 fertilin complex in the appearance of ADAM3 on the sperm surface. *Journal of Biological Chemistry, 279*, 34957–34962.

Nishimura, H., & L'Hernault, S. W. (2010). Spermatogenesis-defective (*spe*) mutants of the nematode *Caenorhabditis elegans* provide clues to solve the puzzle of male germline functions during reproduction. *Developmental Dynamics, 239*, 1502–15014.

Nishimura, H., & L'Hernault, S. W. (2016). Gamete interactions require transmembranous immunoglobulin-like proteins with conserved roles during evolution. *Worm, 5*, e1197485.

Nishimura, H., Tajima, T., Comstra, H. S., Gleason, E. J., & L'Hernault, S. W. (2015). The immunoglobulin-like gene *spe-45* acts during fertilization in *Caenorhabditis elegans* like the mouse *Izumo1* gene. *Current Biology, 25*, 3225–3231.

Nishio, S., Emori, C., Wiseman, B., Fahrenkamp, D., Dioguardi, E., et al. (2024). ZP2 cleavage blocks polyspermy by modulating the architecture of the egg coat. *Cell, 187*, 1440–1459.

Noda, T., Blaha, A., Fujihara, Y., Gert, K. R., Emori, C., et al. (2022). Sperm membrane proteins DCST1 and DCST2 are required for sperm-egg interaction in mice and fish. *Communications Biology, 5*, 332.

Noda, T., Lu, Y., Fujihara, Y., Oura, S., Koyano, T., et al. (2020). Sperm proteins SOF1, TMEM95, and SPACA6 are required for sperm-oocyte fusion in mice. *Proceedings of the National Academy of Sciences of the United States of America, 117*, 11493–11502.

Ohto, U., Ishida, H., Krayukhina, E., Uchiyama, S., Inoue, N., et al. (2016). Structure of IZUMO1-JUNO reveals sperm-oocyte recognition during mammalian fertilization. *Nature, 534*, 566–569.

Olson, J. H., Xiang, X., Ziegert, T., Kittelson, A., Rawls, A., et al. (2001). Allurin, a 21-kDa sperm chemoattractant from *Xenopus* egg jelly, is related to mammalian sperm-binding proteins. *Proceedings of the National Academy of Sciences of the United States of America, 98*, 11205–11210.

Pausch, H., Kolle, S., Wurmser, C., Schwarzenbacher, H., Emmerling, R., et al. (2014). A nonsense mutation in *TMEM95* encoding a nondescript transmembrane protein causes idiopathic male subfertility in cattle. *PLoS Genetics, 10*, e1004044.

Pinello, J. F., Liu, Y., & Snell, W. J. (2021). MAR1 linkes membrane adhesion to membrane merger during cell-cell fusionin *Chlamydomonas*. *Developmental Cell, 56*, 3380–3392.

Pinto, M. R., De Santis, R., D'Alessio, G., & Rosati, F. (1981). Studies on fertilization in the ascidians. Fucosyl sites on vitelline coat of *Ciona intestinalis*. *Experimental Cell Research, 132*, 289–295.

Pinto, M. R., Hoshi, M., Marino, R., Amoroso, A., & DeSantis, R. (1990). Chymotrypsin-like enzymes are involved in sperm penetration through the vitelline coat of *Ciona intestinalis* egg. *Molecular Reproduction and Development, 26*, 319–323.

Putiri, E., Zannoni, S., Kadandale, P., & Singson, A. (2004). Functional domains and temperature-sensitive mutations in SPE-9, an EGF repeat-containing protein required for fertility in *Caenorhabditis elegans*. *Developmental Biology, 272*, 448–459.

Rahban, R., & Nef, S. (2020). CatSper: The complex main gate of calcium entry in mammalian spermatozoa. *Molecular and Cellular Endocrinology, 518*, 110951.

Roe, J. L., Farach, H. A. Jr, Strittmatter, W. J., & Lennarz, W. J. (1988). Evidence for involvement of metalloendoproteases in a step in sea urchin gamete fusion. *The Journal of Cell Biology, 107*, 539–544.

Saito, T., & Sawada, H. (2022). Fertilization of ascidians: Gamete interaction, self/nonself recognition and sperm penetration of egg coat. *Frontiers in Cell and Developmental Biology, 9*, 827214.

Saito, T., Shiba, K., Inaba, K., Yamada, L., & Sawada, H. (2012). Self-incompatiblity response induced by calcium increase in seprm of the ascidian *Ciona intestinalis*. *Proceedings of the National Academy of Sciences of the United States of America, 109*, 4158–4162.

Saitoh, Y., Sawada, H., & Yokosawa, H. (1993). High-molecular-weight protease complexes (proteasomes) of sperm of the ascidian, *Halocynthia roretzi*: Isolation, characterization, and physiological roles in fertilization. *Developmental Biology, 158*, 238–244.

Sakai, N., Sawada, H., & Yokosawa, H. (2003). Extracellular ubiquitin system implicated in fertilization of the ascidian, *Halocynthia roretzi:* Isolation and characterization. *Developmental Biology, 264*, 299–307.

Sakairi, K., & Shirai, H. (1991). Possible MIS production by follicle cells in spontaneous oocyte maturation of the ascidian, *Halocynthia roretzi. Development, Growth & Differentiation, 33*, 155–162.

Saldívar-Hernández, A., González-González, M. E., Sánchez-Tusié, A., Maldonado-Rosas, I., López, P., et al. (2015). Human sperm degradation of zona pellucida proteins contributes to fertilization. *Reproductive Biology and Endocrinology, 13*, 99.

Sasanami, T., Sugiura, K., Tokumoto, T., Yoshizaki, N., Dohra, H., et al. (2012). Sperm proteasome degrades egg envelope glycoprotein ZP1 during fertilization of Japanese quail (*Coturnix japonica). Reproduction (Cambridge, England), 144*, 423–431.

Satoh, N. (1994). *Developmental Biology of Ascidians.*

Sawada, H. (2002). Ascidian sperm lysin system. *Zoological Science, 19*, 139–151.

Sawada, H., Hattori, I., Hashii, N., & Saito, T. (2024). Involvement of metalloproteases in the fertilization of the ascidian *Halocynthia roretzi. Biomolecules, 14*, 1487.

Sawada, H., Morita, M., & Iwano, M. (2014). Self/non-self recognition mechanisms in sexual reproduction: New insight shared by flowering plants and hermaphroditic animals. *Biochemical and Biophysical Research Communications, 450*, 1142–1148.

Sawada, H., Pinto, M. R., & De Santis, R. (1998). Participation of sperm proteasome in fertilization of the Phlebobranch ascidian, *Ciona intestinalis.. Molecular Reproduction and Development, 50*, 493–498.

Sawada, H., & Saito, T. (2022). Mechanisms of sperm-egg interactions: What ascidian ferilization research has taught us. *Cells, 11*, 2096.

Sawada, H., Sakai, N., Abe, Y., Tanaka, E., Takahashi, Y., et al. (2002). Extracellular ubiquitination and proteasome-mediated degradation of the ascidian sperm receptor. *Proceedings of the National Academy of Sciences of the United States of America, 99*, 1223–1228.

Sawada, H., & Someno, T. (1996). Substrate specificity of ascidian sperm trypsin-like proteases, spermosin and acrosin. *Molecular Reproduction and Development, 45*, 240–243.

Sawada, H., Tanaka, E., Ban, S., Yamasaki, C., Fujino, J., et al. (2004). Self/nonself recognition in ascidian fertilization: vitelline coat protein HrVC70 is a candidate allorecognition molecule. *Proceedings of the National Academy of Sciences of the United States of America, 101*, 15615–15620.

Sawada, H., Yamamoto, K., Yamaguchi, A., Yamada, L., Higuchi, A., et al. (2020). Three multi-allelic gene pairs are responsible for self-sterility in the ascidian *Ciona intestinalis. Scientific Reports, 10*, 2514.

Sawada, H., Yokosawa, H., & Ishii, S. (1984a). Purification and characterization of two types of trypsin-like enzymes from sperm of the ascidian (Prochordata) *Halocynthia roretzi*. Evidence for the presence of spermosin, a novel acrosin-like enzyme. *Journal of Biological Chemistry, 259*, 2900–2904.

Sawada, H., Yokosawa, H., Ishii, S., & Hoshi, M. (1982). Evidence for acrosin-like enzyme in sperm extract and its involvement in fertilization of the ascidian, *Halocynthia roretzi. Gamete Research, 5*, 291–301.

Sawada, H., Yokosawa, H., Someno, T., Saino, T., & Ishii, S. (1984b). Evidence for the participation of two sperm proteases, spermosin and acrosin, in fertilization of the ascidian, *Halocynthia roretzi:* Inhibitory effects of leupeptin analogs on enzyme activities and fertilization. *Developmental Biology, 105*, 246–249.

Saxena, D. K., Oh-Oka, T., Kadomatsu, K., Muramatsu, T., & Toshimori, K. (2002). Behaviour of a sperm surface transmembrane glycoprotein basigin during epididymal maturation and its role in fertilization in mice. *Reproduction (Cambridge, England), 123*, 435–444.

Schindl, R., & Romanin, C. (2007). Assembly domains in TRP channels. *Biochemical Society Transactions, 35*, 84–85.

Seifert, R., Flick, M., Bönigk, W., Alvarez, L., Trötschel, C., et al. (2015). The CatSper channel controls chemosensation in sea urchin sperm. *The EMBO Journal, 34*, 379–392.

Shang, X., Shen, C., Liu, J., Tang, L., Zhang, H., et al. (2018). Serine protease PRSS55 is crucial for male mouse fertility via affecting sperm migration and sperm-egg binding. *Cellular and Molecular Life Science, 75*, 4371–4384.

Shen, C., Kuang, Y., Liu, J., Feng, J., Chen, X., et al. (2013). *Prss37* is required for male fertility in the mouse. *Biology of Reproduction, 88*, 123.

Shiba, K., Baba, S. A., Inoue, T., & Yoshida, M. (2008). Ca^{2+} bursts occur around a local minimal concentration of attractant and trigger sperm chemotactic response. *Proceedings of the National Academy of Sciences of the United States of America, 105*, 19312–19317.

Shiba, K., & Inaba, K. (2023). The role of soluble adenylyl cyclase in the regulation of flagellar motility in ascidian sperm. *Biomolecule, 13*, 1594.

Shimada, Y., Kanazawa-Takino, N., & Nishimura, H. (2023). Spermiogenesis in *Caenorhabditis elegans*: An excellent model to explore the molecular basis for sperm activation. *Biomolecules, 13*, 657.

Shur, B. D., Ensslin, M. A., & Rodeheffer, C. (2004). SED1 function during mammalian sperm-egg adhesion. *Current Opinion in Cell Biology, 16*, 477–485.

Shur, B. D., Evans, S., & Lu, Q. (1998). Cell surface galactosyltransferase: Current issues. *Glycoconjugate Journal, 15*, 1537–1548.

Shur, B. D., & Hall, N. G. (1982). A role for mouse sperm surface galactosyltransferase in sperm binding to the egg zona pellucida. *The Journal of Cell Biology, 95*, 574–579.

Shur, B. D., Rodeheffer, C., Ensslin, M. A., Lyng, R., & Raymond, A. (2006). Identification of novel gamete receptors that mediate sperm adhesion to the egg coat. *Molecular and Cellular Endocrinology, 250*, 137–148.

Singaravelu, G., Chatterjee, I., Rahimi, S., Druzhinina, M. K., Kang, L., et al. (2012). The sperm surface localization of the TRP-3/SPE-41 Ca^{2+}-permeable channel depends on SPE-38 function in *Caenorhabditis elegans*. *Developmental Biology, 365*, 376–383.

Singaravelu, G., Rahimi, S., Krauchunas, A., Rizvi, A., Dharia, S., et al. (2015). Forward genetics identifies a requirement for the Izumo-like immunoglobulin superfamily *spe-45* gene in *Caenorhabditis elegans* fertilization. *Current Biology, 25*, 3220–3224.

Singson, A., Mercer, K. B., & L'Hernault, S. W. (1998). The *C. elegans spe-9* gene encodes a sperm transmembrane protein that contains EGF-like repeats and is required for fertilization. *Cell, 93*, 71–79.

Smith, J. R., & Stanfield, G. M. (2011). TRY-5 is a sperm-activating protease in *Caenorhabditis elegans* seminal fluid. *PLoS Genetics, 7*, e1002375.

Sommer, F., Awazu, S., Anton-Erxleben, F., Jiang, D., Klimovich, A. V., et al. (2012). Blood system formation in the urochordate *Ciona intestinalis* requires the variable receptor vCRL1. *Molecular Biology and Evolution, 29*, 3081–3093.

Speer, K. F., Allen-Waller, L., Novikov, D. R., & Barott, K. L. (2021). Molecular mechanisms of sperm motility are conserved in an early-branching metazoan. *Proceedings of the National Academy of Sciences of the United States of America, 118*, e2109993118.

Strünker, T., Goodwin, N., Brenker, C., Kashikar, N. D., Weyand, I., et al. (2011). The CatSper channel mediates progesterone-induced Ca^{2+} influx in human sperm. *Nature, 471*, 382–386.

Su, R., Liu, R., Sun, Y., Su, H., & Xing, W. (2025). Rat copper transport protein 2 (CTR2) is involved in fertilization through interaction with IZUMO1 and JUNO. *Theriogenology, 231*, 160–170.

Sutovsky, P. (2011). Sperm proteasome and fertilization. *Reproduction (Cambridge, England), 142*, 1–14.
Sutovsky, P., Manandhar, G., McCauley, T. C., Caamaño, J. N., Sutovsky, M., et al. (2004). Proteasomal interference prevents zona pellucida penetration and fertilization in mammals. *Biology of Reproduction, 71*, 1625–1637.
Sutton, K. A., Jungnickel, M. K., & Florman, H. M. (2008). A polycystin-1 controls postcopulatory reproductive selection in mice. *Proceedings of the National Academy of Sciences of the United States of America, 105*, 8661–8666.
Suzuki-Toyota, F., Maekawa, M., Cheng, A., & Bleil, J. D. (1995). Immuno-colloidal gold labeled surface replica, and its application to detect sp56, the egg recognition and binding protein, on the mouse spermatozoon. *Journal of Electron Microscopy, 44*, 135–139.
Swanson, W. J., & Vacquier, V. D. (2002). The rapid evolution of reproductive proteins. *Nature Review Genetics, 3*, 137–144.
Takayama, J., & Onami, S. (2016). The sperm TRP-3 channel mediates the onset of a Ca^{2+} wave in the fertilized *C. elegans* oocyte. *Cell Reports, 15*, 625–637.
Takayama, J., Tajima, T., Onami, S., & Nishimura, H. (2021). *C. elegans* spermatozoa lacking *spe-45* are incapable of fusing with the oocyte plasma membrane. *microPublication Biolology, 10*, https://doi.org/10.17912/micropub.biology.000372.
Takizawa, S., Sawada, H., Someno, T., Saitoh, Y., Yokosawa, et al. (1993). Effects of protease inhibitors on binding of sperm to the vitelline coat of ascidian eggs: Implications for participation of a proteasome (multicatalytic proteinase complex). *Journal of Experimental Zoology, 267*, 86–91.
Tanigawa, M., Miyamoto, K., Kobayashi, S., Sato, M., Akutsu, H., et al. (2008). Possible involvement of CD81 in acrosome reaction of sperm in mice. *Molecular Reproduction and Development, 75*, 150–155.
Tardif, S., Wilson, M. D., Wagner, R., Hunt, P., Gertsenstein, M., et al. (2010). Zonadhesin is essential for species specificity of sperm adhesion to the egg zona pellucida. *Journal of Biological Chemistry, 285*, 24863–24870.
Tokuhiro, K., Ikawa, M., Benham, A. M., & Okabe, M. (2012). Protein disulfide isomerase homolog PDILT is required for quality control of sperm membrane protein ADAM3 and male fertility [corrected]. *Proceedings of the National Academy of Sciences of the United States of America, 109*, 3850–3855.
Umeda, R., Satouh, Y., Takemoto, M., Nakada-Nakura, Y., Liu, K., et al. (2020). Structural insights into tetraspanin CD9 function. *Nature Communications, 11*, 1606.
Urayama, S., Harada, Y., Nakagawa, Y., Ban, S., Akasaka, M., et al. (2008). Ascidian sperm glycosylphosphatidylinositol-anchored CRISP-like protein as a binding partner for an allorecognizable sperm receptor on the vitelline coat. *Journal of Biological Chemistry, 283*, 21725–21733.
Urch, U. A., & Patel, H. (1991). The interaction of boar sperm proacrosin with its natural substrate, the zona pellucida, and with polysulfated polysaccharides. *Development (Cambridge, England), 111*, 1165–1172.
Vacquier, V. D., Loza-Huerta, A., García-Rincón, J., Darszon, A., & Beltrán, C. (2014). Soluble adenylyl cyclase of sea urchin spermatozoa. *Biochimica et Biophysica Acta, 1842*, 2621–2628.
Vacquier, V. D., & Swanson, W. J. (2011). Selection in the rapid evolution of gamete recognition proteins in marine invertebrates. *Cold Spring Harbor Perspectives in Biology, 3*, a002931.
Vondrakova, J., Frolikova, M., Ded, L., Cerny, J., Postlerova, P., et al. (2022). MAIA, Fc receptor-like 3, supersedes JUNO as IZUMO1 receptor during human fertilization. *Science Advances, 8*, eabn0047.
Voronina, V. A., Harris, F. M., Schmahl, J., Galligan, C., Oristian, D., et al. (2018). Deletion of *Adam6* in *Mus musculus* leads to male subfertility and deficits in sperm ascent into the oviduct. *Biology of Reproduction, 100*, 686–696.

Ward, G. E., Brokaw, C. J., Garbers, D. L., & Vacquier, V. D. (1985). Chemotaxis of *Arbacia punctulata* spermatozoa to resact, a peptide from the egg jelly layer. *The Journal of Cell Biology, 101*, 2324–2329.

Ward, S., & Carrel, J. S. (1979). Fertilization and sperm competition in the nematode *Caenorhabditis elegans*. *Developmental Biology, 73*, 304–321.

Wassarman, P. M., & Litscher, E. S. (2001). Towards the molecular basis of sperm and egg interaction during mammalian fertilization. *Cells, Tissues, Organs, 168*, 36–45.

Wassarman, P. M., & Litscher, E. S. (2022). Mouse zona pellucida proteins as receptors for binding of sperm to eggs. *Trends in Developmental Biology, 15*, 1–13.

Wessel, G. M., Wada, Y., Yajima, M., & Kiyomoto, M. (2021a). Bindin is essential for fertilization in the sea urchin. *Proceedings of the National Academy of Sciences of the United States of America, 118*, e2109636118.

Wessel, G. M., Wada, Y., Yajima, M., & Kiyomoto, M. (2021b). Sperm lacking Bindin are infertile but are otherwise indistinguishable from wildtype sperm. *Scientific Reports, 11*, 21583.

Wheeler, G. N., & Brändli, A. W. (2009). Simple vertebrate models for chemical genetics and drug discovery screens: Lessons from zebrafish and *Xenopus*. *Developmental Dynamics, 238*, 1287–1308.

Williams, H. L., Mansell, S., Alasmari, W., Brown, S. G., Wilson, S. M., et al. (2015). Specific loss of CatSper function is sufficient to compromise fertilizing capacity of human spermatozoa. *Human Reproduction, 30*, 2737–2746.

Wilson, K. L., Fitch, K. R., Bafus, B. T., & Wakimoto, B. T. (2006). Sperm plasma membrane breakdown during *Drosophila* fertilization requires sneaky, an acrosomal membrane protein. *Development (Cambridge, England), 133*, 4871–4879.

Wilson, L. D., Obakpolor, O. A., Jones, A. M., Richie, A. L., Mieczkowski, B. D., et al. (2018). The *Caenorhabditis elegans* spe-49 gene is required for fertilization and encodes a sperm-specific transmembrane protein homologous to SPE-42. *Molecular Reproduction and Development, 85*, 563–578.

Wyckoff, G. J., Wang, W., & Wu, C. I. (2007). Rapid evolution of male reproductive genes in the descent of man. *Nature, 403*, 304–309.

Xu, H., Liu, F., Srakaew, N., Koppisetty, C., Nyholm, P.-G., et al. (2012). Sperm arylsulfatase A binds to mZP2 and mZP3 glycoproteins in a nonenzymatic manner. *Reproduction (Cambridge, England), 144*, 209–219.

Xu, X.-Z., & Sternberg, P. W. (2003). A *C. elegans* sperm TRP protein required for sperm-egg interactions during fertilization. *Cell, 114*, 285–297.

Xu, J., & Zhang, Y. (2010). How significant is a protein structure similarity with TM-score = 0.5? *Bioinformatics, 26*, 889–895.

Yamada, L., Saito, T., Taniguchi, H., Sawada, H., & Harada, Y. (2009). Comprehensive egg coat proteome of the ascidian *Ciona intestinalis* reveals gamete recognition molecules involved in self-sterility. *Journal of Biological Chemistry, 284*, 9402–9410.

Yamagata, K., Murayama, K., Kohno, N., Kashiwabara, S., & Baba, T. (1998a). *p*-Aminobenzamidine-sensitive acrosomal protease(s) other than acrosin serve the sperm penetration of the egg zona pellucida in mouse. *Zygote (Cambridge, England), 6*, 311–319.

Yamagata, K., Murayama, K., Okabe, M., Toshimori, K., Nakanishi, T., et al. (1998b). Acrosin accelerates the dispersal of sperm acrosomal proteins during acrosome reaction. *Journal of Biological Chemistry, 273*, 10470–10474.

Yamaguchi, R., Fujihara, Y., Ikawa, M., & Okabe, M. (2012). Mice expressing aberrant sperm-specific protein PMIS2 produce normal-looking but fertilization-incompetent spermatozoa. *Molecular Biology of the Cell, 23*, 2671–2679.

Yamaguchi, A., Saito, T., Yamada, L., Taniguchi, H., Harada, Y., et al. (2011). Identification and localization of the sperm CRISP family protein CiUrabin involved in gamete interaction in the ascidian *Ciona intestinalis*. *Molecular Reproduction and Development, 78*, 488–497.

Yamaguchi, R., Yamagata, K., Ikawa, M., Moss, S. B., & Okabe, M. (2006). Aberrant distribution of ADAM3 in sperm from both angiotensin-converting enzyme (*Ace*)- and calmegin (*Clgn*)-deficient mice. *Biology of Reproduction, 75*, 760–766.

Yamashita, M. (1988). A fine structural study of the fertilization process of the jellyfish *Cladonema uchidai*. *Development, Growth & Differentiation, 30*, 81–91.

Yamashita, M., Honda, A., Agura, A., Kashiwabara, S., Fukami, K., et al. (2008). Mouse sperm lacking serine protease Prss21/Tesp5 barely undergo the zona pellucida-induced acrosome reaction, but do fertilize the egg. *Biology of Reproduction, 78*, 301–302.

Yokota, N., & Sawada, H. (2007). Sperm proteasomes are responsible for the acrosome reaction and sperm penetration of the vitelline envelope during fertilization of the sea urchin *Pseudocentrotus depressus*. *Developmental Biology, 308*, 222–231.

Yoshida, M., Hiradate, Y., Sensui, N., Cosson, J., & Morisawa, M. (2013). Species-specificity of sperm motility activation and chemotaxis: a study on ascidian species. *The Biological Bulletin, 224*, 156–165.

Yoshida, M., Ishikawa, M., Izumi, H., De Santis, R., & Morisawa, M. (2003). Store-operated calcium channel regulates the chemotactic behavior of ascidian sperm. *Proceedings of the National Academy of Sciences of the United States of America, 100*, 149–154.

Yoshida, M., Murata, M., Inaba, K., & Morisawa, M. (2002). A chemoattractant for ascidian spermatozoa is a sulfated steroid. *Proceedings of the National Academy of Sciences of the United States of America, 99*, 14831–14836.

Yoshida, K., Shiba, K., Sakamoto, A., Ikenaga, J., Matsunaga, S., et al. (2018). Ca^{2+} efflux via plasma membrane Ca^{2+}-ATPase mediates chemotaxis in ascidian sperm. *Scientific Reports, 8*, 16622.

Yoshida, M., & Yoshida, K. (2011). Sperm chemotaxis and regulation of flagellar movement by Ca^{2+}. *Molecular Human Reproduction, 17*, 457–465.

Yu, X., Zhang, X., Zhao, P., Peng, X., Chen, H., et al. (2021). Fertilized egg cells secrete endopeptidases to avoid polytubey. *Nature, 592*, 433–437.

Zhang, H., Li, Y., Cui, K., Chen, X., Shang, C., et al. (2020). Male fertility in *Mus musculus* requires the activity of TRYX5 in sperm migration into the oviduct. *Journal of Cellular Physiology, 235*, 6058–6072.

Zimmerman, S. W., Manandhar, G., Yi, Y.-J., Gupta, S. K., Sutovsky, M., et al. (2011). Sperm proteasomes degrade sperm receptor on the egg zona pellucida during mammalian fertilization. *PLoS One, 6*, e17256.

Zuo, Y., Mei, X., & Singson, A. (2023). CRISPR/Cas9 mediated fluorescent tagging of *Caenorhabditis elegans* SPE-38 reveals a complete localization pattern in live spermatozoa. *Biomolecules, 13*, 623.

CHAPTER FOUR

Activating the *C. elegans* egg: Molecular players, current knowledge, and unanswered questions☆

Aimee Jaramillo-Lambert and Amber R. Krauchunas*
Department of Biological Sciences, University of Delaware, Newark, DE, United States
*Corresponding author. e-mail address: arkrauch@udel.edu

Contents

Abstract

Egg activation is a global cellular change that, in combination with fertilization, transitions the differentiated, developmentally quiescent oocyte into a totipotent, developmentally active one-cell embryo. In *C. elegans*, key regulators of egg activation include *egg-3, egg-4, egg-5, chs-1,* and *spe-11*. Here we will review our current understanding of how these molecules, and others, ensure the robust activation of the egg by controlling meiosis, formation of the eggshell, and the block to polyspermy.

1. Introduction

The oocyte-to-embryo transition encompasses two interrelated but discrete events: fertilization and egg activation. In *C. elegans*, as in many

☆ *C. elegans* egg activation

Current Topics in Developmental Biology, Volume 162
ISSN 0070-2153, https://doi.org/10.1016/bs.ctdb.2025.01.007

other species, egg activation is triggered by fertilization. Typically, mature eggs are arrested at a species-specific point in the meiotic cell cycle and released from this cell cycle arrest by egg activation. Along with this resumption of the cell cycle are a number of other cellular changes that redefine the cellular state of the egg from fertilization-competent and developmentally quiescent to a fertilization-incompetent, totipotent, one-cell embryo. Because there is no meiotic arrest following oocyte maturation in *C. elegans*, we will describe egg activation in this organism as the changes that take place in the oocyte immediately upon fertilization and through the first embryonic mitosis. Both egg-supplied and sperm-supplied factors have been identified that regulate these changes. Here we will review our current understanding of egg activation in the *C. elegans* model system and the genetic and molecular factors that have thus far been identified for the successful regulation and completion of egg activation in this organism.

2. List of key genes

cbd-1: Chitin binding domain protein involved in eggshell vitelline layer formation. (González et al., 2018; Johnston, Krizus, & Dennis, 2010)
chs-1: Chitin synthase required for the formation of the chitin layer of the eggshell and localization of the EGG complex at the oocyte cortex. (Maruyama et al., 2007; Parry et al., 2009; Zhang, Foster, Nelson, Ma, & Carlow, 2005)
egg-3, -4, -5: Pseudophosphatases required for several aspects of egg activation including eggshell formation and polar body formation; part of the EGG complex. (Cheng, Klancer, Singson, & Seydoux, 2009; Maruyama et al., 2007; Parry et al., 2009)
gskl-1 & gskl-2: Homologs of glycogen synthase kinase, type 3 (GSK3) serine-threonine kinases. GSKL-1 and GSKL-2 are sperm-supplied proteins required for female meiosis II. (Banerjee & Srayko, 2022)
gsp-3 & gsp-4: Sperm-specific PP1 phosphatases required for the MEMI-mediated meiosis II exit. (Ataeian et al., 2016; Banerjee & Srayko, 2022)
let-70: E2 ubiquitin-conjugating enzyme involved in the degradation of maternal plasma membrane proteins and progression past metaphase I of meiosis. LET-70 interacts with MARC-3 in the fast block to polyspermy. (Frazier, Shakes, Hota, & Boyd, 2004; Kawasaki et al., 2024)

marc-3: E3 ubiquitin ligase required for the fast block to polyspermy. It is also involved in the degradation of maternal plasma membrane proteins after fertilization. (Kawasaki et al., 2024)
mbk-2: Dual-specificity YAK-1-related kinase required for degradation of maternal proteins during the oocyte-to-embryo transition; part of the EGG complex. (Pellettieri, Reinke, Kim, & Seydoux, 2003; Stitzel, Pellettieri, & Seydoux, 2006)
memi-1, -2, -3: Meiosis-to-mitosis (MEMI) transition defect proteins consisting of three paralogs. The MEMI proteins are required for the exit from meiosis II. (Ataeian et al., 2016)
perm-2 & *perm-4*: Eggshell proteins that make up the PERM complex. The PERM complex is a component of the eggshell vitelline layer. (González et al., 2018)
rab-11.1: Small GTPase required for cortical granule exocytosis. (Sato et al., 2006)
sep-1: The *C. elegans* homolog of the cysteine protease separase. SEP-1 is required for chromosome segregation during meiosis and also regulates cortical granule exocytosis. (Bembenek et al., 2007; Siomos et al., 2001)
spe-11: Sperm-supplied protein required for polar body formation and proper eggshell formation. (Browning & Strome, 1996; Hill, Shakes, Ward, & Strome, 1989a; McNally & McNally, 2005)
syn-4: Syntaxin 1 homolog. Required for cortical granule exocytosis. (Sato et al., 2006)

3. Calcium waves and zinc fluxes

A rise in Ca^{2+} through the egg is considered a hallmark of egg activation across species and is one of the earliest signaling events that regulates the molecular and cellular changes associated with the onset of development (Kashir, Deguchi, Jones, Coward, & Stricker, 2013; P. Stein, Savy, Williams, & Williams, 2020). The presence of a Ca^{2+} rise in the *C. elegans* oocyte coincident with fertilization was first reported by Samuel, Murthy, and Hengartner (2001). Using a calcium indicator dye injected in the gonad of the worm, this study reported the observation of a single calcium transient when the egg is in the spermatheca, which is the region of the reproductive tract where fertilization occurs (Samuel et al., 2001). More recent studies have described this single calcium rise in *C. elegans* as biphasic, consisting of a rapid local rise at the site of fertilization followed by a slower wave that spreads throughout the egg (Takayama & Onami, 2016; Toperzer, Brennan, Carroll, Guisbert, & Kim Guisbert, 2023). This has been observed by injection of calcium indicator dyes into the worm gonad and by the use of a genetically-encoded GCaMP.

To test whether this rise is dependent on fertilization, or could instead be triggered by entry into the spermatheca, calcium levels were observed in *spe-9* and *spe-42* mutants. Both of these genes encode sperm-specific proteins required for fertilization (Kroft, Gleason, & L'Hernault, 2005; Singson, Mercer, & L'Hernault, 1998). In both *spe-9* and *spe-42* mutants fertilization-competent oocytes enter the spermathecae and encounter fertilization-incompetent sperm. The eggs then exit the spermatheca into the uterus without being fertilized. In the absence of fertilization, the Ca^{2+} rise is not observed indicating that in *C. elegans* fertilization is required to initiate calcium signaling in the egg (Takayama & Onami, 2016; Toperzer et al., 2023).

If fertilization triggers the rise in calcium, what is the mechanism by which this occurs? One piece of the puzzle appears to be the TRP family calcium channel SPE-41 (also known as TRP-3). Present on the plasma membrane of the sperm, sperm-egg fusion places SPE-41 onto the egg plasma membrane (Takayama & Onami, 2016). While SPE-41 is important for the fertilization step itself, such that most sperm lacking SPE-41 fail to fertilize the egg, the small number of sperm from *spe-41* mutants that do manage to fertilize an egg fail to trigger the rapid local calcium rise (Takayama & Onami, 2016; Xu & Sternberg, 2003). Yet, the global Ca^{2+} wave still occurs, albeit with a significantly delayed onset (Takayama & Onami, 2016). This is in contrast to eggs fertilized by sperm lacking SPE-11, a sperm supplied protein that does not play a role in fertilization but is required for several egg activation events. When eggs are fertilized by sperm from a *spe-11* mutant a normal Ca^{2+} rise still occurs (Takayama & Onami, 2016). The only other egg activation gene for which calcium dynamics has been investigated is the oocyte expressed gene *egg-3*. Interestingly, when *egg-3* was knocked down by RNAi, Ca^{2+} rises were observed in eggs present in the uterus (Toperzer et al., 2023). However, whether wild-type Ca^{2+} rises were observed when the eggs were in the spermathecae of these animals was not discussed and the cause and significance of this delayed or secondary Ca^{2+} increase remains to be determined.

Since *spe-11* mutants have multiple egg activation defects, including a weak defective eggshell, but normal calcium dynamics, it appears that a rise in Ca^{2+} is not sufficient to induce full activation of the *C. elegans* egg. This observation also supports the conclusion that SPE-11 is not the "activating factor" introduced into the egg because the Ca^{2+} rise occurs in its absence. Consistent with the conclusion that SPE-11 is not the initiating signal, ectopic expression of *spe-11* in the oocyte does not appear to cause premature activation (Browning & Strome, 1996). In contrast, it was reported that the eggs fertilized without SPE-41 form an eggshell, suggesting that the initial local Ca^{2+} rise that occurs at the site of fertilization is not required for eggshell deposition (Takayama & Onami, 2016). While it is important to keep in mind that the global Ca^{2+} rise still occurs, we can conclude that the SPE-41-mediated local rise is not strictly necessary for complete egg activation as some of these embryos fully developed to the larval stage. However, a reduced hatch rate was observed and the unhatched embryos were described as "rounded" consistent with the phenotypes seen for *spe-11* mutants (Hill, Shakes, Ward, & Strome, 1989b). It would be interesting to more closely examine the phenotype of the eggs fertilized by sperm lacking SPE-41 to better understand how robust eggshell formation, polar body extrusion, protein dynamics, etc. are when the fast local Ca^{2+} rise is absent and the global rise is delayed. Of course, it remains to be discovered what sperm and egg factors are responsible for the slower, global Ca^{2+} wave and whether SPE-41 would be sufficient to induce a Ca^{2+} rise if expressed on the egg membrane in the absence of any additional sperm factors. In other species the calcium rise in the egg is mediated by IP3 (Stein et al., 2020). At present there is no evidence that the *C. elegans* IP3 receptor ITR-1 plays a role in egg activation (Clandinin, DeModena, & Sternberg, 1998). However, it was reported that when *itr-1* was knocked down by RNAi the worms failed to ovulate, precluding any conclusions about the role of the gene in subsequent events such as fertilization and egg activation (Yin, Gower, Baylis, & Strange, 2004). Revisiting the role of ITR-1 with new tools that allow for inducible, tissue-specific degradation of proteins of interest (Dickinson & Goldstein, 2016; Friedland et al., 2013) may now allow us to determine if *C. elegans*, like other species, relies on IP3 for the calcium wave at egg activation.

In addition to calcium dynamics, the importance of zinc dynamics during egg activation was revealed by the discovery of "zinc sparks" that occur at mammalian fertilization (Kim et al., 2011). In *C. elegans*, zinc levels were similarly shown to fluctuate as the egg progresses through meiosis and into embryogenesis. Levels are high at both metaphase I and metaphase II, with levels peaking at metaphase II (Mendoza et al., 2022). They then decrease starting at anaphase II and remain low through the two-cell stage of embryogenesis (Mendoza et al., 2022). In mouse, zinc levels likewise rise from prophase to metaphase II and then decrease when the metaphase II oocyte is fertilized (Kim, Vogt, O'Halloran, & Woodruff, 2010; Kim et al. 2011). It is interesting to note that zinc increases and decreases occur at the same stages of meiosis between these species despite the fact that fertilization occurs at prophase I in *C. elegans* and metaphase II in mouse. This implies that fertilization is not a universal trigger for zinc efflux, but instead closely ties zinc levels to progression through meiosis in both species (Mendoza et al., 2022). When zinc levels in the *C. elegans* egg are limited by the addition of TPEN, defects are observed in polar body extrusion, pronuclear migration, and mitotic spindle positioning demonstrating the importance of these fluctuations for proper development (Mendoza, Woodruff, Wignall, & O'Halloran, 2017; Mendoza et al. 2022). In mouse, there is an interdependency between calcium and zinc oscillations (Akizawa, Lopes, & Fissore, 2023; Kim et al., 2011). Considering the different timing of these two signals in the *C. elegans* egg it would be interesting to investigate whether there is crosstalk between calcium and zinc in this system as well.

4. Completion of meiosis

In many species, oocyte development involves arresting the progression of meiosis at two different steps. The first arrest occurs at prophase I and requires a hormone signal for release. The second arrest typically occurs at either metaphase I or II and requires signaling that occurs during egg activation (Horner & Wolfner, 2008; Von Stetina & Orr-Weaver, 2011). During *C. elegans* oogenesis, oocytes progress through meiotic prophase I but arrest in diakinesis, the final substage of prophase I (McCarter, Bartlett, Dang, & Schedl, 1999). Release from this arrest requires a hormone-like signal that is released from sperm. MSP (major sperm protein) secreted from sperm in the adjacent spermatheca signals for the most proximal oocyte to undergo meiotic maturation and to enter meiotic M-phase (Miller et al., 2001). Meiotic maturation is temporally coupled with ovulation and this oocyte will proceed to enter the spermatheca, which is the site of fertilization. Meiosis proceeds with spindle assembly, rotation, and shortening within 20 min of entering the spermatheca (Yang, McNally, & McNally, 2003). Paired homologous chromosomes congress into a pentagonal array with chromosome segregation perpendicular to the egg cortex (McNally & McNally, 2005; Yang et al., 2003). Anaphase I chromosome segregation mediated by SEP-1 (separase) cleavage of cohesin occurs shortly after spindle rotation (McNally & McNally, 2005; Siomos et al., 2001; Yang et al., 2003). Breakdown of the meiosis I spindle midzone, cytokinesis to form the first polar body, and meiotic progression to metaphase II requires fertilization and some sort of signaling molecule from the sperm (McNally & McNally, 2005). In the absence of fertilization, meiosis II does not take place and both sets of anaphase I chromosomes are eventually incorporated into two female pronuclei (McNally & McNally, 2005).

Egg-supplied proteins are the drivers for meiosis II exit. The MEMI (meiosis-to-mitosis transition defect) proteins are encoded by three paralogous genes (*memi-1*, *memi-2*, and *memi-3*) that are expressed during oogenesis and found in the cytoplasm of oocytes and in fertilized eggs (Ataeian et al., 2016). Loss of MEMI function results in a "skipped meiosis II" phenotype where fertilized eggs complete the first meiotic division but fail to assemble a meiosis II spindle. Nonetheless these eggs are able to proceed to mitosis (Ataeian et al., 2016). This matches the cell cycle progression observed in an unfertilized egg, indicating fertilization activates the MEMI proteins for progression through meiosis II. Analysis of a hypermorphic MEMI allele, *memi-1(sb41)*, provided evidence that proper regulation of the MEMIs is also required for the transition from meiosis II to mitosis. Fertilized *memi-1(sb41)* eggs complete meiosis I normally and are capable of assembling a meiosis II spindle, but do not undergo a meiosis II chromosome segregation event and a second polar body is not formed (Ataeian et al., 2016). Thus, MEMIs are required to proceed from meiosis I to meiosis II, but overactive MEMI can also interfere with the meiosis II to mitosis transition.

Post-meiosis, MEMI proteins are marked for degradation by the CUL-2 E3 ubiquitin ligase complex (Ataeian et al., 2016). Interestingly, the MEMI proteins appear to be regulated, either directly or indirectly, by sperm provided phosphatases and kinases. Either RNAi knockdown or mutations of *gsp-3* and *gsp-4*, which encode PP1 phosphatase subunits, and mutations of *gskl-1* and *gskl-2*, which encode glycogen synthase type 3 kinases, can suppress the meiosis II-arrest phenotype caused by an overactive MEMI protein [*memi-1(sb41)*] (Ataeian et al., 2016; Banerjee & Srayko, 2022). Analysis of a *gsp-4; gskl-1; memi-1(sb41)* triple mutant revealed that the phosphatases and kinases act in a single pathway contributing to a common function. GSP-3 and GSP-4 co-immunoprecipitate with MEMI proteins indicating a possible direct regulation of MEMI through removal of phosphorylation (Ataeian et al., 2016). Direct interaction between MEMI proteins and GSKL-1 and GSKL-2 has not been tested.

GSP-3, GSP-4, GSKL-1, and GSKL-2 are all sperm proteins that have functions in sperm chromosome segregation and sperm motility [sperm functions not reviewed here] (Ataeian et al., 2016; Banerjee & Srayko, 2022; Wu et al., 2012). The regulatory targets of these sperm phosphatases and kinases in the egg upon egg activation are still unknown. One possibility is that GSP-3/4 from the sperm directly dephosphorylate MEMI proteins in the egg for progression through meiosis II. This model would predict that loss of *gsp-3/4* would phenocopy loss of *memi-1/2/3*. Unfortunately, this cannot be directly observed due to the essential role of GSP-3/4 in sperm activation. However, Banerjee and Srayko (2022) argue that the *memi-1(sb41)* suppression data are inconsistent with this simple model of GSP-3/4 dephosphorylating and GSKL-1/2 phosphorylating the MEMI proteins and instead propose that major sperm protein (MSP) is the link between these sperm phosphatases and kinases and the MEMI proteins in the egg. In a small fraction of the observed fertilized eggs from *gsp-4 gskl-2; gskl-1* triple mutants (wild-type *memi*) there was the skipped meiosis II phenotype that is seen in *memi* loss of function mutants consistent with their role in activating the MEMI proteins after fertilization (Banerjee & Srayko, 2022). But this interpretation is complicated by additional phenotypes observed in *gsp-4 gskl-2; gskl-1* fertilized eggs including failure of anaphase I/polar body I formation and egg fragmentation in the spermatheca (Banerjee & Srayko, 2022). It will be interesting and exciting to see what additional work will reveal about the connections between these sperm and egg components and whether they directly regulate more than meiosis II.

Failure to properly segregate chromosomes in anaphase and the failure to produce polar body I are phenotypes observed in *spe-11* mutants. Sperm from *spe-11* mutants can fertilize eggs but in the absence of SPE-11 the fertilized eggs have several defects including a failure to produce a polar body, defective cytokinesis, failure to prevent polyspermy, and defective eggshell formation (Browning & Strome, 1996; Hill et al., 1989a). In the absence of *spe-11*, meiosis I proceeds through anaphase I, however, the segregated chromosomes closest to the cortex collapse back into the egg cytosol and cytokinesis fails to occur, therefore no polar body forms (McNally & McNally, 2005). The 12 pairs of chromosomes proceed to meiosis II undergoing congression, rotation, and segregation at anaphase II however cytokinesis fails a second time and two female pronuclei are formed (McNally & McNally, 2005). This same meiotic defect is seen in fertilized eggs from *chs-1* mutants, *gna-2* mutants, *egg-3* mutants, and *egg-4 egg-5* double mutants (Johnston, Krizus, & Dennis, 2006; Maruyama et al., 2007; Parry et al., 2009).

5. Localization of the EGG complex

Several of the egg-supplied proteins that regulate *C. elegans* egg activation are localized together at the cortex of the oocyte and display dynamic changes upon fertilization. Collectively referred to as the EGG complex these proteins are the pseudophosphatases EGG-3, EGG-4, and EGG-5, the chitin synthase CHS-1, and the kinase MBK-2. EGG-4 and EGG-5 are 99.2 % identical in their amino acid sequence and are functionally redundant (Parry et al., 2009), therefore we will refer to this pair as EGG-4/5. Interactions between EGG complex proteins have been observed by yeast-two-hybrid and immunoprecipitation experiments. Positive interactions have been detected between EGG-3 and EGG-4, EGG-3 and MBK-2, EGG-4 and MBK-2, and EGG-5 and MBK-2 (Cheng et al., 2009; Parry et al., 2009; Stitzel, Cheng, & Seydoux, 2007).

Prior to fertilization, the EGG complex is localized at the cortex of the oocyte. As a presumptive chitin synthase, CHS-1 is expected to be embedded in the plasma membrane (Ren et al., 2022). The other members of the complex, however, are not predicted to contain transmembrane domains and therefore must rely on protein-protein or protein-lipid interactions to maintain their cortical localization. Consistent with this model, loss of CHS-1 leads to loss of EGG-3, EGG-4/5, and MBK-2 at the oocyte cortex suggesting that CHS-1 functions to scaffold these proteins near the plasma membrane (Maruyama et al., 2007; Parry et al., 2009). Interestingly, loss of EGG-3 has the same effect. When *egg-3* is knocked down or knocked out, CHS-1, EGG-4/5, and MBK-2 are all missing from the oocyte cortex (Cheng et al., 2009; Maruyama et al., 2007; Parry et al., 2009; Stitzel et al., 2007). EGG-4/5 are also needed for localizing MBK-2 to the cortex, but EGG-3 and CHS-1 localize to the cortex prior to fertilization independently of EGG-4/5 (Cheng et al., 2009; Parry et al., 2009). In contrast, none of the other proteins rely on MBK-2 for proper localization in the oocyte prior to fertilization (Cheng et al., 2009; Maruyama et al., 2007; Parry et al., 2009; Stitzel et al., 2007). CBD-1, a protein involved in eggshell formation, has also been proposed to interact with the EGG complex at the cell cortex. Knockdown of *cbd-1* by RNAi does not interfere with EGG-3, CHS-1, or MBK-2 localization to the cell cortex, but they do appear less continuous and more punctate in the absence of CBD-1 (González et al., 2018; Johnston et al., 2010).

Post-fertilization the EGG complex leaves the cortex and redistributes throughout the egg. EGG-3, -4, and -5 are then presumed to be degraded while MBK-2 persists until at least the 20-cell stage (Cheng et al., 2009; Maruyama et al., 2007; Parry et al., 2009; Stitzel et al., 2007). EGG-3, CHS-1, and MBK-2 move from the cortex and form foci as the cell cycle progresses from metaphase to anaphase of meiosis I and both CHS-1 and MBK-2 have been shown to colocalize with EGG-3 in these foci suggesting that these proteins remain associated as they move away from the cortex (Maruyama et al., 2007; Stitzel et al., 2007). EGG-4/5 have also been reported to exist in EGG-3-positive subcortical puncta during meiosis II, however these puncta were only ever observed in fixed samples and were not observed in live embryos expressing EGG-4::GFP or EGG-5::GFP (Cheng et al., 2009; Parry et al., 2009). In all cases, this movement away from the cortex does not require fertilization but does depend on cell cycle progression; redistribution away from the cortex occurs in unfertilized eggs as the cell cycle progresses but not in APC/c mutants where the cell cycle remains arrested in meiosis I (Maruyama et al., 2007; Parry et al., 2009; Stitzel et al., 2007). EGG-3 also leaves the cortex early when the cell cycle is precociously activated by knockdown of the kinase *wee-1.3* (Stitzel et al., 2007). And yet, cell cycle progression is not sufficient for the localization change of EGG-3 or CHS-1. In the absence of EGG-4/5 both EGG-3 and CHS-1 persist at the cortex despite progression of meiosis suggesting that EGG-4/5 may couple cell cycle dynamics and EGG-3/CHS-1 relocalization (Parry et al., 2009).

In some studies EGG-1 and EGG-2 have been considered part of the EGG complex (González et al., 2018; Johnston et al., 2010). Though these two proteins also localize to the oocyte cortex, EGG-1 and EGG-2 were first reported to be necessary for fertilization, not egg activation, and whether they play a role in egg activation remains unclear (Kadandale et al., 2005). As such, interactions between EGG-1 and EGG-2 and the egg activation regulators CHS-1, EGG-3, EGG-4/5, and MBK-2 have not been extensively studied. In the one study that did report egg activation defects when *egg-1* and *egg-2* were knocked down by RNAi, both CSH-1 and MBK-2 continued to properly localize at the oocyte cortex prior to fertilization (Johnston et al., 2010).

Thus far, SPE-11 has not been connected to the EGG complex, despite the similarities in mutant phenotypes when SPE-11, EGG-3, EGG-4/5, or CHS-1 are absent. Since SPE-11 is provided by the sperm, and thus does not enter the egg until fertilization, it clearly does not interact with or influence the EGG complex prior to fertilization. However, whether the post-fertilization redistribution of the EGG complex away from the egg cortex occurs in *spe-11* mutants is worth investigating.

6. The *C. elegans* eggshell

Metazoan eggs are covered by a specialized extracellular matrix that helps mediate egg-sperm interactions. After fertilization this extracellular matrix is modified to prevent polyspermy and to protect the embryo as it develops (Wong & Wessel, 2006). The eggshell serves these purposes in nematodes. The *C. elegans* eggshell is a made up of six layers that are assembled in a step-wise manner (Olson, Greenan, Desai, Müller-Reichert, & Oegema, 2012). The outermost layer is the vitelline layer, whose composition is relatively unknown other than that it contains glycoproteins (Johnston et al., 2006; Natsuka et al., 2005). This layer resides on the surface of the oocyte prior to fertilization. The second layer is the chitin layer. The chitin synthase CHS-1 is believed to be activated upon fertilization to polymerize and deposit chitin beneath the vitelline layer, likely contributing to the vitelline layer lifting from the zygote surface (Johnston & Dennis, 2012; Johnston et al., 2006; Maruyama et al., 2007; Wong & Wessel, 2006; Zhang et al., 2005). The third layer is the chondroitin proteoglycan (CPG) layer. Chondroitin proteoglycans found in cortical granules are released during cortical granule exocytosis to form this layer (Olson et al., 2012). Together, these three layers are known as the trilaminar outer eggshell (Stein & Golden, 2018). The fourth layer is the extra-embryonic matrix (EEM), which is an asymmetric layer that sits just below the CPG layer and is thought to be fluid filled and contain several proteins (Johnston et al., 2006; Olson et al., 2012). The fifth layer is the permeability barrier layer. This layer is lipid-rich and serves to limit the passage of small molecules and provide osmotic regulation for the developing embryo (Olson et al., 2012; Stein & Golden, 2018). The gene *seip-1* has been shown to be important for formation of the permeability layer, possibly through the regulation of lipid droplet biogenesis (Bai et al., 2020). The sixth and last layer is the peri-embryonic space. The peri-embryonic space is an amorphous space found between the permeability layer and the embryo (Olson et al., 2012; Stein & Golden, 2018). An extensive review of the *C. elegans* eggshell and all of these layers was previously published in 2018 (Stein & Golden, 2018), thus we will focus on more recent advancements in the field since the last review.

The most recent components of the *C. elegans* eggshell to be discovered make up the PERM complex. CBD-1, along with the PERM complex, which consists of Permeable Eggshell protein 2 (PERM-2) and PERM-4, are the only known proteins of the vitelline layer (González et al., 2018). PERM-2 and PERM-4 were identified as proteins required for eggshell formation in an RNAi screen (Carvalho et al., 2011). PERM-2 and PERM-4 localize to the surface of oocytes before fertilization. PERM-2 and PERM-4 are co-dependent for localization on the vitelline layer of the oocyte and both also require CBD-1 for proper localization (González et al., 2018). Loss of CBD-1 results in a more severe phenotype than loss of the PERM complex, which still assembles in part the vitelline, chitin, and CPG layers (González et al., 2018). CBD-1 contains 12 chitin binding domains, two of which appear to be important for interaction with PERM-2 and PERM-4 (González et al. 2018). While these two N-terminal chitin binding domains appear to be required for PERM-2 and PERM-4 interactions, the structure-function relationships between CBD-1, EGG-1/2 and CHS-1/EGG-3/EGG-4/5 remain to be determined. After fertilization, PERM-2, PERM-4, and CBD-1 are all found localized to the embryonic eggshell (González et al., 2018). Depletion of either PERM-2 or PERM-4 results in a compromised eggshell with embryos becoming permeable to external dyes and probes (e.g., FM4–64 and chitin probe), however the chitin layer of the eggshell still forms (González et al., 2018). In addition, PERM complex depletion results in loss of secreted CPG-2, which is normally retained in the perivitelline space (González et al., 2018; Olson et al., 2012).

The formation of the eggshell is triggered by fertilization; *C. elegans* embryos can be easily distinguished from unfertilized oocytes by their oval shape and presence of an eggshell. However, the molecular connections between fertilization and the eggshell components are still unknown. Loss of function of *spe-11* results in a defective eggshell with only a small "chitin cap" near the site of sperm fusion and when *chs-1*, *egg-3*, or *egg-4/5* are knocked down or knocked out the chitin layer of the eggshell is completely absent (Johnston et al., 2010; Maruyama et al., 2007; Parry et al., 2009; Zhang et al., 2005). The chitin layer is observed as early as metaphase

I in wildtype embryos suggesting that the failure to produce the chitin layer of the eggshell is the earliest defect seen in these mutants (Maruyama et al., 2007). Because of the hierarchical assembly of the *C. elegans* eggshell, it is presumed that the permeability barrier is absent as well. The permeability barrier is normally formed during anaphase II and requires cell cycle progression through meiosis II since APC/c mutants that arrest in metaphase I do not form the permeability barrier despite the presence of the chitin layer (Olson et al., 2012; M. Sato, Grant, Harada, & Sato, 2008). Together this highlights the complex signaling and coordination that is taking place between sperm proteins, egg proteins, meiotic progression and other cellular changes necessary to support the developing embryo.

Further questions regarding formation of the *C. elegans* eggshell remain outstanding. While CBD-1 and the PERM complex have been identified as components of the vitelline layer, ultrastructural determination of this layer pre-fertilization is needed to compare to the post-fertilization structure in order to determine exactly how this membrane is remodeled and how it interacts with downstream eggshell layers such as chitin. Additionally, how the vitelline layer is untethered from the embryo surface is still largely unknown. Johnston and Dennis (2012) proposed that the tether is released upon chitin synthesis rather than protease release as seen in other organisms. In this model, chitin competes for binding domains on CBD-1 which displaces the EGG/CHS-1 interaction releasing the tether (González et al., 2018; Johnston & Dennis, 2012). This substrate competition model is intriguing and requires further testing. How chitin synthase is activated also remains unknown. The chitin cap in *spe-11* mutants suggests that it might be involved in the signal to activate CHS-1 but this has not been demonstrated experimentally. Expressing SPE-11 in an unfertilized oocyte does not cause precocious formation of the chitin layer, indicating that SPE-11 cannot be the sole signal that activates CHS-1 (Browning & Strome, 1996). In addition, the mechanisms of chitin synthesis are not fully elucidated. In *C. elegans*, glutamine synthase (*gln-5*) and glucosamine 6-phosphate N-acetlytransferase (*gna-2*) are required for biosynthesis of UDP-GlcNAc (Johnston et al., 2010; Sönnichsen et al., 2005; Varki & Chrispeels, 1999). While homologs for the enzymes involved in the other five steps of the hexosamine pathway have been identified, they have yet to be validated. Finally, the location of chitin synthesis is unknown. Unlike yeast, which synthesize and store chitin in vesicles prior to incorporation in the plasma membrane, most evidence in *C. elegans* points toward synthesis at the plasma membrane.

7. Block to polyspermy

Polyspermy, fertilization of an egg by more than one sperm, typically leads to aneuploidy and death of the developing embryo. While organisms have developed different mechanisms to prevent fertilization by more than one sperm, most involve physical modifications to the egg. *C. elegans* employ two methods, a fast, early block as well as a slow, more permanent block to prevent polyspermy.

The fast block to polyspermy occurs within seconds of an oocyte-sperm fusion event. Several species including sea urchins, frogs, and mammals utilize a membrane-based mechanism for the fast block. However, the mechanisms are quite varied among the different species. Aquatic species prevent polyspermy with a rapid, transient depolarization of the egg membrane (Cross & Elinson, 1980; Grey, Bastiani, Webb, & Schertel, 1982; Jaffe & Cross, 1986; Kline, 1988; Tyler, Monroy, Kao, & Grundfest, 1956). This "fertilization potential" prevents additional sperm from entering the egg. Mammals do not have an electrical block as depolarization of the egg membrane does not occur (Igusa, Miyazaki, & Yamashita, 1983; Jaffe, Sharp, & Wolf, 1983; McCulloh, Rexroad, & Levitan, 1983; Metz & Monroy, 1985; Miyazaki & Igusa, 1981). Instead, the egg membrane undergoes molecular changes that converts the membrane to a state that is less receptive to sperm (Evans, 2020). One known membrane change that occurs in mice is the loss of Juno, the binding partner of sperm Izumo1, from fertilized mouse eggs (Bianchi, Doe, Goulding, & Wright, 2014). Since Juno-Izumo binding is necessary for fertilization, presumably the loss of Juno prevents any subsequent fertilization events from occurring. The loss of Juno is a more persistent block than the transient electrical block of aquatic species. At this time, it is unknown if other egg membrane changes contribute to the fast block (Evans, 2020).

Presently, very little is known regarding the *C. elegans* fast block to polyspermy. However, like mammals, the block appears to be mediated through modifications of the egg membrane. Thus far the only known mediators of the *C. elegans* fast block are MARC-3 and LET-70. MARC-3 is the homolog of mammalian MARCH3, a Membrane-Associated Ring Finger C3HC4 (MARCH) E3 ubiquitin ligase (Kawasaki et al., 2024). MARC-3 is a transmembrane protein that localizes to the plasma membrane and cortical puncta in oocytes. Some of the puncta colocalize with RAB-5 and RAB-7 endosomal markers. After ovulation, MARC-3 is endocytosed and is degraded during embryogenesis (Kawasaki et al., 2024).

This is similar to what is seen for CHS-1 and EGG-3, although MARC-3 is endocytosed sooner than EGG-3 (Kawasaki et al., 2024). Whether the endocytosis and degradation of MARC-3 requires fertilization or just ovulation remains to be tested. Genetic analysis of *marc-3* mutants or *marc-3* RNAi knockdown results in a polyspermy phenotype (up to 30 % of observed embryos) with most polyspermic eggs containing two paternal nuclei (Kawasaki et al., 2024). In the absence of *marc-3* function, entry of a second sperm occurs rapidly after the first fertilization event (within 12 seconds), but sperm entry no longer occurs after 3–5 min when the slow block is completed (Kawasaki et al., 2024).

Kawasaki et al. (2024) identified LET-70, an E2 ubiquitin conjugating enzyme, as a MARC-3 binding partner and showed that together they form an active E2-E3 ubiquitin ligase complex. Like loss of *marc-3* function, RNAi knockdown of *let-70* also results in a polyspermy phenotype, albeit at a much lower frequency than was seen for *marc-3* (2.4 % compared to 9.3 % for *marc-3* RNAi) (Kawasaki et al., 2024). Lastly, analysis of a strain harboring a mutation of a critical amino acid of the RING-CH domain (Cys38 to Ser), demonstrates that this domain is important for binding to LET-70, ubiquitin ligase activity, and preventing polyspermy (Kawasaki et al., 2024). While the discovery of the role of *marc-3* in the fast block to polyspermy in *C. elegans* is a major advancement in the field, many questions remain regarding how the block is actually achieved. What are the targets of the MARC-3-LET-70 E2-E3 complex and how does their ubiquitination change the plasma membrane to prevent entry of more than one sperm? In addition, the lack of 100 % penetrance in the *marc-3* mutants suggests that there are additional regulators of the fast block to polyspermy that have yet to be discovered.

As the fast block is generally transient, lasting only a few seconds to minutes, a more persistent block also exists, which is termed the slow block to polyspermy. The slow block in many animals is established through changes to the extracellular environment and is sometimes called a fertilization membrane or the egg coat block. In several aquatic species the egg is surrounded by a vitelline envelope that elevates from the egg surface at fertilization and undergoes biochemical changes that prevent polyspermy (Gerton & Hedrick, 1986; Grey, Working, & Hedrick, 1976; Just, 1930; Larabell & Chandler, 1991; Rothschild & Swann, 1949; Wolf, Nishihara, West, Wyrick, & Hedrick, 1976). In mammals, the egg is surrounded by an egg coat called the zona pellucida (ZP). Cortical granules released from the egg during sperm-egg fusion contain peptidases, glycosidases, lectins, and zinc (Evans, 2020; Fahrenkamp, Algarra, & Jovine, 2020). The release of these cargoes onto the extracellular matrix results in structural changes to the zona pellucida that block sperm access to the egg plasma membrane (Fahrenkamp et al., 2020).

C. elegans oocytes do not have a ZP, however sperm-egg fusion triggers events that modify the egg covering. In *C. elegans* the slow block is mediated by the formation of the chitin layer of the eggshell. Biochemical analysis in other organisms shows that chitin is synthesized by the hexosamine pathway, which uses fructose 6-phosphate to synthesize uridine diphosphate-N-acetylglucosamine (UDP-GlcNAc). Then chitin synthase polymerizes chitin using UDP-GlcNAc as a sugar nucleotide donor (Varki & Chrispeels, 1999). Chitin is deposited into the extra cellular space where it provides mechanical support for the developing embryo (Krenger, Burri, Lehnert, Nelson, & Gijs, 2020; Maruyama et al., 2007; Olson et al., 2012; Stein & Golden, 2018; Zhang et al., 2005). The chitin layer is formed normally in *marc-3* mutants (Kawasaki et al., 2024), however polyspermy is also observed in several mutants in which the chitin layer of the eggshell is defective.

In *C. elegans*, CHS-1 is the presumptive chitin synthase required for formation of the chitin layer of the eggshell (Zhang et al., 2005). Depletion of *chs-1* via RNAi or mutation of *gna-2* (glucosamine 6-phosphate N-acetly-transferase), an enzyme involved in UDP-GlcNAc synthesis, results in 40 % and 27 % polyspermy respectively (Johnston et al., 2010). In addition, polyspermy has been observed in *spe-11* (22 %) and *egg-4/5* mutants (25 %) (Johnston et al., 2010; Parry et al., 2009). Since CHS-1 and EGG-4/5 are important for the slow block to polyspermy, and they are all missing from the oocyte cortex prior to fertilization in *egg-3* mutants, it is a logical prediction that polyspermy might also be observed in *egg-3* mutants. Indeed, polyspermy has been reported when *egg-3* is knocked down by RNAi (Johnston et al., 2010; Kawasaki et al., 2024). However, it is worth noting that in many of these experiments the APC/c subunit *mat-1*, which has its own effect on eggshell formation, was simultaneously knocked down by RNAi (Johnston et al., 2010; Kawasaki et al., 2024; Sönnichsen et al., 2005) and polyspermy was not observed when the *egg-3* mutant was initially characterized (Maruyama et al., 2007). Interestingly, yeast-2-hybrid data show that MARC-3 and EGG-3 can physically interact; but when the timing of secondary fertilization events was measured in the *marc-3* and *egg-3* mutants a significant difference was seen between the two mutants, supporting the conclusion that the fast block is still intact in the absence of EGG-3 while the slow block does not require MARC-3 (Kawasaki et al., 2024).

8. Cortical granule exocytosis in *C. elegans*

Cortical granules are large secretory vesicles that contain cargoes required for post-fertilization events (Wessel et al., 2001). In *C. elegans*, cortical granules are 1 μm Golgi-derived, CAV-1 containing vesicles that undergo dynamic trafficking (Bembenek et al., 2007; K. Sato et al., 2006). In oocytes, cortical granules first undergo cortical clustering and then disperse throughout the cortex where they are associated with cortical endoplasmic reticulum (Bembenek et al., 2007). Once the oocyte has entered the spermatheca, a wave of degranulation, or exocytosis, occurs and cortical granules fuse with the plasma membrane releasing their contents into the extracellular space. Embryos in the uterus no longer contain cortical granules (Bembenek et al., 2007; K. Sato et al., 2006).

In *C. elegans* exocytosis of cortical granules is closely linked with cell cycle progression as RNAi knockdown of several cell cycle regulators, including members of the anaphase promoting complex (*apc-2* or *emb-27*), cyclin-dependent kinase 1 (*cdk-1*), and the separase inhibitor *ify-1*, result in retention of cortical granules in arrested eggs in the uterus (Bembenek et al., 2007; K. Sato et al., 2006). In particular, cortical granule exocytosis requires separase (SEP-1), which triggers exocytosis during anaphase I (Bembenek et al., 2007). The recruitment of SEP-1 to the granules requires the Rab GTPase RAB-6.1 (Kimura & Kimura, 2012). Recruitment of SEP-1 by RAB-6.1 is specific to cortical granules and independent of the function of SEP-1 in chromosome segregation. When *rab-6.1* is knocked down by RNAi SEP-1 is not recruited to cortical granules and cortical granule exocytosis is inhibited, but chromosome segregation still occurs, although a polar body fails to form (Kimura & Kimura, 2012). When *cdk-1* or *emb-27* are knocked down, SEP-1 recruitment to cortical granules is also not observed, suggesting that progression into anaphase I is required for this process (Kimura & Kimura, 2012). In addition, cortical granule exocytosis requires RAB-11.1, a small GTPase involved in Golgi-to-plasma membrane transport, and the target-SNARE SYN-4, which is required for docking and fusion of vesicles with their target membranes and is also involved in Golgi-to-plasma membrane transport. RNAi knockdown of either *rab-11.1* or *syn-4* inhibits exocytosis leading to the presence of eggs in the uterus that still contain cortical granules (Sato et al., 2008). In contrast, cortical granule exocytosis occurs independently of fertilization (Sato et al., 2006). By extension, a Ca^{2+} rise is not required for cortical granule exocytosis in *C. elegans* since no Ca^{2+} increase is seen in ovulated unfertilized eggs (Takayama & Onami, 2016). This is different from what is seen in many other organisms where calcium triggers the release of cortical granules (Ducibella et al., 2002; Stein et al., 2020; Wessel et al., 2001).

RNAi knock down of genes involved in cortical granule exocytosis also leads to an osmotic integrity defect indicating that cortical granules contain important cargoes required for eggshell formation (Bembenek et al., 2007). The *C. elegans* eggshell is composed of multiple layers that are assembled in a stepwise fashion (Stein & Golden, 2018). Cortical granule exocytosis is important for the proper formation of the third eggshell layer which consists of proteoglycans. Chondroitin colocalizes with cortical granules and depletion of the chondroitin proteoglycans, CPG-1 and CPG-2, leads to the formation of small cortical granules (Bembenek et al., 2007; Olson et al., 2012; M. Sato et al., 2008). In addition, RNAi of *sqv-4*, a glycosylation enzyme involved in proteoglycan synthesis, also leads to a reduction in cortical granule size (Bembenek et al., 2007).

While some cargoes and mediators of cortical granule exocytosis have been identified, several outstanding questions remain. One such gap is the role of SEP-1. SEP-1 localizes to cortical granules (in addition to chromosomes) in metaphase I and disappears during the exocytic wave in anaphase I, although it remains associated with chromosomes and the cortex near the first polar body (Bembenek et al., 2007). It is not known how SEP-1 functions to ensure exocytosis of the granules. Protein phosphatase 5 (PPH-5) was identified as a negative regulator of SEP-1 function in cortical granule exocytosis (Richie et al., 2011). However, whether PPH-5 directly regulates the phosphorylation status of SEP-1 or whether PPH-5 regulation of SEP-1 is indirect needs further investigation. In addition, while CPG-1 and CPG-2 have been identified as cargoes of cortical granules, whether cortical granules contain additional cargoes needed for modification of the extracellular matrix remains to be determined. Of note, CAV-1, a caveolin, has been used as a marker for cortical granules in *C. elegans* (Sato et al., 2006; M. Sato et al., 2008), but whether caveolins play a role in egg activation is not known. Lastly, whether regulators of other egg activation events such as *egg-3, egg-4/5,* and *chs-1* are required for either formation or exocytosis of cortical granules has not yet been investigated.

9. Regulation of maternal RNAs and proteins

To transition the cellular identity of the newly fertilized cell from "oocyte" to "embryo" the complement of active proteins in the cell must be altered. By inhibiting RNA polymerase II, it was shown that maternal factors are sufficient to support the oocyte-to-embryo transition and embryonic development up to the 28-cell stage (Powell-Coffman, Knight, & Wood, 1996). Just as important as what proteins are present to drive development forward is the regulated degradation of other maternal proteins to allow development to proceed. In addition, there appears to be a significant elimination of maternal mRNAs during the oocyte-to-embryo transition, however the roles of this turnover are not yet well understood (Stoeckius et al., 2014). Here we will focus on a few specific examples of regulation of maternal protein degradation during egg activation.

In addition to preventing polyspermy, MARC-3 and LET-70 are involved in ubiquitination and degradation of maternal proteins during the oocyte-to-embryo transition (Kawasaki et al., 2024). MARC-3 is a E3 ubiquitin ligase that functions in Lys63-linked ubiquitination of target proteins in zygotes (Kawasaki et al., 2024). Upon fertilization there is a transient accumulation of Lys63-linked ubiquitination and this is severely reduced in the absence of MARC-3 (Kawasaki et al., 2024; M. Sato, Konuma, Sato, Tomura, & Sato, 2014). MARC-3 can physically interact with the E2 (ubiquitin-conjugating) enzyme LET-70 based on yeast-2-hybrid assays (Frazier et al., 2004; Kawasaki et al., 2024). Degradation of the maternal proteins CHS-1, CAV-1, and RME-2 is delayed and less efficient in *marc-3* mutants (Kawasaki et al., 2024). Similar results were seen for both CHS-1 and CAV-1 degradation when *let-70* was knocked down by RNAi (Kawasaki et al., 2024). LET-70 (aka UBC-2) also interacts with the APC/c (a multi-subunit ubiquitin ligase) and extended *let-70* knockdown leads to an arrest in metaphase I of meiosis that matches the arrest seen for APC/c mutants (Frazier et al., 2004). Ubiquitination is also important for clearing paternal components of the sperm-derived vesicles known as membranous organelles (MOs) (Molina, Lim, & Boyd, 2019). However, ubiquitination of these paternal factors appears to be differentially regulated from the more global endosomal accumulation of Lys63-linked ubiquitination that occurs in the egg post-fertilization. Arrest of the meiotic cell cycle caused by knockdown of *emb-27* inhibits the accumulation of endosomal ubiquitination but does not appear to affect the ubiquitination of paternal MO components (Sato et al., 2014).

The degradation of other maternal proteins is regulated by the kinase MBK-2 (reviewed in Marcello, Singaravelu, & Singson, 2013; Robertson

& Lin, 2013, 2015). These targets include MEI-1 and MEI-2, the two subunits of the microtubule-severing complex known as katanin (Pellettieri et al., 2003). Degradation of MEI-1 and MEI-2 is critical for the transition from meiosis to mitosis and phosphorylation of MEI-1 at S92 by MBK-2 is both necessary and sufficient for katanin interaction with MEL-26, an E3 ligase adaptor, thus targeting katanin for degradation by the proteasome (Dow & Mains, 1998; Johnson et al., 2009; Joly, Beaumale, Van Hove, Martino, & Pintard, 2020; Lu & Mains, 2007; Pintard et al., 2003; Srayko, Buster, Bazirgan, McNally, & Mains, 2000). Other targets of MBK-2 are OMA-1 and OMA-2 which function in both oocyte maturation and the oocyte-to-embryo transition (Nishi & Lin, 2005). Phosphorylation of OMA-1 by MBK-2 is not only required for its function in the one-cell embryo but also marks it for degradation following the first mitosis (Nishi & Lin, 2005).

It is hypothesized that one function of the EGG complex is to sequester MBK-2 away from its substrates until egg activation is triggered (Cheng et al., 2009). In addition to regulating MBK-2 localization, EGG-4 can also inhibit MBK-2 kinase activity in vitro, indicating an additional layer of regulation when MBK-2 is in complex with the EGG proteins (Cheng et al., 2009). Presumably, when the EGG complex redistributes in the egg after fertilization this allows MBK-2 to enter the cytoplasm and disassociate from EGG-4 and therefore phosphorylate its targets. When *egg-4/5* are knocked down, MEI-1 is prematurely phosphorylated as indicated by its presence in the two most proximal oocytes of some germlines (Cheng et al., 2009). However, knockdown of *egg-4/5* does not noticeably alter the degradation of MBK-2 targets MEI-1 or OMA-1 in fertilized eggs (Parry et al., 2009). In contrast, MEI-1 is prematurely degraded in *egg-3* mutants in a *mbk-2*-dependent manner (Stitzel et al., 2007).

The cellular requirements of an embryo differ greatly from those of an oocyte. Therefore, the regulated and coordinated changes in mRNA and proteins that occur during egg activation are critical to the initiation of development. Discoveries in *C. elegans* have allowed us to mechanistically dissect some of these complicated pathways and connect protein localization, activity, post-translational modification, and degradation. Further analysis of other RNAs and proteins that change in their expression at the time of fertilization offer the potential for additional insights into the molecules, and the connections between them, that drive this developmental transition (Stoeckius et al., 2014).

10. Conclusions

C. elegans is an excellent system in which to explore the genetics and cell biology of egg activation. Several key regulators, from both the egg and the sperm, have been identified and characterized and our molecular understanding of individual events such as eggshell formation and the block to polyspermy continues to expand. However, plenty of questions still remain including what role the Ca^{2+} rise plays in the activation and regulation of these events, how signals from the sperm, such as SPE-11, interact with factors in the egg, like the EGG complex, and how not just progression through the cell cycle, but specific stages of meiosis, coordinate other cellular changes that are required for embryogenesis.

References

Akizawa, H., Lopes, E. M., & Fissore, R. A. (2023). Zn2+ is essential for Ca2+ oscillations in mouse eggs. *eLife, 12*, RP88082. https://doi.org/10.7554/eLife.88082.

Ataeian, M., Tegha-Dunghu, J., Curtis, D. G., Sykes, E. M. E., Nozohourmehrabad, A., Bajaj, M., ... Srayko, M. (2016). Maternal MEMI promotes female meiosis II in response to fertilization in Caenorhabditis elegans. *Genetics, 204*(4), 1461–1477. https://doi.org/10.1534/genetics.116.192997.

Bai, X., Huang, L.-J., Chen, S.-W., Nebenfuehr, B., Wysolmerski, B., Wu, J.-C., ... Wang, C.-W. (2020). Loss of the seipin gene perturbs eggshell formation in Caenorhabditis elegans. *Development (Cambridge, England), 147*(20), dev192997. https://doi.org/10.1242/dev.192997.

Banerjee, R. P., & Srayko, M. (2022). Sperm-specific glycogen synthase kinase 3 is required for sperm motility and the post-fertilization signal for female meiosis II in Caenorhabditis elegans. *Development (Cambridge, England), 149*(10), dev200229. https://doi.org/10.1242/dev.200229. Epub 2022 May 30. https://doi.org/10.1242/dev.200229.

Bembenek, J. N., Richie, C. T., Squirrell, J. M., Campbell, J. M., Eliceiri, K. W., Poteryaev, D., ... White, J. G. (2007). Cortical granule exocytosis in C. elegans is regulated by cell cycle components including separase. *Development (Cambridge, England), 134*(21), 3837–3848. https://doi.org/10.1242/dev.011361.

Bianchi, E., Doe, B., Goulding, D., & Wright, G. J. (2014). Juno is the egg Izumo receptor and is essential for mammalian fertilization. *Nature, 508*(7497), 483–487. https://doi.org/10.1038/nature13203.

Browning, H., & Strome, S. (1996). A sperm-supplied factor required for embryogenesis in C. elegans. *Development (Cambridge, England), 122*(1), 391–404. https://doi.org/10.1242/dev.122.1.391.

Carvalho, A., Olson, S. K., Gutierrez, E., Zhang, K., Noble, L. B., Zanin, E., ... Oegema, K. (2011). Acute drug treatment in the early C. elegans embryo. *PLoS One, 6*(9), e24656. https://doi.org/10.1371/journal.pone.0024656.

Cheng, K. C.-C., Klancer, R., Singson, A., & Seydoux, G. (2009). Regulation of MBK-2/DYRK by CDK-1 and the pseudophosphatases EGG-4 and EGG-5 during the oocyte-to-embryo transition. *Cell, 139*(3), 560–572. https://doi.org/10.1016/j.cell.2009.08.047.

Clandinin, T. R., DeModena, J. A., & Sternberg, P. W. (1998). Inositol trisphosphate mediates a RAS-independent response to LET-23 receptor tyrosine kinase activation in C. elegans. *Cell, 92*(4), 523–533. https://doi.org/10.1016/S0092-8674(00)80945-9.

Cross, N. L., & Elinson, R. P. (1980). A fast block to polyspermy in frogs mediated by changes in the membrane potential. *Developmental bioloID - 0372762, 75*(1), 187–198. https://doi.org/10.1016/0012-1606(80)90154-2.

Dickinson, D. J., & Goldstein, B. (2016). CRISPR-based methods for Caenorhabditis elegans genome engineering. *Genetics, 202*(3), 885–901. https://doi.org/10.1534/genetics.115.182162.

Dow, M. R., & Mains, P. E. (1998). Genetic and molecular characterization of the Caenorhabditis elegans gene, mel-26, a postmeiotic negative regulator of mei-1, a meiotic-specific spindle component. *Genetics, 150*(1), 119–128. https://doi.org/10.1093/genetics/150.1.119.

Ducibella, T., Huneau, D., Angelichio, E., Xu, Z., Schultz, R. M., Kopf, G. S., ... Ozil, J.-P. (2002). Egg-to-embryo transition is driven by differential responses to Ca(2+) oscillation number. *Developmental Biology, 250*(2), 280–291.

Evans, J. P. (2020). Preventing polyspermy in mammalian eggs—Contributions of the membrane block and other mechanisms. *Molecular Reproduction and Development, 87*(3), https://doi.org/10.1002/mrd.23331.

Fahrenkamp, E., Algarra, B., & Jovine, L. (2020). Mammalian egg coat modifications and the block to polyspermy. *Molecular Reproduction and Development, 87*(3), 326. https://doi.org/10.1002/mrd.23320.

Frazier, T., Shakes, D., Hota, U., & Boyd, L. (2004). Caenorhabditis elegans UBC-2 functions with the anaphase-promoting complex but also has other activities. *Journal of Cell Science, 117*(22), 5427–5435. https://doi.org/10.1242/jcs.01417.

Friedland, A. E., Tzur, Y. B., Esvelt, K. M., Colaiácovo, M. P., Church, G. M., & Calarco, J. A. (2013). Heritable genome editing in C. elegans via a CRISPR-Cas9 system. *Nature Methods, 10*(8), 741–743. https://doi.org/10.1038/nmeth.2532.

Gerton, G., & Hedrick, L. (1986). *The vitelline envelope to fertilization envelope conversion in eggs of Xenopus laevis.*

González, D. P., Lamb, H. V., Partida, D., Wilson, Z. T., Harrison, M.-C., Prieto, J. A., ... Olson, S. K. (2018). CBD-1 organizes two independent complexes required for eggshell vitelline layer formation and egg activation in C. elegans. *Developmental Biology, 442*(2), 288–300. https://doi.org/10.1016/j.ydbio.2018.08.005.

Grey, R. D., Bastiani, M. J., Webb, D. J., & Schertel, E. R. (1982). An electrical block is required to prevent polyspermy in eggs fertilized by natural mating of Xenopus levis. *Developmental Biology, 89*, 475–484.

Grey, R. D., Working, P. K., & Hedrick, J. L. (1976). Evidence that the fertilization envelope blocks sperm entry in eggs of Xenopus laevis: Interaction of sperm with isolated envelopes. *Developmental Biology, 54*(1), 52–60. https://doi.org/10.1016/0012-1606(76)90285-2.

Hill, D. P., Shakes, D. C., Ward, S., & Strome, S. (1989a). A sperm-supplied product essential for initiation of normal embryogenesis in Caenorhabditis elegans is encoded by the paternal-effect embryonic-lethal gene, spe-11. *Developmental Biology, 136*(1), 154–166. https://doi.org/10.1016/0012-1606(89)90138-3.

Hill, D. P., Shakes, D. C., Ward, S., & Strome, S. (1989b). A sperm-supplied product essential for initiation of normal embryogenesis in Caenorhabditis elegans is encoded by the paternal-effect embryonic-lethal gene, spe-11. *Developmental Biology, 136*(1), 154–166. https://doi.org/10.1016/0012-1606(89)90138-3.

Horner, V. L., & Wolfner, M. F. (2008). Transitioning from egg to embryo: Triggers and mechanisms of egg activation. *Developmental Dynamics: An Official Publication of the American Association of Anatomists, 237*(3), 527–544. https://doi.org/10.1002/dvdy.21454.

Igusa, Y., Miyazaki, S., & Yamashita, N. (1983). Periodic hyperpolarizing responses in hamster and mouse eggs fertilized with mouse sperm. *The Journal of Physiology, 340*, 633–647. https://doi.org/10.1113/jphysiol.1983.sp014784.

Jaffe, L. A., & Cross, N. L. (1986). Electrical regulation of sperm-egg fusion. *Annual Review of Physiology, 48*, 191–200. https://doi.org/10.1146/annurev.ph.48.030186.001203.

Jaffe, L. A., Sharp, A. P., & Wolf, D. P. (1983). Absence of an electrical polyspermy block in the mouse. *Developmental Biology, 96*(2), 317–323. https://doi.org/10.1016/0012-1606(83)90168-9.

Johnson, J.-L. F. A., Lu, C., Raharjo, E., McNally, K., McNally, F. J., & Mains, P. E. (2009). Levels of the ubiquitin ligase substrate adaptor MEL-26 are inversely correlated with MEI-1/katanin microtubule-severing activity during both meiosis and mitosis. *Developmental Biology, 330*(2), 349–357. https://doi.org/10.1016/j.ydbio.2009.04.004.

Johnston, W. L., & Dennis, J. W. (2012). The eggshell in the C. elegans oocyte-to-embryo transition. *Genesis (New York, N. Y.: 2000), 50*(4), 333–349. https://doi.org/10.1002/dvg.20823.

Johnston, W. L., Krizus, A., & Dennis, J. W. (2006). The eggshell is required for meiotic fidelity, polar-body extrusion and polarization of the C. elegansembryo. *BMC Biology, 4*(1), https://doi.org/10.1186/1741-7007-4-35.

Johnston, W. L., Krizus, A., & Dennis, J. W. (2010). Eggshell chitin and chitin-interacting proteins prevent polyspermy in C. elegans. *Current Biology, 20*(21), 1932. https://doi.org/10.1016/j.cub.2010.09.059.

Joly, N., Beaumale, E., Van Hove, L., Martino, L., & Pintard, L. (2020). Phosphorylation of the microtubule-severing AAA+ enzyme Katanin regulates C. elegans embryo development. *Journal of Cell Biology, 219*(6), e201912037. https://doi.org/10.1083/jcb.201912037.

Just, E. E. (1930). The "fertilization" membrane of echinid ova. *Science (New York, N. Y.), 71*(1835), 243. https://doi.org/10.1126/science.71.1835.243.

Kadandale, P., Stewart-Michaelis, A., Gordon, S., Rubin, J., Klancer, R., Schweinsberg, P., ... Singson, A. (2005). The egg surface LDL receptor repeat-containing proteins EGG-1 and EGG-2 are required for fertilization in Caenorhabditis elegans. *Current Biology: CB, 15*(24), 2222–2229. https://doi.org/10.1016/j.cub.2005.10.043.

Kashir, J., Deguchi, R., Jones, C., Coward, K., & Stricker, S. A. (2013). Comparative biology of sperm factors and fertilization-induced calcium signals across the animal kingdom. *Molecular Reproduction and Development, 80*(10), 787–815. https://doi.org/10.1002/mrd.22222.

Kawasaki, I., Sugiura, K., Sasaki, T., Matsuda, N., Sato, M., & Sato, K. (2024). MARC-3, a membrane-associated ubiquitin ligase, is required for fast polyspermy block in Caenorhabditis elegans. *Nature Communications, 15*(1), https://doi.org/10.1038/s41467-024-44928-6.

Kim, A. M., Bernhardt, M. L., Kong, B. Y., Ahn, R. W., Vogt, S., Woodruff, T. K., & O'Halloran, T. V. (2011). Zinc sparks are triggered by fertilization and facilitate cell cycle resumption in mammalian eggs. *ACS Chemical Biology, 6*(7), 716–723. https://doi.org/10.1021/cb200084y.

Kim, A. M., Vogt, S., O'Halloran, T. V., & Woodruff, T. K. (2010). Zinc availability regulates exit from meiosis in maturing mammalian oocytes. *Nature Chemical Biology, 6*(9), 674–681. https://doi.org/10.1038/nchembio.419.

Kimura, K., & Kimura, A. (2012). Rab6 is required for the exocytosis of cortical granules and the recruitment of separase to the granules during the oocyte-to-embryo transition in Caenorhabditis elegans. *Journal of Cell Science, 125*(Pt 23), 5897–5905. https://doi.org/10.1242/jcs.116400.

Kline, D. (1988). *Calcium-dependent events at fertilization of the frog egg: Injection of a calcium buffer blocks ion channel opening, exocytosis, and formation of pronuclei.*

Krenger, R., Burri, J. T., Lehnert, T., Nelson, B. J., & Gijs, M. A. M. (2020). Force microscopy of the Caenorhabditis elegans embryonic eggshell. *Microsystems & Nanoengineering, 6*(1), 1–11. https://doi.org/10.1038/s41378-020-0137-3.

Kroft, T. L., Gleason, E. J., & L'Hernault, S. W. (2005). The spe-42 gene is required for sperm-egg interactions during C. elegans fertilization and encodes a sperm-specific transmembrane protein. *Developmental Biology, 286*(1), 169–181. https://doi.org/10.1016/j.ydbio.2005.07.020.

Larabell, C., & Chandler, D. E. (1991). Fertilization-induced changes in the vitelline envelope of echinoderm and amphibian eggs: Self-assembly of an extracellular matrix. *Journal of Electron Microscopy Technique, 17*(3), 294–318. https://doi.org/10.1002/jemt.1060170305.

Lu, C., & Mains, P. E. (2007). The *C. elegans* anaphase promoting complex and MBK-2/DYRK kinase act redundantly with CUL-3/MEL-26 ubiquitin ligase to degrade MEI-1 microtubule-severing activity after meiosis. *Developmental Biology, 302*(2), 438–447. https://doi.org/10.1016/j.ydbio.2006.09.053.

Marcello, M. R., Singaravelu, G., & Singson, A. (2013). Fertilization. In T. Schedl (Ed.). *Germ cell development in C. elegans* (pp. 321–350) Springer. https://doi.org/10.1007/978-1-4614-4015-4_11.

Maruyama, R., Velarde, N. V., Klancer, R., Gordon, S., Kadandale, P., Parry, J. M., ... Singson, A. (2007). EGG-3 regulates cell-surface and cortex rearrangements during egg activation in Caenorhabditis elegans. *Current Biology: CB, 17*(18), 1555–1560. https://doi.org/10.1016/j.cub.2007.08.011.

McCarter, J., Bartlett, B., Dang, T., & Schedl, T. (1999). On the control of oocyte meiotic maturation and ovulation in Caenorhabditis elegans. *Developmental Biology, 205*(1), 111–128. https://doi.org/10.1006/dbio.1998.9109.

McCulloh, D. H., Rexroad, C. E. J., & Levitan, H. (1983). Insemination of rabbit eggs is associated with slow depolarization and repetitive diphasic membrane potentials. *Developmental Biology, 95*(2), 372–377. https://doi.org/10.1016/0012-1606(83)90038-6.

McNally, K. L., & McNally, F. J. (2005). Fertilization initiates the transition from anaphase I to metaphase II during female meiosis in C. elegans. *Developmental Biology, 282*(1), 218–230. https://doi.org/10.1016/j.ydbio.2005.03.009.

Mendoza, A. D., Sue, A., Antipova, O., Vogt, S., Woodruff, T. K., Wignall, S. M., & O'Halloran, T. V. (2022). Dynamic zinc fluxes regulate meiotic progression in Caenorhabditis elegans†. *Biology of Reproduction, 107*(2), 406–418. https://doi.org/10.1093/biolre/ioac064.

Mendoza, A. D., Woodruff, T. K., Wignall, S. M., & O'Halloran, T. V. (2017). Zinc availability during germline development impacts embryo viability in Caenorhabditis elegans. *Comparative Biochemistry and Physiology. Toxicology & Pharmacology: CBP, 191*, 194–202. https://doi.org/10.1016/j.cbpc.2016.09.007.

Metz, C. B., & Monroy, A. (1985). The fertilization response of the egg. In C. B. Metz (Ed.). *Alberto monroy*. Academic Press.

Miller, M. A., Nguyen, V. Q., Lee, M. H., Kosinski, M., Schedl, T., Caprioli, R. M., & Greenstein, D. (2001). A sperm cytoskeletal protein that signals oocyte meiotic maturation and ovulation. *Science (New York, N. Y.), 291*(5511), 2144–2147. https://doi.org/10.1126/science.1057586.

Miyazaki, S., & Igusa, Y. (1981). Fertilization potential in golden hamster eggs consists of recurring hyperpolarizations. *Nature, 290*. https://doi.org/10.1038/290702a0.

Molina, P., Lim, Y., & Boyd, L. (2019). Ubiquitination is required for the initial removal of paternal organelles in *C. elegans*. *Developmental Biology, 453*(2), 168–179. https://doi.org/10.1016/j.ydbio.2019.05.015.

Natsuka, S., Kawaguchi, M., Wada, Y., Ichikawa, A., Ikura, K., & Hase, S. (2005). Characterization of wheat germ agglutinin ligand on soluble glycoproteins in Caenorhabditis elegans. *Journal of Biochemistry, 138*(2), 209–213. https://doi.org/10.1093/jb/mvi117.

Nishi, Y., & Lin, R. (2005). DYRK2 and GSK-3 phosphorylate and promote the timely degradation of OMA-1, a key regulator of the oocyte-to-embryo transition in *C. elegans*. *Developmental Biology, 288*(1), 139–149. https://doi.org/10.1016/j.ydbio.2005.09.053.

Olson, S. K., Greenan, G., Desai, A., Müller-Reichert, T., & Oegema, K. (2012). Hierarchical assembly of the eggshell and permeability barrier in *C. elegans*. *Journal of Cell Biology, 198*(4), 731–748. https://doi.org/10.1083/jcb.201206008.

Parry, J. M., Velarde, N. V., Lefkovith, A. J., Zegarek, M. H., Hang, J. S., Ohm, J., ... Singson, A. (2009). EGG-4 and EGG-5 link events of the oocyte-to-embryo transition with meiotic progression in C. elegans. *Current Biology: CB, 19*(20), 1752–1757. https://doi.org/10.1016/j.cub.2009.09.015.

Pellettieri, J., Reinke, V., Kim, S. K., & Seydoux, G. (2003). Coordinate activation of maternal protein degradation during the egg-to-embryo transition in C. elegans. *Developmental Cell, 5*(3), 451–462. https://doi.org/10.1016/s1534-5807(03)00231-4.

Pintard, L., Willis, J. H., Willems, A., Johnson, J.-L. F., Srayko, M., Kurz, T., ... Peter, M. (2003). The BTB protein MEL-26 is a substrate-specific adaptor of the CUL-3 ubiquitin-ligase. *Nature, 425*(6955), 311–316. https://doi.org/10.1038/nature01959.

Powell-Coffman, J. A., Knight, J., & Wood, W. B. (1996). Onset of *C. elegans* gastrulation is blocked by inhibition of embryonic transcription with an RNA polymerase antisense RNA. *Developmental Biology, 178*(2), 472–483. https://doi.org/10.1006/dbio.1996.0232.

Ren, Z., Chhetri, A., Guan, Z., Suo, Y., Yokoyama, K., & Lee, S.-Y. (2022). Structural basis for inhibition and regulation of a chitin synthase from Candida albicans. *Nature Structural & Molecular Biology, 29*(7), 653–664. https://doi.org/10.1038/s41594-022-00791-x.

Richie, C. T., Bembenek, J. N., Chestnut, B., Furuta, T., Schumacher, J. M., Wallenfang, M., & Golden, A. (2011). Protein phosphatase 5 is a negative regulator of separase function during cortical granule exocytosis in C. elegans. *Journal of Cell Science, 124*(Pt 17), 2903–2913. https://doi.org/10.1242/jcs.073379.

Robertson, S., & Lin, R. (2013). The oocyte-to-embryo transition. In T. Schedl (Ed.). *Germ cell development in C. elegans* (pp. 351–372) Springer. https://doi.org/10.1007/978-1-4614-4015-4_12.

Robertson, S., & Lin, R. (2015). The maternal-to-zygotic transition in *C. elegans*. In H. D. Lipshitz (Vol. Ed.), *Current topics in developmental biology: Vol. 113*, (pp. 1–42). Academic Press. https://doi.org/10.1016/bs.ctdb.2015.06.001.

Rothschild, L., & Swann, M. M. (1949). The fertilization reaction in the sea-urchin egg: A propagated response to sperm attachment. *Journal of Experimental Biology, 26*(2), 164–176. https://doi.org/10.1242/jeb.26.2.164.

Samuel, A. D., Murthy, V. N., & Hengartner, M. O. (2001). Calcium dynamics during fertilization in C. elegans. *BMC Developmental Biology, 1*, 8. https://doi.org/10.1186/1471-213X-1-8.

Sato, K., Sato, M., Audhya, A., Oegema, K., Schweinsberg, P., & Grant, B. D. (2006). Dynamic regulation of caveolin-1 trafficking in the germ line and embryo of Caenorhabditis elegans. *Molecular Biology of the Cell, 17*(7), 3085–3094. https://doi.org/10.1091/mbc.e06-03-0211.

Sato, M., Grant, B. D., Harada, A., & Sato, K. (2008). Rab11 is required for synchronous secretion of chondroitin proteoglycans after fertilization in Caenorhabditis elegans. *Journal of Cell Science, 121*(Pt 19), 3177–3186. https://doi.org/10.1242/jcs.034678.

Sato, M., Konuma, R., Sato, K., Tomura, K., & Sato, K. (2014). Fertilization-induced K63-linked ubiquitylation mediates clearance of maternal membrane proteins. *Development (Cambridge, England), 141*(6), 1324–1331. https://doi.org/10.1242/dev.103044.

Singson, A., Mercer, K. B., & L'Hernault, S. W. (1998). The C. elegans spe-9 gene encodes a sperm transmembrane protein that contains EGF-like repeats and is required for fertilization. *Cell, 93*(1), 71–79. https://doi.org/10.1016/s0092-8674(00)81147-2.

Siomos, M. F., Badrinath, A., Pasierbek, P., Livingstone, D., White, J., Glotzer, M., & Nasmyth, K. (2001). Separase is required for chromosome segregation during meiosis I in Caenorhabditis elegans. *Current Biology: CB, 11*(23), 1825–1835. https://doi.org/10.1016/s0960-9822(01)00588-7.

Sönnichsen, B., Koski, L. B., Walsh, A., Marschall, P., Neumann, B., Brehm, M., ... Echeverri, C. J. (2005). Full-genome RNAi profiling of early embryogenesis in Caenorhabditis elegans. *Nature, 434*(7032), 462–469. https://doi.org/10.1038/nature03353.

Srayko, M., Buster, D. W., Bazirgan, O. A., McNally, F. J., & Mains, P. E. (2000). MEI-1/MEI-2 katanin-like microtubule severing activity is required for Caenorhabditis elegans meiosis. *Genes & Development, 14*(9), 1072–1084. https://doi.org/10.1101/gad.14.9.1072.

Stein, K. K., & Golden, A. (2018). The C. elegans eggshell. *WormBook: The Online Review of C. Elegans Biology, 2018,* 1–36. https://doi.org/10.1895/wormbook.1.179.1.

Stein, P., Savy, V., Williams, A. M., & Williams, C. J. (2020). Modulators of calcium signalling at fertilization. *Open Biology, 10*(7), 200118. https://doi.org/10.1098/rsob.200118.

Stitzel, M. L., Cheng, K. C.-C., & Seydoux, G. (2007). Regulation of MBK-2/Dyrk kinase by dynamic cortical anchoring during the oocyte-to-zygote transition. *Current Biology: CB, 17*(18), 1545–1554. https://doi.org/10.1016/j.cub.2007.08.049.

Stitzel, M. L., Pellettieri, J., & Seydoux, G. (2006). The C. elegans DYRK kinase MBK-2 marks oocyte proteins for degradation in response to meiotic maturation. *Current Biology: CB, 16*(1), 56–62. https://doi.org/10.1016/j.cub.2005.11.063.

Stoeckius, M., Grün, D., Kirchner, M., Ayoub, S., Torti, F., Piano, F., ... Rajewsky, N. (2014). Global characterization of the oocyte-to-embryo transition in Caenorhabditis elegans uncovers a novel mRNA clearance mechanism. *The EMBO Journal, 33*(16), 1751–1766. https://doi.org/10.15252/embj.201488769.

Takayama, J., & Onami, S. (2016). The sperm TRP-3 channel mediates the onset of a Ca (2+) wave in the fertilized C. elegans oocyte. *Cell Reports, 15*(3), 625–637. https://doi.org/10.1016/j.celrep.2016.03.040.

Toperzer, K. M., Brennan, S. J., Carroll, D. J., Guisbert, E. A., & Kim Guisbert, K. S. (2023). Visualization of the biphasic calcium wave during fertilization in Caenorhabditis elegans using a genetically encoded calcium indicator. *Biology Open, 12*(9), bio059832. https://doi.org/10.1242/bio.059832.

Tyler, A., Monroy, A., Kao, C. Y., & Grundfest, H. (1956). Membrane potential and resistance of the starfish egg before and after fertilization. *Marine Biological Laboratory Bulletin, 111*, 153–177.

Varki, A., & Chrispeels, M. J. (1999). *Essentials of glycobiology*. Cold Spring Harbor Laboratory Press. https://books.GoogleScholar.com/books?id=lH72FFWIIpgC.

Von Stetina, J. R., & Orr-Weaver, T. L. (2011). Developmental control of oocyte maturation and egg activation in metazoan models. *Cold Spring Harbor Perspectives in Biology, 3*(10), a005553. https://doi.org/10.1101/cshperspect.a005553.

Wessel, G. M., Brooks, J. M., Green, E., Haley, S., Voronina, E., Wong, J., & Conner, S. (2001). *The biology of cortical granules. International review of cytology*, Academic Press, *Vol. 209*, 117–206. https://doi.org/10.1016/S0074-7696(01)09012-X.

Wolf, D. P., Nishihara, T., West, D. M., Wyrick, R. E., & Hedrick, J. L. (1976). Isolation, physicochemical properties, and the macromolecular composition of the vitelline and fertilization envelopes from Xenopus laevis eggs. *Biochemistry, 15*(17), 3671–3678. https://doi.org/10.1021/bi00662a005.

Wong, J. L., & Wessel, G. M. (2006). Defending the zygote: Search for the ancestral animal block to polyspermy. *Current Topics in Developmental Biology, 72*, 1–151. https://doi.org/10.1016/S0070-2153(05)72001-9.

Wu, J., Go, A. C., Samson, M., Cintra, T., Mirsoian, S., Wu, T. F., ... Chu, D. S. (2012). Sperm development and motility are regulated by PP1 phosphatases in Caenorhabditis elegans. *Genetics, 190*(1), 143–157. https://doi.org/10.1534/genetics.111.135376.

Xu, X.-Z. S., & Sternberg, P. W. (2003). A C. elegans sperm TRP protein required for sperm-egg interactions during fertilization. *Cell, 114*(3), 285–297. https://doi.org/10.1016/s0092-8674(03)00565-8.

Yang, H., McNally, K., & McNally, F. J. (2003). MEI-1/katanin is required for translocation of the meiosis I spindle to the oocyte cortex in C elegans. *Developmental Biology, 260*(1), 245–259. https://doi.org/10.1016/s0012-1606(03)00216-1.

Yin, X., Gower, N. J. D., Baylis, H. A., & Strange, K. (2004). Inositol 1,4,5-trisphosphate signaling regulates rhythmic contractile activity of myoepithelial sheath cells in Caenorhabditis elegans. *Molecular Biology of the Cell, 15*(8), 3938–3949. https://doi.org/10.1091/mbc.e04-03-0198.

Zhang, Y., Foster, J. M., Nelson, L. S., Ma, D., & Carlow, C. K. S. (2005). The chitin synthase genes chs-1 and chs-2 are essential for C. elegans development and responsible for chitin deposition in the eggshell and pharynx, respectively. *Developmental Biology, 285*(2), 330–339. https://doi.org/10.1016/j.ydbio.2005.06.037.

CHAPTER FIVE

Fertilization and the fast block to polyspermy in the African Clawed Frog, *Xenopus laevis*: A historical perspective

Kayla M. Komondor and Anne E. Carlson*
Department of Biological Sciences, University of Pittsburgh, Pittsburgh, PA, United States
*Corresponding author. e-mail address: acarlson@pitt.edu

Contents

Abstract

The African clawed frog, *Xenopus laevis*, has long been a model organism for studying fertilization due to its large and abundant eggs that are easily manipulated and rapidly undergo embryonic development. Research on this model organism has provided significant insights into the mechanisms that ensure successful fertilization, including the prevention of polyspermy. Polyspermy, the fertilization of an egg by multiple sperm, poses a significant threat to successful embryonic development in most sexually reproducing animals. To counter this, eggs have evolved mechanisms known as polyspermy blocks, which prevent additional sperm from entering once fertilization has occurred. This review focuses on fertilization research in general, and specifically on studies of the fast block to polyspermy in *X. laevis*. We trace key discoveries and experimental advancements that have shaped our current understanding. Indeed, studies on *X. laevis* have revealed that fertilization triggers a depolarization of the egg membrane mediated by an efflux of Cl^- through the Ca^{2+}-activated Cl^- channel TMEM16A, effectively preventing polyspermy. Despite these advances, several questions remain regarding the precise molecular interactions and signaling pathways involved. Continued research on *X. laevis* promises to uncover further details about the earliest events in embryogenesis and the voltage-dependent mechanisms of fertilization, offering broader insights into reproductive biology across species.

Current Topics in Developmental Biology, Volume 162
ISSN 0070-2153, https://doi.org/10.1016/bs.ctdb.2024.12.003

1. Introduction

For most sexually reproducing animals, successful embryonic development requires that an egg is fertilized by only one sperm (Hassold et al., 1980; Rojas et al., 2021). Eggs have evolved mechanisms to prevent polyspermy, the fertilization by multiple sperm, by stopping additional sperm from entering once fertilization has occurred. Scientists call these processes polyspermy blocks, which are conserved in the eggs of diverse animals (Evans, 2020; Fahrenkamp, Algarra, & Jovine, 2020; Wong & Wessel, 2006). Polyspermy blocks take two forms, named based on their timing (Wong & Wessel, 2006). The fast block occurs immediately after fertilization (Wong & Wessel, 2006) and is used by externally fertilizing animals. In contrast, the slow block to polyspermy occurs minutes later and is conserved in both internal and external fertilization animals (Evans, 2020). Much of our understanding of the fast block to polyspermy has been derived from studies on the African clawed frog, *Xenopus laevis* (Wozniak & Carlson, 2020), which will be the focus of this review.

X. laevis is a widely studied amphibian species in developmental biology and embryology. Native to sub-Saharan Africa, *X. laevis* thrives in various aquatic environments, from ponds to rivers (Tinsley, Loumont, & Kobel, 1996; Toft-Bertelsen, Krizaj, & MacAulay, 2017). This species is particularly valued in scientific research due to its large, easily manipulated eggs and rapid developmental processes (Heasman, Holwill, & Wylie, 1991; Nieuwkoop & Faber, 1994; Sive, Grainger, & Harland, 2010). Over the years, *X. laevis* has become a crucial model organism, providing insights into cellular processes and developmental mechanisms applicable across vertebrates.

The use of *X. laevis* as a model organism in scientific research dates back to the early 20th century, but its popularity surged in the 1950s and 1960s. One of the landmark studies that highlighted the utility of *X. laevis* in developmental biology was conducted by John Gurdon in the early 1960s. Gurdon's pioneering work involved the successful cloning of frogs using the technique of nuclear transplantation in *X. laevis*, demonstrating that the nucleus from a differentiated cell could revert to a totipotent state when injected into an enucleated egg (Gurdon, Elsdale, & Fischberg, 1958; Gurdon, 1962). This experiment, published in 1962, provided profound insights into cellular differentiation and the reversibility of the differentiated state, laying the groundwork for future research in developmental biology, cloning, and stem cell research (Gurdon, 2013; Yamanaka & Blau, 2010).

Before Gurdon's experiments, *X. laevis* had been used in endocrinology research in the lab (Malmros, 1948) and for pregnancy tests in the clinic

(Elkan, 1938). These *Xenopus*-based pregnancy tests known as the "Hogben test", and were considered to be more humane than the mouse and rabbit-based tests used before, which required slaughter of these animals and careful dissection of their ovaries to assess for ovulation (Nuwer, 2013). The pregnancy test was rooted in the observation that *X. laevis* females ovulate when injected with urine from pregnant women due to the presence of human chorionic gonadotropin (hCG) hormone (Oliver & Miller, 1946; Sanders, 1946). This led to the widespread use of *X. laevis* in pregnancy testing until more modern methods were developed.

X. laevis also serves as a vital model system in fertilization research. These frogs belong to the class Amphibia and order Anura. As external fertilizers, *X. laevis* eggs are abundantly and readily available for laboratory research. The large size of the eggs allows for easier electrophysiology recordings and manipulation during recordings (Bhatt et al., 2022). As opposed to mammalian fertilization, *X. laevis* fertilization is readily accessible in the lab as the animals naturally fertilize outside of the mother. Additionally, *X. laevis* females lay thousands of eggs in a single experimental day in the lab when induced to ovulate, and these eggs natively use the fast block to polyspermy to keep sperm out of an already fertilized egg. Altogether, these characteristics have made *X. laevis* a widely used model organism for over 70 years for studying fertilization. Specifically, studies on *X. laevis* have been instrumental in elucidating the fast block to polyspermy, a critical process that prevents multiple sperm from fertilizing an egg. This review will focus on the fast block to polyspermy in *X. laevis*, highlighting the key discoveries and their implications for broader biological research.

2. Mating system

Like external fertilization in other aquatic and marine animals, *X. laevis* fertilization occurs outside the mother. *X. laevis* mating involves a unique courtship behavior known as amplexus, which involves the male grasping the female around her midsection using specialized structures called nuptial pads. Nuptial pads are located on the forearms of males from various amphibians and provide a firm grip on the female during amplexus to ensure close contact during mating (Fisher et al., 2023). During this process, the female lays eggs, and the male simultaneously releases sperm for external fertilization. This synchronization ensures successful reproduction and is a common practice among freely swimming animals like frogs and fish (Wootton & Smith, 2014). In the *X. laevis* mating system, gametes mix shortly after release, resulting in a high sperm-to-egg ratio.

By contrast, broadcast spawning is a common reproductive strategy in stationary marine invertebrates, such as sea urchins, corals, mollusks, and marine sponges (Barton, Willis, & Hutson, 2017; Evans & Lymbery, 1813; Ponder, Lindberg, & Ponder, 2019). In this process, both eggs and sperm are released into the water column, often in synchrony with environmental cues such as lunar cycles (Glese & Kanatani, 1987; Mercier, Ycaza, & Hamel, 2007; Morgan, 2020), water temperature (Bernard et al., 2016; Eads, Evans, & Kennington, 2016; Heyward & Negri, 2010), or tides (Crimaldi & Zimmer, 2014; Monteiro et al., 2016). The synchrony is crucial, as it maximizes the chances of gametes encountering one another in the vast ocean environment (Babcock et al., 1992; Levitan, 2020). Unlike species with direct mating behaviors, these invertebrates rely on external cues and chemical signals to ensure gamete viability and successful fertilization. This method increases genetic diversity but also requires the production of large quantities of gametes to compensate for the high dilution and potential predation in the open water (Evans & Lymbery, 1813; Evans & Sherman, 2013). The ratio of sperm to eggs at the site and time of fertilization can be much higher than in mammalian fertilization, as large quantities of sperm are needed to increase the likelihood of encountering and fertilizing the eggs in the dispersed aquatic environment.

Following mating, *X. laevis* sperm rely on chemotaxis to locate the egg (Burnett et al. 2008, 2011). Chemotaxis involves cells moving toward a chemical gradient. The jelly surrounding the *X. laevis* egg releases allurin (Sugiyama et al., 2009), a potent chemoattractant that forms a concentration gradient (al-Anzi & Chandler, 1998). Allurin creates a chemical map that ensures sperm find and fertilize eggs in the aquatic environment. Chemotaxis also plays a role in the fertilization of stationary marine invertebrates, with eggs releasing specific chemical attractants that guide sperm towards them, enhancing the likelihood of fertilization amidst the dispersed gametes (Evans & Sherman, 2013). While efficient for species anchored to substrates, broadcast fertilization presents challenges such as polyspermy and fertilization failure, necessitating mechanisms to enhance reproductive success and offspring survival in these environments (Evans & Lymbery, 1813; Evans & Sherman, 2013; Marshall & Bolton, 2007).

By contrast, for mammals, the sperm-to-egg ratio at the site and time of fertilization is closer to 1 to 1 (Coy et al., 2012), and this lower ratio is likely the result of the requirement that sperm travel through the anatomical maze of the female reproductive tract (Suarez & Pacey, 2006). A higher sperm-to-egg ratio increases the potential for polyspermy, requiring mechanisms to prevent additional sperm from entering a fertilized egg. Scientists have proposed that the excess of sperm relative to eggs necessitates the fast block to prevent polyspermy (Gilbert, 2003). Future comparative studies are needed to test this hypothesis.

3. Sperm and egg interaction

Fertilization involves the fusion of two cells: the sperm and the egg. Sperm are terminally differentiated cells designed solely for delivering their genetic material, whereas eggs are prepared to develop into a new organism upon fertilization. The union of these two cells creates new life and represents an active area of research.

Sperm egg binding is facilitated by proteins localized to the plasma membrane of each gamete. The sperm proteins Izumo1, SPACA6, and TMEM81 form a heterotrimer required for sperm-egg binding (Deneke et al., 2023). The egg receptor, in contrast to the highly conserved sperm complex, shows much greater variability across species. In mammals, this complex binds to the egg protein Juno (Bianchi et al., 2014), while in fish, it binds to Bouncer (Deneke et al., 2023). Interestingly, despite advances in understanding fertilization, the specific sperm receptor in frog eggs has not yet been identified. So far, researchers have identified Izumo1 and Juno as the only interacting proteins essential for fertilization in mammals (Deneke & Pauli, 2021). Several fundamental questions about the precise molecular mechanisms remain unanswered.

Neither the process of sperm-egg fusion nor the proteins required for this fusion have been characterized for frogs or mammals. Previous research suggested that the fusion between sperm and egg in *X. laevis* was analogous to external fertilization in fish and sea urchins, where actin polymers from the egg were responsible for bringing the sperm into the egg (Limatola, Chun, & Santella, 2022). However, inhibition of actin polymerization during *X. laevis* fertilization has shown that these protrusions are unnecessary for *X. laevis* sperm-egg fusion at fertilization (Tembo et al., 2021). How sperm and egg physically unite in frogs and mammals remains an active area of research.

Fertilization signals an elevation in the egg's cytoplasmic Ca^{2+}. This phenomenon is conserved in nearly all animals and several plants (Whitaker, 2006).

This intracellular Ca^{2+} signals the polyspermy blocks, initiates embryonic development, and regulates the early embryo's cell division and patterning process (Whitaker, 2006). In addition to its importance in activating the egg to prevent polyspermy, the increase in Ca^{2+} in *X. laevis* eggs at fertilization is also essential for initiating embryonic development by triggering the exit of the condition of meiotic arrest that *X. laevis* eggs are in before fertilization (Tokmakov et al., 2014). Several years ago, an early review on Ca^{2+} sources in egg activation among external fertilizers proposed that Ca^{2+} is the primary activation mechanism in the egg (Jaffe, 1983). Since its publication, subsequent research has confirmed that Ca^{2+} is crucial for fertilization, polyspermy blocks, and development in nearly all sexual reproducers studied, and these ideas continue to be an active area of study.

4. Polyspermy is embryonic lethal for most animals

Polyspermy, or fertilization of an egg by more than one sperm, is generally detrimental to successful embryonic development for nearly all sexual reproducers. For most animals, polyspermy leads to chromosomal abnormalities and dysregulation of cellular processes such as cell division, ultimately resulting in cell death (Elinson, 1986). However, some animals, including certain birds, reptiles, and even some amphibians, not only tolerate polyspermy but require it to initiate development (Mizushima, 2017). For these species, a few to hundreds of sperm are necessary to trigger successful embryonic development (Hemmings & Birkhead, 1818; Snook, Hosken, & Karr, 2011). For example, salamanders, which are amphibians like frogs but use internal fertilization, are known to have eggs that are inseminated by multiple sperm (Elinson, 1986). Monospermic animals, on the other hand, have evolved strict regulatory mechanisms to ensure that only one sperm fertilizes the egg, employing various methods to prevent polyspermy.

For most animals, embryos cannot survive the excess genetic material resulting from polyspermic fertilization. This excess genetic material disrupts DNA replication in the embryo, leading to cell death because the egg cannot regulate its cell cycle with multiple nuclei present (Skinner, 2018). To prevent this catastrophic outcome, eggs have evolved methods to stop additional sperm from entering once fertilization has occurred. Two commonly used polyspermy blocks are the fast block and the slow block to polyspermy.

The slow block to polyspermy occurs in the eggs of most sexually reproducing animals (Wong & Wessel, 2006) and is triggered by an increase in the egg's cytoplasmic Ca^{2+} levels following fertilization. Secretory vesicles called cortical granules docked at the plasma membrane of these eggs play a crucial role in this process (Berg & Wessel, 1997; Liu, 2011; Wyrick, Nishihara, & Hedrick, 1974). Although the exact contents of cortical granules are still being studied, it is known that their release includes a Zn^{2+} spark, which is conserved in both external and internal fertilizers (Wozniak et al., 2020). During the slow block, fertilization prompts the cortical granules to release their contents, which modify the extracellular matrix and the plasma membrane to prevent additional sperm from entering. This release also hardens the extracellular matrix (known as the zona pellucida in mammals and the vitelline envelope in *X. laevis*) to protect the nascent zygote (Nishio et al., 2023; Wessel & Wong, 2009). This process is highly conserved (Swanson & Vacquier, 2002). Upon fertilization, the membrane structure is altered, and proteins that interact with sperm are cleaved, ensuring no more sperm can enter. While the slow block plays a critical role in preventing polyspermy, the fast block will be the focus of this review.

5. Fast block to polyspermy

The first description of polyspermy blocking mechanisms was published in 1919 by Ernest Everett Just, who reported a physical lifting of the membrane of sea urchin eggs following fertilization (Just, 1919). During his experiments, Just observed a wave-like lifting of the vitelline envelope around the sea urchin egg, starting from the point of sperm entry, which acted as a barrier to additional sperm. He also noticed sperm near the egg stopped swimming and no longer attempted to enter the egg once this lifting occurred (Just, 1919). Although the underlying mechanism was not understood at the time, Just was observing the effects of the fast block to polyspermy, as evidenced by the immobilized sperm and the prevention of further fertilization near the membrane before the envelope lifting.

Forty years later, the first electrophysiology recording on a frog egg was made. Using eggs from the common toad, *Bufo vilagris*, Takashi Maéno recorded membrane potentials of eggs and oocytes in solutions with varying Na^+, K^+, and Ca^{2+} solutions (Maeno, 1959). Maéno observed differences in the membrane potential of immature oocytes versus fertilization-competent eggs, concluding that ion permeability changes as oocytes mature into eggs. Together, findings from the Just and Maéno experiments laid the conceptual and technical groundwork needed to discover the fast block.

Laurinda Jaffe used voltage clamping to demonstrate that fertilization-triggered depolarization is a polyspermy block in sea urchin eggs (Jaffe, 1976). She performed whole-cell recordings during the fertilization of eggs from the purple sea urchin, *Strongylocentrotus purpuratus.* Jaffe observed that before sperm addition, the resting potential of the eggs averaged −70 mV. Within seconds after sperm addition, she recorded a rapid depolarization to a less negative fertilization potential, averaging −10 mV eggs (Jaffe, 1976). She made these observations by passively recording the egg's membrane potential.

To test whether fertilization-triggered depolarization affects the ability of sperm to fertilize, Jaffe clamped sea urchin eggs at either the resting potential (−10 mV and lower) or the fertilization potential (over +5 mV) and determined the incidence of fertilized eggs (Jaffe, 1976). Whereas Maéno used passive recordings, Jaffe actively induced a positive potential in the sea urchin eggs similar to that observed after fertilization. This was only possible following the invention of the voltage clamp (Hille, 2001). When eggs were held at a negative potential, multiple sperm fertilized because this interfered with the depolarization-induced polyspermy block. Conversely, fertilization stopped altogether when eggs were held at the positive potential typically reached after fertilization. Sperm bound to but did not enter depolarized eggs (Jaffe, 1976).

In a well-controlled experiment, Jaffe allowed the membrane potential to return to rest, at which point fertilization occurred. Indeed, fertilization did not occur until after the positive hold was released, and the egg could return to the typical negative resting potential. Interestingly, Jaffe also observed a delay in lifting the fertilization membrane in eggs held at a positive potential compared to surrounding eggs not clamped at a positive potential (Jaffe, 1976). The surrounding eggs showed membrane lifting within the expected timeframe after sperm addition. In eggs clamped at a negative potential, although there was no membrane depolarization, the fertilization membrane lifted, indicating that the slow block to polyspermy occurred independently of the fast block depolarization eggs (Jaffe, 1976).

Since the initial recordings of egg membrane depolarizations, the fast block has been observed in many external fertilizers, including amphibians, echinoderms, and teleost fish. In amphibians, this includes the clouded salamander *Hynobius nebulosus* (Iwao, 1989), the American toad *Bufo americanus* (Cross & Elinson, 1980), the northern leopard frog *Rana pipiens*

(Cross & Elinson, 1980; Cross, 1981), and the African clawed frog *Xenopus laevis* (Busa et al., 1985; Grey et al., 1982; Jaffe, Cross, & Picheral, 1983a; Webb & Nuccitelli, 1985). In echinoderms, the fast block has been observed in sea urchins (Jaffe, 1976; Whitaker & Steinhardt, 1983), starfish (Miyazaki & Hirai, 1979; Moccia et al., 2004), and sand dollars (Steinhardt, Lundin, & Mazia, 1971). Among teleost fish, eggs from the medaka *Oryzas latipes* (Nuccitelli, 1980) exhibit a membrane depolarization, although this is not necessary to prevent polyspermy and might suggest evolutionary differences in polyspermy blocks (Nuccitelli, 1980). Notably, the fast block in *X. laevis* represents the most thoroughly understood signaling pathway among the amphibian species studied (Cross & Elinson, 1980; Glahn & Nuccitelli, 2003; Kline & Nuccitelli, 1985; Kline et al., 1991; Komondor et al., 2023; Tembo et al., 2019; Webb & Nuccitelli, 1985; Wozniak et al., 2018a, 2018a).

To explore whether the fast block exists in mammalian eggs, whole-cell recordings were made to determine if fertilization triggered a depolarization of the membrane or if altering the membrane potential of the eggs could alter the egg's ability to prevent fertilization by multiple sperm. Experiments performed in mice (Jaffe, Sharp, & Wolf, 1983b), rabbits (McCulloh, Rexroad, & Levitan, 1983), and hamsters (Miyazaki & Igusa, 1981, 1982) do not support the existence of a membrane potential-reliant polyspermy-blocking mechanism. These data suggest that mammalian eggs do not use the fast block.

6. Fast block mechanism

The exact mechanism by which egg membrane depolarization prevents sperm entry is not fully understood. However, cross-fertilization experiments suggest that a sperm-derived factor plays a crucial role. A 1983 study demonstrated that the voltage required to prevent sperm entry depends on the species of sperm rather than the egg (Jaffe et al., 1983a). In these experiments, *X. laevis* eggs clamped at voltages that blocked *X. laevis* sperm did not prevent fertilization by eastern newt (*Notophthalmus viridescens*) sperm. Similarly, when *Notophthalmus* sperm and *Rana pipiens* eggs were used, sperm was only blocked at voltages typical of *Notophthalmus* egg depolarization (Jaffe et al., 1983a). These findings suggest that sperm recognize species-specific membrane potentials, indicating a role for sperm in polyspermy prevention.

Further comparative studies on different amphibians support the role of fertilization-triggered depolarization as a polyspermy block. Eggs from polyspermic species, such as *Pleurodeles* and *Ambystoma*, do not depolarize upon sperm entry and consistently exhibit polyspermic fertilizations, even with monospermic species' sperm like *Rana temporaria* (Charbonneau et al., 1983). This research suggests that eggs from monospermic species actively use depolarization to block additional sperm, whereas polyspermic species do not rely on this mechanism (Charbonneau et al., 1983).

7. Signaling pathway of the fast block in *X. laevis*

To understand the signaling pathway that triggers the fast block to polyspermy, scientists first needed to identify the ion responsible for the depolarization. Depolarization of a cell membrane occurs when there is a change in the cell's membrane potential, making it less negative (Hille, 2001). This process is typically achieved through the movement of ions across the cell membrane. When positively charged ions such as Na^+ or Ca^{2+} flow into the cell, or negatively charged ions such as Cl^- flow out of the cell, the membrane potential becomes less negative (Bainbridge & Carlson, 2023). These ionic movements are facilitated by ion channels, which open in response to specific signals, allowing ions to pass through and alter the membrane potential. This change is crucial for various cellular processes, including the fast block to polyspermy during fertilization. Notably, different animal species use various ions to mediate this process.

In *X. laevis* eggs, the fast block to polyspermy is mediated by an efflux of Cl^-. Fertilization recordings of *X. laevis* eggs conducted in solutions with varying Cl^- concentrations revealed that reducing extracellular Cl^- resulted in larger depolarizations, as expected for a Cl^- conducted current, due to the increased driving force behind Cl^- movement out of the egg. Conversely, increasing extracellular Cl^- led to smaller depolarizations (Grey et al., 1982; Webb & Nuccitelli, 1985). Scientists leveraged the ability of Cl^- channels to conduct other halides (Wright & Diamond, 1977) and found that when Cl^- was replaced with other halides like I^- and Br^-, fertilization did not change the membrane potential in Br^-, whereas it triggered hyperpolarization in I^--enriched solutions (Grey et al., 1982; Watabe et al., 2019). These conditions also led to polyspermic embryos (Grey et al., 1982; Watabe et al., 2019). These findings demonstrate that fertilization opens a Cl^- channel in *X. laevis* eggs during the fast block to polyspermy.

Fertilization triggers an increase in intracellular Ca^{2+} to trigger the Cl^- current that depolarizes *X. laevis* eggs during the fast block (Grey et al., 1982; Kline, 1988). A requirement for elevated intracellular Ca^{2+} was demonstrated in experiments using electrophysiology recordings on *X. laevis* eggs during fertilization, that had been loaded with the Ca^{2+} chelator BAPTA (Kline, 1988). BAPTA binds Ca^{2+} with a high affinity (Tsien, 1980) and would, therefore, quickly bind labile Ca^{2+} within the egg to prevent an elevation of cytoplasmic Ca^{2+}. Fertilization did not trigger a depolarization in these BAPTA-loaded eggs, and these eggs displaced an increased incidence of polyspermy (Kline, 1988). Additional experiments documented that applying a Ca^{2+} ionophore to *X. laevis* eggs triggered a depolarization (Grey et al., 1982). These data demonstrated that elevated intracellular Ca^{2+} is both necessary and sufficient for triggering the fast block in *X. laevis* eggs. These findings gave rise to the hypothesis that a Ca^{2+}-activated Cl^- channel mediates the depolarizing Cl^- current in *X. laevis* eggs (Hartzell et al., 2009).

To uncover the molecular identity of the Cl^- channel responsible for the fast block to polyspermy in *X. laevis*, published proteomics and RNA-sequencing datasets were interrogated (Wozniak et al., 2018b). Two candidate Cl^- channels emerged that both localized to the plasma membrane and could potentially conduct the depolarizing Cl^- current: TransMEMbrane Member 16 A (TMEM16A) and Bestrophin 2 A (BEST2A) (Wozniak et al., 2018a). Notably, both TMEM16A and BEST2A are Ca^{2+}-activated Cl^- channels (Caputo et al., 2008; Kane Dickson, Pedi, & Long, 2014; Nguyen & Chen, 2024; Owji et al., 2021; Qu & Hartzell, 2004; Schroeder et al., 2008; Yang et al., 2008). Moreover, both TMEM16A and BEST2A channels are present in fertilization-competent *X. laevis* eggs.

After identifying two candidate Cl^- channels, Wozniak and colleagues exogenously expressed BEST2A in axolotl (*Ambystoma mexicanum*) oocytes and performed two-electrode voltage clamp electrophysiology recordings. Axolotl oocytes were chosen because these cells lack other Ca^{2+}-activated currents (Schroeder et al., 2008). Using a photolabile caged-IP_3, axolotl oocytes expressing either TMEM16A or BEST2A exhibited large currents with light application, substantiating that both channels act as Ca^{2+}-activated Cl^- channels. Another research group has previously demonstrated similar results for the exogenous expression of TMEM16A in axolotl oocytes (Schroeder et al., 2008).

Using the axolotl oocytes expressing either TMEM16A or BEST2a, various inhibitors were screened to identify compounds capable of discriminating currents conducted by the two channels to determine then which channel is used during the fast block. Ani9 and MONNA significantly inhibited TMEM16A-conducted currents of the five inhibitors screened while nominally acting on BEST2A channels (Wozniak et al., 2018b). When Ani9 and MONNA were used during whole-cell recordings during fertilization of *X. laevis* eggs, there was a significant slowing or complete lack of depolarization in each inhibitor, respectively, and polyspermic development occurred. These findings demonstrate that fertilization signals the Ca^{2+}-activated Cl^- channel TMEM16A opening, which conducts an efflux of Cl^- to depolarize *X. laevis* eggs during the fast block to polyspermy (Fig. 1) (Wozniak et al., 2018b).

Additional experiments uncovered the source of Ca^{2+} that activates TMEM16A. To do so, Wozniak et al. examined whether *X. laevis* eggs need Ca^{2+} to initiate the fast block. Using Ca^{2+} channel inhibitors, they examined whether the fast block depolarization rate and incidence of polyspermy were different relative to control conditions. This study established that IP_3 is necessary for membrane depolarization as the fast block did not occur when using IP_3 receptor inhibitors during whole-cell recordings (Wozniak et al., 2018c). In addition to needing functional IP_3 receptors, phospholipase C (PLC) is also required for the fast block, as inhibiting it prevents the fast block and led to polyspermy (Fig. 1) (Komondor et al., 2023; Wozniak et al., 2018c). This research showed that the elevated Ca^{2+} that is responsible for activating TMEM16A in *X. laevis* eggs, which has been confirmed by multiple research groups (Grey et al., 1982; Kline, 1988; Nuccitelli, Yim, & Smart, 1993; Wagner et al., 1998; Watabe et al., 2019; Wozniak et al., 2018b, 2018c), comes from a PLC/IP_3-mediated cytoplasmic Ca^{2+} increase. Early experiments examined the speed and concentration of Ca^{2+} released at fertilization in *X. laevis* eggs. They showed that the Ca^{2+} is rapidly released in a wave around the egg from the point of sperm entry at a rate as quick as 17.2 mM/s with a concentration as high as 1.2 mM (Fontanilla & Nuccitelli, 1998).

During the fast block to polyspermy in *X. laevis*, fertilization triggers the activation of PLC, which cleaves PIP_2 into its two secondary messengers, diacylglycerol (DAG) and IP_3. IP_3 binds to the IP_3 receptor in the ER, initiating an increase in cytoplasmic Ca^{2+} that activates TMEM16A in the egg membrane. TMEM16A activation causes an efflux of Cl^- that depolarizes the egg membrane to prevent further sperm entry (Fig. 1).

Different PLC subtypes are activated by different signaling pathways. Therefore, to uncover how fertilization activates PLC, it is useful to understand which PLCs are present. To identify which PLC isoforms are present in the *X. laevis* eggs, Komondor and colleagues interrogated two previously published

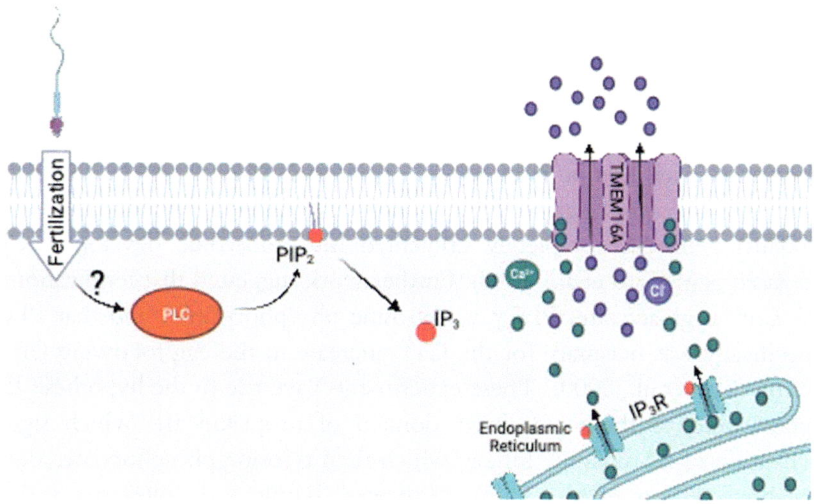

Fig. 1 *The fast block to polyspermy pathway in X. laevis.* Fertilization activates phospholipase C (PLC) through an unknown mechanism. PLC cleaves phosphatidylinositol 4,5-bisphosphate (PIP_2) to create the second messenger inositol trisphosphate (IP_3). IP_3 binds to the IP_3 receptor in the endoplasmic reticulum (ER) to release Ca^{2+} from intracellular stores. Ca^{2+} activates TransMEMbrane 16A (TMEM16A), which releases Cl^- to depolarize the membrane and prevent further sperm entries in the egg.

high-throughput expression datasets (Komondor et al., 2023): the proteome of fertilization-competent eggs (Wuhr et al., 2014) and RNA-sequencing dataset from developing oocytes (Session et al., 2016). These datasets were queried for all proteins encoded by known PLC genes and three PLC proteins are present in the egg: PLCγ1 (encoded by the PLCG1 gene), PLCβ1 and PLCβ3 (encoded by the PLCB1 and PLCB3 genes, respectively). PLCγ1 was the most abundant in the *X. laevis* egg, present at 85.2 nM, an approximately 20-fold higher concentration than either PLCβ1 (3.9 nM), or PLCβ3 (4.3 nM) (Komondor et al., 2023). An RNA-sequencing dataset acquired during different stages of development of *X. laevis* oocytes and eggs (Session et al., 2016) was used as a control, reasoning that if these three PLC proteins are present in the fertilization-competent eggs, the RNA encoding these enzymes should be present in the developing gamete. Indeed, mRNA for all three PLC types was present in the developing oocytes. The PLCγ enzyme is typically activated by phosphorylation of a critical tyrosine residue (Y776 in *X. laevis*) near its catalytic domain (Gresset, Sondek, & Harden, 2012). In contrast, PLCβ isoforms are activated by G-protein pathways requiring members of the $G\alpha_{q/11}$ family (Gresset et al., 2012).

Conflicting data has been published regarding whether fertilization in X. laevis eggs, triggers canonical activation of PLCγ or PLCβ. For example, by imaging cytoplasmic Ca^{2+}, Glahn and colleagues reported that injecting *X. laevis* eggs with tyrosine kinase inhibitors lavendustin A or tyrphostin B46 stopped the fertilization evoked Ca^{2+} wave (Glahn et al., 1999). Similarly, Sato and colleagues reported that inhibiting tyrosine phosphorylation with 100 μM genistein completely prevented the embryonic development of *X. laevis* eggs (Sato et al., 1998). Further work suggested that fertilization in *X. laevis* eggs activates PLCγ via tyrosine phosphorylation and that PLCγ specifically was necessary for the Ca^{2+} increase in the egg following fertilization (Sato et al., 2000). These experiments gave rise to the hypothesis that sperm bind to the extracellular domain of uroplakin III, which signals activation of a Src family kinase, which then tyrosine phosphorylates PLCγ (Mahbub Hasan et al., 2005). However, Runft and colleagues imaged cytoplasmic Ca^{2+} in *X. laevis* eggs. They reported that stopping either PLCγ activity by overexpressing the Src-homology 2 (SH2) domain did not interfere with the fertilization-signaled increase of cytoplasmic Ca^{2+} (Runft, Watras, & Jaffe, 1999). Similarly, Komondor et al. reported that inhibiting tyrosine phosphorylation with 100 μM genistein or 100 nM dasantinib, did not stop a fertilization-signaled depolarization. However, both inhibitors effectively inhibited the phosphorylation of the critical active-site tyrosine in oocytes exogenously expressing the receptor tyrosine kinase for the platelet-derived growth factor. Moreover, western blots probing for phosphorylated Y776 in PLCγ1 suggested that this residue is not phosphorylated at fertilization. Although fertilization may activate PLCγ1 by a mechanism independent of the phosphorylation of the critical tyrosine residue, these data suggest that a pathway involving the canonical activation of PLCγ1 is likely not required.

Additional experiments probed for a fertilization-signaled activation of PLCβ. Typically, PLCβ isoforms are activated by G-protein pathways requiring members of the $G\alpha_{q/11}$ family (Rhee, 2001; Smrcka et al., 1991). *X. laevis* eggs express the $G\alpha_{11}$ subunit, capable of activating PLCβ (Komondor et al., 2023). For example, Runft and colleagues loaded *X. laevis* eggs with calcium green dextran to observe the fertilization-signaled Ca^{2+} wave (Runft et al., 1999). They reported that stopping PLCβ activation with a $G\alpha_{q/11}$ targeting antibody did not disrupt the fertilization-signaled elevation of intracellular Ca^{2+} (Runft et al., 1999). Komondor et al. reported that using YM-254890 to inhibit the $G\alpha_{11}$-mediated activation of PLCβ, did not alter the fast block. Specifically, they inseminated *X. laevis* eggs in the presence of 10 μM YM-254890; using whole-cell recordings, they reported normal fertilization-signaled depolarizations and monospermic fertilizations (Komondor et al., 2023). Altogether, these data suggest that neither tyrosine phosphorylation of PLCγ nor $G\alpha_{q/11}$ activation of PLCβ is needed for the fast block.

How fertilization activates a PLC to trigger the fast block is yet to be determined. However, a few possibilities remain. For example, it is possible that the *X. laevis* sperm release a PLC in the egg after fertilization. Indeed, the soluble and sperm-derived PLCζ is transferred to the egg during fertilization in birds and mammals (Bedford-Guaus et al., 2011; Cox et al., 2002; Ito et al., 2008; Mizushima et al. 2007, 2014; Ross et al., 2008; Saunders et al., 2002; Yoneda et al., 2006). Notably, the *X. laevis* genome lacks an annotated PLCZ1, the gene encoding PLCζ. It is possible that a different sperm-derived PLC is released into the egg. Alternatively, it is possible that a different mechanism at fertilization activates an egg-derived PLC.

8. Concluding remarks

In this review, we have explored the intricacies of fertilization and the fast block to polyspermy in *X. laevis*, tracing the historical developments and key discoveries that have shaped our current understanding. The pioneering work of researchers like Ernest Everett Just and Laurinda Jaffe laid the foundation for elucidating the mechanisms underlying the fast block. Studies on *X. laevis* have demonstrated that fertilization triggers a depolarization of the egg membrane mediated by an efflux of Cl^- through the Ca^{2+}-activated Cl^- channel TMEM16A, effectively preventing polyspermy. Despite significant advances, questions remain about the precise molecular interactions and signaling pathways that activate PLC and initiate the fast block. Addressing these questions will not only enhance our comprehension of the earliest events in embryogenesis but also provide broader insights into the voltage-dependent mechanisms of fertilization across species. The continued use of *X. laevis* as a model organism will undoubtedly yield further breakthroughs in this fundamental area of reproductive biology.

References

al-Anzi, B., & Chandler, D. E. (1998). A sperm chemoattractant is released from Xenopus egg jelly during spawning. *Developmental Biology, 198*(2), 366–375.

Babcock, R., et al. (1992). Predictable and unpredictable spawning events: In situ behavioural data from free-spawning coral reef invertebrates. *Invertebrate Reproduction & Development, 22*(1-3), 213–227.

Bainbridge, R. E., & Carlson, A. E. (2023). Ion channels in sperm and eggs. In J. Zheng, & M. C. Trudeau (Eds.). *Handbook of ion channels*. Boca Raton, FL: CRC Press.

Barton, J. A., Willis, B. L., & Hutson, K. S. (2017). Coral propagation: A review of techniques for ornamental trade and reef restoration. *Reviews in Aquaculture, 9*(3), 238–256.

Bedford-Guaus, S. J., et al. (2011). Molecular cloning and characterization of phospholipase C zeta in equine sperm and testis reveals species-specific differences in expression of catalytically active protein. *Biology of Reproduction, 85*(1), 78–88.

Berg, L. K., & Wessel, G. M. (1997). Cortical granules of the sea urchin translocate early in oocyte maturation. *Development (Cambridge, England), 124*(9), 1845–1850.

Bernard, I., et al. (2016). In situ spawning in a marine broadcast spawner, the Pacific oyster Crassostrea gigas: Timing and environmental triggers. *Limnology and Oceanography, 61*(2), 635–647.

Bhatt, M., et al. (2022). The "www" of Xenopus laevis oocytes: The why, when, what of Xenopus laevis oocytes in membrane transporters research. *Membranes (Basel), 12*(10).

Bianchi, E., et al. (2014). Juno is the egg Izumo receptor and is essential for mammalian fertilization. *Nature, 508*(7497), 483–487.

Burnett, L. A., et al. (2008). Crisp proteins and sperm chemotaxis: Discovery in amphibians and explorations in mammals. *The International Journal of Developmental Biology, 52*(5-6), 489–501.

Burnett, L. A., et al. (2011). Egg jelly proteins stimulate directed motility in Xenopus laevis sperm. *Molecular Reproduction and Development, 78*(6), 450–462.

Busa, W. B., et al. (1985). Activation of frog (Xenopus laevis) eggs by inositol trisphosphate. I. Characterization of Ca2+ release from intracellular stores. *The Journal of Cell Biology, 101*(2), 677–682.

Caputo, A., et al. (2008). TMEM16A, a membrane protein associated with calcium-dependent chloride channel activity. *Science (New York, N. Y.), 322*(5901), 590–594.

Charbonneau, M., et al. (1983). Fertilization of amphibian eggs: A comparison of electrical responses between anurans and urodeles. *Developmental Biology, 98*(2), 304–318.

Cox, L. J., et al. (2002). Sperm phospholipase Czeta from humans and cynomolgus monkeys triggers Ca2+ oscillations, activation and development of mouse oocytes. *Reproduction (Cambridge, England), 124*(5), 611–623.

Coy, P., et al. (2012). Roles of the oviduct in mammalian fertilization. *Reproduction (Cambridge, England), 144*(6), 649–660.

Crimaldi, J. P., & Zimmer, R. K. (2014). The physics of broadcast spawning in benthic invertebrates. *Annual Review of Marine Science, 6*(1), 141–165.

Cross, N. L., & Elinson, R. P. (1980). A fast block to polyspermy in frogs mediated by changes in the membrane potential. *Developmental Biology, 75*(1), 187–198.

Cross, N. L. (1981). Initiation of the activation potential by an increase in intracellular calcium in eggs of the frog, Rana pipiens. *Developmental Biology, 85*(2), 380–384.

Deneke, V. E., & Pauli, A. (2021). The fertilization enigma: How sperm and egg fuse. *Annual Review of Cell and Developmental Biology, 37*, 391–414.

Deneke, V. E., et al. (2023). A conserved fertilization complex of Izumo1, Spaca6, and Tmem81 mediates sperm-egg interaction in vertebrates. *bioRxiv* p. 2023.07.27.550750.

Eads, A. R., Evans, J. P., & Kennington, W. J. (2016). Plasticity of fertilization rates under varying temperature in the broadcast spawning mussel, Mytilus galloprovincialis. *Ecology and evolution, 6*(18), 6578–6585.

Elinson, R. P. (1986). Fertilization in amphibians: The ancestry of the block to polyspermy. *International Review of Cytology, 101*, 59–100.

Elkan, E. R. (1938). The xenopus pregnancy test. *British Medical Journal, 2*(4067), 1253–1274 2.

Evans, J. P., & Sherman, C. D. H. (2013). Sexual selection and the evolution of egg-sperm interactions in broadcast-spawning invertebrates. *The Biological Bulletin, 224*(3), 166–183.

Evans, J. P., & Lymbery, R. A. (1813). Sexual selection after gamete release in broadcast spawning invertebrates. *Philosophical Transactions of the Royal Society of London. Series B, Biological Sciences, 2020*(375), 20200069.

Evans, J. P. (2020). Preventing polyspermy in mammalian eggs-Contributions of the membrane block and other mechanisms. *Molecular Reproduction and Development, 87*(3), 341–349.

Fahrenkamp, E., Algarra, B., & Jovine, L. (2020). Mammalian egg coat modifications and the block to polyspermy. *Molecular Reproduction and Development, 87*(3), 326–340.

Fisher, M., et al. (2023). Xenbase: Key features and resources of the Xenopus model organism knowledgebase. *Genetics, 224*(1).

Fontanilla, R. A., & Nuccitelli, R. (1998). Characterization of the sperm-induced calcium wave in Xenopus eggs using confocal microscopy. *Biophysical Journal, 75*(4), 2079–2087.

Gilbert, S. F. (2003). *Developmental biology* (7th ed.). Sunderland, MA: Sinauer Associates838 (xvii).

Glahn, D., & Nuccitelli, R. (2003). Voltage-clamp study of the activation currents and fast block to polyspermy in the egg of Xenopus laevis. *Development, Growth & Differentiation, 45*(2), 187–197.

Glahn, D., et al. (1999). Tyrosine kinase inhibitors block sperm-induced egg activation in Xenopus laevis. *Developmental Biology, 205*(1), 171–180.

Glese, A. C., & Kanatani, H. (1987). Maturation and spawning. In A. C. Giese, J. S. Pearse, & V. B. Pearse (Eds.). *Roproduction of marine invertebrates*. Palo Alto: Blackwell Scientific Press.

Gresset, A., Sondek, J., & Harden, T. K. (2012). The phospholipase C isozymes and their regulation. *Sub-Cellular Biochemistry, 58*, 61–94.

Grey, R. D., et al. (1982). An electrical block is required to prevent polyspermy in eggs fertilized by natural mating of Xenopus laevis. *Developmental Biology, 89*(2), 475–484.

Gurdon, J. B., Elsdale, T. R., & Fischberg, M. (1958). Sexually mature individuals of Xenopus laevis from the transplantation of single somatic nuclei. *Nature, 182*(4627), 64–65.

Gurdon, J. B. (1962). The developmental capacity of nuclei taken from intestinal epithelium cells of feeding tadpoles. *Journal of Embryology and Experimental Morphology, 10*, 622–640.

Gurdon, J. B. (2013). The egg and the nucleus: A battle for supremacy. *Development (Cambridge, England), 140*(12), 2449–2456.

Hartzell, H. C., et al. (2009). *Anoctamin/TMEM16 family members are Ca^{2+}-activated Cl^- channels. The Journal of Physiology, 587*(Pt 10), 2127–2139.

Hassold, T., et al. (1980). A cytogenetic study of 1000 spontaneous abortions. *Annals of Human Genetics, 44*(2), 151–178.

Heasman, J., Holwill, S., & Wylie, C. C. (1991). Fertilization of cultured Xenopus oocytes and use in studies of maternally inherited molecules. *Methods in Cell Biology, 36*, 213–230.

Hemmings, N., & Birkhead, T. R. (1818). Polyspermy in birds: Sperm numbers and embryo survival. *Proceedings of the Royal Society B: Biological Sciences, 2015*(282), 20151682.

Heyward, A. J., & Negri, A. P. (2010). Plasticity of larval pre-competency in response to temperature: Observations on multiple broadcast spawning coral species. *Coral Reefs, 29*, 631–636.

Hille, B. (2001). *Ion channels of excitable membranes* (3rd ed.). Sinauer Associates.

Ito, M., et al. (2008). Difference in Ca2+ oscillation-inducing activity and nuclear translocation ability of PLCZ1, an egg-activating sperm factor candidate, between mouse, rat, human, and medaka fish. *Biology of Reproduction, 78*(6), 1081–1090.

Iwao, Y. (1989). An electrically mediated block to polyspermy in the primitive urodele Hynobius nebulosus and phylogenetic comparison with other amphibians. *Developmental Biology, 134*(2), 438–445.

Jaffe, L. A., Sharp, A. P., & Wolf, D. P. (1983b). Absence of an electrical polyspermy block in the mouse. *Developmental Biology, 96*(2), 317–323.

Jaffe, L. A. (1976). Fast block to polyspermy in sea urchin eggs is electrically mediated. *Nature, 261*(5555), 68–71.

Jaffe, L. A., Cross, N. L., & Picheral, B. (1983a). Studies of the voltage-dependent polyspermy block using cross-species fertilization of amphibians. *Developmental Biology, 98*(2), 319–326.

Jaffe, L. F. (1983). Sources of calcium in egg activation: A review and hypothesis. *Developmental Biology, 99*(2), 265–276.

Just, E. (1919). The fertilization reaction in Echinarachnius parma: I. Cortical response of the egg to insemination. *The Biological Bulletin, 36*(1), 1–10.

Kane Dickson, V., Pedi, L., & Long, S. B. (2014). Structure and insights into the function of a Ca2+-activated Cl- channel. *Nature, 516*(7530), 213–218.

Kline, D., & Nuccitelli, R. (1985). The wave of activation current in the Xenopus egg. *Developmental Biology, 111*(2), 471–487.

Kline, D. (1988). Calcium-dependent events at fertilization of the frog egg: Injection of a calcium buffer blocks ion channel opening, exocytosis, and formation of pronuclei. *Developmental Biology, 126*(2), 346–361.

Kline, D., et al. (1991). Evidence for the involvement of a pertussis toxin-insensitive G-protein in egg activation of the frog, Xenopus laevis. *Developmental Biology, 143*(2), 218–229.

Komondor, K. M., et al. (2023). TMEM16A activation for the fast block to polyspermy in the African clawed frog does not require conventional activation of egg PLCs. *The Journal of General Physiology, 155*(10).

Levitan, D. R. (2020). *The ecology of fertilization in free-spawning invertebrates. Ecology of marine invertebrate larvae.* CRC Press 123–156.

Limatola, N., Chun, J. T., & Santella, L. (2022). Regulation of the actin cytoskeleton-linked Ca(2+) signaling by intracellular pH in fertilized eggs of sea urchin. *Cells, 11*(9).

Liu, M. (2011). The biology and dynamics of mammalian cortical granules. *Reproductive Biology and Endocrinology: RB&E, 9*, 149.

Maeno, T. (1959). Electrical characteristics and activation potential of Bufo eggs. *The Journal of General Physiology, 43*(1), 139–157.

Mahbub Hasan, A. K., et al. (2005). Uroplakin III, a novel Src substrate in Xenopus egg rafts, is a target for sperm protease essential for fertilization. *Developmental Biology, 286*(2), 483–492.

Malmros, H. (1948). The South African frog Xenopus laevis as a test object in hormone analyses. *Acta Medica Scandinavica, 131*(206), 512–515.

Marshall, D. J., & Bolton, T. F. (2007). Sperm release strategies in marine broadcast spawners: The costs of releasing sperm quickly. *Journal of Experimental Biology, 210*(21), 3720–3727.

McCulloh, D. H., Rexroad, C. E., Jr., & Levitan, H. (1983). Insemination of rabbit eggs is associated with slow depolarization and repetitive diphasic membrane potentials. *Developmental Biology, 95*(2), 372–377.

Mercier, A., Ycaza, R. H., & Hamel, J. F. (2007). Long-term study of gamete release in a broadcast-spawning holothurian: Predictable lunar and diel periodicities. *Marine Ecology Progress Series, 329*, 179–189.

Miyazaki, S., & Hirai, S. (1979). Fast polyspermy block and activation potential. Correlated changes during oocyte maturation of a starfish. *Developmental Biology, 70*(2), 327–340.

Miyazaki, S., & Igusa, Y. (1982). Ca-mediated activation of a K current at fertilization of golden hamster eggs. *Proceedings of the National Academy of Sciences of the United States of America, 79*(3), 931–935.

Miyazaki, S., & Igusa, Y. (1981). Fertilization potential in golden hamster eggs consists of recurring hyperpolarizations. *Nature, 290*(5808), 702–704.
Mizushima, S., et al. (2007). Possible role of calcium on oocyte development after intracytoplasmic sperm injection in quail (Coturnix japonica). *Journal of Experimental Zoology Part A: Ecological Genetics and Physiology, 307*(11), 647–653.
Mizushima, S., et al. (2014). The birth of quail chicks after intracytoplasmic sperm injection. *Development (Cambridge, England), 141*(19), 3799–3806.
Mizushima, S. (2017). Fertilization 2: Polyspermic fertilization. *Advances in Experimental Medicine and Biology, 1001*, 105–123.
Moccia, F., et al. (2004). NAADP triggers the fertilization potential in starfish oocytes. *Cell Calcium, 36*(6), 515–524.
Monteiro, C. A., et al. (2016). Temporal windows of reproductive opportunity reinforce species barriers in a marine broadcast spawning assemblage. *Scientific Reports, 6*(1), 29198.
Morgan, S. G. (2020). *The timing of larval release. Ecology of marine invertebrate larvae.* CRC Press157–191.
Nguyen, D. M., & Chen, T. Y. (2024). Structure and function of calcium-activated chloride channels and phospholipid scramblases in the TMEM16 family. *Handbook of Experimental Pharmacology, 283*, 153–180.
Nieuwkoop, P. D., & Faber, J. (1994). *Normal table of Xenopus laevis (Daudin): A systematical and chronological survey of the development from the fertilized egg till the end of metamorphosis.*
Nishio, S., et al. (2023). Architecture of the vertebrate egg coat and structural basis of the ZP2 block to polyspermy. *bioRxiv* p. 2023.06.21.544075.
Nuccitelli, R., Yim, D. L., & Smart, T. (1993). The sperm-induced Ca2+ wave following fertilization of the Xenopus egg requires the production of Ins(1, 4, 5)P3. *Developmental Biology, 158*(1), 200–212.
Nuccitelli, R. (1980). The fertilization potential is not necessary for the block to polyspermy or the activation of development in the medaka egg. *Developmental Biology, 76*(2), 499–504.
Nuwer, R. (2013). Doctors used to use live African frogs as pregnancy tests. *Smithsonian Magazine.*
Oliver, J. A., & Miller, M. F. (1946). The frog test in the diagnosis of early pregnancy. *Arizona Medicine, 3*, 31–35.
Owji, A. P., et al. (2021). Structure and Function of the Bestrophin family of calcium-activated chloride channels. *Channels (Austin), 15*(1), 604–623.
Ponder, W. F., Lindberg, D. R., & Ponder, J. M. (2019). *Biology and evolution of the mollusca, Vol. 1.* CRC Press.
Qu, Z., & Hartzell, C. (2004). Determinants of anion permeation in the second transmembrane domain of the mouse bestrophin-2 chloride channel. *The Journal of General Physiology, 124*(4), 371–382.
Rhee, S. G. (2001). Regulation of phosphoinositide-specific phospholipase C. *Annual Review of Biochemistry, 70*, 281–312.
Rojas, J., et al. (2021). Knockin' on egg's door: Maternal control of egg activation that influences cortical granule exocytosis in animal species. *Frontiers in Cell and Developmental Biology, 9*, 704867.
Ross, P. J., et al. (2008). Parthenogenetic activation of bovine oocytes using bovine and murine phospholipase C zeta. *BMC Developmental Biology, 8*, 16.
Runft, L. L., Watras, J., & Jaffe, L. A. (1999). *Calcium release at fertilization of Xenopus eggs requires type I IP_3 receptors, but not SH2 domain-mediated activation of PLCγ or G_q-mediated activation of PLCβ. Developmental Biology, 214*(2), 399–411.
Sanders, C. B. (1946). Frog test (Xenopus laevis) for pregnancy. *Texas State Journal of Medicine, 42*, 375.

Sato, K., et al. (1998). Involvement of protein-tyrosine phosphorylation and dephosphorylation in sperm-induced Xenopus egg activation. *FEBS Letters, 424*(1–2), 113–118.

Sato, K., et al. (2000). Tyrosine kinase-dependent activation of phospholipase Cgamma is required for calcium transient in Xenopus egg fertilization. *Developmental Biology, 224*(2), 453–469.

Saunders, C. M., et al. (2002). PLC zeta: A sperm-specific trigger of Ca(2+) oscillations in eggs and embryo development. *Development (Cambridge, England), 129*(15), 3533–3544.

Schroeder, B. C., et al. (2008). Expression cloning of TMEM16A as a calcium-activated chloride channel subunit. *Cell, 134*(6), 1019–1029.

Session, A. M., et al. (2016). Genome evolution in the allotetraploid frog Xenopus laevis. *Nature, 538*(7625), 336–343.

Sive, H. L., Grainger, R. M., & Harland, R. M. (2010). *Early development of Xenopus laevis: A laboratory manual* (1st ed.). Cold Spring Harbor Laboratory Press338.

Skinner, M. K. (2018). (2nd ed.). *Encyclopedia of reproduction Vol. 6*. Amsterdam; Boston: Elsevier, Academic Press.

Smrcka, A. V., et al. (1991). Regulation of polyphosphoinositide-specific phospholipase C activity by purified Gq. *Science (New York, N. Y.), 251*(4995), 804–807.

Snook, R. R., Hosken, D. J., & Karr, T. L. (2011). The biology and evolution of polyspermy: Insights from cellular and functional studies of sperm and centrosomal behavior in the fertilized egg. *Reproduction (Cambridge, England), 142*(6), 779.

Steinhardt, R. A., Lundin, L., & Mazia, D. (1971). Bioelectric responses of the echinoderm egg to fertilization. *Proceedings of the National Academy of Sciences of the United States of America, 68*(10), 2426–2430.

Suarez, S. S., & Pacey, A. A. (2006). Sperm transport in the female reproductive tract. *Human Reproduction Update, 12*(1), 23–37.

Sugiyama, H., et al. (2009). Purification and multimer formation of allurin, a sperm chemoattractant from Xenopus laevis egg jelly. *Molecular Reproduction and Development, 76*(6), 527–536.

Swanson, W. J., & Vacquier, V. D. (2002). The rapid evolution of reproductive proteins. *Nature Reviews. Genetics, 3*(2), 137–144.

Tembo, M., et al. (2021). Actin polymerization is not required for the fast block to polyspermy in the African clawed frog, Xenopus laevis. *microPublication Biology,* 2021.

Tembo, M., et al. (2019). Phosphatidylinositol 4,5-bisphosphate (PIP2) and Ca(2+) are both required to open the Cl(-) channel TMEM16A. *The Journal of Biological Chemistry, 294*(33), 12556–12564.

Tinsley, R., Loumont, C., & Kobel, H. (1996). Geographical distribution and ecology. *The Biology of Xenopus,* 35–60.

Toft-Bertelsen, T. L., Krizaj, D., & MacAulay, N. (2017). When size matters: Transient receptor potential vanilloid 4 channel as a volume-sensor rather than an osmo-sensor. *The Journal of Physiology, 595*(11), 3287–3302.

Tokmakov, A. A., et al. (2014). Calcium signaling and meiotic exit at fertilization in Xenopus egg. *International Journal of Molecular Sciences, 15*(10), 18659–18676.

Tsien, R. Y. (1980). New calcium indicators and buffers with high selectivity against magnesium and protons: Design, synthesis, and properties of prototype structures. *Biochemistry, 19*(11), 2396–2404.

Wagner, J., et al. (1998). Simulation of the fertilization Ca2+ wave in Xenopus laevis eggs. *Biophysical Journal, 75*(4), 2088–2097.

Watabe, M., et al. (2019). The electrical block to polyspermy induced by an intracellular Ca (2+) increase at fertilization of the clawed frogs, Xenopus laevis and Xenopus tropicalis. *Molecular Reproduction and Development, 86*(4), 387–403.

Webb, D. J., & Nuccitelli, R. (1985). Fertilization potential and electrical properties of the Xenopus laevis egg. *Developmental Biology, 107*(2), 395–406.

Wessel, G. M., & Wong, J. L. (2009). Cell surface changes in the egg at fertilization. *Molecular Reproduction and Development, 76*(10), 942–953.

Whitaker, M. (2006). Calcium at fertilization and in early development. *Physiological Reviews, 86*(1), 25–88.

Whitaker, M. J., & Steinhardt, R. A. (1983). Evidence in support of the hypothesis of an electrically mediated fast block to polyspermy in sea urchin eggs. *Developmental Biology, 95*(1), 244–248.

Wong, J. L., & Wessel, G. M. (2006). Defending the zygote: Search for the ancestral animal block to polyspermy. *Current Topics in Developmental Biology, 72*, 1–151.

Wootton, R. J., & Smith, C. P. (2014). *Reproductive biology of teleost fishes*. John Wiley & Sons.

Wozniak, K. L., & Carlson, A. E. (2020). Ion channels and signaling pathways used in the fast polyspermy block. *Molecular Reproduction and Development, 87*(3), 350–357.

Wozniak, K. L., et al. (2018a). PLC and IP3-evoked Ca(2+) release initiate the fast block to polyspermy in Xenopus laevis eggs. *The Journal of General Physiology*.

Wozniak, K. L., et al. (2018c). PLC and IP3-evoked Ca(2+) release initiate the fast block to polyspermy in Xenopus laevis eggs. *The Journal of General Physiology, 150*(9), 1239–1248.

Wozniak, K. L., et al. (2018a). The TMEM16A channel mediates the fast polyspermy block in Xenopus laevis. *The Journal of General Physiology*.

Wozniak, K. L., et al. (2018b). The TMEM16A channel mediates the fast polyspermy block in Xenopus laevis. *The Journal of General Physiology, 150*(9), 1249–1259.

Wozniak, K. L., et al. (2020). Zinc protection of fertilized eggs is an ancient feature of sexual reproduction in animals. *PLoS Biology, 18*(7), e3000811.

Wright, E. M., & Diamond, J. M. (1977). Anion selectivity in biological systems. *Physiological Reviews, 57*(1), 109–156.

Wuhr, M., et al. (2014). Deep proteomics of the Xenopus laevis egg using an mRNA-derived reference database. *Current Biology: CB, 24*(13), 1467–1475.

Wyrick, R. E., Nishihara, T., & Hedrick, J. L. (1974). Agglutination of jelly coat and cortical granule components and the block to polyspermy in the amphibian Xenopus laevis. *Proceedings of the National Academy of Sciences of the United States of America, 71*(5), 2067–2071.

Yamanaka, S., & Blau, H. M. (2010). Nuclear reprogramming to a pluripotent state by three approaches. *Nature, 465*(7299), 704–712.

Yang, Y. D., et al. (2008). TMEM16A confers receptor-activated calcium-dependent chloride conductance. *Nature, 455*(7217), 1210–1215.

Yoneda, A., et al. (2006). Molecular cloning, testicular postnatal expression, and oocyte-activating potential of porcine phospholipase Czeta. *Reproduction (Cambridge, England), 132*(3), 393–401.

CHAPTER SIX

No transcription, no problem: Protein phosphorylation changes and the transition from oocyte to embryo

Jonathon M. Thomalla[a,b] and Mariana F. Wolfner[a,*]
[a]Department of Molecular Biology and Genetics, Cornell University, Ithaca, NY, United States
[b]Department of Biomedical Sciences, Cornell University, College of Veterinary Medicine, Ithaca, NY, United States
*Corresponding author. e-mail address: mfw5@cornell.edu

Contents

Current Topics in Developmental Biology, Volume 162
ISSN 0070-2153, https://doi.org/10.1016/bs.ctdb.2025.01.001

Abstract

Although mature oocytes are arrested in a differentiated state, they are provisioned with maternally-derived macromolecules that will start embryogenesis. The transition to embryogenesis, called 'egg activation', occurs without new transcription, even though it includes major cell changes like completing stalled meiosis, translating stored mRNAs, cytoskeletal remodeling, and changes to nuclear architecture. In most animals, egg activation is triggered by a rise in free calcium in the egg's cytoplasm, but we are only now beginning to understand how this induces the egg to transition to totipotency and proliferation. Here, we discuss the model that calcium-dependent protein kinases and phosphatases modify the phosphorylation landscape of the maternal proteome to activate the egg. We review recent phosphoproteomic mass spectrometry analyses that revealed broad phospho-regulation during egg activation, both in number of phospho-events and classes of regulated proteins. Our interspecies comparisons of these proteins pinpoints orthologs and protein families that are phospho-regulated in activating eggs, many of which function in hallmark events of egg activation, and others whose regulation and activity warrant further study. Finally, we discuss key phospho-regulating enzymes that may act apically or as intermediates in the phosphorylation cascades during egg activation. Knowing the regulators, targets, and effects of phospho-regulation that cause an egg to initiate embryogenesis is crucial at both fundamental and applied levels for understanding female fertility, embryo development, and cell-state transitions.

1. Background, outstanding questions, and emerging research

1.1 What is egg activation?

Eggs emerge from the ovary as terminally differentiated cells: in the ovary, the oocyte (the egg's precursor) completes its differentiation and development and stalls in a taxon-specific stage of meiosis (e.g. Metaphase I in Drosophila (Fellmeth & McKim, 2022), Metaphase II in mouse (Kubiak, Weber, de Pennart, Winston, & Maro, 1993) and humans (Diluigi et al., 2008)). During its development, the oocyte is loaded with maternal factors, including proteins, RNAs, and organelles. In order for the oocyte to transition to a proliferating embryo, it must reinitiate meiosis, upregulate translation, and other cellular events. The oocyte-to-embryo transition, known as "egg activation" is universal and essential. Without it, the terminally-differentiated oocyte cannot transition to the totipotent "ultimate stem cell" that is the zygote, which will give rise to all cells and cell types of the embryo.

Understanding the mechanism of egg activation is important at the fundamental level for overarching questions like germ cell biology and how major changes in cell potential occur, and at a more granular level to understand how a cell can kickstart meiosis and translation, and rapidly modify its eggshell and cytoskeleton on cue. It is also of applied importance for human health and fertility: individuals whose oocytes cannot activate cannot currently be diagnosed for a failure in egg activation ("Oocyte Activation Deficiency") until in vitro fertilization (IVF) fails (Kashir, Ganesh, Jones, & Coward, 2022; Yestes, Jones, Amdani, Patel, & Coward, 2015). Knowing the molecules that mediate egg activation could provide diagnostic markers to identify these individuals, as well as markers to ensure that IVF conditions are optimal.

Despite its importance for early development, we know little about egg activation's mechanisms and molecules. The process is usually fast and often occurs within the female reproductive tract (Coy, García-Vázquez, Visconti, & Avilés, 2012; Heifetz, Yu, & Wolfner, 2001), making it difficult to observe. Moreover, egg activation relies on molecules and organelles that are already present in the oocyte rather than involving changes to gene expression. There is no new transcription during this transition; thus, transcriptomic analyses only reveal dynamics of existing pools of stored maternal mRNAs (Chousal et al., 2018; Vastenhouw, Cao, & Lipshitz, 2019). Despite the challenges that accompany the study of egg activation, fundamental players have been uncovered in several organisms. For example, frogs and marine invertebrates that produce large, or copious, oocytes permitted discovery of phosphorylated proteins during egg activation (i.e. meiotic exit) through biochemical assays. In species like mouse, genetic studies have allowed for testing essentiality of specific genes. In other systems, like Drosophila, synergized application of both biochemical and genetic approaches allowed for discovery of essential players and their mechanism of action during egg activation. In this review we highlight several species whose specific advantages have together contributed to our ongoing understanding of egg activation.

1.2 How does a cell/oocyte regulate cell state change without transcription?

Changes in cell state often rely on transcriptional alteration (Weinreb, Rodriguiez-Fraticelli, Camargo, & Klein, 2020; White, Rifkin, Hurba, & Hogness, 1999). However, oocytes lack the ability to actively transcribe, as their sole nucleus is locked in metaphase of meiosis in many species (Madgwick & Jones, 2007). In spite of this, activating oocytes undergo a dramatic change in state very quickly. The rapid and coordinated need to turn on or off the function of materials stored in the egg may explain the oocyte's bias away from transcription-based regulation.

When fully mature, the egg retains molecules that built and maintained it as an oocyte. It also contains molecules that will be needed for the initial steps of embryogenesis, and are presumably kept inactive in the mature oocyte. Parthenogenetically-reproducing species, and haplodiploid species, highlight the ability of maternally-stored proteins to facilitate development with no male contribution, even beyond certain stages of embryogenesis. For example, in haplodiploid insects such as wasps and bees, males develop from unfertilized activated eggs (Went, 1982).

Non-parthenogenetic species, like the mouse, can be made to undergo parthenogenesis by modifications to the egg: for example, adult mice can develop from unfertilized eggs by epigenetically modifying some genes to mimic normal imprinting. This was initially accomplished by mutagenesis of the H19 locus, a maternally-biased imprinted lncRNA, in reconstructed oocytes (Kono et al., 2004). More recently, an elegant study used catalytically-dead Cas9 fused to the DNA methyltransferase Dnmt3a and directed its activity to several imprinting control regions, in artificially-diploid oocytes (Wei, Yang, & Zhao, 2022). Intriguingly, mouse oocytes can switch to partially-parthenogenetic outcomes if their ploidy can match that of a fertilized egg: inhibition of polar body extrusion with actin depolymerizing drugs in metaphase of the 2nd meiotic division results in retaining the DNA that would usually be expelled and lost in a polar body (Kishigami & Wakayama, 2007; Ma et al., 2005; Surani & Barton, 1983). By this method, in artificially activated eggs and fertilized eggs in which the paternal DNA was removed, embryos can develop to blastocyst stage or day 10 of gestation. Importantly, these studies reveal that although the paternal genome is normally required for successful development, with a little help to overcome imprinting issues, the egg contains all the materials needed to drive development beyond zygotic genome activation (ZGA), including the potential to produce live offspring.

A recent study in Drosophila uncovered a genetic "recipe" to achieve consistent, albeit low, rates of parthenogenesis in the normally non-parthenogenetic *melanogaster* species. When centriole biogenesis, cell cycle, and proliferative factors are overexpressed in unfertilized, but activated, eggs, parthenogenetic development to adulthood increased from 0% to 1.4% (Sperling, Fabian, Garrison, & Glover, 2023). This excitingly showed that changing gene dosage for select proteins in the developing oocyte, and therefore raising their protein levels in the mature oocyte, is sufficient to induce parthenogenesis. Thus, maternal proteins alone can permit the initiation of embryo development.

Recent proteomic analyses in mouse revealed that egg proteins are remarkably long-lived, and that maternally-provided proteins remain stably detectable in eggs across generations (Harasimov et al., 2024). Additionally, protein levels were seen to remain stable during egg activation (Zhang et al., 2023). The stability is hypothesized to be accomplished by storing proteins on a cytoplasmic protein lattice containing PADI6; loss of PADI6 depletes levels of chromatin remodeling proteins, ribosomal proteins, and others, leading to embryonic arrest (Esposito et al., 2007; Jentoft et al., 2023). This dramatic protein longevity suggests that these stable proteins are competent to facilitate cellular events from oocyte maturation through ZGA, a stretch of time that can span decades in mammals, with egg activation serving as the major "switching point" between the terminally-differentiated oocyte and newly-reprogrammed zygote. Together, these data suggest that maternal proteins are able to survive long stretches without turnover, and raises questions about how this stability is maintained. While some proteome changes accompany oocyte maturation and egg activation in many species, degradation of oocyte factors or biosynthesis of embryonic-promoting factors are only part of the mechanism controlling early development. These studies strengthen the notion that oocytes/eggs come 'pre-packed' with the materials needed to manifest their own biological outcomes (or, at least, to support themselves through early development), even without male contributions.

2. Post-translational modifications as regulators of egg activation

Without available transcription, eggs are restricted in mechanisms available to transition from oocyte to embryo. Post-transcriptional regulation of molecular events is a common developmental motif in germ cell biology, occurring during spermatogenesis after post-meiotic spermatids cease transcription and later, during capacitation and acrosome reaction (Morgan, Kumar, Li, & Baptissart, 2021). Oocytes similarly complete their development and coordinate the events before and after fertilization in the absence of transcription.

Early analyses of egg activation focused heavily on the roles of protein kinases and phosphatases that are activated by elevated calcium levels, the conserved first step in egg activation (Ducibella & Fissore, 2008; Mahbub Hasan, Matsumoto, Kihira, Yoshida, & Sato, 2012a; Mahbub Hasan, Matsumoto, Kihira, Yoshida, & Sato, 2012b) (discussed in Section 4). Indeed, analysis of phospho-regulating enzymes and particular phosphorylated substrates established a need for protein phosphorylation during egg activation, especially in the context of release from meiotic arrest. However, the scale of phospho-events that occur during egg activation was only discovered in the last 15 years. Meiotic proteins are among phospho-regulated proteins, but the classes of proteins extend far beyond cell cycle regulation, thus implicating phosphorylation as a critical post-translational modification (PTM) for several signaling pathways and processes during egg activation.

2D gel electrophoresis studies in sea urchins and Drosophila first identified that numerous proteins shifted their electrophoretic mobility and charge as eggs activate. Mass spectrometry on the shifting protein spots revealed changes in phosphorylated peptides between mature oocytes and activated eggs. In sea urchin, most of the phospho-changing proteins were metabolic, cytoskeletal, or calcium-interacting proteins (Roux, Radeke, Goel, Mushegian, & Folz, 2008); in Drosophila, the phospho-changing proteins related to calcium-binding, proteolysis, and translation (Krauchunas, Horner, & Wolfner, 2012). These results suggested that phospho-modulation of maternal proteins could drive egg activation, potentially by turning "off" oocyte maintenance proteins in the egg while turning "on" maternally-provided early embryogenesis proteins. We will review the evidence for this hypothesis here.

Phospho-state controlling protein function or activity is a well-documented and common biological mechanism to change a protein's activity: (de)phosphorylation may change a protein's conformation, changing its affinity for binding partners or ligands, promote its degradation, or change its subcellular localization. Relatedly, aberrant cellular phospho-profiles or incorrectly-"wired" phosphorylation networks underlie disease states like cancer and neurodegeneration (Drake et al., 2016; Harsha & Pandey, 2010; Liu et al., 2021; Xiao, Kim, Pang, & Yang, 2022; Yilmaz et al., 2024). Understanding the nature and effects of differential phosphorylation (and other PTMs) of maternal proteins during egg activation may be the missing link to understanding the regulation of early development, including how disruption of normal phosphorylation events, such as by specific mutations in important phospho-proteins, could lead to infertility. Given the need for particular protein kinases or phosphatases for successful egg activation, phospho-regulation of stored maternal proteins seems likely to "pull the strings" during the transition to embryogenesis.

Multiple methods have been employed to identify proteins with dynamically-regulated phospho-sites (Delom & Chevet, 2006). Early studies relied on incorporation of radioactive phosphate isotopes followed by gel electrophoresis; development of phosphate-binding tag 'phos-tag' gels allowed for detection of phospho-changes on proteins analyzed by SDS-PAGE (Kinoshita, Kinoshita-Kikuta, Takiyama, & Koike, 2006). Additional tools, like monoclonal antibodies raised against a phosphorylated peptide, allowed for identification of phospho-isoforms' levels in protein extracts or for localization of those isoforms by cytological examination. Further advancements in 2D electrophoresis and mass spectrometry widened the throughput of such analyses, but required manual inspection of spectra to identify phosphorylation sites, and were not totally quantitative. Within the last decade, phosphoproteomic mass spectrometry (hereby shortened to 'phospho-mass spectrometry') has allowed substantial advances in the understanding of how maternal proteins are handled. In phospho-mass spectrometry analyses, phosphorylated proteins are biochemically enriched from a protein lysate before mass spectrometry, allowing for quantitative analysis of these otherwise low-abundance modifications (~0.1 % of the total proteome; while a significant fraction of proteins represented by the proteome are phosphorylated, phosphorylated peptides are scarce (Gerritsen & White, 2021; Vlastaridis et al., 2017)). In the species in which phosphoproteomic analysis of activating eggs has been possible (fruit flies, frogs, sea urchins, sea stars), phosphorylation modifications on maternal proteins were found to change dramatically between mature oocyte and activated egg. In three of those studies (Drosophila, Xenopus, sea star), tandem mass tag (TMT)-multiplexing technology was used and allowed for quantitative analysis of phosphorylated peptides across biological samples. These high throughput datasets have provided a list of candidate proteins and pathways whose phospho-state is modulated during egg activation, and might be important in the control of early developmental events.

Even before comprehensive analyses of phospho-events during egg activation were possible, specific examples have been tested and point to an important role for phosphorylation changes to maternal proteins at egg activation. For example, in mouse, Wee1B/WEE2 activity is controlled by its phosphorylation state, and mimicking an 'active' phospho-state is sufficient to support completion of stalled meiosis and pronucleus formation in lieu of calcium (Oh, Susor, & Conti, 2011). In flies, numerous phospho-regulated proteins are critical for oogenesis or fertility (Zhang, Krauchunas, Huang, & Wolfner, 2018), although it is as yet unknown whether phospho-regulation dictates the effects of these proteins. However, specific examples like the Drosophila Gnu protein, whose dephosphorylation during egg activation allows activation of the Pan Gu Kinase (PNG) (Hara, Petrova, & Orr-Weaver, 2017), or the Drosophila Young Arrest protein, whose dephosphorylation during egg activation correlates with its ability to enter the nucleus (Yu, Song, Turner, & Wolfner, 1999) support this hypothesis.

In this review, we compare trends in phosphorylation to maternal proteins as identified by phospho-mass spectrometry in fruit flies (*Drosophila melanogaster* (Zhang, Ahmed-Braimah, Goldberg, & Wolfner, 2019)), frogs (*Xenopus laevis* (Presler et al., 2017)), sea urchin (*Strongylocentrotus purpuratus* (Guo et al., 2015)), and sea star (*Patiria miniata* (Swartz et al., 2021)). Each study compared proteins and phospho-proteins in mature oocytes to activated eggs, but with different details for each: in Drosophila, mature oocytes were compared to unfertilized, naturally-activated eggs that were 30 min or younger. In Xenopus, the phosphoproteome was surveyed in mature oocytes and artificially-activated (electric shock activated, unfertilized) eggs every 2 min post-activation for 18 min. In sea urchin, the unfertilized egg phosphoproteome was compared to the phosphoproteome 2 min and 5 min post-fertilization (activation). For both Xenopus and sea urchin, we discuss proteins with changes to phospho-dynamics during the timecourses analyzed. Finally, in sea star, the phosphoproteome of maturing oocytes and early embryos were analyzed. Compared to the other species, sea stars' eggs are unusual in that they are competent for fertilization and egg activation at any time between germinal vesicle breakdown and female pronucleus formation (Stricker, 1999). The phospho-mass spectrometry analysis used sea star eggs fertilized during Meiosis I (MI); as fertilization induces calcium rise and downstream egg activation events in sea star, for the purposes of this review we consider sea star 'egg activation' to be the progression from MI to Meiosis II (MII).

Although differences in biology, phospho-mass spectrometry, and statistical methods make inter-organism comparisons imperfect, we summarize here phospho-regulation during egg activation in the species where high-throughput phospho-mass spectrometry has been performed. We contextualize the results with examples in which phosphorylation controls the function of specific maternal proteins. We also discuss the kinases and phosphatases implicated in regulating the phospho-proteome during the oocyte to embryo transition.

3. Conserved events related to phospho-regulation during egg activation

3.1 General trends in protein/phospho-site dynamics

Phospho-mass spectrometry identified abundant phosphoproteome changes during egg activation in all four model systems studied (Guo et al., 2015;

Presler et al., 2017; Swartz et al., 2021; Zhang et al., 2019). In each system, these changes predominate over other proteome changes during this developmental transition. For example, in Xenopus, about 15 % of detectable phospho-sites are reported to be dynamic, but only 1 % of proteins change in abundance (Table 1, Row 3, Column 4) (Presler et al., 2017). In sea urchin, a striking 93.71 % of phospho-sites are reported to be dynamic whereas only 10.9 % of proteins change in abundance (Table 1, Row 4, Column 4) (Guo et al., 2015). The trend remains, but is less dramatic, in sea star, where 55.35 % of phospho-sites are dynamically regulated during the oocyte-embryo transition but only 25.50 % of proteins change in abundance (Table 1, Row 5. Column 4) (Swartz et al., 2021). In Drosophila, the trend is more mild: 19 % of the phosphoproteome is remodeled, but only 14.65 % of proteins change in abundance (Table 1, Row 2, Column 4) (Kronja, Whitefield et al., 2014; Zhang et al., 2019). Where examined (sea stars and sea urchins), serine or threonine phosphorylation dominated the population of regulated substrates: 79 % of the (de)phosphorylated amino acids were serines in sea star.

These studies additionally revealed a general bias towards dephosphorylation (versus phosphorylation) of substrates during activation. In the species that we were able to compare, dephosphorylation occurs 1.4-13.03X more than phosphorylation during activation (Table 1, Column 6). When we averaged the phosphorylation versus dephosphorylation events reported across the four species, we found that, on average, ~69 % of dynamic phospho-sites get dephosphorylated, whereas ~31 % of sites get phosphorylated.

While the functional relevance of differentially regulated phosphorylation sites is largely unknown, some studies have shown the critical nature of these phospho-events or phospho-regulated proteins for egg activation events such as relieving meiosis arrest or translating new proteins (Sections 3.2–3.5). Genetic analyses in Drosophila have shown that some of the phospho-regulated proteins are required for, or immediately after, egg activation, but many others had pleiotropic roles, (including during oogenesis) preventing their study during egg activation (Zhang et al., 2018). An interesting possibility, given this pleiotropy is that the proteins' phosphorylation states dictate their activities and functions so that maternal proteins can be "recycled" for different activities during oogenesis, egg activation, and early embryo development; however this has not yet been tested so it remains a (tempting) hypothesis at present.

These studies suggest broad importance for phosphorylation of maternal proteins during egg activation. Below, we highlight groups of phospho-regulated proteins identified by phospho-mass spectrometry analyses, whose

Table 1 Trends in phosphoproteome and proteome regulation during egg activation. Species-specific details about egg activation and accompanying trends in phosphoproteome and proteome changes during egg activation. Calculations used to determine these numbers can be found in Appendix II.

1. Species	2. Samples compared	3. Oocyte arrest stage	4. % Proteome changes vs. % phosphoproteome changes	5. % Phosphorylated vs. dephosphorylated	6. Ratio dephosphorylation:-phosphorylation	7. References
2. Fruit fly (*Drosophila melanogaster*)	Mature oocyte vs. activated, unfertilized egg (<30 min)	Metaphase I	14.65 % proteome, 19 % phosphoproteome	38.14 % phosphorylated, 61.86 % dephosphorylated	1.62	Zhang et al. (2019), PMID: 30478224; Kronja, Whitfield et al. (2014), PMID: 2534940
3. Frog (*Xenopus laevis*)	Mature oocyte vs, activated, unfertilized egg (2–18 min)	Metaphase II	1.41 % proteome, 15.66 % phosphoproteome	36.09 % phosphorylated, 63.91 % dephosphorylated	1.77	Presler et al. (2017), PMID: 29183978
4. Sea urchin (*Strongylocentrotus purpuratus*)	Mature oocyte vs. activated, fertilized embryo (2 and 5 min)	G1	10.9 % proteome, 93.71 % phosphoproteome	43.67 % phosphorylated, 56.33 % dephosphorylated	1.40	Guo et al. (2015), PMID: 26227301
5. Sea star (*Patiria miniata*)	Meiosis I vs. Meiosis II (unfertilized vs. activated, fertilized embryo)	Pronucleus (if unfertilized)	25.50 % proteome, 55.35 % phosphoproteome	7.13 % phosphorylation, 92.87 % dephosphorylation	13.03	Swartz et al. (2021), PMID: 34342579

phospho-regulation seems likely to mediate major egg activation events. Additionally, we present examples where experimental analysis has shown important roles for phospho-regulation in mammalian egg activation.

3.2 Translation and RNA handling

One characteristic of egg activation is a change in the translation of stored maternal mRNAs and in some cases related changes to those RNAs (e.g. polyadenylation). Phospho-mass spectrometry has revealed changes in phospho-state of proteins critical for these processes, suggesting that their phospho-modulation underlies translational upregulation during egg activation and pointing to specific factors required for it. Specifically, in comparing the available phospho-mass spectrometry datasets for egg activation, we find differentially-regulated phospho-sites on proteins related to translational regulation and RNA processing in Drosophila, Xenopus, sea urchin, and sea star. Among these phospho-regulated proteins are factors whose (de)phosphorylation matches experimentally-determined events, like the Gnu subunit of PNG in Drosophila.

Table 2/Appendix 1 lists differentially-regulated phospho-proteins during egg activation (Translation: Column 2). Examples of phospho-regulated proteins include translation initiation factors (Jackson, Hellen, & Pestova, 2010) (e.g. eIF4G1 in all species examined; eIF4B in fruit fly, frog, and sea star) and ribosomal subunits (Petibon, Ghulam, Catala, & Elela, 2020) (e.g. RPS10, in frog and sea star; RPS2 in sea urchin and sea star; RPL34 in fruit fly and frog), suggesting phospho-regulation controls multiple discrete steps of translation. Phospho-regulation has functional ties to the control of such proteins in other systems, but for many factors the functional relevance of phosphorylation has not been established. For particular cases, there are hints, including studies of ribosomal subunit assembly during Xenopus oocyte maturation and sea urchin fertilization, in which 40S ribosomal subunit proteins undergo phosphorylation (Ballinger & Hunt, 1981; Martin-Perez, Rudkin, Siegmann, & Thomas, 1986).

In addition to factors that may directly influence the localization or complexing of translational machinery, RNA-binding proteins were also represented among phospho-regulated substrates, including mRNA decapping proteins (Cassani & Seydoux, 2024) like DCP1 and Edc3 in fruit fly, or polyadenylation-promoting factors like CPEB1 (Rouhana et al., 2023) in sea star. In developing oocytes, stable mRNAs are stored in 'P-bodies' (Lin et al., 2008). The localization, translation, or destruction of mRNAs relies on the proteins that interact with P-bodies (Cassani &

Seydoux, 2024). During Drosophila egg activation, P-bodies disperse (Weil et al., 2012; York-Andersen et al., 2015), which is thought to allow for translation of mRNAs that had been stored within the P-bodies. A well-studied example is *bicoid*, whose transcripts are prematurely lost in conditions with altered P-body structure (Sankaranarayanan et al., 2021). There are also examples in other systems where phospho-state controls function of RNA-binding proteins. For example, in cancer cell lines, phosphorylation of EDC3 controls its ability to interact with P-bodies (Bearss et al., 2021). Thus, phospho-modulation of RNA-interacting factors during egg activation may underly their ability to contact or disperse P-bodies.

In Drosophila, translational activation during the oocyte-to-embryo transition is controlled in part by phospho-regulation of the Pan Gu Kinase. PNG is made up of three subunits, which must assemble to trigger the kinase's activity (Lee, Hoewyk, & Orr-Weaver, 2003); specifically, dephosphorylation of PNG's subunit Gnu during egg activation is needed to assemble PNG, allowing the kinase to act (Hara et al., 2017). Active PNG phosphorylates translational repressors like Pumilio and TRAL, inactivating them, and thus promoting translation of crucial proteins related to meiotic completion and DNA modification (Hara et al., 2018). Phospho-mass spectrometry identified TRAL and Pumilio/PUM2 as phospho-regulated in fruit fly and sea star. Two additional confirmed (Me31B) or likely (Cup/EIF4ENIF1) PNG targets (Hara et al., 2018) were found by phospho-mass spectrometry to change phospho-state during egg activation. In fruit fly embryos, TRAL, Cup, and Me31B function in a complex (Wang et al., 2017). We raise the possibility that phospho-regulation might control the complexing of these three proteins during egg activation.

During egg activation, the bulk of translated mRNAs are related to RNA processing and chromatin organization (Kronja, Yuan et al., 2014). For example, in Drosophila, the RNA-binding protein Smaug is among the first protein translated from stored maternal mRNA (Tadros et al., 2007). Smaug's translation is triggered at egg activation in a PNG Kinase-dependent manner; Smaug protein then destabilizes maternally-stored mRNAs that were needed for oogenesis, but whose presence would inhibit normal embryogenesis. Activated eggs deficient for *smaug* fail to degrade maternal transcripts, leading to cell cycle defects later in embryogenesis (Tadros et al., 2007).

In sea stars, phospho-mass spectrometry uncovered phospho-state changes in numerous translational regulators during the MI→MII transition (Table 2, Row 5, Column 2). Although the function of these phospho-changes has not

Table 2 Selected phospho-regulated proteins identified by phosphoproteomic mass spectrometry during egg activation. Curated phospho-proteins related to Translation/RNA handling, Meiosis/Spindle regulation, Cytoskeletal remodeling, Nuclear architecture, and Kinases/phosphatases are displayed. The fruit fly proteins listed have aberrant phospho-regulation when Calcineurin is genetically depleted or overactivated; however, they are not all known to be direct substrates of Calcineurin. Bolded entries are orthologs that undergo phospho-regulation in at least two of the four species. We note that similar protein families are represented across species, even when not direct orthologs (e.g. KIFs, CDCs, Myosins, NUPs, EIFs). Orthologs were identified by name and/or by manual inspection of predicted orthologs using Flybase or UniProt. Kinases may be double listed (e.g. 'MASTL' is listed in 'Meiosis' and 'Kinases' columns because it is a kinase with roles in Meiosis). Additionally, motor proteins or interactors (e.g. KIFs) are listed in both 'Meiosis' and 'Cytoskeleton' columns because of their possible roles in both processes. The full list of analyzed phospho-proteins can be found in Appendix I.

1. Species	2. Translation and RNA handling	3. Meiosis and spindle regulation	4. Cytoskeletal remodeling	5. Nuclear architecture	6. Kinases and phosphatases
2. Fruit fly (*Drosophila melanogaster*)	**Cup/EIF4ENIF1, eIF-4B, eIF4G1, Pumilio/PUM2, RpL34b/RPS34, Tral/LSM14B,** DCP1, Edc3, eIF2A, eIF5B, Gnu, Me31B, RpL3, RpL31, RpLP1, RpS30, RpS3A, RpS6	**Asp/ASPM, BubR1/BUB1, Dlic/DYNC1LI1, Gwl/MASTL, INCENP, polo/PLK1, Spd-2/CEP192,** Cdc16, Endos/ENSA, Fzy/CDC20, Klp10A/KIF2A, Map205, Mtrm, SRPK	**Cib/TMSB4X, CLIP190/CLIP1, Coronin/CORO1C, Diaphanous/DIAPH2, Dlic/DYNC1LI2, Hts/ADD1, Jupiter/JPT1,** Feo/PRC1, Klp10A/KIF2A, Map205, Moesin/MSN, Patronin/CAMSAP1, RhoGEF2/ARHGEF12, Sofe/PRC1, Sqh/MYL9	**Nup50, Nup98, Nup153, Nup358/RANBP2, Pendulin/KPNA2** Nup133, Nup154	**BubR1/BUB1, Gwl/MASTL, polo/PLK1, Sgg/GSK3,** PNK, SRPK
3. Frog (*Xenopus laevis*)	**EIF4ENIF1, EIF4B, EIF4G1, RPL34, RPS2, RPS10, SF3B1,** EDC4, EIF2AK2, EIF3A, EIF4EBP2, EIF4H, RPL24, RPL7A, RPS28, SF3B2, SLU7	**INCENP, MASTL, CSPP1, DYNC1LI1,** ANAPC1, CCNE1, CDC23, CDC27, CDC5L, CDCA3, CDCA7, DKA3, ESPL1, KIF4A, MAP4, MAP7D1, MAPRE1, MAST4, WEE2	**ADD1, CLIP1, DYNC1LI1, ECT2, TMSB4X,** ARHGAP25, CKAP2, EML4, KIF4A, LIMA1, LIMCH1, MAP4, MAP7D1, MAPRE1, MAST4, MYH10, SVIL, WASL, WIPF2	**NUP50, NUP98, NUP133, NUP153, NUP188, NUP205, RANBP2,** ELYS, KPNA3, KPNA4, LEMD3, LMNB2, NPM1, NUP35, NUP107, NUP155, NUP214,	**MASTL,** EIF2AK2, MAP4K4, MARCKS, MAST4, MELK, NEK3, PAK2, PHACTR4, PPM1H, PTPN12, RPS6KB1, SSH3, STK31, WEE2

(continued)

Table 2 Selected phospho-regulated proteins identified by phosphoproteomic mass spectrometry during egg activation. Curated phospho-proteins related to Translation/RNA handling, Meiosis/Spindle regulation, Cytoskeletal remodeling, Nuclear architecture, and Kinases/phosphatases are displayed. The fruit fly proteins listed have aberrant phospho-regulation when Calcineurin is genetically depleted or overactivated; however, they are not all known to be direct substrates of Calcineurin. Bolded entries are orthologs that undergo phospho-regulation in at least two of the four species. We note that similar protein families are represented across species, even when not direct orthologs (e.g. KIFs, CDCs, Myosins, NUPs, EIFs). Orthologs were identified by name and/or by manual inspection of predicted orthologs using Flybase or UniProt. Kinases may be double listed (e.g. 'MASTL' is listed in 'Meiosis' and 'Kinases' columns because it is a kinase with roles in Meiosis). Additionally, motor proteins or interactors (e.g. KIFs) are listed in both 'Meiosis' and 'Cytoskeleton' columns because of their possible roles in both processes. The full list of analyzed phospho-proteins can be found in Appendix I. (*cont'd*)

1. Species	2. Translation and RNA handling	3. Meiosis and spindle regulation	4. Cytoskeletal remodeling	5. Nuclear architecture	6. Kinases and phosphatases
4. Sea urchin (*Strongylocentrotus purpuratus*)	**EIF4ENIF1, EIF4G1, RPS2,** DDX24, EEF1B2, EIF3L, QKI, RPS12, RPS6KA1, SRSF1	**BUB1, CEP192, SPDL1,** CASC5, CCNY, CDC43BPA, CDC6, CDCA7L, CDK13, CEP131, CEP85, CHK1, DNS1, KIF16B, KIF17, KIF20A, KIFAP3, MAP7, MAP7D3, MAP9, NUSAP1, PCNT, SPICE1, STAG1, TUBGCP5, TUBGCP6	**ADD1,** ARAP1, ARHGEF11, ESPN, KIF16B, KIF17, KIF20A, KIFAP3, MAP7, MAP7D3, MAP9, MYO9B, SPIRE1, STMN1, TUBGCP5, TUBGCP6	**NOP58, NUP153, NUP205, RANBP2,** KPNA4, NUP98, RANBP3	**BAZ1B, BUB1,** CDK13, CHK1, PAK1, PDPK1, PPP1R37, PPP4R2, PRPF4B, RIO2, STK33, TEP1, TTBK2, VRK1
5. Sea star (*Patiria miniata*)	**EIF4B, EIF4G1, LSM14B, PUM2, RPS2, RPS10, SF3B1,** CPEB1, DDX42, EIF1, EIF2A, EIF2B, EIF3A, EIF3G, EIF3J, EIF6, RPL39,	**ASPM, CSPP1, INCENP, MASTL, PLK1, SPDL1,** CCNH, CCNK, CCNT2, CDK12, CDKL1, CDKL4, CEP19, CSM3-like, DC1L2, DCTN1, DYHC2, KIF14, KIF18A, KIF21A, KIF23, KIF24, KIF26A, KIF28P, KIF3C, KIF4A, MAP10, MAST3, MND1, MNS1, NUF2, SPC24, TUBGCP2	**CLIP1, CORO1C, DIAPH2, ECT2, JPT1,** ADDA, CARMIL1, DC1L2, DCTN1, DYHC2, EPN2, JKIP1, KIF14, KIF18A, KIF21A, KIF23, KIF24, KIF26A, KIF28P, KIF3C, KIF4A, MAP10, MAST3, MPRIP, MYH4, MYLK, TUBGCP2, WIPF1, XIRP2	**ELYS, KPNA2, NOP58, NUP98, NUP133, NUP153, NUP188, RANBP2,** ANO39, IPO8, LAP2A, LBR, MAN1, NUCL, NUP53, NUP93, NUP214	**BAZ1B, GSK3, MASTL, PLK1,** CAMK1, CDC42BPA, CHK2, JKIP1, MAPK1, MAST3, MLTK, MPRIP, MTOR, MYLK, MYPT1, NEK10, NEK9, PKC, PKN2, PP14C, PTN12, SSH1

yet been elucidated, experiments have shown that there is a general need for translation for successful transition to MII: when treated with translational inhibitor emetine, sea star eggs prematurely decondense their DNA into an apparent maternal pronucleus, but this nucleus is unable to fuse with the paternal pronucleus (Swartz et al., 2021).

Interestingly, mouse eggs do not require translation for egg activation as a whole: when unfertilized mouse eggs are treated with cycloheximide, a pronucleus can form, indicating completed meiosis (Siracusa, Whittingham, Molinaro, & Vivarelli, 1973). Further analysis found that translation inhibition lowers levels of Cyclin B, reducing active M-phase promoting factor (MPF = Cyclin-B-cdc2/CDK1) in the egg and leading to release of meiotic arrest (Moos, Kopf, & Schultz, 1996). Recent analyses in mouse eggs show that translational activity increases during egg activation. Interestingly, translation-related proteins decrease in their translation during activation, and cell cycle proteins increase in translation (Iyyappan et al., 2023). Analysis of polysome-bound mRNAs suggest some proteins are translated at constant levels during activation whereas others are increased or decreased in their translation (Chen et al., 2011). Of the mRNAs that increase in translation during mouse egg activation, many contained binding regions recognized by the RNA-binding proteins DAZL (deleted in azoospermia-like) and CPEB1 (cytoplasmic polyadenylation element-binding protein). CPEB1 and DAZL function in the same pathway to support translation of meiotic spindle proteins. CPEB1 polyadenylates *Dazl* mRNA, which allows translation of DAZL protein. DAZL's activity is thought to be important to bind to and localize mRNAs that must be translated to support meiotic spindle integrity. In agreement, morpholino-based knockdown of DAZL results in stalled development (Chen et al., 2011). A recent study showed that CPEB-phosphorylation is required for its activity during oocyte maturation (Kunitomi, 2024; Kunitomi, Romero, Daldello, Schindler, & Conti, 2024). Additional work showed that CPEB-phosphorylation is Mitogen Activated Protein Kinase (MAPK)-regulated during oocyte maturation (Sha et al., 2017). While CPEB-phosphorylation function has not been studied during egg activation in mouse, CPEB1 was detected as a phospho-regulated substrate in activating sea star oocytes, suggesting phospho-regulation of CPEB1 might be used during egg activation.

Phospho-regulation of similar translation regulators is observed in somatic cells, and specific examples described above indicate phospho-regulation may allow for the burst of translation that accompanies egg activation. Taken together, phospho-mass spectrometry of activating eggs reveals that translation initiation may be regulated at the level of ribosome assembly, recruitment of initiation factors, or proteins that handle stored mRNAs.

3.3 Meiosis and spindle regulation

Perhaps the most visible and important cellular event during egg activation is the completion of stalled meiosis. For meiosis to reinitiate and complete properly, the spindle machinery must be properly regulated and chromosomes must segregate equally. Failed spindle assembly or integrity is a major cause of aneuploidy and is mostly lethal, and in cases of viability, genomic instability, developmental abnormality, and cancer risk increase (Orr, Godek, & Comtpon, 2015).

Numerous proteins related to spindle, meiosis, or cell cycle are represented among proteins with differentially-regulated phospho-sites during egg activation in all species analyzed (Table 2, Column 3). Phospho-regulated proteins include spindle regulators such as factors that interact with the kinetochore (e.g. kinesin interacting factors/KIFs found in all species) (Camlin, McLaughlin, & Holt, 2017) or centrosomal proteins (e.g. CEP192 found in fruit fly and sea urchin; CSPP1 found in frog and sea star) (Athwal, Kochiyanil, Bhat, Allan, & Parsyan, 2024). Other microtubule motor proteins (dyneins or regulators, e.g. Dlic/DYNC1LI1 in fruit flies and frogs)(Roberts, Kon, Knight, Sutoh, & Burgess, 2013) or non-motor microtubule-binding proteins (e.g. MAPs in all species) (Bodakuntla, Jijumon, Villablanca, Gonzalez-Billault, & Janke, 2019) were found to be phospho-regulated and could influence spindle integrity, stability, and/or division. Additional spindle regulators such as Abnormal spindle (Asp/ASPM) (Saunders, Avides, Howard, Gonzalez, & Glover, 1997) or BubR1/BUB1 (Bolanos-Garcia & Blundell, 2011) were found. Cell cycle proteins (CDKs, Cdcs, and Cyclins) (Marston & Amon, 2004) were also highly represented among phospho-regulated proteins.

An example of a phospho-regulated protein is Inner Centromere Protein (INCENP), which was found to be phospho-regulated in sea star, fruit fly, and frog. In sea star, INCENP dephosphorylation by PP2A-B55 is required for efficient INCENP translocation from the centromeres to the spindle midbody. Phospho-mutation of INCENP's T61 to S disrupted INCEP's translocation efficiency, presumably due to failed dephosphorylation by PP2A-B55 (Swartz et al., 2021). This phospho-mutation uncovered a cellular consequence for INCENP phosphorylation, but its functional importance for egg activation has not yet been resolved.

Meiosis is arrested before completion in mature oocytes of most species; egg activation relieves this meiotic arrest. A common stage of arrest is metaphase of MI or MII; thus, a critical molecular task in many systems is handling, stability, and disjunction of chromosomes on the meiotic spindle. The proteins that are responsible for holding eggs in a metaphase-arrested state are important phospho-regulated targets, many of which have been shown to be phospho-modulated during egg activation. For example, active MPF (CDK1-Cyclin B) keeps eggs arrested in M phase by maintaining phosphorylation of spindle proteins and nuclear envelope proteins (Enserink & Kolodner, 2010; Jones, 2004). CDK1 itself must be dephosphorylated to remain complexed with Cyclin B and active, and a regulatory kinase Wee1B/WEE2 must be phosphorylated to deactivate CDK1 through phosphorylation of the latter's Y15 residue (Fig. 1A1) (Oh et al., 2011; Potapova, Daum, Byrd, & Gorbsky, 2009). This phospho-cascade, triggered by egg activation, dissociates MPF and leads to rapid degradation of Cyclin B (Fig. 1C6). Another important phospho-loop that regulates meiosis resumption and completion involves the anaphase promoting complex/cyclosome (APC/C) (Pesin & Orr-Weaver, 2008). In frogs, Emi2 (early meiotic induction) is a direct inhibitor of the APC/C (Sako et al., 2014). Emi2 binds directly to APC/C subunit APC10 to competitively inhibit binding of Ube2S (Sako et al., 2014). When Ube2S is bound, APC/C promotes degradation of Cyclin B by ubiquitinating it. To activate Emi2, and thereby disable APC/C, Emi2 is phosphorylated at its T195 site, which creates a docking site for Polo-like kinase (PLK1) to subsequently phosphorylate Emi2's S32 degron, which induces Emi2's ubiquitination and degradation via the SCF degradation complex (Hansen, Tung, & Jackson, 2006; Heim, Tischer, & Mayer, 2018; Rauh, Schimdt, Bormann, Nigg, & Mayer, 2005). Phospho-mass spectrometry showed that PLK1 itself is phospho-regulated in activating fruit fly and sea star oocytes, and experiments in Xenopus egg extracts showed that phosphorylation of PLK1 can control its activity (Kelm, Wind, Lehmann, & Nigg, 2002).

In summary, meiosis or spindle proteins whose phospho-sites were found to change could be important for the activity of these proteins or prime them for their activity later as the embryo begins its mitotic divisions.

3.4 Cytoskeletal remodeling

Phospho-regulation of proteins that regulate the oocyte's cytoskeleton were commonly seen in the mass spectrometry datasets of activating eggs. Many microtubule-binding proteins were found (their roles were discussed in

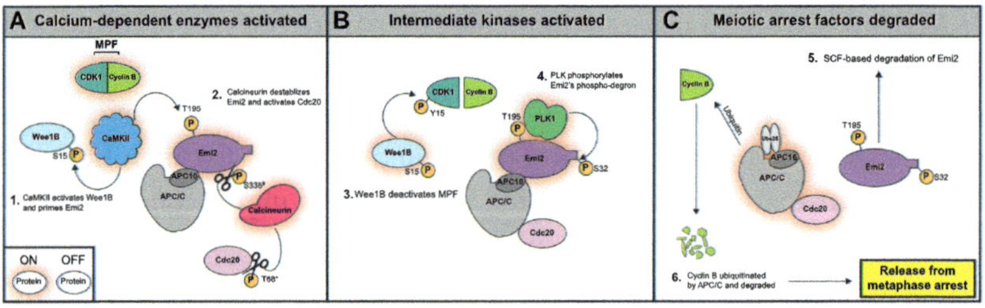

Fig. 1 Overview of calcium-mediated phosphorylation events that lead to release from meiotic arrest. (A) Calcium rise at the start of egg activation stimulates calcium-dependent enzymes (CaMKII and Calcineurin; Calmodulin not depicted). CDK1-Cyclin B (MPF) activity is high, while APC/C activity is off. Emi2 is active and competitively inhibits APC/C subunit APC10 from binding its activator. (A1) CaMKII phosphorylates Emi2 at T195 to create a docking site for PLK1. CaMKII also phosphorylates Wee1B kinase at S15 to activate it. (A2) Calcineurin dephosphorylates Emi2 at S335, which destabilizes its interaction with PP2A. Calcineurin also dephosphorylates Cdc20 at T68, which allows Cdc20 to perform its role as a cofactor for APC/C. (B) Intermediate kinases propagate phospho-signaling started by CaMKII and Calcineurin. (B3) Wee1B phosphorylates CDK1 at its inhibitory Y15 site, promoting CDK1's disassociation from Cyclin B and deactivation of MPF. (B4) PLK1 recognizes the binding site on Emi2 resulting from CaMKII phosphorylation. PLK1 then phosphorylates Emi2 at its phospho-degron site S32. (C) Emi2 is degraded and inactivated, relieving its inhibition of APC/C; thus, APC/C is activated to release metaphase arrest. (C5) Emi2 phosphorylation at its phospho-degron promotes degradation and subsequent deactivation of Emi2 via the SCF degradation complex. (C6) APC/C is activated by association with its cofactor Cdc20 and activator Ube2S and marks Cyclin B for degradation, along with components of the metaphase arrest machinery like Securin (not shown). Orange glow around a protein represents that its activity is on. # = MAPK-regulated sites. * = CDK1 sites. A-C represent temporal progression of metaphase arrest release, but 1–6 do not imply order of events (e.g. CaMKII and Calcineurin are simultaneously activated by calcium/Calmodulin). Phospho-sites depicted were all identified in Xenopus except for Wee1B S15 (found in mouse). As discussed in Section 4.2, CaMKII and Calcineurin are not both known to be involved in all organisms, but are both needed in Xenopus.

Section 3.3 in the context of the meiotic spindle regulation). We also found actin-binding proteins among proteins phospho-regulated during egg activation. For example, non-muscle myosin components or myosin regulators (Vicente-Manzanares, Ma, Adelstein, & Horwitz, 2009) were found in fruit fly (Sqh), frog (MYH10), sea urchin (MYO9B), and sea star (MPRIP, MYH4, MLK). Actin nucleating proteins were also found to change phospho-states during egg activation (e.g. Diaphanous/DIAPH2 (Afshar, Stuart, & Wasserman, 2000) in fruit flies and sea stars; SPIRE1 in sea urchins, WASL and WIPF2 in frogs (Kramer, Piper, & Chen, 2022)). Further, actin binding proteins like Adducin, which stabilize the actin cytoskeleton through interactions with the Spectrin cytoskeleton, were found in all four species to change in phospho-state (e.g. Hts/ADD1 in fruit fly, frog, and sea urchin; ADDA in sea star) (Naydenov & Ivanov, 2011).

There are numerous possible roles for phospho-regulation of actin-related processes during egg activation, including egg-envelope modifications, polar body extrusion, and regulation of the meiotic spindle. However, experimental evidence tying phospho-regulation to these actin-based functions has not yet been broadly explored; the following text explores candidates.

The actin cytoskeleton plays important roles in meiosis in activating oocytes. For example, actin interacts with microtubules in the metaphase spindle and is required for faithful chromosome segregation. In mouse, the F-actin network promotes kinetochore fiber formation and ensures chromosomes align and separate properly (Mogessie & Schuh, 2017). In Drosophila, F-actin dictates the position and shape of the meiotic spindle and ensures proper maintenance of spindle microtubules (Wood, Shi, & Weil, 2024). An interesting phospho-regulated candidate identified by phospho-mass spectrometry, CLIP190/CLIP1 (fruit fly and frog), is thought to support linkage of microtubule and actin networks, and could function in actin spindle integrity (Lantz &Miller, 1998; Sanghavi, Lu, & Gonsalvez, 2012).

Actin is also required for cytokinesis events following the meiotic divisions, that result in only one haploid set of chromosomes being retained in the oocyte (Maddox, Azoury, & Dumont, 2012). In vertebrates, echinoderms, and nematodes, asymmetric cell divisions extrude products of MI or MII into small "polar bodies" that are eventually destroyed (in Drosophila, all four meiotic products remain within the large oocyte, but only one is positioned to associate with the male pronucleus (Loppin, Dubruille, & Horard, 2015)). In mouse oocytes, actin is required for the cytokinesis that leads to polar body extrusion (Duan & Sun, 2019). The regulators of polar body division have been partially characterized; more general developmental studies indicate a role for actin in their extrusion, as treatment with actin depolymerizing drugs prevent polar body extrusion (Kubiak, Paldi, Weber, & Maro, 1991). Further, mouse eggs require the activity of myosin-light chain kinase (MYLK2) for efficient polar body extrusion, hinting that phosphorylation of myosin (or others) may be an important way to switch on the actin-interacting machinery necessary for extrusion (Matson, Markoulaki, & Ducibella, 2006). Another factor, the RhoA-GTPase activator ECT2, is required for polar body extrusion in mouse and is phospho-regulated (Elbaz, Reizel, Nevo, Galiani, & Dekel, 2010); ECT2 was detected as a phospho-regulated protein in sea star and Xenopus. In Xenopus, the polar bodies are extruded by contractile F-actin/actomyosin, and phosphorylated myosin-II was found to be present at contractile rings, indicating phospho-regulation of myosin-II may apply to its activity here (Zhang et al., 2008). Phospho-regulation of myosin was detected in the mass spectrometry datasets; interestingly, factors that regulate myosin phosphorylation, such as MYLK and myosin phosphatase (MPRIP (Surks, Riddick, & Ohtani, 2005)) in sea star, were also found, suggesting that myosin activity may be phospho-regulated at multiple steps.

The cytoskeleton is also important in remodeling the egg's coverings. In organisms in which sperm-entry stimulates egg activation, an early egg activation event (after membrane depolarization) is release of the contents of the egg's cortical granules (CGs) into the perivitelline space, modifying the vitelline envelope/zona pellucida to prevent binding by additional sperm. In mouse, the actin cytoskeleton supports the anchoring of cortical granules to the oocyte cortex, which is required for their efficient exocytosis during fertilization (Rojas et al., 2021). In zebrafish, F-actin and myosin-II localize to the site of CG exocytosis, and depolymerizing actin expedites CG exocytosis, while stabilization of F-actin prevents efficient CG exocytosis (Becker & Hart, 1999). When Xenopus eggs are treated with actin depolymerizing drugs, CG exocytosis can occur (Tembo et al., 2021). Mouse eggs also require MYLK2 for CG exocytosis and subsequent polyspermy block (Matson et al., 2006); this exocytosis is impaired when oocytes are treated with blebbistatin, which prevents phosphorylation of myosin light chain (Vogt et al., 2019). When actin is depleted in in vitro-fertilized mouse eggs, their polyspermy block is inefficient (McAvey, Wortzman, Williams, & Evans, 2002). Together, these findings imply actin's importance, and potential involvement of phospho-regulated actin-binding proteins, in CG exocytosis. In Drosophila, CG exocytosis and polar body extrusion do not occur; however, the cortical actin cytoskeleton is rapidly remodeled during egg activation, including by changes to actin binding protein dynamics, such as Moesin, which links the actin cytoskeleton to the plasma membrane (York-Andersen, Hu, Wood, Wolfner, & Weil, 2019). Moesin was identified as a phospho-regulated substrate in activating fruit fly eggs, hinting that phospho-regulation might control its change in dynamics.

Although the functional consequences of modulation of specific phospho-sites in cytoskeletal proteins during egg activation have yet to be established, identification of differentially-regulated phospho-sites in cytoskeletal proteins suggests that phospho-modifications may be used to control multiple important cytoskeletal processes during egg activation.

3.5 Nuclear architecture

Among the phospho-regulated proteins detected across the queried organisms were proteins related to nuclear envelope dynamics and nuclear transport (Table 2, Column 5). These proteins included various components of the nuclear pore complex (Otsuka & Ellenberg, 2018), nuclear import factors (Importins) (Stewart, 2007), and structural proteins like lamins (Liu & Ikegami, 2020). The function of phospho-regulation during egg activation for these groups of proteins is largely undetermined; however, differential phosphorylation may point to these proteins' activities being turned on or off during egg activation.

The completion of meiosis that is triggered by egg activation culminates with formation of the haploid female pronucleus that will later fuse or combine with the male pronucleus (Loppin et al., 2015; Meaders & Burgess, 2020) (except in cases of parthenogenetic/haplodiploid development). Pronucleus formation requires both decondensing maternal and paternal DNA and biogenesis of nuclear envelopes. In sea urchins, the decondensation of sperm DNA is phospho-regulated (Porter & Vacquier, 1983): sperm specific histones are phosphorylated to facilitate their exchange with canonical histones. In mouse, in vitro experiments showed that Nucleophosmin (NMP) proteins, which are expressed in the oocyte, must be phospho-regulated for efficient sperm nuclear decondensation (Okuwaki et al., 2012). Mass spectrometry also detected NMP-family proteins being phospho-regulated in activating frog and sea star oocytes (NMP1 and ANO39, respectively). To generate a nuclear envelope for the pronuclei and later embryonic nuclei, gametes store pre-assembled nuclear envelope components. For example, nuclear pore complexes (NPCs) pre-assemble into tightly folded (ER) sheets termed "annulate lammelae that are thought to support the rapid nuclear assembly needed for early mitotic divisions (Hampoelz et al., 2019; Otsuka & Ellenberg, 2018).

NPC components are among the most common and most dynamically-regulated phospho-proteins identified by phospho-mass spectrometry (Table 2, Column 5), although the function of their phospho-regulation during egg activation has yet to be determined. For example, Nup358/RANBP2, which regulates annulate lamella formation in Drosophila (Hampoelz et al., 2019), is phospho-regulated during egg activation in all four species. RANBP2's function in human cell lines is controlled by its phospho-state (Gloerich et al., 2011), but the effect of its phospho-regulation during egg activation is unknown. In timed examination of the Xenopus phosphoproteome during egg activation, NPC components related to the inner basket and nuclear ring undergo very fast phosphorylation changes, whereas the core scaffold proteins have slower phosphorylation regulation (Presler et al., 2017). This temporal pattern echoes that of NPC assembly post-mitosis, hinting that there is egg activation-induced assembly of new NPCs from stored components in addition to inclusion of pre-assembled NPCs in oocytes (Otsuka & Ellenberg, 2018).

Nuclear envelope assembly/disassembly also requires availability of lamins in their proper forms. Lamins are phospho-regulated during frog and sea star egg activation (e.g. frog: LMNB2; sea star: LAP2A, LBR). Another example is the Drosophila Young Arrest protein, a nuclear lamina protein that is needed to initiate mitosis in the early embryo (Sackton, Lopez, Berman, & Wolfner, 2009). After being dephosphorylated during egg activation, YA's localization changes from cytosolic to nuclear; this relocalization is thought to depend on YA's phosphorylation state (Yu et al., 1999). In sea urchin, phosphorylation of paternal nuclear lamin is PKC- and calcium-regulated and is required for efficient sperm membrane breakdown prior to pronucleus formation (Collas, Thompson, Fields, Poccia, & Courvalin, 1997).

Finally, nuclear import proteins were identified by phospho-mass spectrometry. One example, Pendulin/KPNA2/Importin-alpha-2, is phospho-regulated in both fruit fly and sea star; similar family members were found in frog (KPNA3 and KPNA4) and sea urchin (KPNA4). In Drosophila, mutation of Pendulin's S56 phosphorylation site causes reduction in embryonic viability, delayed mitotic divisions, and severe disorganization of early mitotic spindles, despite displaying normal meiotic divisions (Gorjanacz et al., 2006). Thus, Pendulin phosphorylation during egg activation may be important for temporal and/or spatial organization of the mitotic divisions that occur in the early embryo, soon after egg activation.

Functional dissections of phospho-regulated nuclear proteins during egg activation are scarce. However, the phospho-mass spectrometry analyses suggest that phospho-regulation could be an important mechanism to control NPC assembly or incorporation into new nuclei. Phospho-regulation could also affect the decompaction of maternal and paternal DNA or their ability to fuse together, in species where pronuclear fusion occurs.

Finally, phosphorylation state of lamins or other nuclear architecture proteins could be important for the assembly of nuclear envelopes during egg activation.

4. Kinases and phosphatases that regulate maternal phospho-proteins during egg activation

4.1 What triggers egg activation?

The results discussed above suggest that egg activation is driven by phospho-changes that turn on or off the activities of stored maternal proteins. But this raises the question: what triggers the changes in protein phosphorylation state? If proteins are initially all in their "oocyte" state, what switches them suddenly and coordinately to their "egg activation" state?

Despite the mystery that surrounds many egg activation mechanisms, its most upstream trigger is known. In virtually all sexually-reproducing metazoans, including humans, egg activation is initiated by a rise in free cytoplasmic calcium in the oocyte. This calcium increase is tightly regulated in time, place and level. It can occur via uptake of calcium from the environment, release of calcium from internal stores (endoplasmic reticulum, ER), or careful coordination of both. This important process has been reviewed in detail (Saunders et al., 2002; Stein, Savy, Williams, & Williams, 2020; Swann & Lai, 2016), so we will only present it in outline here, referring calcium-curious readers to the excellent reviews just noted. In general, calcium levels first rise locally, at the site of sperm entry or site of mechanical stress, and then spread in a "wave" across the ooplasm. This wave of increased calcium concentration is due to calcium release from stores by IP3R (triggered by IP3 and/or calcium levels) causing successive activation of IP3R as calcium spreads across the oocyte.

In mammals, frogs, echinoderms, and nematodes, the free-calcium increase in the oocyte is induced by the fertilizing sperm. However, calcium levels first rise at the site of sperm entry by different mechanisms in different species. The fertilizing sperm introduces into the oocyte cytoplasm sperm-soluble factors (e.g. PLC-zeta, in mammals (Saunders et al., 2002)) that induce calcium release from stores, or provides calcium-permeant channels that become part of the cell membrane of the fertilized egg (TRP channel in C. elegans (Takayama & Onami, 2016)). The increased calcium and IP3 act on the ER-located IP3 receptors to cause release of calcium from stores. This occurs progressively, spreading a wave of calcium across the oocyte. In some organisms (frogs, marine invertebrates, nematodes) this single calcium wave is enough to trigger egg activation. In contrast, mammalian oocytes undergo a subsequent series of "oscillations" to induce events of egg activation, and the frequency and amplitude of oscillations correlate with discrete egg activation events (Ducibella et al., 2002); the molecular control of oscillatory calcium rises and their cellular consequences are being investigated.

Interestingly, eggs of some species, including insects and some fish, activate naturally without fertilization. In Hymenoptera (ants, wasps, bees), unfertilized eggs not only activate but also develop fully into adults - haploid males (drones) (Went, 1982). Went and Krause showed that for Pimpla wasps, the trigger to start development is passage of the egg through the narrow ovipositor, but the mechanism by which this causes egg activation has not been investigated (Went & Krause, 1973). In Drosophila, unfertilized eggs can also activate, completing meiosis, translating stored maternal mRNAs, hardening their eggshell, and modifying their phosphoproteomes normally. However, these activated eggs cannot develop into embryos because they lack centrioles (which are donated by the sperm in this species) (Blachon, Khire, & Avidor-Reiss, 2014). Examination of the cell state of Drosophila oocytes in the reproductive tract showed that ovulation, the movement of the mature oocyte from the ovary into the oviduct, is the signal for egg activation (Heifetz et al., 2001). In vitro experiments showed that eggs activate upon pressure on, or swelling of, their membranes – if the egg is also in an environment that contains calcium (Kaneuchi et al., 2015; York-Andersen et al., 2015). This suggested that egg activation involves the opening of mechanically gated ion channels that allow calcium to enter the oocyte, a hypothesis that was confirmed by studies with chemical inhibitors and genetic mutants of such channels (Hu & Wolfner, 2019).

4.2 How does elevated calcium trigger egg activation?

While much is known about the necessity for, and characteristics of, the calcium rise that triggers egg activation, how this sparks the cell biological changes (cell cycle, translation, etc.) was initially a mystery. Genetic studies showed that egg activation requires calcium-dependent kinases and phosphatases, suggesting a simple model (Fig. 2) that is substantiated in Drosophila (Zhang et al., 2019): the calcium trigger is transduced to the cellular machinery through the action of phosphorylation-modifying enzymes that calcium activates. Here, we will discuss these enzymes' role in egg activation and protein phosphorylation changes.

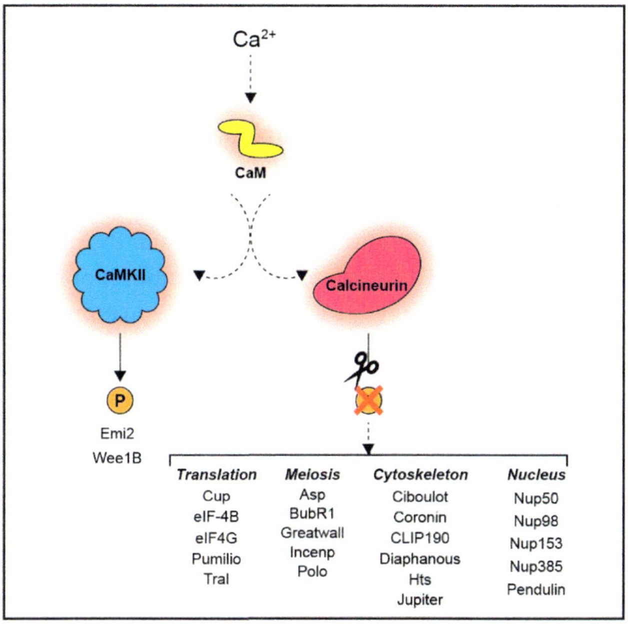

Fig. 2 **Calcium rise at egg activation simulates phospho-protein regulation by CaMKII and Calcineurin.** Raised free cytoplasmic calcium (Ca^{2+}) activates CaMKII and/or Calcineurin via Calmodulin (CaM). Orange haze indicates active proteins. CaMKII's known targets during egg activation are Emi2 and Wee1B, as identified in vertebrates (Xenopus and mouse, respectively). Calcineurin regulates (directly or indirectly, hence the dotted line beneath Calcineurin) the phosphorylation states of dozens of maternal proteins. Those shown are the Drosophila proteins whose orthologs were identified in multiple species' phospho-mass spectrometry screens (bolded proteins in the Drosophila row of Table 2). For technical reasons, CaM activity has not been directly assessed during egg activation; a dotted line therefore connects Ca^{2+} to CaM, and CaM to CaMKII/Calcineurin.

4.2.1 Calcium-dependent enzymes

4.2.1.1 CaMKII

CaMKII(g) is a calmodulin/Ca^{2+}-dependent kinase that has been shown to be essential for egg activation in mouse and frog oocytes (Backs et al., 2010; Hansen et al., 2006). Mouse eggs contain two CaMKII isoforms, and the CaMKIIg isoform is found in high levels (Backs et al., 2010). *CaMKIIg* null animals are female sterile, due to the inability of their oocytes to respond to calcium; the oocytes remain stalled in MII (Backs et al., 2010). Moreover, expression of constitutively active CaMKII in mouse eggs is sufficient to drive embryogenesis without a calcium rise (e.g. after fertilization by PLCz-deficient sperm) (Knott et al., 2006), indicating that CaMKIIg is the major transducer of the calcium signal for egg activation in mice. In activating mouse oocytes, CaMKII activity mirrors the pattern of calcium oscillations (Markoulaki, Matson, Abbott, & Ducibella, 2003; Markoulaki, Matson, & Ducibella, 2004). In sea urchin, CaMKII activity is also stimulated by fertilization/egg activation (Tombes & Peppers, 1995).

Despite its importance for egg activation in several organisms, most of CaMKII's targets in oocytes are as yet undetermined. Its few known targets point to a role for CaMKII in regulating the resumption of meiosis. In mice CaMKII regulates the Wee1B/WEE2-CDK1 regulation loop by phosphorylating Wee1B at S15 (Figs. 1A1 and 2) (Oh et al., 2011). In Xenopus, CaMKII directly phosphorylates Emi2 at T195 to allow for PLK1 binding and subsequent degradation of Emi2 and release from meiotic arrest (Figs. 1 and 2) (Hansen et al., 2006). In mouse, Emi2-degradation is also CaMKII-dependent; however, it has not yet been determined whether Emi2 is a direct target of mouse CaMKII (Medvedev, Stein, & Schultz, 2014). CaMKII also appears to be required for translational efficiency of activating mouse oocytes, because recruitment of mRNAs to polysomes is impaired when *CaMKIIg*-null eggs are artificially activated with strontium (Backs et al., 2010). Consistent with the idea that CaMKII has numerous downstream targets whose activities drive egg activation, phospho-motif searches have identified CaMKII target sites among phospho-regulated proteins in sea urchin egg activation (Guo et al., 2015), and in Xenopus 25% of phospho-sites that increase in their phosphorylation during egg activation contain the CaMKII-preferred motif (Presler et al., 2017). Some phospho-regulated maternal proteins in Drosophila also contain predicted CaMKII sites, hinting at a possible role for CaMKII in Drosophila egg activation (Zhang et al., 2018), but this has not yet been tested.

4.2.1.2 Calcineurin

Dephosphorylation of maternal proteins is common during egg activation (Table 1, Column 5). The calcium-regulated phosphatase Calcineurin has been implicated in egg activation in Drosophila and Xenopus, so far. In Drosophila, initial genetic studies found that Calcineurin's regulator Sarah/calcipressin is essential for egg activation (Horner et al., 2006); Calcineurin itself was subsequently shown to be essential (Takeo, Hawley, & Aigaki, 2010). In Xenopus, transient activity of Calcineurin was also shown to be essential for egg activation (Mochida & Hunt, 2007; Nishiyama, Yoshizaki, Kishimoto, & Ohsumi, 2007). Further studies in Drosophila integrated genetic and biochemical approaches to uncover the Calcineurin-dependent phosphorylation events during egg activation and found Calcineurin to regulate many targets (Table 2, Row 2 and Fig. 2) (Zhang et al., 2019).

Comparison of phosphoproteome changes between wild type and Calcineurin-deficient Drosophila oocytes demonstrated ~100+ Calcineurin-dependent phospho-changes during egg activation (Zhang et al., 2019). Some of these sites were in proteins required for cell cycle progression or for translation initiation, pointing to Calcineurin's role in regulating these processes during egg activation (Fig. 2). Some sites were found in regulators of Calcineurin activity, consistent with a negative feedback loop leading to transient Calcineurin activity, as in Xenopus. Among the Calcineurin-regulated sites were three in Fzy, a cofactor for APC/C whose binding is required for APC/C activation, indicating that Fzy's release from, and subsequent deactivation of, APC/C is Calcineurin-controlled (Swan & Schupbach, 2007; Zhang et al., 2019). Additional cell cycle regulators, like the Greatwall kinase and Endosulfine, were found to be mis-regulated in Calcineurin knockdown oocytes. Greatwall, which is normally phosphorylated and activated by MPF, phosphorylates and activates Endosulfine, which regulates MPF by inhibiting its inhibitor PP2A-B55 (Mochida, Maslen, Skehel, & Hunt, 2010; Yu, Zhao, Li, Galas, & Goldberg, 2006). When Calcineurin is depleted, the phospho-regulation of these factors is abnormal, likely underlying the failure of Calcineurin-deficient eggs to complete meiosis. Dephosphorylation of Gnu, the component of the PNG kinase that activates translation, is also dependent on Calcineurin.

Results in Xenopus are also consistent with Calcineurin regulating meiosis resumption/completion. As in Drosophila, cdc20/Fzy is Calcineurin-regulated in Xenopus egg extracts, and it has been shown to be a direct Calcineurin substrate there (Fig. 1A2) (Heim et al., 2018; Mochida & Hunt, 2007). Calcineurin was also shown to regulate the phosphorylation of Emi2 (Xerp1) (Fig. 1A2) (Heim et al., 2018). Interestingly, Emi2 appears to require both phosphorylation and dephosphorylation events for it to efficiently release meiotic arrest.

Calcineurin's role in mammalian oocytes is less clear. When porcine oocytes are treated with Calcineurin-inhibitors, the rate of overall successful egg activation decreases and cortical granule exocytosis is disrupted, but activity levels of MPF are unaltered (Tumova et al., 2016). The role of Calcineurin in egg activation is unexplored in other mammals, including mouse.

Table 2, Row 2 highlights Calcineurin-regulated substrates in Drosophila. As seen by the bolded entries, orthologous proteins are found to undergo similar phospho-regulation during egg activation in all species examined, suggesting that Calcineurin may transduce calcium signaling to downstream events of egg activation in species beyond the fruit fly. For the full list of Calcineurin-dependent phospho-proteins identified in Drosophila, see Appendix I.

4.2.2 Other enzymes

Calcium-stimulated kinases or phosphatases represent the obvious candidates to facilitate the beginning of calcium-mediated phosphorylation cascades – i.e., it is likely that these are the most apical kinases activated during egg activation. However, mass spectrometry identified phospho-regulated sites during egg activation that represent binding motifs for kinases or phosphatases that are not calcium-dependent. The most common such motif is the phospho-S/T-P motif targeted by CDK1/MAPK. Additional enrichments were found for Casein Kinase II (ShiYang et al., 2020), PKC (Halet, 2004), and others. Thus, phospho-motif enrichment implicates these kinases may act as intermediates, downstream of the calcium-activated kinases/phosphatases discussed above.

4.2.2.1 MAPK and CDK1

MAPK and CDK1 play multiple roles during oocyte maturation, including the establishment and release of meiotic arrest. In short, the MAPK pathway holds oocytes in arrest, in part through stabilizing Emi2 (Inoue, Kanemore, Nobui, & Sagata, 2007) (Fig. 1A). CDK1-Cyclin B (i.e. MPF) phosphorylates meiotic proteins, like spindle proteins, to promote M-phase arrest (Cdc20 (Labit et al., 2012) shown Fig. 1A). Both kinases play roles in the processes leading up to egg activation, and must be rapidly inhibited upon egg activation (see reviews (Dupre, Haccard, & Jessus, 2011; Jones, 2004)). Indeed, MAPK activity is seen to plummet during egg activation in a number of animals, including Drosophila (Sackton, Buehner, & Wolfner, 2007), sea urchin (Swartz et al., 2021), and mouse (Gonzalez-Garcia et al., 2014; Moos, Xu, Schultz, & Kopf, 1996): this loss of activity is due to dephosphorylation of MAPK. Phospho-mass spectrometry in Xenopus oocytes showed that the minimal phospho-S/T-P site preferred by CDK1 and MAPK comprised ~70% of sites that were dephosphorylated during egg activation (Presler et al., 2017). This supports the idea that MPF and MAPK are major players in establishing the phospho-signatures of proteins before egg activation, and their deactivation through CaMKII (Wee1B activation to disable MPF (Oh et al., 2011)) or yet-to-be fully described calcium-dependent mechanisms (Gonzalez-Garcia et al., 2014) causes their activity to drop, and their targets are dephosphorylated.

4.2.3 *Other phospho-regulated kinases*

Other kinases are likely required to propagate phospho-based signaling cascades. As noted above, in Xenopus, CaMKII primes Emi2 by creating a binding site for the kinase PLK1, which then phosphorylates a phospho-degron and leads to Emi2's destruction (Fig. 1A–C) (Hansen et al., 2006). In another example, among phospho-proteome changes in activating Drosophila oocytes were ones that were less phosphorylated in Calcineurin-depleted Drosophila oocytes relative to controls (Zhang et al., 2019). This result suggests that Calcineurin activates intermediate kinases, which normally phosphorylate some target proteins; in the absence of the Calcineurin phosphatase, this phosphorylation cannot occur.

We note many kinases and phosphatases that display differential phosphorylation between these phospho-mass spec datasets. For example, BubR1/BUB1 (Bolanos-Garcia & Blundell, 2011) was found to be phospho-modulated during egg activation and is required in Drosophila for efficient spindle separation during both MI and MII (Malmanche et al., 2007). Another intermediary kinase is Akt1; in Drosophlia, Akt1 is phosphorylated during egg activation in a Calcineurin-dependent manner (Zhang et al., 2019). Akt1's activation disables the GSK3 kinase by phosphorylating its S9 residue (Takeo et al., 2012); this does not occur when Calcineurin is depleted. In Table 2, Column 6, we list notable phospho-regulated kinases or phosphatases that may act downstream of CaMKII, Calcineurin, or other possible apical enzymes.

5. Summary and future directions

5.1 Conservation of phospho-regulation during egg activation

The studies discussed above highlight the importance of protein phospho-regulation during egg activation, although further studies are required to understand its full functional importance and to identify the most important players in the processes. We highlight the dynamics of phospho-regulation vs. protein level regulation during egg activation, and discuss commonly-phospho-regulated processes. An intriguing observation from these data is that there are common pathways and proteins phospho-regulated during egg activation across organisms, even though their oocytes have unique methods of accomplishing the goal of successfully starting embryogenesis. Differences across species include arresting in MI vs. MII vs. G1 (or no arrest);

requirement for egg activation in different environments; different requirements for block to polyspermy; different handling of genetic material (polar body extrusion vs. fusion/degradation). Many conserved proteins are phospho-regulated during egg activation (bolded entries in Table 2), and in cases where orthologous proteins were not detected, activating eggs in the species examined showed similar protein families being phospho-regulated. We hypothesize that phospho-control of egg activation is a deeply conserved developmental module. A recent phospho-mass spectrometry analysis of oocyte maturation in mouse compared its identified phospho-sites to egg activation phospho-sites from fruit fly, frog, and sea star; 76 conserved sites were found (Sun et al., 2024). Computational analyses of phospho-proteins regulated during egg activation across species will be fruitful in identifying conserved proteins and sites to reveal the most deeply conserved signals and whether/how they respond to calcium.

The deep conservation of phospho-regulated processes and proteins described in this review supports the idea that the invertebrates and frogs whose results were discussed here serve as excellent models for studying phospho-protein regulation and its importance for fertility and development in a manner that is applicable to mammalian systems and ultimately human health. However, a high throughput investigation of phospho-dynamics in a mammalian system will be crucial to understanding the level of conservation, and thus relevance, of phospho-modulated proteins in invertebrates/frogs.

5.2 Development of new tools to assess phospho-dynamics

The study of phosphorylation dynamics in a high-throughput capacity requires the use of phospho-enriched peptides followed by mass spectrometry (Gerritsen & White, 2021). This experimental workflow requires substantial amounts of starting protein, limiting available systems to use for such analysis and hampering the temporal sensitivity of the assay in many systems. Further, the phospho-mass spectrometry analyses thus far have relied on bulk samples, potentially missing low-abundance or non-uniform phospho-changes (e.g., where only a small population of a phospho-protein, perhaps in a specific location, undergoes phospho-regulation). Thus, single cell phosphoproteomics will be critical for understanding the true complexity of phospho-signaling during early development. Additionally, systems like sea urchin and frog, which allowed very detailed timecourses, reveal how quickly phosphorylation events can change and thus advancement of single cell (or, at least, "low input") phosphoproteomics will allow for such analyses in species with small oocytes. Recent developments in phospho-mass spectrometry (e.g. nano-phospho-enrichment + near single cell input (Tsai et al., 2023)) show promise that such studies will soon be possible.

Additional technologic advancements are needed at the genetic/protein engineering level. Even excellent genetic models like Drosophila require finer spatiotemporal control than current tools permit, to uncover mechanistic details of phospho-control. For example, the widely used UAS-GAL4 "binary" system (Brand & Perrimon, 1993), which can allow expression of proteins specifically in the female germline, cannot give precise temporal control during egg activation because transcription is turned off in maturing and activating oocytes. One exciting alternative is to use optogenetics to manipulate protein levels or activity without the need for new transcription. For example, upon blue light activation of an optogenetically-controlled SOS protein, MAPK activity can be turned on (Toettcher, Weiner, & Lim, 2013). This tool has allowed for imaging-based studies of MAPK activity and has been coupled with high-throughput technologies, like phospho-mass spectrometry, to uncover MAPK targets (Yang et al., 2024). While not yet applied to the study of egg activation, recent studies with 'Opto-SOS' suggest that optogenetics may be an informative future tool in the study of egg activation.

5.3 Functional dissection of phospho-regulated proteins

Functional dissection of the effects of the phospho-changes during egg activation is a major area for future study. While studies in multiple species strongly suggest widespread functional roles for the protein phosphorylation changes during egg activation, and specific cases have been validated, a large-scale investigation is needed. Moreover, unless a phospho-protein plays an apical role in the signaling cascade, it is likely that a system that allows combinatorial manipulation will be critical to understanding the broader importance of phospho-regulation. Additionally, only in Drosophila has it been possible to link protein phosphorylation changes to the calcium rise (Zhang et al., 2019). Further analysis of calcium's tie to phospho-regulation is needed in other species.

Finally, phosphorylation is just one of numerous post-translations modifications that could potentially be triggered by calcium during egg activation to modulate the function of target proteins. Comprehensive analyses for PTMs and their consequences during egg activation are currently lacking; they will be an interesting avenue for future study.

5.4 Broader applications of the study of protein phosphorylation in egg activation

Elucidating the mechanism by which the oocyte transitions to totipotency is important in understanding germline biology and, more generally, mechanisms that mediate changes in cell state. However, potential biomedical applications could result from understanding egg activation. As noted earlier, individuals whose oocytes cannot activate cannot currently be diagnosed until failed IVF is ameliorated by addition of a calcium ionophore, a blunt tool that introduces excess calcium into the egg to robustly activate it (Kashir et al., 2022). The use of calcium ionophores in fertility medicine is controversial (Anifandis et al., 2019), and recent studies suggest negative consequences of ionophores and excess calcium on embryo development (Chen et al., 2020; Savy, Stein, Shi, & Williams, 2022). If protein phosphorylation regulation is a conserved mechanism by which calcium acts, identifying the network of calcium-dependent phospho-regulated proteins would be of critical importance. Understanding the connection between calcium and phospho-regulation could allow for diagnostic tools to identify abnormal versions of maternal proteins that prevent normal development. It could also allow for optimizing IVF conditions to produced ones that result in the most physiologically-normal response.

Acknowledgments

We thank Drs. Paul Wassarman and Eveline Litscher for the invitation to write this article. We are grateful to Drs. Carmen Williams and Marcus Kilwein, and Melissa Mychalczuk for their helpful insights and comments on the manuscript. We appreciate support for this work from NIH grants R03HD101732 and R01HD114777 (to MFW), and R21HD105230 (to MFW and John Schimenti). JMT is a Cornell CVG Scholar who is also grateful for partial support from Cornell's Harry and Samuel Mann Award.

Appendix A. Supporting information

Supplementary data associated with this article can be found in the online version at https://doi.org/10.1016/bs.ctdb.2025.01.001.

References

Afshar, K., Stuart, B., & Wasserman, S. A. (2000). Functional analysis of the drosophila diaphanous FH protein in early embryonic development. *Development, 127*, 1887–1897.

Anifandis, G., Michopoulos, A., Daponte, A., Chatzimeletiou, K., Simopoulou, M., Messini, C. I., ... Goulis, D. G. (2019). Artificial oocyte activation: Physiological, pathophysiological and ethical aspects. *Systems Biology in Reproductive Medicine, 65*(1), 3–11.

Athwal, H., Kochiyanil, A., Bhat, V., Allan, A. L., & Parsyan, A. (2024). Centrosomes and associated proteins in parthenogenesis and treatment of breast cancer. *Frontiers in Oncology, 14*.

Backs, J., Stein, P., Backs, T., Duncan, F. E., Grueter, C. E., McAnally, J., ... Olson, E. N. (2010). The γ isoform of CaM kinase II controls mouse egg activation by regulating cell cycle resumption. *Proceedings of the National Academy of Sciences, 107*, 81–86.

Ballinger, D. G., & Hunt, T. (1981). Fertilization of sea urchin eggs is accompanied by 40S ribosomal subunit phosphorylation. *Developmental Biology, 87*, 277–285.

Bearss, J. J., Padi, S. K. R., Singh, N., Cardo-Vila, M., Song, J. H., Mouneimne, G., ... Okumura, K. (2021). EDC3 phosphorylation regulates growth and invasion through controlling P-body formation and dynamics. *EMBO Reports, 22*.

Becker, K. A., & Hart, N. H. (1999). Reorganization of filamentous actin and myosin-II in Zebrafish eggs correlates temporally and spatially with cortical granule exocytosis. *Journal of Cell Science, 112*, 97–110.

Blachon, S., Khire, A., & Avidor-Reiss, T. (2014). The Origin of the Second Centriole in the Zygote of Drosophila melanogaster. *Genetics, 197*(1), 199–205. https://doi.org/10.1534/genetics.113.160523 Epub2014 Feb 13. PMID: 24532732.

Bodakuntla, S., Jijumon, A. S., Villablanca, C., Gonzalez-Billault, C., & Janke, C. (2019). Microtubule-associated proteins: Structuring the cytoskeleton. *Trends in Cell Biology, 29*(10), 804–819.

Bolanos-Garcia, V. M., & Blundell, T. L. (2011). BUB1 and BUBR1: Multifaceted kinases of the cell cycle. *Trends in Biochemical Sciences, 36*(3), 141–150.

Brand, A. H., & Perrimon, N. (1993). Targeted gene expression as a means of altering cell fates and generating dominant phenotypes. *Development, 118*(2), 401–415. https://doi.org/10.1242/dev.118.2.401.

Camlin, N. J., McLaughlin, E. A., & Holt, J. E. (2017). Motoring through: The roles of kinesin superfamily proteins in female meiosis. *Human Reproduction Update, 23*(4), 409–420.

Cassani, M., & Seydoux, G. (2024). P-body-like condensates in the germline. *Seminars in Cell & Developmental Biology, 157*, 24–32.

Chen, J., Melton, C., Suh, N., Oh, J. S., Horner, K., Xie, F., ... Conti, M. (2011). Genome-wide analysis of translation reveals a critical role for deteled in azoospermia-like (Dazl) at the oocyte-to-zygote transition. *Genes and Development, 25*, 755–766.

Chen, C., Sun, T., Yin, M., Yan, Z., Yu, W., Long, H., ... Lyu, Q. (2020). Ionomycin-induced mouse oocyte activation can disrupt preimplantation embryo development through increased reactive oxygen species reaction and DNA damage. *Molecular Human Reproduction, 26*(10), 773–783.

Chousal, J. N., Cho, K., Ramaiah, M., Skarbrevik, D., Mora-Castilla, S., Stumpo, D. J., ... Cook-Andersen, H. (2018). Chromatin modification and global transcriptional silencing in the oocyte mediated by the mRNA decay activator ZFP36L2. *Develompental Cell, 44*, 392–402.

Collas, P., Thompson, L., Fields, A. P., Poccia, D. L., & Courvalin, J. (1997). Protein kinase C-mediated interphase lamin B phosphorylation and solubilization. *Journal of Biological Chemistry, 272*(34), 21274–21280.

Coy, P., García-Vázquez, F. A., Visconti, P. E., & Avilés, M. (2012). Roles of the oviduct in mammalian fertilization. *Reproduction, 144*(6), 649–660. https://doi.org/10.1530/REP-12-0279 Retrieved Feb 4, 2025.

Delom, F., & Chevet, E. (2006). Phosphoprotein analysis: From proteins to proteomes. *Proteome Science, 4*(15).

Diluigi, A., Weitzman, V. N., Pace, M. C., Siano, L. J., Maier, D., & Mehlmann, L. M. (2008). Meiotic arrest in human oocytes is maintained by a GS signaling pathway. *Biology of Reproduction, 78*, 667–672.

Drake, J. M., Paull, E. O., Graham, N. A., Lee, J. K., Smith, B. A., Titz, B., ... Stuart, J. M. (2016). Phosphoproteome integration reveals patient-specific networks in prostate cancer. *Cell, 166*, 1041–1054.

Duan, X., & Sun, S. (2019). Actin cytoskeleton dynamics in mammalian oocyte meiosis. *Biology of Reproduction, 100*(1), 15–24.

Ducibella, T., & Fissore, R. (2008). The roles of Ca2+, downstream protein kianses, and oscillatory signaling in regulating fertilization and the activation of development. *Developmental Biology, 315*, 257–279.

Ducibella, T., Huneau, D., Angelichio, E., Xu, Z., Schultz, R. M., Kopf, G. S., ... Ozil, J. (2002). Egg-to-embryo transition is driven by differential responses to Ca2+ oscillation number. *Developmental Biology, 250*, 280–291.

Dupre, A., Haccard, O., & Jessus, C. (2011). Mos in the oocyte: How to use MAPK independently of growth factors and transcription to control meiotic divisions. *Journal of Signal Transduction.*

Elbaz, J., Reizel, Y., Nevo, N., Galiani, D., & Dekel, N. (2010). Epithelial cell transforming protein 2 (ECT2) depletion blocks polar body extrusion and generates mouse oocytes containing two metaphase II spindles. *Endocrinology, 151*(2), 755–765.

Enserink, J. M., & Kolodner, R. D. (2010). An overview of Cdk1-controlled targets and processes. *Cell Division, 5*(11).

Esposito, G., Vitale, A. M., Leijten, F. P. J., Strik, A. M., Koonen-Reemst, A. M. C. B., Yurttas, P., ... Gossen, J. A. (2007). Peptidylarginine deiminase (PAD) 6 is essential for oocyte cytoskeletal sheet formation and female fertility. *Molecular and Cellular Endocrinology, 273*, 25–31.

Fellmeth, J. E., & McKim, K. S. (2022). Brief history of drosophila (female) meiosis. *Genes, 13*, 775.

Gerritsen, J. S., & White, F. M. (2021). Phosphoproteomics: A valuable tool for uncovering molecular signaling in cancer cells. *Expert Review of Proteomics, 18*(8), 661–674.

Gloerich, M., Vliem, M. J., Prummel, E., Meijer, L. A. T., Rensen, M. G. A., Rehmann, H., & Bos, J. L. (2011). The nucleoporin RanBP2 tethers the cAMP effector Epac1 and inhibits its catalytic activity. *Journal of Cell Biology, 193*(6), 1009–1020.

Gonzalez-Garcia, J. R., Bradley, J., Nomikos, M., Paul, L., Machaty, Z., Lai, F. A., & Swann, K. (2014). The dynamics of MAPK inactivation at fertilization in mouse eggs. *Journal of Cell Science, 127*, 2749–2760.

Gorjanacz, M., Torok, I., Pomozi, I., Garab, G., Szlanka, T., Kiss, I., & Mechler, B. M. (2006). Domains of Importin-a2 required for ring canal assembly during Drosophila oogenesis. *Journal of Structural Biology, 154*, 27–41.

Guo, H., Garcia-Vedrenne, A. E., Isserlin, R., Lugowski, A., Morada, A., Sun, A., ... Emili, A. (2015). Phosphoproteomic network analysis in the sea urchin Strongylocentrotus purpuratus reveals new candidates in egg activation. *Proteomics, 15*, 4080–4095.

Halet, G. (2004). PKC signaling at fertilization in mammalian eggs. *Biochimica et Biophysica Acta, 1742*, 185–189.

Hampoelz, B., Schwarz, A., Ronchi, P., Bragulat-Teixidor, H., Tischer, C., Gaspar, I., ... Beck, M. (2019). Nuclear pores assemble from nucleoporin condensates during oogenesis. *Cell, 179*, 671–686.

Hansen, D. V., Tung, J. J., & Jackson, P. K. (2006). CaMKII and Polo-like kinase 1 sequentially phosphorylate the cytostatic factor Emi2/XErp1 to trigger its destruction and meiotic exit. *Proceedings of the National Academy of Sciences, 103*, 608–613.

Hara, M., Lourido, S., Petrova, B., Lou, H. J., Von Stetina, J. R., Kashevsky, H., ... Orr-Weaver, T. L. (2018). Identification of PNG kinase substrates uncovers interactions with the translational repressor TRAL in the oocyte-to-embryo transition. *eLife, 7*.

Hara, M., Petrova, B., & Orr-Weaver, T. L. (2017). Control of PNG kinase, a key regulator of mRNA translation, is coupled to meiosis completion at egg activation. *eLife, 6*.

Harasimov, K., Gorry, R. L., Welp, L. M., Penir, S. M., Horokhovskyi, Y., Cheng, S., ... Schuh, M. (2024). The maintenance of oocytes in the mammalian ovary involves extreme protein longevity. *Nature Cell Biology, 26*, 1124–1138.

Harsha, H. C., & Pandey, A. (2010). Phosphoproteomics in cancer. *Molecular Oncology, 4*, 482–495.

Heifetz, Y., Yu, J., & Wolfner, M. F. (2001). Ovulation triggers activation of drosophila oocytes. *Developmental Biology, 234*(2), 416–424.

Heim, A., Tischer, T., & Mayer, T. U. (2018). Calcineurin promotes APC/C activation at meiotic exit by acting on both XErp1 and Cdc20. *EMBO Reports, 19*.

Horner, V. L., Czank, A., Jang, J. K., Singh, N., Williams, B. C., Puro, J., ... Goldberg, M. L. (2006). The Drosophila calcipressin sarah is required for several aspects of egg activation. *Current Biology, 15*, 1441–1446.

Hu, Q., & Wolfner, M. F. (2019). The Drosophila Trpm channel mediates calcium influx during egg activation. *Proceedings of the National Academy of Sciences, 116*, 18994–19000.

Inoue, D. O., Kanemore, M., Nobui, Y., & Sagata, T. N. (2007). A direct link of the Mos-MAPK pathway to Erp1/Emi2 in meiotic arrest of Xenopus laevis eggs. *Nature, 446*, 1100–1104.

Iyyappan, R., Aleshkina, D., Ming, H., Dvoran, M., Kakavand, K., Jansova, D., ... Susor, A. (2023). The translational oscillation in oocyte and early embryo development. *Nucleic Acids Research, 51*(22), 12076–12091.

Jackson, R. J., Hellen, C. U. T., & Pestova, T. V. (2010). The mechanism of eukaryotic translation initiation and principles of its regulation. *Nature Reviews, 10*, 113–127.

Jentoft, I. M. A., Bauerlein, F. J. B., Welp, L. M., Cooper, B. H., Petrovic, A., So, C., ... Schuh, M. (2023). Mammalian oocytes store proteins for the early embryo on cytoplasmic lattices. *Cell, 186*, 5308–5327.

Jones, K. T. (2004). Turning it on and off: M-phase promoting factor during meiotic maturation and fertilization. *Molecular Human Reproduction, 10*(1), 1–5.

Kaneuchi, T., Sartain, C. V., Takeo, S., Horner, V. L., Buehner, N. A., Aigaki, T., & Wolfner, M. F. (2015). Calcium waves occur as Drosophila oocyte activate. *Proceedings of the National Academy of Sciences, 112*, 791–796.

Kashir, J., Ganesh, D., Jones, C., & Coward, K. (2022). Oocyte activation deficiency and assisted oocyte activation: mechanisms, obstacles and prospects for clinical application. *Human Reproduction Open*, 1–29.

Kelm, O., Wind, M., Lehmann, W. D., & Nigg, E. A. (2002). Cell cycle-regulated phosphrylation of the Xenopus polo-like kinase Plx1. *Journal of Biological Chemistry, 277*(28), 25247–25256.

Kinoshita, E., Kinoshita-Kikuta, E., Takiyama, K., & Koike, T. (2006). Phosphate-binding tag, a new tool to visualize phosphorylated proteins. *Molecular & Cellular Proteomics. 5.4*, 749–757.

Kishigami, S., & Wakayama, T. (2007). Efficient strontium-induced activation of mouse oocytes in standard culture media by chelating calcium. *Journal of Reproduction and Development, 53*(6), 1207–1215.

Knott, J., Gardner, A. J., Madgwick, S., Jones, K. T., Williams, C. J., & Schultz, R. M. (2006). Calmodulin-dependent protein kinase II triggers mouse egg activation and embryo development in the absence of Ca2+ oscillations. *Developmental Biology, 296*, 388–395.

Kono, T., Obata, Y., Wu, Q., Niwa, K., Ono, Y., Yamamoto, Y., ... Ogawa, H. (2004). Birth of parthenogenetic mice that can develop to adulthood. *Nature, 428*, 860–864.

Kramer, D. A., Piper, H. K., & Chen, B. (2022). WASP family proteins: Molecular mechanisms and implications in human disease. *European Journal of Cell Biology*.

Krauchunas, A. R., Horner, V. L., & Wolfner, M. F. (2012). Protein phosphorylation changes reveal new candidates in the regulation of egg activation and early embryogenesis in D. melanogaster. *Developmental Biology, 370*, 125–134.

Kronja, I., Whitefield, Z. J., Yuan, B., Dzeyk, K., Kirkpatrick, J., Krijgsveld, J., & Or-Weaver, T. L. (2014). Quantitative proteomics reveals the dynamics of protein changes during Drosophila oocyte maturation and the oocyte-to-embryo transition. *Proceedings of the National Academy of Sciences, 111*(45), 16023–16028.

Kronja, I., Yuan, B., Eichhorn, S. W., Dzeyk, K., Krijgsveld, J., Bartel, D. P., & Orr-Weaver, T. L. (2014). Widespread changes in the posttranscriptional landscape at the drosophila oocyte-to-embryo transition. *Cell Reports, 7*, 1495–1508.

Kubiak, J., Paldi, A., Weber, M., & Maro, B. (1991). Genetically identical parthenogeneticm ouse embryos produced by inhibition of the first meiotic cleavage with cytochalasin D. *Development (Cambridge, England), 111*, 763–769.

Kubiak, J. Z., Weber, M., de Pennart, H., Winston, N. J., & Maro, B. (1993). The metaphase II arrest in mouse oocytes is controlled through microtubule-dependent destrubtion of cyclin B in the presence of CSF. *The EMBO Journal, 12*(10), 3773–3778.

Kunitomi, C., Romero, M., Daldello, E. M., Schindler, K., & Conti, M. (2024). Multiple intersecting pathways are involved in CPEB1 phosphorylation and regulation of translation during mouse oocyte meiosis. *Development (Cambridge, England), 151*.

Labit, H., Fujimitsu, K., Bayin, N. S., Takaki, T., Gannon, J., & Tamano, H. (2012). Dephosphorylation of Cdc20 is required for its C-box-dependent activation of the APC/C. *The EMBO Journal, 31*, 3351–3362.

Lantz, V. A., & Miller, K. G. (1998). A class VI unconventional myosin is associated with a homologue of a microtubule-binding protein, cytoplasmic linker protein-170, in neurons and at the posterior pole of Drosophila embryos. *Journal of Cell Biology, 140*(4), 897–910.

Lee, L. A., Hoewyk, D. V., & Orr-Weaver, T. L. (2003). The Drosophila cell cycle kianse PAN GU forms an active complex with PLUTONIUM and GNU to regulate embryonic divisions. *Genes and Development, 17*, 2979–2991.

Lin, M., Jiao, X., Grima, D., Newbury, S. F., Kiledjian, M., & Chou, T. (2008). Drosophila processing bodies in oogenesis. *Developmental Biology, 322*, 276–288.

Liu, S. Y., & Ikegami, K. (2020). Nuclear lamin phosphorylation: An emerging role in gene regulation and pathogenesis of laminopathies. *Nucleus, 11*(1), 299–314.

Liu, Y., Zhao, Q., Xu, F., Wang, K., Zhao, Y., Chen, H., ... Zhang, J. (2021). Dysregulation of phosphoproteins in hepatocellular carcinoma revealed via quantitative analysis of the phosphoproteome. *Oncology Letters, 21*(117).

Loppin, B., Dubruille, R., & Horard, B. (2015). The intimate genetics of Drosophila fertilization. *Open Biology, 5*, 150076.

Ma, S., Liu, X., Miao, D., Han, Z., Zhang, X., Miao, Y., ... Tan, J. (2005). Parthenogenetic activation of mouse oocytes by strontium chloride: A search for the best conditions. *Theriogenology, 64*, 1142–1157.

Maddox, A. S., Azoury, J., & Dumont, J. (2012). Polar body cytokinesis. *Cytoskeleton, 69*, 855–868.

Madgwick, S., & Jones, K. T. (2007). How eggs arrest at metaphase II: MPF stabilization plus APC/C inhibition equals Cytostatic Factor. *Cell Division, 2*(4).

Mahbub Hasan, A. K. M., Matsumoto, T., Kihira, S., Yoshida, J., & Sato, K. (2012a). Phospho-signaling at oocyte maturation and fertilization: Set up for embryogenesis and beyond Part I. Protein kinases. In K. Sato (Ed.). *Embryogenesis* (pp. 447–498). InTech.

Mahbub Hasan, A. K. M., Matsumoto, T., Kihira, S., Yoshida, J., & Sato, K. (2012b). Phospho-signaling at oocyte maturation and fertilization: Set up for embryogenesis and beyond Part II. Kinase regulators and substrates. In K. Sato (Ed.). *Embryogenesis* (pp. 499–554). InTech.

Malmanche, N., Owen, S., Gegick, S., Steffensen, S., Tomkiel, J. E., & Sunkel, C. E. (2007). Drosophila BubR1 is essential for meiotic sister-chromatid cohesion and maintenance of synaptonemal complex. *Current Biology, 17*, 1489–1497.

Markoulaki, S., Matson, S., & Ducibella, T. (2004). Fertilization stimulates long-lasting oscillations of CaMKII activity in mouse eggs. *Developmental Biology, 272*, 15–25.

Markoulaki, S., Matson, S., Abbott, A. L., & Ducibella, T. (2003). Oscillatory CaMKII activity in mouse egg activation. *Developmental Biology, 258*, 464–474.

Marston, A. L., & Amon, A. (2004). Meiosis: Cell-cycle controls shuffle and deal. *Nature Reviews, 5*, 983–997.
Martin-Perez, J., Rudkin, B. B., Siegmann, M., & Thomas, G. (1986). Activation of ribosomal protein S6 phosphorylation during meiotic maturation of Xenopus laevis oocytes: In vitro ordered appearance of S6 phosphopeptides. *The EMBO Journal, 5*(4), 725–731.
Matson, S., Markoulaki, S., & Ducibella, T. (2006). Antagonists of myosin light chain kinase and of myosin II inhibit specific events of egg activation in fertilized mouse eggs. *Biology of Reproduction, 74*, 167–176.
McAvey, B. A., Wortzman, G. B., Williams, C. J., & Evans, J. P. (2002). Involvement of calcium signaling and the actin cytoskeleton in the membrane block to polyspermy in mouse eggs. *Biology of Reproduction, 67*, 1342–1352.
Meaders, J. L., & Burgess, D. R. (2020). Mucrotubule-based mechanisms of pronuclear positioning. *Cells, 9*(505).
Medvedev, S., Stein, P., & Schultz, R. M. (2014). Specificity of calcium/calmodulin-dependent protein kinases in mouse egg activation. *Cell Cycle (Georgetown, Tex.), 13*(9), 1482–1488.
Mochida, S., & Hunt, T. (2007). Calcineurin is required to release Xenopus egg extracts from meiotic M phase. *Nature Letters, 449.*
Mochida, S., Maslen, S. L., Skehel, M., & Hunt, T. (2010). Greatwall phosphorylates an inhibitor of protein phosphatase 2A that is essential for mitosis. *Science, 330*, 1670–1673.
Mogessie, B., & Schuh, M. (2017). Actin protects mammalian eggs against chromosome segregation errors. *Science, 357.*
Moos, J., Kopf, G. S., & Schultz, R. M. (1996). Cycloheximide-induced activation of mouse eggs: Effects on cdc2/cyclin B and MAP kinase activities. *Journal of Cell Science, 109*, 739–748.
Moos, J., Xu, Z., Schultz, R. M., & Kopf, G. S. (1996). Regulation of nuclear envelope assembly/disassembly by MAP kinase. *Developmental Biology, 175*, 358–361.
Morgan, M., Kumar, L., Li, Y., & Baptissart, M. (2021). Post-transcriptional regulation in spermatogenesis: All RNA pathways lead to healthy sperm. *Cellular and Molecular Life Sciences, 78*, 8049–8071.
Naydenov, N. G., & Ivanov, A. I. (2011). Spectrin-adducin membrane skeleton. *BioArchitechture, 1*(4), 186–191.
Nishiyama, T., Yoshizaki, N., Kishimoto, T., & Ohsumi, K. (2007). Transient activation of calcineurin is essential to initiate embryonic development in Xenopous laevis. *Nature, 449*, 341–346.
Oh, J. S., Susor, A., & Conti, M. (2011). Protein tyrosine kinase Wee1B Is essential for metaphase II exit in mouse oocytes. *Science, 332*, 462–465.
Okuwaki, M., Sumi, A., Hisaoka, M., Saotome-Nakamura, A., Akashi, S., Nishimura, Y., & Nagata, K. (2012). Function of homo- and hetero-oligomers of human nucleoplasmin/nucleophosmin family proteins NPM1, NPM2, and NPM3 during sperm chromatin remodeling. *Nucleic Acids Research, 40*(11), 4861–4878.
Orr, B., Godek, K. M., & Comtpon, D. (2015). Aneuploidy. *Current Biology, 25*, R523–R548.
Otsuka, S., & Ellenberg, J. (2018). Mechanisms of nuclear pore complex assembly – Two different ways of building one molecular machine. *FEBS Letters, 592*, 475–488.
Pesin, J. A., & Orr-Weaver, T. L. (2008). Regulation of APC/C activators in mitosis and meiosis. *Annual Review of Cell and Developmental Biology, 24*, 475–499.
Petibon, C., Ghulam, M. M., Catala, M., & Elela, S. A. (2020). Regulation of ribosomal protein genes: An ordered anarchy. *WIREs RNA, 12.*
Porter, D. C., & Vacquier, V. D. (1983). Phosphorylation of sperm histone H1 Is induced by the egg jelly layer in the sea urchin strongylocentrotus purpuratus. *Developmental Biology, 116*, 203–212.

Potapova, T. A., Daum, J. R., Byrd, K. S., & Gorbsky, G. J. (2009). Fine tuning the cell cycle: Activation of the Cdk1 inhibitory phosphorylation pathway during mitotic exit. *Molecular Biology of the Cell, 20*, 1737–1748.

Presler, M., Van Itallie, E., Klein, A. M., Kunz, R., Coughlin, M. L., Peshkin, L., ... Kirschner, M. W. (2017). Proteomics of phosphorylation and protein dynamics during fertilization and meiotic exit in the Xenopus egg. *Proceedings of the National Academy of Sciences*, E10838–E10847.

Rauh, N. R., Schimdt, A., Bormann, J., Nigg, E. A., & Mayer, T. U. (2005). Calcium triggers exit from meiosis II by targeting the APC/C inhibitor XErp1 for degradation. *Nature, 437*(13), 1048–1052.

Roberts, A. J., Kon, T., Knight, P. L., Sutoh, K., & Burgess, S. A. (2013). Functions and mechnics of dynein motor proteins. *Nature Reviews, 14* 713-276.

Rojas, J., Hinostroza, F., Vergara, S., Pinto-Borguero, I., Aguilera, F., Fuentes, R., & Carvacho, I. (2021). Knockin' on egg's door: Maternal control of egg activation that influences cortical granule exocytosis in animal species. *Frontiers in Cell and Developmental Biology, 9*.

Rouhana, L., Edgar, A., Hugosson, F., Dountcheva, V., Martindale, M. Q., & Ryan, J. F. (2023). Cytoplasmic polyadeylation is an ancestral hallmark of early development in animals. *Molecular Biology and Evolution, 40*(6).

Roux, M. M., Radeke, M. J., Goel, M., Mushegian, A., & Folz, K. R. (2008). 2DE identification of proteins exhibiting turnover and phosphorylation dynamics during sea urchin egg activation. *Developmental Biology, 313*, 630–647.

Sackton, K. L., Buehner, N. A., & Wolfner, M. F. (2007). Modulation of MAPK activities during egg activation in drosophila. *Fly, 1*(4), 222–227.

Sackton, K. L., Lopez, J. M., Berman, C. L., & Wolfner, M. F. (2009). YA is needed for proper nuclear organization to transition between meiosis and mitosis in Drosophila. *BMC Developmental Biology, 9*, 43.

Sako, K., Suzuki, K., Isoda, M., Yoshikai, S., Senoo, C., Nakajo, N., ... Sagata, N. (2014). Emi2 mediates meiotic MII arrest by competitively inhibiting the binding of Ube2S to the APC/C. *Nature Communications, 5*.

Sanghavi, P., Lu, S., & Gonsalvez, G. B. (2012). A functional link between localized Oskar, dynamic microtubules, and endocytosis. *Developmental Biology, 367*, 66–77.

Sankaranarayanan, M., Emenecker, R. J., Wilby, E. L., Jahnel, M., Trussina, I. R. E. A., Wayland, M., ... Weil, T. T. (2021). Adaptable P body physical states differentially regulate bicoid mRNA storage during early Drosophila development. *Develompental Cell, 56*, 2886–2901.

Saunders, R. D. C., Avides, M. C., Howard, T., Gonzalez, C., & Glover, D. M. (1997). The Drosophila gene abnormal spindle encodes a novel microtubule-associated protein that associates with the polar regions of the mitotic spindle. *Journal of Cell Biology, 137*(4), 881–891.

Saunders, C. M., Larman, M. G., Parrington, J., Cox, L. J., Royse, J., Blayney, L. M., ... Lai, F. A. (2002). PLCζ: a sperm-specific trigger of Ca2+ oscillations in eggs and embryo development. *Development, 129*, 3533–3544.

Savy, V., Stein, P., Shi, M., & Williams, C. J. (2022). PMCA1 depletion in mouse eggs amplifies calcium signaling and impacts offspring growth. *Biology of Reproduction, 107*(6), 1439–1451.

Sha, Q., Dai, X., Dang, Y., Tang, F., Liu, J., Zhang, Y., & Fan, H. (2017). A MAPK cascade couples maternal mRNA translation and degradation to meiotic cell cycle progression in mouse oocytes. *Development, 144*, 452–463.

ShiYang, X., Miao, Y., Cui, Z., Lu, Y., Zhou, C., Zhang, Y., & Xiong, B. (2020). Casein kinase 2 modulates the spindle assembly checkpoint to orchestrate porcine oocyte meiotic progression. *Journal of Animal Science and Biotechnology, 11*, 31.

Siracusa, G., Whittingham, D. G., Molinaro, M., & Vivarelli, E. (1973). Parthenogenetic activation of mouse oocytes induced by inhibitors of protein synthesis. *Journal of Embryology and Experimental Morphology, 43*, 157–166.

Sperling, A. L., Fabian, D. K., Garrison, E., & Glover, D. M. (2023). A genetic basis for facultative parthenogenesis in Drosophila. *Current Biology, 33*, 3545–3560.

Stein, P., Savy, V., Williams, A. M., & Williams, C. J. (2020). Modulators of calcium signalling at fertilization. *Open Biology, 10.*

Stewart, M. (2007). Molecular mechanism of nuclear protein import cycle. *Nature Reviews, 8*, 195–208.

Stricker, S. A. (1999). Comparative biology of calcium signaling during fertilization and egg activation in animals. *Developmental Biology, 211*, 157–176.

Sun, H., Han, L., Guo, Y., An, H., Wang, B., Zhang, X., ... Wang, Q. (2024). The global phosphorylation landscape of mouse oocytes during meiotic maturation. *The EMBO Journal, 43*(20), 4752–4785.

Surani, M. A. H., & Barton, S. C. (1983). Development of gynogenetic eggs in the mouse: Implications for parthenogenetic embryos. *Science, 222*, 1034–1036.

Surks, H. K., Riddick, N., & Ohtani, K. (2005). M-RIP targets myosin phosphatase to stress fibers to regulate myosin light chain phosphorylation in vascular smooth muscle cells. *Journal of Biological Chemistry, 280*(52), 42543–42551.

Swan, A., & Schupbach, T. (2007). The Cdc20 (Fzy)/Cdh1-related protein, cort, cooperates with Fzy in cyclin destruction and anaphase progression in meiosis I and II in Drosophila. *Development, 134*, 891–899.

Swann, K., & Lai, F. A. (2016). Egg activation at fertilization by a soluble sperm protein. *Physiological Reviews, 96*, 127–149.

Swartz, S. Z., Nguyen, H. T., McEwan, B. C., Adamo, M. E., Cheeseman, I. M., & Kettenback, A. N. (2021). Selective dephosphorylation by PP2A-B55 directs the meiosis I-meiosis II transition in oocytes. *eLife, 10*, e70588.

Tadros, W., Goldman, A. L., Babak, T., Menzies, F., Vardy, L., Orr-Weaver, T., ... Lipshitz, H. D. (2007). SMAUG is a major regulator of maternal mRNA destabilization in drosophila and its translation is actiated by the PAN GU kinase. *Developmental Cell, 12*, 143–155.

Takayama, J., & Onami, S. (2016). The sperm TRP-3 channel mediates the onset of a Ca2+ wave in the fertilized C. elegans oocyte. *Cell Reports, 15*, 625–637.

Takeo, S., Hawley, R. S., & Aigaki, T. (2010). Calcineurin and its regulation by Sra/RCAN is required for completion of meiosis in Drosophila. *Developmental Biology, 344*, 957–967.

Takeo, S., Swanson, S. K., Nandanan, K., Nakai, Y., Aigaki, T., Washburn, M. P., ... Hawley, R. S. (2012). Shaggy/glycogen synthase kinase 3B and phosphorylation of sarah/regulator of calcineurin are essential for completion of Drosophila female meiosis. *Proceedings of the National Academy of Sciences, 109*(17), 6382–6389.

Tembo, M., Sauer, M. L., Wisner, B. W., Beleny, D. O., Napolitano, M. A., & Carlson, A. E. (2021). Actin polymerization is not required for the fast block to polyspermy in the African clawed frog, Xenopus laevis. *microPublication Biology.*

Toettcher, J. E., Weiner, O.,D., & Lim, W. A. (2013). Using optogenetics to interrogate the dynamic control of signal transmission by the ras/erk module. *Cell, 155*, 1422–1434.

Tombes, R. M., & Peppers, L. S. (1995). Sea urchin fertilization stimulates CaM kinase-II (multifunctional [type II] Ca2+/CaM kinase) activity and association with p34cdc2. *Development, Growth & Differentiation, 37*, 589–596.

Tsai, C., Wang, Y., Hsu, C., Kitata, R. B., Chu, R. K., Velickovic, M., ... Liu, T. (2023). A streamlined tandem tip-based workflow for sensitive nanoscale phosphoproteomics. *Communications Biology, 6*, 70.

Tumova, L., Chmelikova, E., Zalmanova, T., Kucerova-Chrpova, V., Romar, R., Dvorakova, M., ... Petr, J. (2016). Calcineurin role in porcine oocyte activation. *Animal, 10*(12), 1998–2007.

Vastenhouw, N. L., Cao, W. X., & Lipshitz, H. D. (2019). The maternal-to-zygotic transition revisited. *Development, 146.*

Vicente-Manzanares, M., Ma, X., Adelstein, R. S., & Horwitz, A. R. (2009). Non-muscle myosin II takes centre stage in cell adhesion and migration. *Nature Reviews, 10*, 778–790.

Vlastaridis, P., Kyriakidou, P., Chaliotis, A., Van De Peer, Y., Oliver, S. G., & Amoutzias, G. D. (2017). Estimating the total number of phosphoproteins and phosphorylation sites in eukaryotic proteomics. *GigaScience, 6*, 1–11.

Vogt, E., Tokuhiro, K., Guo, M., Dale, R., Yang, G., Shin, S., ... Dean, J. (2019). Anchoring cortical granules in the cortex ensures trafficking to the plasma membrane for post-fertilization exocytosis. *Nature Communications, 10.*

Wang, M., Ly, M., Lugowski, A., Laver, J. D., Lipshitz, H. D., Smibert, C. A., & Rissland, O. S. (2017). ME31B globally represses maternal mRNAs by two distinct mechanisms during the Drosophila maternal-to-zygotic transition. *eLife, 6.*

Wei, Y., Yang, C., & Zhao, Z. (2022). Vialbe offspring derived from single unfertilized mammalian oocytes. *Proceedings of the National Academy of Sciences, 119*(12).

Weil, T. T., Parton, R. M., Herpers, B., Soetaert, J., Veenendaal, T., Xanthakis, D., ... Davis, I. (2012). Drosophila patterning is established by differential association of mRNAs with P bodies. *Nature Cell Biology, 14*(12), 1305–1313.

Weinreb, C., Rodriguiez-Fraticelli, A., Camargo, F. D., & Klein, A. M. (2020). Lineage tracing on transcriptional landscapes links state to fate during differentiation. *Science, 367*(755).

Went, D. F. (1982). Egg activation and parthenogenetic reproduction in insects. *Biological Reviews, 57*, 319–344.

Went, D. F., & Krause, G. (1973). Normal development of mechanically activated, unlaid eggs of an endoparasitic hymenopteran. *Nature, 244*, 454–455.

White, K. P., Rifkin, S. A., Hurba, P., & Hogness, D. S. (1999). Microarray znalysis of Drosophila development during metamorphosis. *Science, 286*, 2179–2184.

Wood, B. W., Shi, X., & Weil, T. T. (2024). F-actin coordinates spindle morphology and function in Drosophila meiosis. *PLoS Genetics, 20*(1).

Xiao, D., Kim, H. J., Pang, I., & Yang, P. (2022). Functional analysis of the stable phosphoproteome reveals cancer vulnerabilities. *Bioinformatics, 38*(7), 1956–1963.

Yang, L., Zhu, A., Aman, J. M., Denberg, D., Kilwein, M. D., Marmion, R. A., ... Shvartsman, S. Y. (2024). ERK synchronizes embryonic cleavages in Drosophila. *Developmental Cell, 59*, 3061–3071.

Yestes, M., Jones, C., Amdani, S. N., Patel, S., & Coward, K. (2015). Oocyte activation deficiency: A role for an oocyte contribution? *Human Reproduction Update, 22*, 23–47.

Yilmaz, S., Lopes, F. B. T. P., Schlatzer, D., Wang, R., Qi, X., Koyuturk, M., & Chance, M. R. (2024). Exploring temporal and sex-linked dysregulation in Alzheimer disease phosphoproteome. *iScience, 27.*

York-Andersen, A. H., Hu, Q., Wood, B. W., Wolfner, M. F., & Weil, T. T. (2019). A calcium-mediated actin redistribution at egg activation in Drosophila. *Molecular Reproduction & Development, 87*, 293–304.

York-Andersen, A. H., Parton, R. M., Bi, C. J., Bromley, C. L., Davis, I., & Weil, T. T. (2015). A single and rapid calcium wave at egg activation in Drosophila. *Biology Open, 4*, 553–560.

Yu, J. L. J., Song, K., Turner, S. G., & Wolfner, M. F. (1999). Nuclear entry of the Drosophila melanogaster nuclear lamina protein YA correlates with developmentally regulated changes in its phosphorylation state. *Developmental Biology, 210*, 124–134.

Yu, J., Zhao, Y., Li, Z., Galas, S., & Goldberg, M. L. (2006). Greatwall kinase participates in the Cdc2 autoregulatory loop in Xenopus egg extracts. *Molecular Cell, 22*, 83–91.
Zhang, Z., Ahmed-Braimah, Y. H., Goldberg, M. L., & Wolfner, M. F. (2019). Calcineurin-dependent protein phosphorylation changes during egg activation in Drosophila melanogaster. *Molecular & Cellular Proteomics, 18*, S145–S158.
Zhang, H., Ji, S., Zhang, K., Chen, Y., Ming, J., Kong, F., ... Xie, W. (2023). Stable maternal proteins underlie distinct transcriptome, translatome, and proteome reprogramming during mouse oocyte-to-embryo transition. *Genome Biology, 24*(166).
Zhang, Z., Krauchunas, A. R., Huang, S., & Wolfner, M. F. (2018). Maternal proteins that are phosphoregulated upon egg activation include crucial factors for oogenesis, egg activation and embryogenesis in Drosophila melanogaster. *Genes Genomes Genetics, 8*, 3005–3018.
Zhang, X., Ma, C., Miller, A. L., Katbi, H. A., Bement, W. M., & Liu, X. J. (2008). Polar body emission requires a RhoA contractile ring and Cdc42-mediated membrane protrusion. *Developmental Cell, 15*, 386–400.

CHAPTER SEVEN

The mammalian egg's zona pellucida, fertilization, and fertility☆

Eveline S. Litscher* **and Paul M. Wassarman***
Department of Cell, Developmental, and Regenerative Biology, Icahn School of Medicine at Mount Sinai, New York, NY, United States
*Corresponding authors. e-mail address: eveline.litscher@mssm.edu; paul.wassarman@mssm.edu

Contents

☆ *Current Topics in Developmental Biology* "Fertilization and Activation of Development", Volume 164, 2025

Current Topics in Developmental Biology, Volume 162
ISSN 0070-2153, https://doi.org/10.1016/bs.ctdb.2024.10.008

Abstract

The zona pellucida (ZP) is a relatively thick extracellular matrix (ECM) that surrounds all mammalian eggs and plays vital roles during oogenesis, fertilization, and preimplantation development. The ZP is a semi-permeable, viscous ECM that consists of three or four glycosylated proteins, called ZP1–4, that differ from proteoglycans and proteins of somatic cell ECM. Mammalian ZP proteins are encoded by single-copy genes on different chromosomes and synthesized and secreted by growing oocytes arrested in meiosis. Secreted ZP proteins assemble in the extracellular space into long fibrils that are crosslinked polymers of ZP proteins and exhibit a structural repeat. Several regions of nascent ZP proteins, the signal-sequence, ZP domain, internal and external hydrophobic patches, transmembrane domain, and consensus furin cleavage-site regulate secretion and assembly of the proteins. The ZP domain is required for assembly of ZP fibrils, as well as for assembly of other kinds of ZP domain-containing proteins. ZP proteins adopt immunoglobulin (Ig)-like folds that resemble C- and V-type Ig-like domains, but represent new immunoglobulin-superfamily subtype structures. Interference with synthesis, processing, or secretion of ZP proteins by either gene-targeting in mice or mutations in human ZP genes can result in failure to assemble a ZP and female infertility. ZP2 and ZP3 must be present to assemble a ZP during oocyte growth and both serve as receptors for binding of free-swimming sperm to ovulated eggs. Acrosome-reacted sperm bind to ZP2 polypeptide by inner-acrosomal membrane and acrosome-intact sperm bind to ZP3 oligosaccharides by plasma membrane overlying the sperm head. Binding of acrosome-intact sperm to ZP3 induces them to undergo cellular exocytosis, the acrosome reaction. Only acrosome-reacted sperm can penetrate the ZP, bind to, and then fuse with the egg's plasma membrane to produce a zygote. Following sperm-egg fusion (fertilization) the ZP undergoes structural and functional changes (zona reaction) induced by cortical granule components (cortical reaction) deposited into the ZP. The latter include zinc and ovastacin, a metalloendoprotease that cleaves ZP2 near its amino-terminus and hardens the egg's ZP. The changes prevent penetration of bound sperm through and binding of supernumerary sperm to the ZP of fertilized eggs as part of a secondary or slow block to polyspermy. Therefore, ZP proteins act as structural proteins and sperm receptors, and help to prevent fertilization by more than one sperm. Here we review some of this information and provide details about several key features of ZP proteins, ZP matrix, and mammalian fertilization.

1. Introduction to the ZP

All plant and animal cells are surrounded by an extracellular matrix (ECM) that supports cell structure and regulates cell behavior (Fig. 1). The Baltic German embryologist Karl Ernst von Baer (1792–1876) was first to use the term *zona pellucida* (ZP) in ca. 1827 to describe the ECM surrounding human eggs. By the 1840s the term ZP had gained widespread use among embryologists and clinicians.

The roles of the ZP during fertilization of mammalian eggs were recognized at the beginning of the 1970s when investigators proposed that the ZP was the site of species-restricted receptors for sperm (Gwatkin, 1977). Since then the ZP has been a subject of considerable research. Whereas only 50 or so papers with the tag ZP were cited on PubMed for the 45-year period 1930–1974, starting in the mid-1970s the number of papers increased about 120-fold over the next 45 years. Approximately 1100 papers with the tag ZP were cited for the 15-year period 1975–1989, 2700 for 1990–2004, and 2200 for 2005–2019, resulting in a total of about 6000 papers cited in 45 years (≃130 papers/year). Since 2020 an additional 670 papers or more with the tag ZP have been cited on PubMed.

During the past half-century information about the synthesis, structure, and functions of the ZP during oogenesis and fertilization has increased enormously. Research has led to (1) identification and characterization of ZP proteins from a variety of mammals, from marsupials and monotremes to human beings; (2) identification of the site of synthesis of ZP proteins and

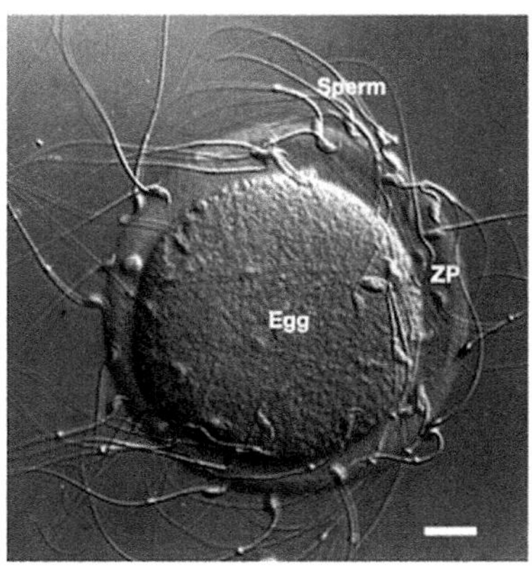

Fig. 1 Photographic image of a light micrograph (Nomarski differential interference contrast) of an unfertilized mouse egg incubated in the presence of free-swimming mouse sperm. Sperm are bound to the zona pellucida (ZP). Scale bar: ≃0.7 cm = ≃14 μm.

cis-acting gene sequences and *trans*-acting factors that regulate expression of ZP genes; (3) identification of regions of ZP proteins that regulate secretion and assembly of nascent proteins into crosslinked fibrils; (4) determination of the three-dimensional structure of ZP proteins, fibrils, and related proteins such as uromodulin; (5) cloning of ZP genes and establishment of transgenic and knockout mouse lines; (6) identification and characterization of mutations in human ZP genes that result in female infertility; (7) identification and characterization of ZP proteins that serve as receptors for binding of sperm to unfertilized eggs; and (8) characterization of changes in ZP proteins and ZP structure following sperm-egg fusion (fertilization).

In terms of human health, mutations in human ZP genes can have deleterious effects on synthesis, processing, or secretion of ZP proteins, result in failure to produce Graafian follicles and ovulated eggs, and cause female infertility. Although contraceptive vaccines based on native or recombinant ZP proteins are effective in non-human primates, they cause a variety of ovarian pathologies (e.g., oophoritis) making this an unacceptable method of contraception in human females at the present time (Gupta, 2023). On the other hand, contraceptive vaccines based on ZP proteins have proven to be very effective in controlling wildlife populations, such as rabbits, deer, and wild horses, where either long-term or permanent infertility is the goal (Kirkpatrick et al. 2009).

2. Mammalian oogenesis

Our body is made up of a large variety of cell types, including blood, bone, fat, muscle, and nerve cells. These cells reproduce by cell division or *mitosis* that consists of four phases, prophase, metaphase, anaphase, and telophase. Mitosis ensures that each cell division results in daughter cells whose nuclei have the same number of chromosomes (diploid) as the parental cell's nucleus; for example, 46 chromosomes in 23 pairs for human beings and 40 chromosomes in 20 pairs for mice. In this manner tissues grow, with each cell giving rise to two genetically identical daughter cells.

2.1 Meiosis

Of the tens-of-trillions of cells in our body, only male and female gametes, sperm and eggs, undergo *meiosis* (John & Lewis, 1973; Barresi & Gilbert, 2019; Cole, 2023). Meiosis results in reductive cell divisions to produce gametes containing one-half the number of chromosomes (haploid) of the parental cell's nucleus. Meiosis ensures an increase in genotypic variation due to crossing-over and recombination and a decrease in gamete ploidy from diploid to haploid. When sperm and egg fuse, the fertilized egg or *zygote* is returned to a diploid state. It has been found that errors in meiosis are a leading cause of infertility, miscarriage, and birth defects in human beings (Biswas et al., 2021).

Meiosis in mammals occurs during *spermatogenesis* in the male's testis and during *oogenesis* in the female's ovary. Following a final doubling of DNA in *spermatogonia* and *oogonia* during mitosis, the gametes enter meiotic prophase as *spermatocytes* or *oocytes* rather than going through another mitotic division. During meiotic prophase homologous chromosomes pair, cross-over, and recombine resulting in a reshuffling of genetic information. In males, each spermatocyte undergoes two reductive divisions, giving rise to four haploid *spermatids*, each of which will become mature *spermatozoa* during *spermiogenesis*. In females, oocytes await ovulation to undergo separation of homologous chromosomes, the first reductive division, that is an unequal cytokinesis with emission of the first polar body and become *unfertilized eggs*. This is followed by separation of sister chromatids, the second reductive division, at the time of sperm-egg fusion (fertilization), accompanied by an unequal cytokinesis with emission of a second polar body. Following fertilization, the zygote undergoes extensive rounds of mitosis and differentiation and, in time, a fully-developed organism representative of the species is produced.

2.2 Oogenesis

The unfertilized egg or ovum is the link between one generation and the next. It is the process of oogenesis in females that is responsible for production of unfertilized eggs (Austin & Short, 1982; Byskov, 1982; Baker, 1982; Claw & Swanson, 2012; Albertini, 2015; Telfer et al., 2023). During each reproductive cycle in sexually mature females one or more ovarian oocytes grows, undergoes *meiotic maturation* (first meiotic division with separation of homologous chromosomes and emission of a first polar body), is ovulated, and becomes an unfertilized egg that upon fusion with a single haploid sperm (second meiotic division with separation of chromatids and emission of a second polar body) produces a diploid zygote (see above).

2.2.1 Primordial germ cells

Oogenesis begins during fetal development with formation of *primordial germ cells* (PGC) that migrate to the genital ridges, at day-10 in mice and week-5 in humans, the site of the ovary in females (Fig. 2). Migration of

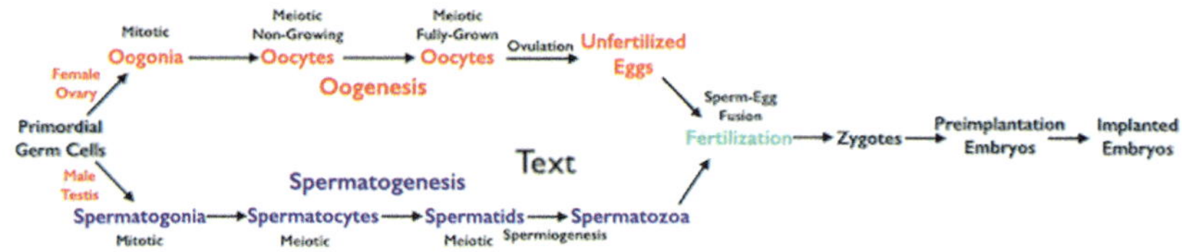

Fig. 2 Schematic diagram of some steps involved in mammalian oogenesis (*red*) and spermatogenesis (*blue*). In the female ovary, primoridial germ cells (PGC) progress from oogonia (mitotic) to non-growing oocytes (meiotic), to fully-grown oocytes, to ovulated unfertilized eggs. In the male testis, PGC progress from spermatogonia (mitotic) to spermatocytes (meiotic), to spermatids, to mature spermatozoa. Sperm-egg fusion or fertilization gives rise to a diploid zygote and ultimately to an adult.

PGC to other organs, such as the adrenal glands rather than to the genital ridges, is not unusual in mice and ectopic germ cells follow the same timetable of meiotic events characteristic of ovarian germ cells (Upadhyay & Zamboni, 1982). Specification of PGC in mice occurs ≃6–7 days *post coitum*, beginning with ≃125 cells and increasing to ≃15,000 cells in each ovary over ≃7–8 days (Findlay et al., 2015). PGC become mitotic oogonia that in time cluster into groups and become meiotic oocytes at various stages of prophase. Entrance into meiosis is triggered by exposure to retinoic acid, a metabolite of vitamin A (retinol) produced by somatic cells of the ovary.

At birth oocytes are arrested in late diplotene of meiosis, the *dictyate stage*, where they remain until stimulated to grow. This pool of non-growing oocytes is the sole source of unfertilized eggs in sexually mature adults (Krarup et al., 1969; Peters et al., 1973; Yoshihara et al., 2023). The number of oocytes entering their growth phase is a function of the pool size of non-growing oocytes. In female mice nearly half of the oocyte population at birth, ≃7000–8000 oocytes, is lost 2–3 weeks after birth and less than ≃1000 oocytes remain at 40 weeks after birth. Human females have ≃1–2 million non-growing oocytes at birth, but by puberty only ≃300,000–500,000 remain (Findlay et al., 2015). The number of oocytes decreases continuously with age, reaching ≃5000–10,000 at age 40 and zero-1000 at menopause, age 45–55. Since human females ovulate only one egg during each reproductive cycle, with only ≃300–400 eggs ovulated over ≃30 years, the substantial decrease in oocytes is attributable to massive degeneration of oocytes over time. This pattern of oocyte loss with age is characteristic of female mammals in general.

2.2.2 Follicle growth

Oocytes are contained in *ovarian follicles* that grow concomitantly with growth of oocytes under control of the hypothalamic-pituitary-gonadal axis (Fig. 3). Non-growing oocytes are surrounded initially by a single layer of a few somatic cells that becomes several layers of *granulosa* or *cumulus* cells by the time oocytes complete growth. Oocytes undergo tremendous enlargement during growth. For example, mouse oocytes increase ≃7-fold

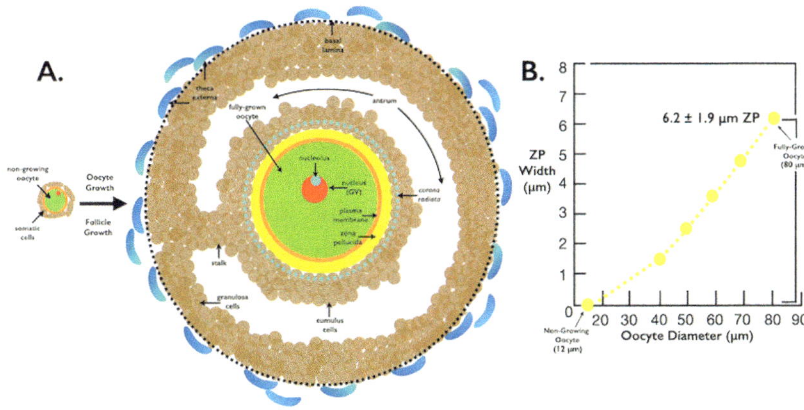

Fig. 3 Diagrammatic representation of the growth of the mammalian ovarian follicle. **Panel A.** A non-growing oocyte (≃12 μm diameter) surrounded by a few somatic cells (*brown*) on the left and a fully-grown oocyte (≃80 μm diameter) in a Graafian follicle containing a large antrum on the right. Non-growing oocytes do not possess a ZP. Fully-grown ocytes possess a nucleus or germinal vesicle (GV) (*red*), nucleolus (*turquoise*), plasma membrane (*orange*), and ZP (*yellow*). In the Graafian follicle, two-to-three layers of cumulus cells surround the oocyte (*cumulus oophorous; brown*) with the innermost cumulus cells (*corona radiata*) indicated by a dashed *turquoise* circle. The oocyte is connected to granulosa cells (*brown*) by a stalk and is located in a fluid filled cavity, the antrum. At the outermost region of the Graafian follicle is a basal lamina (dashed *black* circle) and outside of this are the theca externa cells (*blue*). At ovulation, oocytes arrested in metaphase II of meiosis are ovulated from the Graafian follicle surrounded by the *cumulus oophorous.* Oocytes resume meiosis, complete the first meiotic reductive division or meiotic maturation with separation of homologous chromosomes and emission of the first polar body, and become unfertilized eggs. At fertilization, eggs complete meiosis with separation of chromatids and emission of a second polar body. The sperm's genome restores the fertilized egg to a diploid state. **Panel B.** A graph depicting changes in the width of the mouse oocyte's ZP (*yellow*) at various stages of oocyte growth. Mouse oocytes grow from ≃12 μm (non-growing oocytes) to ≃80 μm in diameter (fully-grown oocytes) over 2–3 weeks; this corresponds to a more than 300-fold increase in oocyte volume, from ≃0.9 to ≃270 pl, during oocyte growth. Concomitant with oocyte growth, the ZP increases in width from zero (non-growing oocytes) to 6.2 ± 1.9 μm (fully-grown oocytes).

in diameter, from ≃12 to 80 μm, and ≃300-fold in volume, from ≃0.9 to 270 pl. Growth is accompanied by intense metabolic activity, marked changes in oocyte morphology and ultrastructure, and relatively high rates of transcription and translation (Wassarman & Josefowicz, 1978; Wassarman & Albertini, 1994). A *thecal layer* is first distinguishable, outside of and separated by a basement membrane from the cumulus cells, when the cumulus layer is a few layers thick. Granulosa cells proliferate extensively once oocyte growth is completed and produce a *Graafian follicle* with a large cavity or *antrum* filled with follicular fluid. As the antrum expands, the oocyte takes up an acentric position surrounded by a few layers of cumulus cells, the *cumulus oophorous*, with the innermost layer constituting the *corona radiata*. Fully-grown oocytes in Graafian follicles undergo ovulation and meiotic maturation with emission of a first polar body in response to a surge of luteinizing hormone (LH). Oocytes acquire the ability to undergo meiotic maturation, with separation of chromosomes, emission of a polar body, nuclear envelope breakdown, and meiotic spindle assembly during their growth phase (Sorensen & Wassarman, 1976). Following ovulation, the empty follicle that remains in the ovary becomes an endocrine gland, the *corpus luteum*, that secretes high levels of progesterone and moderate levels of estradiol to support pregnancy by maintaining a uterine environment that allows implantation.

The ovarian follicle is a functional syncytium responsible for bidirectional communication between oocytes and follicle cells; follicle cells regulate oocyte growth and oocytes regulate follicle development (Eppig et al., 2002; Robert, 2021; Doherty et al., 2022; Marchais et al., 2022; Crozet et al., 2023). *Gap junctions* are present between all follicle cells as well as between oocytes and surrounding follicle cells at sites where follicle cell processes traverse the ZP and contact the plasma membrane (Matzuk et al., 2002; Wassarman, 2002; Li & Albertini, 2013; Baena & Terasaki, 2019). Innermost follicle cells (*corona radiata*) of the *cumulus oophorous* send long transzonal projections or filopodia through the ZP surrounding growing oocytes, form gap junctions with the oocyte's plasma membrane, and provide a route for movement of a variety of small molecules (< 1500 MW) needed by growing oocytes (Goodenough & Paul, 2009; Alexander & Goldberg, 2003; El-Hayek & Clarke, 2016; El-Hayek et al., 2018; Clarke, 2018). The number of gap junctions and oocyte-follicle cell communication increase throughout oocyte growth.

In the absence of ZP formation, the number of gap junctions between oocytes and follicle cells is severely reduced, compromising electrical and metabolic coupling necessary for oocyte growth, follicle development, and fertility (Carabatsos et al., 2000; Kidder & Mhawi, 2002; Gittens & Kidder, 2005). This is consistent with the phenotype of female mice that are homozygous-nulls for gap junction proteins, such as connexin-37 that is associated with gap junctions between oocytes and innermost cumulus cells and connexin-43 that is associated with gap junctions between granulosa cells. Ovaries of these null-mice are deficient in growing oocytes and multi-layered follicles and females are infertile. Failure to form a sufficient number of gap junctions leads to a substantial decrease in the production of Graafian follicles and results in female infertility (Simon et al., 1997; Ackert et al., 2001; Wassarman & Litscher, 2012, 2022a). It has been suggested that the ZP may serve as a kind of "glue" with which to stabilize gap junctions and other contacts between oocytes and cells of the *corona radiata* (Wassarman & Litscher, 2022a).

3. Features of ZP matrix

The thickness of the mammalian ZP varies from <1 to >20 μm (≃1–30 ng protein) depending on the animal species (mouse ≃6.2 μm, human ≃18 μm, and cow ≃16 μm thick). Unfertilized mouse eggs contain ≃25 ng of total protein and ≃3–4 ng of ZP proteins. A ZP first appears as an extremely thin and patchy ECM around early-stage growing oocytes and then increases in thickness as oocytes increase in diameter (Wassarman & Mortillo, 1991). For example, as mouse oocytes grow from ≃12 to 80 μm in diameter over ≃2–3 weeks, a thin ZP appears and then increases in thickness as oocytes grow, reaching a final width of 6.2 ± 1.9 μm (Fig. 3).

Electron micrographs of the ZP reveal that it is an amorphous, fibrillar network, not unlike other types of ECM, that has a surface covered with pores (≃50 pores per mouse ZP) giving it a sponge-like or Swiss cheese appearance (Phillips & Shalgi, 1980; Familiari et al., 2006) (Fig. 4). Consequently, the ZP is a semi-porous matrix that can be penetrated by relatively large macromolecules, such as enzymes, antibodies, and small viruses (Hastings et al., 1972; Gwatkin, 1977; Legge, 1995) and is recognized by a variety of lectins, some of which can inhibit binding of sperm to unfertilized eggs and fertilization.

The ZP is composed of relatively long, crosslinked fibrils ≃7–8 nm wide with a ≃14–15 nm structural repeat (Greve & Wassarman, 1985; Wassarman & Mortillo, 1991; Wassarman et al., 1996; Familiari et al., 2006; Matassa et al., 2023). Innermost ZP fibrils are oriented radially to the

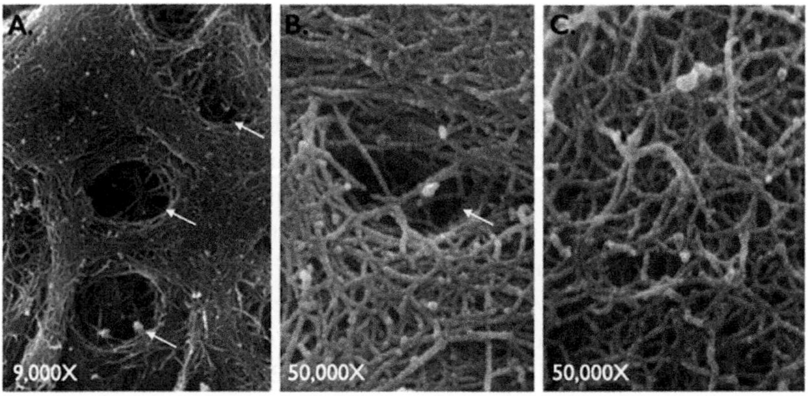

Fig. 4 Scanning electron micrographs of the surface of human and mouse oocytes. **Panel A.** Human oocyte showing the presence of many pores (9000-times magnification) on the outer surface of the ZP. **Panel B.** Higher magnification of a human oocyte showing the fibrillar organization of the ZP (50,000-times magnification). Fibrils are 0.1–0.4 μm long and 10–14 nm wide. **Panel C.** Outer surface of a mouse oocyte showing the fibrillar organization of the ZP (50,000-times magnification). Samples were treated with saponin-ruthenium red-osmium-thiocarbohydrazide to reveal ZP fibrils. This figure was adapted with permission from Professor G. Familiari (see Familiari et al., 2006).

oocyte's plasma membrane and are closely packed, whereas outermost fibrils are oriented tangentially to the plasma membrane and are more loosely packed (Keefe et al., 1997; Pelletier et al., 2004). The reorientation of fibrils during oocyte growth and loose packing of fibrils in the outer region of the ZP may be due to considerable stretching of the outer matrix to accommodate the large increase in circumference (≃6–7-fold), during oocyte growth (Litscher & Wassarman, 2020a). The different orientation of ZP fibrils may be related to early descriptions of the ZP of eggs from different mammalian species suggesting that the ZP is composed of two distinct concentric layers, an inner and outer layer (Dickmann, 1965).

It is noteworthy that, prior to fertilization, the ZP can be completely dissolved by a variety of agents that either break (e.g., proteases or reducing agents) or do not break (e.g., moderately high or low pH or mild heat) covalent bonds. The latter suggests that non-covalent bonding stabilizes ZP protein interactions in fibrils that make up the matrix. Following fertilization the ZP is considerably less soluble (i.e., harder) in the presence of these agents than it is prior to fertilization (see Sections 10.5.1 and 10.5.2). Hardening of the ZP should enhance protection of cleavage-stage embryos by the ZP as they traverse the female reproductive tract on their way to the uterus to implant.

4. ZP proteins and genes

4.1 ZP proteins

The ZP surrounding mammalian eggs is a unique ECM composed of only a few glycosylated proteins, called ZP1–4, that differ from proteoglycans and proteins found in somatic cell ECM (Bleil & Wassarman, 1980a; Wassarman, 1988a, 2008; Wassarman & Litscher, 2018; Litscher & Wassarman, 2020a,b). Mouse (*Mus musculus*) ZP1–3 were first identified in 1980 (Bleil & Wassarman, 1980a) and ZP4 in human ZP in 2004 (Lefievre et al., 2004); ZP4 is a pseudogene in *Mus musculus*, but not in other murine species, such as *Mus pahari* (Izquierdo-Rico et al., 2021). Very similar glycosylated proteins are found in the ZP, or vitelline envelope, of mollusk, ascidian, fish, amphibian, and bird eggs; this represents an evolutionary time-span of more than 600 million years (Litscher & Wassarman, 2014, 2015).

Each ZP protein consists of a unique polypeptide that is heterogeneously glycosylated with complex-type asparagine (N)-linked and serine/threonine (O)-linked oligosaccharides (Fig. 5). For example, mouse ZP1 has four complex-type N-linked oligosaccharides, ZP2 six, and ZP3 three or four, as well as an undetermined number of O-linked oligosaccharides. In this context, it has been estimated that ≃50–70% of eukaryotic proteins are glycosylated. ZP2–4 are monomers and ZP1 is a dimer of identical polypeptides connected by a single intermolecular disulfide. ZP2 and ZP3 are present in nearly equimolar amounts and account for the bulk of protein in the ZP. ZP1 and ZP4 are relatively minor components of the ZP compared to ZP2 and ZP3.

A comparison of nascent ZP polypeptide lengths for fifteen different mammalian species, from monotremes and marsupials to primates, reveals that they vary from 609–689 amino acids (aa) for ZP1 (monomer), 694–784 aa for ZP2, 418–462 aa for ZP3, and 526–570 aa for ZP4 (Table 1). Primary structures of ZP2 and ZP3 from different mammals are relatively well conserved and exhibit ≃65–95% identity, whereas ZP1 and ZP4 exhibit only ≃40% identity. Comparison of ZP proteins from various mammals compared to human ZP proteins reveals that they can be divided into three groups, ranging from 46–56% (Group 1; monotremes and marsupials), 58–73% (Group ll; rodents and domesticated animals), and 92–99% identical (Group lll; primates) (Table 2).

The molecular weights of mature ZP proteins on denaturing gels are considerably higher than those of their polypeptides. For example, mouse (*Mus musculus*) ZP1–3 have molecular weights of ≃200 (dimer), ≃120, and

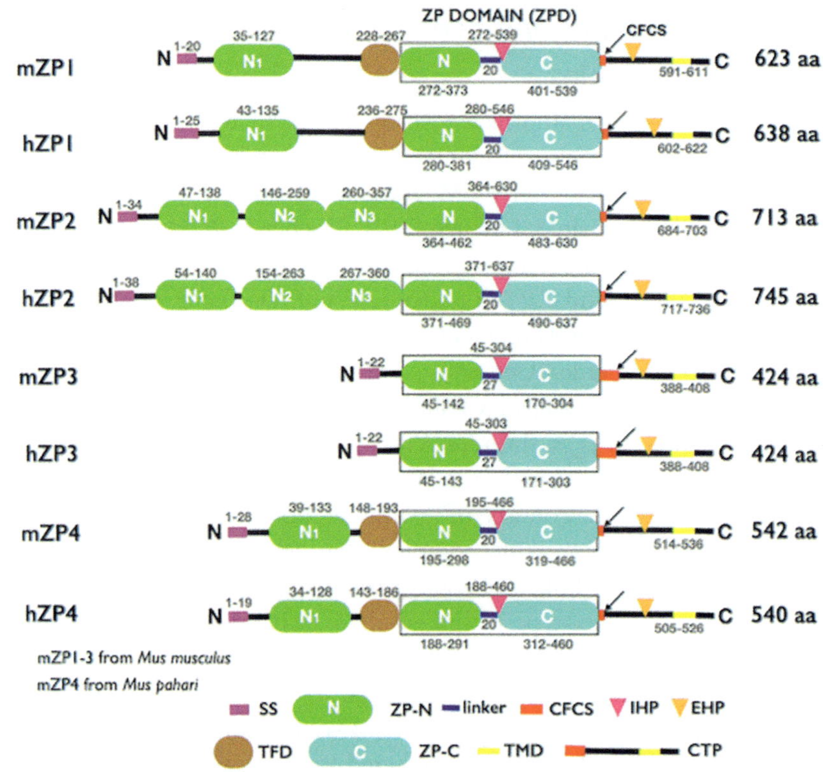

Fig. 5 Schematic representation of the overall organization of mouse (m) and human (h) ZP proteins. The polypeptides have an N-terminal signal-sequence (SS; *magenta*); a zona pellucida domain (ZPD) that consists of ZP-N (*green*) and ZP-C (*cyan*) subdomains, and a linker-region (*blue*); a consensus furin cleavage-site (CFCS; *black* arrow); an internal hydrophobic peptide (IHP; *pink* triangle) and external hydrophobic peptide (EHP; *orange* triangle); a transmembrane domain (TMD; *yellow*); and a short C-terminal tail (CT; *black*) at the C-terminal peptide (CTP; *black*). mZP1, hZP1, and hZP4 also have a trefoil domain (TFD; *brown*). mZP2 (713 aa) and hZP2 (745 aa) have three external ZP-N subdomains (N1–N3; *green*) between the N-terminus of the polypeptide and the ZPD. mZP1 (623 aa), hZP1 (638 aa), mZP4 (542 aa), and hZP4 (540 aa) have one external ZP-N subdomain (N1; *green*), between the N-terminus of the polypeptide and the TFD. mZP3 (424 aa) and hZP3 (424 aa) are the smallest of the ZP proteins and consist primarily of a ZPD. mZP1, hZP1, mZP2, and hZP2 have only 3 or 4 aa between the C-terminus of the ZPD and the CFCS, whereas mZP3 and hZP3 have >45 aa (see Fig. 6). The latter is a region of positive Darwinian selection during evolution.

≃83 kD, respectively, whereas their polypeptides have molecular weights of ≃116 (dimer), ≃66, and ≃36 kD. This is largely due to the type of sugar linkage (e.g., complex N- or O-linked oligosaccharides), sugar composition, and extent of glycosylation of the polypeptides, as well as to oligosaccharide

Table 1 Polypeptide lengths for mammalian ZP proteins.

	Polypeptide Length (number aa)			
Organism	**ZP1**	**ZP2**	**ZP3**	**ZP4**
Platypus (*Ornithorhynchus anatinus*)	625	712	462	565
Possum (*Trichosurus vulpecula*)	609	712	418	527
Koala (*Phascolarctos cinereus*)	656	709	419	526
Mouse (*Mus musculus*)	623	713	424	
Mouse (*Mus pahari*)	623	714	424	542
Rat (*Rattus rattus*)	614	694	422	545
Hamster (*Mesocricetus auratus*)	616	708	422	543
Cow (*Bos taurus*)		713	421	534
Pig (*Sus domesticus*)		716	421	536
Cat (*Felis catus*)	652	716	424	570
Dog (*Canis lupus familiaris*)		715	426	552
Marmoset (*Callithrix jacchus*)	689	747	424	541
Monkey (*Macaca mulatta*)	640	784	424	538
Chimpanzee (*Pan troglodytes*)	638	745	424	540
Human (*Homo sapiens*)	638	745	424	540

modifications such as sialylation and sulfation. Mature ZP proteins are acidic (pIs 4.1–5.5), with oligosaccharides, not the polypeptides (pIs > 6.5), accounting for their relatively low isoelectric points. Removal of sialic acid from oligosaccharides of ZP proteins by treatment with neuraminidase results in a significant increase in their isoelectric points.

4.2 ZP genes

For a wide variety of animal species, from lampreys to humans, ZP proteins are encoded by three to five genes, called *ZP1, ZP2, ZP3, ZP4,* and *ZPax* (Hughes & Barratt, 1999; Spargo & Hope, 2003; Hughes, 2007; Goudet et al., 2008; Feng et al., 2018; Nishio et al., 2018; Litscher & Wassarman, 2007, 2014) (Table 3). One of the four genes, *ZP1, ZP2, ZP4,* and *ZPax,*

Table 2 ZP Proteins from various mammals compared to human ZP proteins.

Organism (ZP Proteins)	% Identity: % Similarity to Human ZP1-4	
Platypus (ZP1–4)	46–52: 73–78	**Group I**
Possum (ZP1–4)	48–56: 70–81	46–56: 70–81
Mouse (ZP1–3)	58–68: 80–87	
Rat (ZP1–4)	58–67: 80–87	
Hamster (ZP1–4)	58–68: 77–86	**Group II**
Cow (ZP2–4)	61–72: 84–86	58–73: 77–87
Pig (ZP2–4)	63–73: 83–87	
Dog (ZP2–4)	67–71: 84–87	
Monkey (ZP1–4)	92–94: 98	**Group III**
Chimpanzee (ZP1–4)	99: 99–100	92–99: 98–100

Table 3 ZP Genes in vertebrates.

Organism	ZP1	ZP2	ZP3	ZP4	ZPax
Fish	−	−	+	+	+
Amphibians	−	+	+	+	+
Reptiles	+	+	+	+	+
Birds	+	+	+	+	+
Mammals	+	+	+	+/−	−

can be missing from the genome or be present as non-functional pseudogenes. The ZP of some mammalian organisms consist of ZP1–4, while others of only ZP2–4 or ZP1–3. ZP of rodents (e.g., mouse, rat, and hamster), except *Mus musculus*, tend to consist of ZP1–4 and ZP of domesticated species (e.g., cow, pig, and dog) of ZP2–4. The ZP of non-mammalian organisms (e.g., fish, amphibians, reptiles, and birds) also consists of various combinations of ZP1–4, but can have as an additional protein ZPax. Marsupials have *ZP3*-related genes, *ZP3-A, -B,* and *-C;*

ZP3-C is an orthologue and *ZP3-A* and *-B* are paralogues of eutherian *ZP3* (Moros-Nicolás et al., 2018; Frankenberg & Renfree, 2018). While ZP proteins are synthesized in ovaries of mammals, fish, amphibians, and birds, they are also synthesized in the liver of fish and birds and transported via the circulation to the ovary for assembly into the egg's ZP.

Among mammals, ZP proteins are encoded by single-copy genes located on different chromosomes (Table 4). In the mouse, *Mus musculus*, *ZP1–3* are located on chromosomes 19 (12 exons, 623 aa), 7 (18 exons, 713 aa), and 5 (8 exons, 424 aa), respectively; *ZP4* is present as a pseudogene on chromosome 13. In humans ZP1–4 are located on chromosomes 11 (12 exons, 638 aa), 16 (19 exons, 745 aa), 7 (8 exons, 424 aa), and 1 (12 exons, 540 aa), respectively.

ZP genes are expressed coordinately and exclusively by growing oocytes (Bleil & Wassarman, 1980a; Shimizu et al., 1983; Chamberlin & Dean, 1989, 1990; Lira et al., 1990; Epifano et al., 1995a,b; Hoodbhoy et al., 2006; Gupta, 2018). Genomic organization of ZP loci is identical in oocytes and somatic cells, but ZP genes are hypomethylated in oocytes compared to somatic cells (Ringuette et al., 1986). ZP genes are highly conserved among mammals. Although intron sizes of ZP genes vary greatly between species, the number and lengths of exons are highly conserved and exon/intron boundaries define distinct domains (see Section 9). There is a high degree of conservation between coding regions and between the first

Table 4 Mouse (m) and human (h) ZP genes.

ZP Gene	Chromosome Number	Gene Length (kb)	Number Exons
mZP1	19	6.5	12
mZP2	7	18.5	18
mZP3	5	8.6	8
mZP4	13	-	-
hZP1	11	8.1	12
hZP2	16	14	19
hZP3	7	18.3	8
hZP4	1	17	12

Table adapted from Lunsford et al., 1990; Epifano et al., 1995b; Gupta, 2018.

300 base pairs (bp) of the 5'-flanking regions of ZP genes, such that the human ZP promoter can utilize transcriptional machinery of mouse oocytes (Liang & Dean, 1993). Mutation of a 12 bp sequence (element 4) located ≃200 bp upstream of the transcription start-site of ZP genes (TATA box) reduces reporter gene activity to ≃1–4 % of wild-type activity (Millar et al., 1991).

4.3 ZP gene expression

ZP proteins are synthesized by oocytes throughout their growth phase and represent ≃1.5–2.5 % of total protein synthesis in growing oocytes (Haddad & Nagai, 1977; Bleil & Wassarman, 1980b; Greve et al., 1982; Salzmann et al., 1983; Shimizu et al., 1983; Philpott et al., 1987; Chamberlin & Dean, 1989, 1990; El-Mestrah et al., 2002; Hoodbhoy et al., 2006). The number of copies of ZP messenger-RNA (mRNA) per mouse oocyte or egg increases from undetectable levels (<1000 copies) in non-growing oocytes, to hundreds-of-thousands of copies in mid-stage growing and fully-grown oocytes, to almost undetectable levels in fertilized eggs and cleavage-stage embryos (Roller et al., 1989). The amount of ZP3 mRNA declines dramatically when transcription is terminated during meiotic maturation and ovulation (≃10–12 h), from ≃240,000 copies per oocyte to ≃5000 copies per unfertilized egg (≃98 % reduction), and ZP3 mRNA is undetectable in fertilized eggs (<1000 copies) (Roller et al., 1989).

Cis-acting sequence elements in the 5'-flanking region of ZP genes and *trans*-acting factors in growing oocytes regulate ZP gene expression during oocyte growth (Lira et al., 1990, 1993; Millar et al., 1991, 1993; Schickler et al., 1992; Liang et al., 1997; Wassarman & Litscher, 2021). The sequence CANNTG, an E-box domain located ≃200 bp upstream of the transcription start-site of ZP genes, is identical to the consensus sequence required for binding of proteins that belong to the basic helix-loop-helix (bHLH) family of DNA binding-proteins (Liang et al., 1997; Soyal et al., 2000; Millar et al., 1993). ZP genes have a potential TATA-box just upstream of the transcription start-site that is involved in oocyte-specific expression of ZP genes when ZAP-1, a complex of proteins, binds to DNA (Millar et al., 1993). Similarly, OSP-1 binds to a minimal sequence TGATAA located within the first 100 bp of the ZP3 promoter (Schickler et al., 1992). Like ZAP-1, the abundance of OSP-1 during oogenesis mirrors the pattern of ZP gene expression. The minimal sequence recognized by OSP-1 is identical to the consensus sequence A/TGATAG/A that is the binding motif for the GATA multigene family of DNA binding-proteins. As little as

153 nucleotides of ZP3 5'-flanking sequence is sufficient to target expression of foreign proteins to growing oocytes (Lira et al., 1993) and ≃6.5 kilobases regulates *cre* expression to specifically target genes of mouse oocytes (Lewandoski et al., 1997).

5. Assembly of ZP proteins

Nascent ZP proteins are secreted in membrane-bound, doughnut-shaped secretory vesicles, ≃2–3 μm in diameter (i.e., ≃10-times larger than secretory vesicles of somatic cells), that originate from the Golgi (Qi et al., 2002) and/or in large, multivesicular aggregates, ≃1–5 μm in diameter, that also originate from the Golgi (Hoodbhoy et al., 2006). Since nascent ZP proteins assemble solely on the innermost layer of the ZP, the ZP thickens from the inside to the outside during oocyte growth (Qi et al., 2002). In mice, as much as one-half of newly synthesized ZP protein passes through the matrix unassembled and the ZP of fully-grown oocytes varies considerably in thickness, from ≃4.3 to ≃8.1 μm, suggesting that assembly of the ZP is stochastic.

Polypeptides of nascent ZP1–4 have several features in common (Table 5 and 6). They possess an N-terminal signal-sequence (SS) that targets nascent polypeptide to the secretory pathway; a ZP domain (ZPD) or module close to the C-terminus, composed of two subdomains, ZP-N and ZP-C, required for protein assembly into fibrils; a consensus furin cleavage-site (CFCS; Arg-X-X-Arg or Arg-X-Arg/Lys-Arg) close to the ZPD that is cleaved by a furin-like protease during secretion (Tables 5 and Table 6); internal (IHP) and external (EHP) hydrophobic patches that function as control switches in regulating nascent protein assembly (Tables 5 and 6); a transmembrane domain (TMD) composed of hydrophobic aa that anchors ZP proteins in secretory vesicle and plasma membranes (Tables 5, 6 and Table 7); a short cytoplasmic tail (CT) involved in incorporation of ZP proteins into the ZP; and a trefoil domain (TFD) in ZP1 and ZP4 that may be involved in crosslinking of ZP fibrils. The CFCS, EHP, TMD, and CT are located in a carboxy-terminal propeptide (CTP) downstream of the ZPD. The IHP and EHP are integral components located on the A (IHP) and G (EHP) strands of the ZP-C fold (see Section 8.1) and are in intimate contact prior to cleavage of ZP proteins at the CFCS, a cleavage event that disengages the EHP from the IHP.

Table 5 Primary structure of mouse (m) and human (h) ZP proteins.

	Polypeptide Amino Acid Position							
	Nascent Polypeptide Length (aa)	**SS**	**ZPD**	**CFCS**	**TMD**	**TFD**	**IHP**	**EHP**
mZP1	623	1–20	272–539	545–548	591–611	228–267	394–400	566–572
hZP1	638	1–25	280–546	552–555	602–622	236–275	402–408	573–579
mZP2	713	1–34	364–630	632–635	684–703	–	483–489	648–654
hZP2	745	1–38	371–637	639–642	717–736	–	490–496	657–663
mZP3	424	1–22	45–304	350–353	388–408	–	170–177	357–369
hZP3	424	1–22	45–303	349–352	388–408	–	171–178	356–368
mZP4	542	1–28	195–466	470–473	514–536	148–193	319–325	487–493
hZP4	540	1–19	188–460	464–67	505–526	143–186	312–318	482–488

mZP1-3 from *Mus musculus*
mZP4 from *Mus pahari*
aa, amino acid; **SS**, signal sequence; **ZPD**, ZP domain; **CFCS**, consensus furin cleavage-site; **TMD**, transmembrane domain; **TFD**, trefoil domain; **IHP**, internal hydrophobic patch; **EHP**, external hydrophobic patch.

Table 6 Sequence of mouse (m) and human (h) IHP, EHP, CFCS, and TMD.

	IHP	EHP	CFCS	TMD
mZP1	394SGPLRLE400	566PGAVGFE572	545RRRR548	591LLLLLAITLALAAGIFVGLI611
hZP1	402PGPLRLE408	573PGPVGFE579	552RQRR555	602LLWAVLLLPAVALVLGFGVFV622
mZP2	483PGPLVLV489	648PGPILLL654	632RSKR635	684LGAVAALVGSAVILGFICYL703
hZP2	490LGPFTLI496	657PGPILLL663	639RHRR642	717VAAVAAFAGVVATLGFIYYL736
mZP3	170AFSLRLME177	357DEADVTVGPLIFL369	350RNRR353	388VALGLGLATVAFLTLAAIVLAV408
hZP3	171TFSLRLME17868	356EEADVTVGPLIFL3	349RNRR352	388VLLGVGLAVVVSLTLTAVILV408
mZP4	319PGPLSLK325	487KGPVILL493	470RRRK473	514PALWVMGLSATMIIIGVLVVSYL536
hZP4	312PGPLTLE318	482KGPMILL488	464RRRN467	506KVLWVAGLSGTLILGALLVSYL526

mZP1, mZP2 and mZP3 from *Mus musculus* and mZP4 from *Mus pahari*
IHP, internal hydrophobic patch; **EHP**, external hydrophobic patch; **CFCS**, consensus furin cleavage-site;
TMD, transmembrane domain

Table 7 TMD of mouse (m) and human (h) ZP proteins.

	Transmembrane Domain (TMD) Position and Sequence	No. Hydrophobic aa (L,I,A,V,F,W,M,P)
mZP1	591**LLLLLAITLALALAGIFVGLI**611	18 (+2 G)
mZP2	684**LGAVAALVGSAVILGFICYL**703	14 (+3 G)
mZP3	388**VALGLGLATVAFLTLAAIVLA**408	17 (+2 G)
mZP4	514**PALWVMGLSATMIIIGVLVVSYL**536	17 (+2 G)
hZP1	602**LLWAVLLLPAVALVLGFGVFV**622	19 (+2 G)
hZP2	717**VAAVAAFAGVVATLGFIYYL**736	15 (+2 G)
hZP3	388**VLLGVGLAVVVSLTLTAVILV**408	16 (+2 G)
hZP4	505**KVLWVAGLSGTLILGALLVSYL**526	14 (+3 G)

mZP1 and mZP2 from *Mus musculus*
mZP3 and mZP4 from *Mus pahari*
A, Ala; **F**, Phe; **G**, Gly; **I**, Ileu; **L**, Leu; **M**, Met; **P**, Pro; **S**, Ser; **T**, Thr; **V**, Val; **W**, Tryp; **Y**, Tyr

The presence of a CT, IHP, and EHP prevents assembly of nascent ZP proteins within growing oocytes (Litscher et al., 1999; Williams & Wassarman, 2001; Zhao et al., 2002, 2003; Qi et al., 2002; Jovine et al., 2004, 2007; Jimenez-Movilla & Dean, 2011). Assembly of ZP proteins into fibrils in the oocyte's extracellular space is dependent on proteolysis of nascent ZP proteins at their CFCS which disrupts IHP-EHP interactions; failure to do so reduces secretion ≃20-fold and results in accumulation of uncleaved ZP proteins in the endoplasmic reticulum (Williams & Wassarman, 2001; Qi et al., 2002; Jovine et al., 2007). Cleavage of inhibitory sequences of protein precursors controls assembly of many other types of proteins as well (Jovine et al., 2007).

6. ZP domain

A hallmark feature of all ZP proteins is their ZPD, a ≃260-amino acid sequence that contains eight conserved Cys residues present as four intramolecular disulfides (Bork & Sander, 1992; Jovine et al., 2005; Plaza et al., 2010; Litscher & Wassarman, 2015). The ZPD is essential for assembly of extracellular proteins into fibrils (Jovine et al., 2002, 2005).

A comparison of ZPD sequences for ZP1–4 from platypus to humans reveals an average ≃73 % identity and ≃90 % similarity. The ZPD arose more than 600 million years ago and is found in hundreds of other proteins (e.g., cuticlin, endoglin, hensin, muclin, nompA, oikosin, tectorin, and uromodulin) in a variety of animal species from invertebrates (e.g., jellyfish) to vertebrates (e.g., humans), and in a wide variety of tissues and organs (Jovine et al., 2005; Litscher & Wassarman, 2015) (Table 8). These proteins include secreted proteins that function as receptors or mechanical transducers between adjacent cells, as intermediators during differentiation, morphogenesis, and signaling, and as ECM proteins.

6.1 ZPD subdomains and linker

The ZPD is a bipartite structure consisting of an N-terminal subdomain (ZP-N; ≃100 aa) and a C-terminal subdomain (ZP-C; ≃135–150 aa) separated by a short, protease-sensitive, linker region (≃25–30 aa). It has been suggested that the ZP-N and ZP-C should be treated as independent domains and that the term ZPD should be changed to ZP module (Wilburn & Swanson, 2017; Rivera et al., 2022). The linker between ZP-N and ZP-C can be flexible as in ZP1–4 or rigid (short) as in uromodulin and differences in flexibility of the linker may modulate assembly of structurally similar ZPD proteins (Bokhove et al., 2016). The two ZPD subdomains are always found in the sequential order ZP-N→ZP-C and never ZP-C→ZP-N and ZP-N, but not ZP-C, is found by itself in some proteins (e.g., in Oosp1, Plac1, and papillote).

Proteins possessing a ZPD form polymers and isolated ZPDs can polymerize into fibrils on their own, with ZP-N required for polymerization (Jovine et al., 2002, 2006; Bokhove et al., 2016). However, there is evidence to suggest that both ZP-N and ZP-C may be required for polymerization of some ZPD-containing proteins (Stsiapanava et al., 2020). Mutation of the ZP-N subdomain can lead to severe pathologies, such as deafness, vascular disease, renal disease, infertility, or cancer.

6.2 ZPD Cys residues

Each ZPD subdomain has four conserved Cys residues present as two intramolecular disulfides, with ZP-N disulfides linked Cys1→Cys4 and Cys2→Cys3 and ZP-C disulfides linked Cys5→Cys7 and Cys6→Cys8. ZP1, ZP2, and ZP4 have two additional Cys residues, a and b, in their ZP-C subdomain that form an a→b linked disulfide. Additional copies of the ZP-N are found in single or multiple copies in the N-terminal regions

Table 8 ZP Domain proteins: from jellyfish to mammals.

Mammals	**SS**	**ZPD**	**CFCS**	**TMD**	**Additional Domains**
Ebnerin (tongue)	+	+	+	+	SRCR, CUB
Endoglin (blood vessels)	+	+	+	+	None
Hensin (brain)	+	+	+	−	SRCR, CUB
Muclin (intestine)	+	+	+	+	SRCR, CUB
α-Tectorin (inner ear)	+	+	+	−	GI, vWF, EGF
Uromodulin (kidney)	+	+	+	+	EGF, DBC
Vomeroglandin (nose)	+	+	+	+	SRCR, CUB
Worms (*C. elegans*)					
Cuticlins (cuticle)	+	+	+	+	Cys-rich
Dyf−7 (dendrites)	+	+	+	+	None
Ram−5 (rays)	+	+	−	+	None
Flies (*D. melanogaster*)					
Dumpy (trachia, wings)	+	+	+	+	EGF/DPY
NompA (dendritic cap)	+	+	+	+	PAN
Quasimodo (eye)	+	+	+	+	None
Tunicates (*O. dioca*)					
Oikosins	−	+	−	−	CUB, Cys-rich
Jellyfish (*A. aurita*)					
Mesoglein	−	+	−	−	DSL

SS, signal sequence; **ZPD**, zona pellucida domain; **CFCS**, consensus furin cleavage-site; **TMD**, transmembrane domain; **SRCR**, scavenger receptor cysteine-rich; **CUB,** complement C1r/C1s, Uegf, Bmp1; **vWF**, von Willebrand factor; **EGF**, epidermal growth factor; **PAN**, plasminogen; **DSL**, delta/serrate/lag-2; **DBC**, deleted in breast cancer.

of ZP1 (1 copy, ZP-N1), ZP2 (3 copies, ZPN1–3), and ZP4 (1 copy, ZP-N1) (Callebaut et al., 2007; Bokhove & Jovine, 2018). According to structure-based sequence alignments, ZP-N1 has four conserved Cys residues linked Cys1→Cys4 and Cys2→Cys3, whereas N2 and N3 have two conserved Cys residues linked Cys1→Cys4; ZP-N3 has four Cys residues, but only two are conserved. Perhaps a single Cys1→Cys4 linkage is sufficient for structural stability of the ZP-N subdomain. ZP3 has only ZP-N and not additional copies as extensions at its N-terminus.

In general, the ZPD is a conserved evolutionary unit essential for polymerization of proteins, whereas adjoining regions contribute functional diversification that may be caused by selective pressure related to species-specific functions. The strong structural similarity between ZP-N and ZP-C suggests that the ZPD may have arisen by duplication of an ancestral gene encoding a protein containing a single ZP-N. Certain molecular features must regulate whether ZPD proteins assemble into homopolymers (e.g., uromodulin fibrils) or heteropolymers (e.g., ZP fibrils). It has been suggested that ZP2 and ZP3 are amyloidogenic, that ZP fibrils have biophysical properties similar to amyloid fibrils, and that the ZP is a functional amyloid (Louros et al., 2013; Egge et al., 2015; Hewetson et al., 2017; Litscher & Wassarman, 2020a).

7. Primary structure of ZP proteins

7.1 ZP2 and ZP3

A major difference between ZP2 and ZP3 is the length and composition of the aa sequence that follows the ZP-C subdomain and abuts the CFCS (Fig. 6). For ZP2 from different species this region includes only 3–4 aa, whereas for ZP3 it includes 32–48 aa. For ZP2 a Pro residue always follows the last Cys of ZP-C (Cys-Pro), whereas for ZP3 the last Cys is always followed by Ser-Phe (Cys-Ser-Phe) and then either Ser or Asn. Eight aa downstream is the consensus sequence Trp-X-Pro-Val-Glu-Gly followed three aa downstream by the consensus sequence Ile/Val-Cys-X-Cys-Cys-X-X-Gly-X-Cys-X.

The C-terminal region of ZP3 exhibits a high degree of divergence compared to other regions of the protein. It is a site of positive Darwinian selection, perhaps having to do with species-restricted fertilization (Rosiere & Wassarman, 1992; Wassarman & Litscher, 1995; Kinloch et al., 1995; Litscher & Wassarman, 1996; Swanson et al., 2001, 2003; Williams et al., 2006; Claw & Swanson, 2012; Killingbeck & Swanson, 2018; Duarte et al., 2021). Possibly selective pressure drives mutations in this region and promotes divergence and evolution of ZP3.

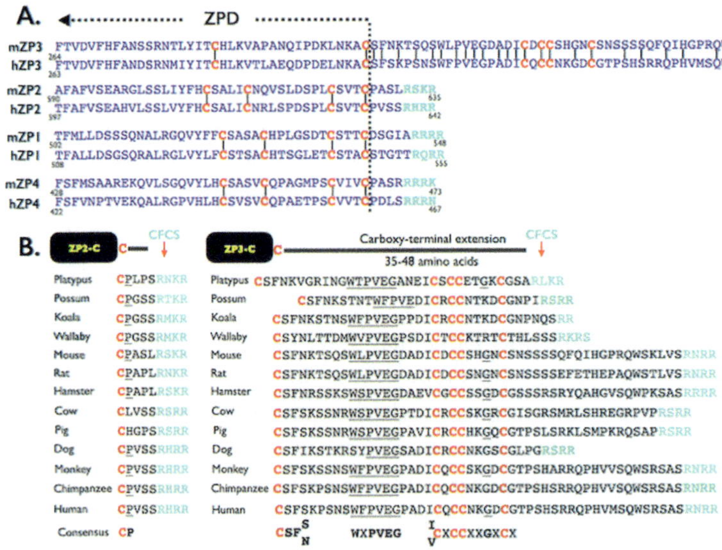

Fig. 6 Primary structures of the region following the terminal Cys residue of the ZPD. **Panel A.** Shown are primary structures for mZP1-mZP4 and hZP1-hZP4 downstream of the terminal Cys residue of the ZPD. Cys residues are in *red* and the CFCS is in *cyan* (mZP4 from Mus pahari). **Panel B.** Shown are primary structures for the carboxy-terminal extensions downstream of the terminal Cys residue of the ZPD for mZP2 and mZP3 from 13 different species. Cys residues are in *red*, the CFCS is in *cyan*, and consensus sequences are indicated.

7.2 ZP1 and ZP4

ZP1 and ZP4 are paralogues that share ≃72% similarity and ≃43% identity, with the major difference being the length of the polyP sequence upstream of the TFD and adjacent ZP-N1 (Fig. 7). Each protein consists of a ZPD close to the C-terminus, a TFD adjacent to the ZPD, and an additional ZP-N subdomain (ZP-N1) close to the SS.

Pro generally accounts for ≃6.3% of aa in proteins, of which ≃4.4% are present in trimers or longer repeats (Morgan & Rubenstein, 2013). Vertebrate ZP1 contains an N-terminal region that is rich in Pro with long spans of multiple, but not necessarily consecutive Pro residues. Pro-rich regions (PRRs) are referred to as low-complexity domains and often form left-handed helical structures called polyproline type II (PPII) helices (Adzhubei et al., 2013). PPII is not a standard secondary structure, and predictions of the structure of PRR-containing proteins without high-resolution structural data are only suggestive (Rath et al., 2005). For ZP1 the PRRs stretch over 100 aa or more and Pro accounts for ≃20% and

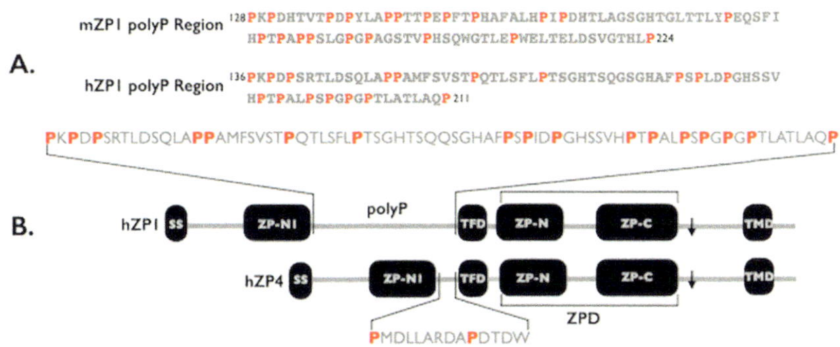

Fig. 7 Schematic representation of the organization of hZP1 and hZP4 and the primary structures of the mZP1 and hZP1 polyP regions. **Panel A.** Shown are the sequences of the polyP regions of mZP1(aa 128–224) and hZP1 (aa 136–211); Pro residues are in *red*. **Panel B.** Shown is the overall organization of hZP1 and hZP4 together with the primary sequences between ZP-N1 and the TFD; Pro residues are in *red*.

≃22% of the residues in this stretch for mouse and human ZP1, respectively. The PRRs are located between the ZP-N1 and the adjacent TFD. ZP4 proteins are shorter versions of ZP1 that lack the relatively long PRR, but otherwise have the same domain organization.

A PPII helix is a short structure consisting of 3–6 aa and consecutive helices can form a relatively flexible rod in an extended helical structure or at terminal regions of a protein. This flexibility may be important for creating elasticity, supporting self-assembly, and promoting intermolecular interactions. It is relevant to note that squamate reptiles, such as geckos, lizards, and snakes produce eggs with either flexible or hard eggshells. The two groups of eggshells can be distinguished from one another by a significant difference in the number of Pro residues present. Flexible eggshells have much higher levels of Pro (> 3-fold) than hard eggshells (Sexton et al., 2005).

ZP1 and ZP4 have a single ZP-N1 subdomain and the ZP-N1 of ZP1 is unusual in that it has five Cys residues, as opposed to the ZP-N subdomain of ZP4 that has four Cys residues. However, ZP-N1 adopts the same immunoglobulin (Ig)-like fold as ZP-N, with two disulfides linked Cys1→Cys4 and Cys2→Cys3. The fifth Cys residue of ZP-N1 (Cys69 in *Mus musculus* and Cys77 in human) is likely used to covalently link the two polypeptides of the ZP1 homodimer (Fig. 8). If the ZP1 homodimer crosslinks ZP fibrils, the two flexible PRR rods of ZP1 could serve as elastic regions between ZP fibrils and give the matrix its stretchability. ZP-N1 of ZP4 does not have a fifth Cys residue and may not have the elasticity presumably present in ZP1 since it is missing an N-terminal PRR.

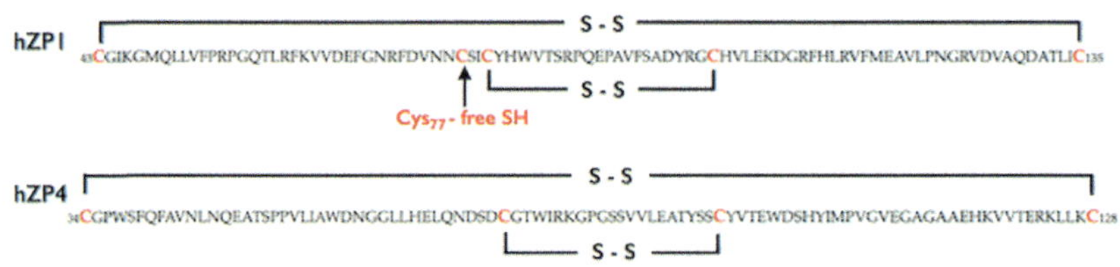

Fig. 8 Schematic representation of the positions of the Cys residues (*red*) and intramolecular disulfide linkages (*black*) of subdomain ZP-N1of hZP1 and hZP4. The free sulfhydryl group of Cys77 in hZP1 (Cys69 in mZP1) is shown in *red*.

7.3 Trefoil domain

A TFD was first identified in a rabbit ZP protein (Bork, 1993) and has since been found in ZP1 and ZP4 from a variety of animal species (Litscher & Wassarman, 2015). TFDs consist of three loops formed by three intramolecular disulfide bonds, Cys1→Cys5, Cys2→Cys4, and Cys3→Cys6, and are present upstream of the ZPD and downstream of ZP-N1 of ZP1 and ZP4 (Fig. 9). For both *hZP1* and *hZP4* the TFD is encoded by exon-4. TFD aa sequences of mouse and human ZP1 and ZP4 are ≃70% identical and of chicken and mouse ZP1 are ≃48% identical.

Trefoil factors (TF) of the gastrointestinal tract consist of ≃45 aa, are secreted by mucosal epithelium, bind oligosaccharides of large mucus glycoproteins and affect viscoelastic properties of gastric mucins (Polshakov et al., 1997; Thim & May, 2005; Braga Emidio et al., 2019; Järva et al., 2020; Notova et al., 2020). It is possible that the TFD of ZP1 and ZP4 participate in crosslinking of ZP fibrils and at least partly account for the increase in ZP stiffness and viscosity following fertilization of eggs.

8. Three-dimensional structure of ZP proteins

8.1 ZPD subdomains

X-ray diffraction studies of ZPD-containing proteins have revealed that both ZP-N and ZP-C adopt an Ig-like fold despite the absence of sequence identity (Monné et al., 2008; Monné & Jovine, 2011; Raj et al., 2017; Bokhove & Jovine, 2018; Nishio et al., 2024). The structures of ZP-N and ZP-C resemble C-type and V-type Ig-like domains, respectively. However, certain structural features make the ZPDs distinct from other known Ig-like domains such that they represent new Ig superfamily subtype structures.

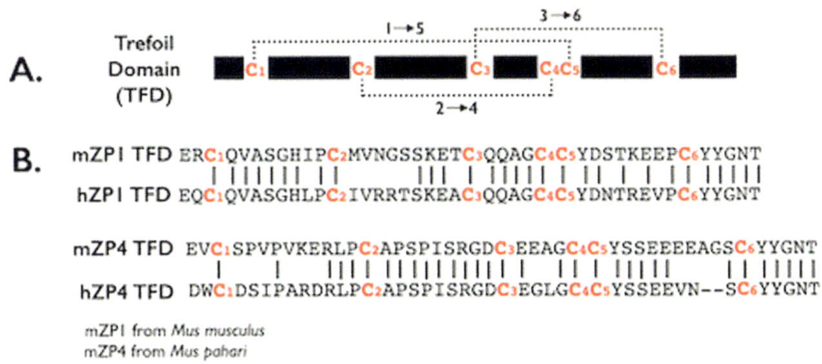

Fig. 9 Schematic representation of the organization of the TFD and primary structures of the TFDs of mZP1, hZP1, mZP4 (*Mus pahari*), and hZP4. **Panel A.** Shown is the overall organization of the TFD with the three intramolecular disulfides indicated. The six Cys residues are in *red*. **Panel B.** Shown are primary structures of the TFDs of mZP1, hZP1, mZP4, and hZP4. The six Cys residues are in *red*.

The ZP-N subdomain consists of an antiparallel sandwich of two β-sheets made up of eight strands of polypeptide, designated as A, B, D, and E and C, E', F, and G strands. The strands enclose a hydrophobic core of buried side-chain residues, with two disulfides that have Cys1→Cys4 and Cys2→Cys3 connectivity and clamp both sides of the sandwich (Fig. 10). Similar to the structure of ZP-N, the ZP-C subdomain consists of a β-sandwich comprising two stacked β-sheets, one with four strands and the other with six strands, that approximate a Greek key–like motif which is characteristic of Ig-like domains (Hutchinson & Thornton, 1993) (Fig. 10).

The ancestral gene for the Ig superfamily may have originated ≃750 million years ago in sponges as a primitive sandwich-like fold used later in vertebrate extracellular recognition systems (Williams & Barclay, 1988). The strong structural similarity between the ZP-N and ZP-C suggests that the ZPD may have arisen by duplication of an ancestral gene encoding a protein containing a single ZP-N.

8.2 ZP3 dimers

In ZP3 crystals two molecules are arranged as homodimers in antiparallel orientation to form an asymmetric structure (Han et al., 2010; Raj et al., 2017; Bokhove & Jovine, 2018). The two ZPDs are bonded by electrostatic interactions between the ZP-N and ZP-C of opposing molecules, ZP-N(1):ZP-C(2) and ZP-N(2):ZP-C(1). There are no ZP-N(1):ZP-N(2) or ZP-C(1):ZP-C(2) contacts within the homodimer. The structures of

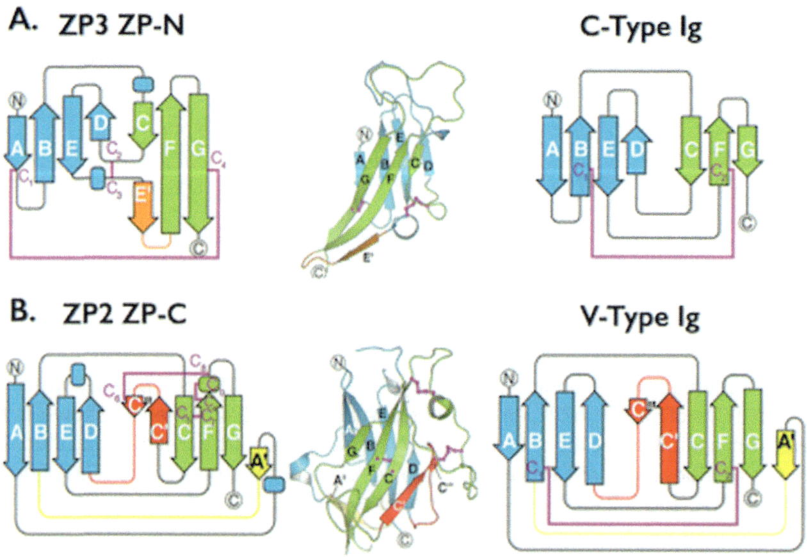

Fig. 10 Three-dimensional structures of ZPD subdomains ZP-N and ZP-C that are related to C-type and V-type Ig-like domains. **Panel A.** ZP3 subdomain ZP-N and C-type Ig-like domains. β-strands are labeled using Ig terminology, helices are indicated by rectangles. Opposing β-sheets 1 and 2 are *blue* and *green,* respectively, with termini circled. The E′ strand is *orange* and disulfides *magenta.* **Panel B.** ZP2 ZP-C and V-type Ig-like domains. As in panel A, except for the additional A′ and C′/C′′ strands that are *yellow* and *red,* respectively. This figure was adapted with permission from Professor L. Jovine (see Bokhove & Jovine, 2018).

ZP-N and ZP-C of uromodulin, another ZPD-containing protein, are similar to those of ZP3 having the same fold, disulfide connectivities, and a conserved Tyr residue in the ZP-N subdomain (aa111 in mouse ZP3). Mutation of this Tyr residue in mouse ZP2 reduces incorporation of the mutated protein into the ZP and prevents its secretion from transfected cell lines (Jovine et al., 2002). Mutation of Tyr (aa1870) to Cys in human tectorin, a ZPD-containing protein associated with the tectorial membrane, results in malformation of the membrane and nonsyndromal deafness (Legan et al., 1997, 2005).

8.3 ZP fibrils

Early structural evidence suggested that each ZP fibril consists of two intertwined filaments, with each filament composed of ZP2-ZP3 dimers interposed with ZP1 and ZP4. The filaments wound around each other to form a double-helix having an axial repeat of $\simeq$12 nm, diameter of

≃9–14 nm, and structural repeat of ≃15 nm (Greve & Wassarman, 1985; Wassarman & Mortillo, 1991; Jovine et al., 2002). However, more recent structural evidence suggests a different arrangement of ZP proteins in filaments, with ZP1, ZP2, and ZP4 assembled into one filament (heteropolymer) that pairs with a second filament composed solely of ZP3 (homopolymer) to form a left-handed helical structure (Nishio et al., 2024). ZP1 present in one fibril connects via its intermolecular disulfide with another fibril. Regardless of the arrangement of individual ZP proteins in filaments, the extended N-terminal regions of ZP1, ZP2, and ZP4 protrude from the filaments and give rise to the ≃15 nm structural repeat (see Fig. 7A–D, Nishio et al., 2024).

9. Evolution of ZP proteins

Proteins are the product of evolution and their molecular phylogeny provides a natural basis for hierarchical classification. Several mechanisms have been described for domain acquisitions or rearrangements; they include exon-deletion, exon-shuffling, exon-extension into a non-coding region, intron-recombination, and other possibilities. To gain or rearrange domains without disrupting reading frames of genes, exons must be shuffled symmetrically (Liu & Grigoriev, 2004; Marsh & Teichmann, 2010; Patthy, 2021).

It has been suggested that *ZP3* evolved from an ancestral *ZP3*-like gene and subsequently duplicated several times and became the other ZP genes, *ZP1, ZP2, ZP4,* and *ZPax* (Fig. 11) (Swanson et al., 2001; Spargo & Hope, 2003; Conner & Hughes, 2003; Smith et al., 2005; Litscher & Wassarman, 2007, 2014; Goudet et al., 2008; Claw & Swanson, 2012; Killingbeck & Swanson, 2018; Feng et al., 2018; Rivera et al., 2022). Over time *ZP3* gained three ZP-N subdomains, ZP-N1, -N2, and -N3, to become *ZP2* (Fig. 11). Again, over time *ZP2* was shortened by the loss of exons 5, 6, 7, 9, and 10, and gained a TFD (Fig. 11; see Fig. 9) perhaps by intron-recombination, and two-three Pro residues by extension of exon-3 to become *ZP4*. *ZP4* then gained a long stretch of Pro residues (polyP) to become *ZP1* (Fig. 11; see Fig. 7). Comparison of exon-boundaries and lengths of aa sequences suggests that the ZP-N1 subdomain of ZP1 and ZP4 is a homolog of ZP-N1 of ZP2. The gain of an additional Cys residue in ZP1 (5th Cys in ZP-N1; see Fig. 8), possibly by a mutation in exon-2 of *ZP1*, enabled formation of an intermolecular disulfide bond between two ZP1 monomers and covalent crosslinking of ZP fibrils.

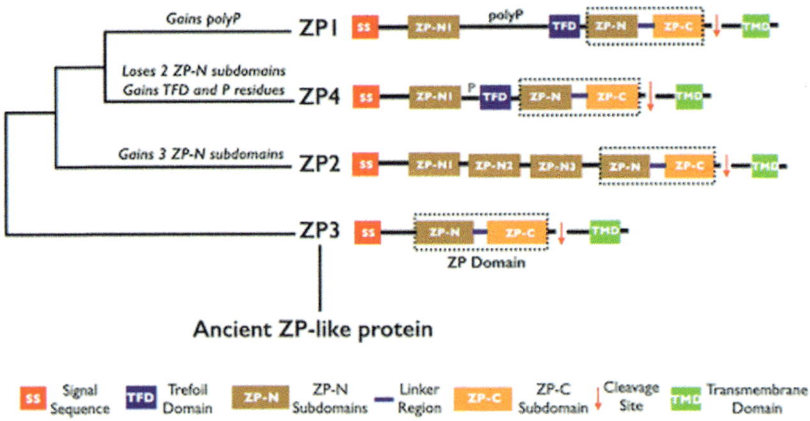

Fig. 11 Schematic representation of the evolution of ZP proteins. Shown is a scheme for the evolution of ZP proteins from an ancient ZP-like protein to ZP3, the simplest of the four ZP proteins, to ZP2 (gain of three ZP-N subdomains; *brown*), to ZP4 (loss of two ZP-N subdomains and gain of P residues and a TFD; *blue*), to ZP1 (a gain of a polyP stretch; *black*). All ZP proteins have an SS, ZPD, CFCS, and TMD.

The ZP of most mammalian species consists of ZP1–4, however, some domesticated animals (e.g., *Mus musculus*, cow, and dog) have *ZP1* or *ZP4* as pseudogenes. *ZP1* and *ZP4* are paralogs, but ZP1 appeared later than ZP4 as a component of the ZP during species evolution. The appearance of ZP4 in early vertebrate egg envelopes apparently was followed later on by the appearance of ZP1 with an added polyP stretch and Cys residue (described above). In more ancient vertebrates (e.g., teleost fish, sharks, and lampreys) the ZP does not have ZP1, but does have ZP4. However, *ZP1* is found in birds, turtles, and snakes.

10. Mammalian fertilization and the ZP

Virtually all multicellular organisms are capable of sexual reproduction. It involves two kinds of gametes, eggs and sperm, that fuse with each other (*fertilization*) to produce a one-cell embryo or zygote. It was Karl Ernst von Baer (1792–1876) who established that mammals develop from eggs and concluded that "every animal which springs from the coition of male and female is developed from an ovum". Fertilization of eggs by sperm was first proposed by the Italian biologist Lazzaro Spallanzani (1729–1799) in ca. 1786 and was accurately described by the German embryologist Oscar Hertwig (1849–1922) and the Swiss biologist Hermann Fol (1845–1892) in ca. 1876.

10.1 Mammalian fertilization

Among mammalian species, few eggs are found in female oviducts (≃1–10) and relatively few sperm are found at the site of fertilization (≃100–150) compared to the number of sperm deposited into the female reproductive tract (≃100–200 million). Eggs released from Graafian follicles enter the opening (*ostium*) of the oviduct (*Fallopian tube*) and move to the lower *ampulla* region where fertilization takes place. The final and essential maturation of sperm, called *capacitation*, occurs following ejaculation of sperm into the female's genital tract (Florman & Ducibella, 2006). During capacitation, efflux of cholesterol from the sperm's plasma membrane takes place and results in a change in membrane fluidity, rearrangement and/or modification of membrane lipids and proteins, and activation of signal transduction by protein kinases. As a result, sperm become hyperactivated in terms of motility, acquire the ability to complete the *acrosome reaction* (ARx), and the ability to bind and penetrate the egg's ZP. Binding of sperm to eggs occurs due to a chance encounter in the oviduct, but is promoted by a chemical gradient stimulus (*chemotaxis*) mediated by an egg follicular factor (Eisenbach & Giojalas, 2006).

Fertilization begins when free-swimming sperm bind to unfertilized eggs in the ampulla region of the female's oviduct (Yanagimachi, 1994; Wassarman, 1987; Wassarman & Litscher, 1998, 1999; Jungnickel et al., 2003; Florman & Ducibella, 2006; Visconti & Florman, 2010; Florman & Fissore, 2015). Sperm bind in a species-restricted manner to the ZP and penetrate it using enzymes associated with the sperm's inner-acrosomal membrane (IAM) and forward thrust due to flagellar hyperactivity. Following fertilization, the ZP undergoes physical and functional changes that prevent fusion of eggs with more than one sperm (*polyspermy*) (Fahrenkamp et al., 2020). The process of fertilization involves three membrane fusion events, the ARx, cortical reaction (CRx), and sperm-egg fusion (Wassarman & Litscher, 2008).

10.2 Acrosome reaction

Mammalian sperm possess an organelle, the *acrosome*, that is derived from Golgi and overlies the sperm's head (Toshimori & Eddy, 2015). For capacitated sperm to penetrate the ZP and fuse with the egg's plasma membrane they must undergo the ARx, a form of cellular exocytosis (Florman et al., 2008; Balbach et al., 2020). The ARx involves fusion of sperm's plasma membrane with outer-acrosomal membrane (OAM) and leads to exposure of the IAM with its associated contents that include proteases and hyaluronidases (Fig. 12). Both acrosome-intact (AI) and acrosome-reacted (AR) sperm are able to bind to the ZP. AI-sperm's plasma membrane and AR-sperm's IAM recognize and bind to different epitopes on the ZP. However, only AR-sperm with an exposed IAM can penetrate the ZP and fuse with the egg's plasma membrane.

Depending on the animal species, the ARx may occur in the female reproductive tract, upon interaction with cumulus cells surrounding ovulated eggs, or after binding of AI-sperm to the ZP (Avella & Dean, 2011). In the latter case, AI-sperm bind to ZP3 and undergo the ARx on the surface of the ZP (Florman & Storey, 1982; Bleil & Wassarman, 1983; Bleil et al., 1988; Vazquez et al., 1989). ZP3 drives the sperm ARx by opening low voltage-activated Ca^{2+} channels (T-type) that causes a transient Ca^{2+}

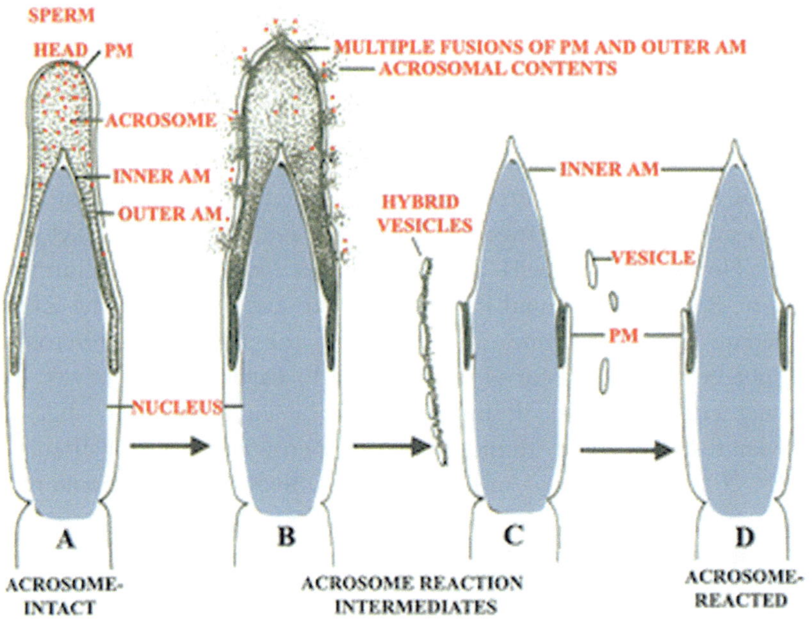

Fig. 12 Diagrammatic representation of some morphological features of a mammalian sperm undergoing and completing the acrosome reaction (ARx). The course of the AR is indicated by stages (A)–(D). An acrosome-intact (AI) sperm head and sperm nucleus (pale *blue*) is shown in (A). In (B), fusion between the sperm's outer acrosomal membrane and plasma membrane is indicated. Hybrid membrane vesicles, composed of plasma and outer acrosomal membrane, are shown in (C) and (D). pm, plasma membrane; am, acrosomal membrane. This figure is a modified version of Figure 16 in Yanagimachi (1994) that appeared as Fig. 5 in Wassarman (1999).

influx followed by activation of heterotrimeric G proteins and phospholipase C, elevation of intracellular pH, and a sustained Ca^{2+} influx (Evans & Florman, 2002).

10.3 Sperm-binding to ZP

AI-sperm bind to ZP3 and AR-sperm bind to ZP2 of unfertilized eggs (Bleil & Wassarman, 1980c, 1986; Florman & Storey, 1982; Bleil et al., 1988; Mortillo & Wassarman, 1991; Arnoult et al., 1996; Jungnickel et al., 2007; Wassarman & Litscher, 2022b) (Fig. 13). AI-sperm bind to ZP3 oligosaccharides (Florman et al., 1984; Lambert, 1984; Florman & Wassarman, 1985; Wassarman et al., 1985; Bleil & Wassarman, 1988; Shalgi et al., 1990; Kinloch et al., 1991; Litscher et al., 1995; Johnston et al., 1998; Dell et al., 2003; Kerr et al., 2004; Hanna et al., 2004; Pang et al., 2011; Wassarman & Litscher, 2022b), whereas AR-sperm bind to ZP-N1 of ZP2, not to ZP2 oligosaccharides (Gahlay et al., 2010; Baibakov et al., 2012; Avella et al., 2014; Tokuhiro & Dean, 2018). However, recent evidence obtained with mice lacking the ZP-N1 region of ZP2 suggests that this region may not be

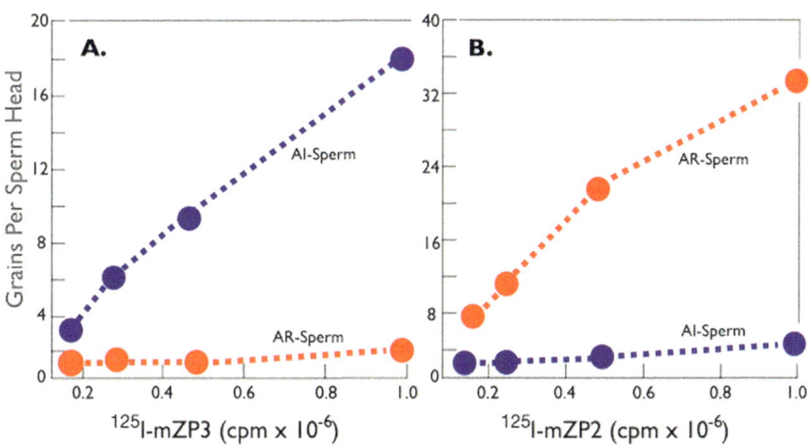

Fig. 13 Quantitation of mZP2, mZP3, and fetuin radiolabeled-probes bound to acrosome-intact (AI-) and acrosome-reacted (AR-) mouse sperm. **Panel A.** Binding of ^{125}I-mZP3 to AI- (*blue* circles) and AR- (*red* circles) sperm as a function of probe concentration. **Panel B.** Binding of ^{125}I-mZP2 to AI- (*blue* circles) and AR- (*red* circles) mouse sperm as a function of probe concentration. ^{125}I-fetuin binding to AI- and AR-sperm occurred at insignificant levels. These results demonstrate that mZP3 binds preferentially to AI-sperm and mZP2 binds preferentially to AR-sperm. For further information see Bleil and Wassarman (1986). Similar results were obtained using colloidal gold-labeled mZP2 and mZP3; see Mortillo and Wassarman, (1991).

required or essential for binding of AR-sperm to eggs (Nishio et al., 2024). Sperm bind but fail to penetrate ZP lacking the ZP-N1 region of ZP2 since the ZP resembles that of fertilized eggs that have undergone hardening and the mice are subfertile. In any case, apparently there are two pathways for initiation of fertilization in mammals (Fig. 14).

Involvement of ZP3 oligosaccharides in AI-sperm binding is akin to binding of sperm to eggs from marine animals and amphibia, sexual agglutination in budding yeast, binding of pollen to the plant stigma, and binding of bacteria, animal viruses, parasites, and other pathogens to their cellular hosts. The region of ZP3 proteins to which sperm bind, called the sperm combining-site, may be a site of positive Darwinian selection, perhaps having to do with species-restricted fertilization (Wassarman & Litscher, 1995; Litscher & Wassarman, 1996; Swanson et al., 2001; Killingbeck & Swanson, 2018). Results of X-ray crystallographic analysis of the structure of avian ZP3 suggests that the sperm's combining-site is relatively close to a conserved O-linked glycan essential for sperm binding (Han et al., 2010).

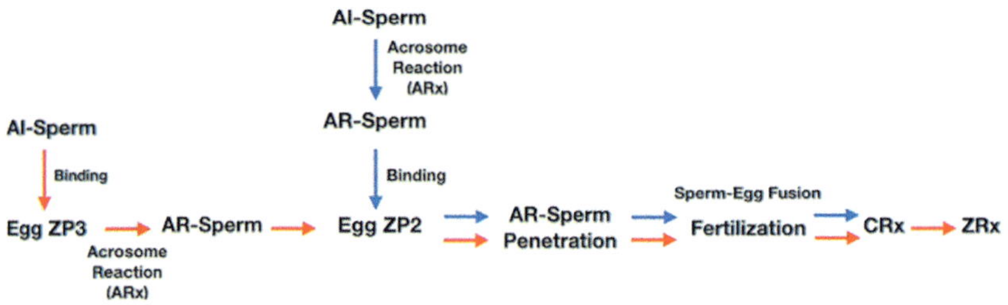

Fig. 14 Schematic diagram of some steps involved in two different pathways to fertilization in mice. In the first pathway (*red* arrows), acrosome-intact (AI-) sperm bind to mZP3 in the egg's zona pellucida (ZP). As a result of binding to mZP3, AI-sperm undergo the acrosome-reaction (ARx; multiple fusions between the sperm's outer acrosomal membrane and plasma membrane leading to exposure of the inner acrosomal membrane), and acrosome-reacted (AR-) sperm then bind to mZP2. In the second pathway (*blue* arrows), AI-sperm undergo the ARx prior to reaching the ZP, perhaps in the *cumulus oophorous* of ovulated eggs, and then bind to mZP2 rather than to mZP3. In both pathways AR-sperm penetrate the ZP and fuse with the egg's plasma membrane (fertilization). Fusion of sperm and egg induces the cortical reaction (CRx) in eggs (fusion of cortical granule membrane and the egg's plasma membrane) that results in exocytosis of cortical granule components, including Zn^{2+} and ovastacin, into the ZP and induces the zona reaction (ZRx). The ZRx consists of inactivation of mZP2 and mZP3 as sperm receptors (i.e., AI- and AR-sperm cannot bind to the ZP of fertilized eggs) and hardening of the ZP. Changes in the ZP assist in the prevention of polyspermy (slow block) and protection of cleavage-stage embryos as they traverse the female reproductive tract.

10.4 Sperm-egg fusion

Sperm-egg fusion involves at least two AR-sperm proteins, Izumo and SPACA6, and two egg plasma membrane proteins, Juno and CD9 (Miyado et al., 2000; Inoue et al., 2005; Bianchi et al., 2014; Bianchi & Wright, 2016; Okabe, 2018; Noda et al., 2020; Siu et al., 2021). The former are testis-specific, Ig-like, and localized on the equatorial segment of the sperm's plasma membrane that is involved in sperm-egg fusion. Juno is a membrane-anchored, folate receptor-like protein present only in mammals, whereas CD9 is a member of the tetraspanin family of proteins. Izumo and SPACA6 homozygous-null male mice are infertile, Juno homozygous-null female mice are infertile, and CD9 homozygous-null female mice are subfertile. Although binding of AR-sperm to the egg's plasma membrane is mediated by Izumo and Juno, apparently their interaction alone is not sufficient to induce fusion of gametes. In this context, a recent report suggests that formation of a pentameric complex, called a "fusion synapse", consisting of sperm's Izumo1, SPACA6, and TMEM81 and egg's Juno and CD9, is responsible for gamete fusion (Elofsson et al., 2024).

10.5 Fertilized Egg's ZP

Fertilization and activation of eggs leads to the *cortical reaction* (CRx) and *zona reaction* (ZRx). These reactions involve changes in the ionic composition of eggs, changes in the egg's plasma membrane, and release of Ca^{2+} from the egg's intracellular stores. It is likely that the sperm's oocyte-activator factor (SOAF), phospholipase C-zeta (PLCz), is involved in egg activation (Swann et al., 2006; Saleh et al., 2020).

The primary block to polyspermy following fertilization occurs at the level of the egg's plasma membrane so that unfused sperm accumulate in the perivitelline space between the ZP and plasma membrane. While the primary block in mammalian eggs apparently is not due to a change in electrical potential at the plasma membrane, a depolarization found with marine animals, the actual cause remains unclear.

10.5.1 Cortical and zona reactions

Fertilization and activation of eggs causes completion of meiosis with emission of a second polar body, formation of male and female pronuclei, DNA synthesis, and the CRx (Xiong et al., 2017; Vogt et al., 2019). The CRx requires translocation of *cortical granules* to the inner surface of the egg's plasma membrane using myosin IIA and elevation of Ca^{2+} concentration (1 μM in mouse eggs) due to release of Ca^{2+} stores from endoplasmic reticulum by inositol triphosphate (IP3). The CRx leads to the release of proteases, glycosidases, lectins, Zn^{2+}, and other molecules into the ZP. This results in accumulation of Zn^{2+} in the ZP, called Zn^{2+}-sparks, and release of ovastacin, a member of the astacin family of metalloendoproteases, that cleaves ZP2 at ZP-N2. High concentrations of Zn^{2+} have a negative effect on sperm motility and helps to prevent penetration of bound sperm through the ZP. Proteolytic cleavage occurs without release of a peptide (mouse ZP2 cleavage-site 166Leu-Ala↓Asp-Glu169; human ZP2 cleavage-site 171Leu-Ala↓Asp-Asp174) (Bleil et al., 1981; Moller & Wassarman, 1989; Quesada et al., 2004; Gahlay et al., 2010; Burkart et al., 2012; Avella et al., 2013).

The CRx induces changes in the ZP that constitute the ZRx. Following fertilization and activation of eggs, supernumerary sperm are no longer able to bind to either ZP2 or ZP3, thereby contributing to prevention of polyspermy. The changes restrict penetration of bound sperm through and binding of additional sperm to the ZP (Que et al., 2017; Tokuhiro & Dean, 2018; Evans, 2020; Fahrenkamp et al., 2020). The ZRx causes a significant decrease in the solubility of the ZP (hardening) together with a ≃2–4-fold increase in the stiffness and viscosity of the ZP (Kim & Kim, 2013). ECM elasticity, or stiffness, has been shown to dramatically affect a variety of cellular processes, including adhesion of cells to ECM (Chaudhuri et al., 2020). Fusion of a single sperm with the egg's plasma membrane is sufficient to prevent fusion by additional sperm present in the perivitelline space and thereby prevents adverse effects of polyspermy on development of zygotes.

10.5.2 Proteolytic cleavage ZP2

A recent three-dimensional structural analysis of the ZP revealed that cleavage of ZP2 within ZP-N2 by ovastacin results in homodimerization of ZP2 and causes extensive crosslinking of ZP fibrils; this accounts for the hardening of the ZP (Nishio et al., 2024). The enhanced crosslinking brings ZP fibrils very close to one another (within < 11 nm), ≃7-fold closer than mediated by the intermolecular ZP1 disulfide crosslinks present in the unfertilized egg's ZP. This change significantly increases the rigidity of the ZP matrix, hinders sperm penetration, and is part of the secondary or slow block to polyspermy.

11. Inactivation and mutation of ZP genes

Results of experiments in which antisense oligonucleotides directed against ZP2 or ZP3 mRNAs were injected into growing mouse oocytes strongly suggest that ZP2 and ZP3 are dependent upon each other for incorporation into the ZP (Tong et al., 1995). To confirm and extend this finding, gene-targeting was used to establish mouse lines in which ZP genes were inactivated by either homologous recombination or insertional mutagenesis and the fertility of female mice was assessed. Results of these experiments are described below and summarized in Table 9.

11.1 Homozygous-null mice

Male mice that are homozygous-nulls for mouse ZP1–3 are as fertile as wild-type males. However, female mice that are homozygous-nulls for ZP1 ($ZP1^{-/-}$) are fertile, but exhibit reduced fertility compared to wild-type females due to early loss of cleavage-stage embryos in oviducts (Rankin et al., 1999). Such loss is due to a ZP that is insufficiently crosslinked and extremely fragile as cleavage-stage embryos traverse the female reproductive tract.

On the other hand, female mice that are homozygous-nulls for either ZP2 ($ZP2^{-/-}$) or ZP3 ($ZP3^{-/-}$) produce eggs that lack a ZP and these females are completely infertile (Liu et al., 1996; Rankin et al., 1996, 2001). Infertility is attributable to a scarcity of growing oocytes, Graafian follicles, and ovulated eggs. The paucity of growing oocytes and follicles in ovaries of $ZP3^{-/-}$ mice is reflected in weight differences of ovaries from 20 day-old wild-type females (1.0 ± 0.17 mg per ovary) and ovaries from homozygous-null females the same age (0.26 ± 0.1 mg per ovary). This difference in weight ($\simeq$4-fold) is due to significantly retarded oocyte growth and follicle development (Wassarman & Litscher, 1998, 2012). These findings are consistent with results of antisense experiments (Tong et al., 1995) and strongly suggest that the presence of both ZP2 and ZP3 is absolutely required for assembly of a ZP around growing oocytes and development of Graafian follicles.

Table 9 Phenotypes of ZP null female mice.

Genotype	ZP	Fertility
Wild-type	Normal	Fertile
mZP1 homozygous-null	Abnormal	Low fertility
mZP2 homozygous-null	None	Infertile
mZP3 homozygous-null	None	Infertile
mZP3 heterozygous-null	Thin	Fertile

11.2 Heterozygous-null mice

Female mice that are heterozygous-nulls for ZP3 ($ZP3^{+/-}$) are as fertile as wild-type females ($ZP3^{+/+}$), but their eggs have a thin ZP (ave. width $2.7 \pm 1.2\,\mu m$) compared to the ZP of eggs from wild-type females (ave. width $6.2 \pm 1.9\,\mu m$) (Wassarman et al., 1997). The thin ZP contains about one-half the amount of ZP2 and ZP3 found in ZP of eggs from wild-type mice. These findings suggest that the width of the ZP is not a critical parameter for binding of free-swimming sperm to the ZP or for fertilization of eggs. They also suggest that as long as both nascent ZP2 and ZP3 are present, even at reduced amounts, a functional ZP can be assembled.

11.3 Mutated human ZP genes

Today it is estimated that approximately one-in-ten married women worldwide are infertile and infertility can have a genetic component (Deshpande & Gupta, 2019). In this context, it was found that there is a causal relationship between gene sequence variations (mutations) in human ZP genes and female fertility. Some women unsuccessful in IVF trials had significantly more mutations in their ZP genes than women with proven fertility (Männikkö et al., 2005; Pökkylä et al., 2011; Margalit et al., 2012).

In recent years more than thirty cases of human female infertility seen at IVF clinics have been linked to mutations in ZP genes encoding ZP1–4 (e.g., Huang et al., 2014; Liu et al., 2017, 2023; Zhou et al., 2019; Cao et al., 2020; Litscher & Wassarman, 2020b; Wassarman & Litscher, 2021, 2022a; Jia et al., 2022; Kong et al., 2023; Sun et al., 2023). Such mutations often result in empty follicle syndrome (EFS); a condition in which no oocytes were retrieved from stimulated, growing ovarian follicles. For several of the mutated genes insertion of a premature stop-codon led to synthesis of truncated forms of ZP proteins and the absence of a ZP around oocytes. For example, a heterozygous missense mutation in *ZP2* exon-18 Arg 698 > X with insertion of a stop-codon at aa698 and a heterozygous frameshift mutation in *ZP3* exon-7 Arg 349 > X followed by a stop-codon were identified (Liu et al., 2017). Both mutations resulted in synthesis of truncated ZP proteins, ZP2 lacking a TMD and ZP3 lacking a CTP. More than a dozen mutations have been identified in exons encoding the ZPD of *ZP1–4*; mutations in the ZP-N subdomain, ZP-C subdomain, or linker prevented construction of a ZP around growing oocytes. For *ZP1* and *ZP4* several mutations occurred in exons encoding the N1 subdomain before the TFD and resulted in either the presence of an abnormal ZP or no ZP around oocytes. As expected, the absence of ZP polypeptide sequence elements necessary for proteolytic processing, secretion, or assembly of nascent ZP proteins during oogenesis resulted in female infertility. Infertile patients often had either abnormal oocytes, eggs lacking a ZP, or no eggs at all. As found with homozygous-null female mice, these results confirm that the presence of a normal ZP around growing oocytes is essential for production of ovulated eggs and establishment of human female fertility.

12. Final comments

Reproduction of the species is a fundamental and essential property of all living things. Consequently, involvement of meiosis during development of eggs (oogenesis) and sperm (spermatogenesis) and species-restricted interactions between haploid gametes (binding of sperm to eggs) that culminate in fertilization (sperm-egg fusion) and restoration of a diploid state to the fertilized egg are all highly regulated. Fertilization activates mammalian eggs to initiate a complex program of development, over time transforming a single cell, the fertilized egg, into a multicellular organism; in the case of human adults, consisting of tens-of-trillions of cells, hundreds of differentiated cell-types, and a few tissue-types. Fertilization also activates mechanisms at the level of both the egg's plasma membrane (primary or fast block to polyspermy) and ZP (secondary or slow block to polyspermy) that prevent fertilization by more than one sperm, a lethal event. Many egg and sperm molecules, macromolecules, and organelles that participate before, during, and after fertilization have now been identified and, in many cases, signaling cascades and mechanisms of action have been clarified in some detail.

Formation of a ZP is an essential activity of growing oocytes during mammalian oogenesis. Failure to do so results in an inability to produce Graafian follicles and ovulated eggs during oogenesis and in female infertility. The ZP stabilizes interactions between innermost cumulus cells and oocytes essential for oocyte growth and follicle development during oogenesis, provides specific receptors for binding of sperm to ovulated eggs prior to fertilization, assists in prevention of polyspermy following fertilization, and protects cleavage-stage embryos as they traverse the female's reproductive tract. The participation of ZP proteins ZP1–4 in these various roles is fairly well understood today, however, several aspects of mammalian fertilization still remain unresolved, vague, or controversial. It is likely that continued basic and clinical research directed at fundamental questions about mammalian fertilization will lead to practical applications for both human conception and contraception in the future.

Acknowledgments

We are grateful to past members of the Wassarman laboratory at Harvard Medical School, Roche Institute of Molecular Biology, and Icahn School of Medicine at Mount Sinai. We thank them all for their valuable contributions to research on the zona pellucida and fertilization. We thank Professor Giuseppe Familiari, University of Rome, Italy, and Professor Luca Jovine, Karolinska Institute, Sweden, for kindly providing images from their publications. We apologize to investigators whose contributions were not cited directly here. Our research was supported largely by the National Institutes of Health (NICHD-12275 and -35105) and F. Hoffmann-La Roche AG.

References

Ackert, C. L., Gittens, J. E., O'Brien, M. J., Eppig, J. J., & Kidder, G. M. (2001). Intercellular communication via connexin43 gap junctions is required for ovarian folliculogenesis in the mouse. *Developmental Biology, 233*, 258–270.

Adzhubei, A. A., Sternberg, M. J. E., & Makarov, A. A. (2013). Polyproline-II helix in proteins: Structure and function. *Journal of Molecular Biology, 425*, 2100–2132.

Albertini, D. F. (2015). *The mammalian oocyte. Knobil and Neill's physiology of reproduction, 1*, San Diego, CA: Academic Press/Elsevier59–97.

Alexander, D. B., & Goldberg, G. S. (2003). Transfer of biologically important molecules between cells through gap junction channels. *Current Medicinal Chemistry, 10*, 2045–2058.

Arnoult, C., Zeng, Y., & Florman, H. M. (1996). ZP3-dependent activation of sperm cation channels regulates acrosomal secretion during mammalian fertilization. *Journal of Cell Biology, 134*, 637–645.

Austin, C. R., & Short, R. V. (Vol. Eds.), (1982). *Reproduction in mammals 1, germ cells and fertilization: 177*. Cambridge, UK: Cambridge University Press.

Avella, M. A., & Dean, J. (2011). Fertilization with acrosome-reacted mouse sperm: Implications for the site of exocytosis. *Proceedings of the National Academy of Sciences, USA, 108*, 19843–19844.

Avella, M. A., Xiong, B., & Dean, J. (2013). The molecular basis of gamete recognition in mice and humans. *Molecular Human Reproduction, 19*, 279–289.

Avella, M. A., Baibakov, B., & Dean, J. (2014). A single domain of the ZP2 zona pellucida protein mediates gamete recognition in mice and humans. *Journal of Cell Biology, 205*, 801–809.

Baibakov, B., Boggs, N. A., Yauger, B., Baibakov, G., & Dean, J. (2012). Human sperm bind to the N-terminal domain of ZP2 in humanized zonae pellucidae in transgenic mice. *Journal of Cell Biology, 197*, 897–905.

Baena, V., & Terasaki, M. (2019). Three-dimensional organization of transzonal projections and other cytoplasmic extensions in the mouse ovarian follicle. *Scientific Reports, 9*, 1262.

Baker, T. G. (1982). *Oogenesis and ovulation. Reproduction in mammals 1, germ cells and fertilization*. Cambridge, UK: Cambridge University Press17–45.

Balbach, M., Hamzeh, H., Jikeli, J. F., Brenker, C., Schiffer, C., Hansen, J. N., & Wachten, D. (2020). Molecular mechanisms underlying the action of zona pellucida glycoproteins on mouse sperm. *Frontiers in Cell and Developmental Biology, 8*, 572735.

Barresi, M., & Gilbert, S. (2019). *Developmental biology* (12th ed.). Sunderland, MA: Sinauer Associates.

Bianchi, E., Doe, B., Goulding, D., & Wright, G. J. (2014). Juno is the egg Izumo receptor and is essential for mammalian fertilization. *Nature, 508*, 483–487.

Bianchi, E., & Wright, G. (2016). Sperm meets egg: The genetics of mammalian fertilization. *Annual Review of Genetics, 50*, 93–111.

Biswas, L., Tyc, K., El Yakoubi, W., Morgan, K., Xing, J., & Schindler, K. (2021). Meiosis interrupted: The genetics of female infertility via meiotic failure. *Reproduction (Cambridge, England), 161*, R13–R35.

Bleil, J. D., & Wassarman, P. M. (1980a). Structure and function of the zona pellucida: Identification and characterization of the proteins of the mouse oocyte's zona pellucida. *Developmental Biology, 76*, 185–202.

Bleil, J. D., & Wassarman, P. M. (1980b). Synthesis of zona pellucida proteins by denuded and follicle-enclosed mouse oocytes during culture in vitro. *Proceedings of the National Academy of Sciences, USA, 77*, 1029–1033.

Bleil, J. D., & Wassarman, P. M. (1980c). Mammalian sperm-egg interaction: Identification of a glycoprotein in mouse egg zonae pellucidae possessing receptor activity for sperm. *Cell, 20*, 873–882.

Bleil, J. D., Beall, C. F., & Wassarman, P. M. (1981). Mammalian sperm-egg interaction: Fertilization of mouse eggs triggers modification of the major zona pellucida glycoprotein, ZP2. *Developmental Biology, 86*, 189–197.

Bleil, J. D., & Wassarman, P. M. (1983). Mammalian sperm-egg interaction: Sequence of events and induction of the acrosome reaction by a zona pellucida glycoprotein. *Developmental Biology, 95*, 317–324.

Bleil, J. D., & Wassarman, P. M. (1986). Autoradiographic visualization of the mouse egg's sperm receptor bound to sperm. *Journal of Cell Biology, 102*, 1363–1371.

Bleil, J. D., & Wassarman, P. M. (1988). Galactose at the nonreducing terminus of O-linked oligosaccharides of mouse egg zona pellucida glycoprotein ZP3 is essential for the glycoprotein's sperm receptor activity. *Proceedings of the National Academy of Sciences, USA, 85*, 6778–6782.

Bleil, J. D., Greve, J. M., & Wassarman, P. M. (1988). Identification of a secondary sperm receptor in the mouse egg zona pellucida: Role in maintenance of binding of acrosome-reacted sperm to eggs. *Developmental Biology, 128*, 376–385.

Bokhove, M., Nishimura, K., Brunati, M., Han, L., de Santis, D., Rampoldi, L., & Jovine, L. (2016). A structured interdomain linker directs self-polymerization of human uromodulin. *Proceedings of the National Academy of Sciences, USA, 113*, 1552–1557.

Bokhove, M., & Jovine, L. (2018). Structure of zona pellucida module proteins. *Current Topics in Developmental Biology, 130*, 413–442.

Bork, P., & Sander, C. (1992). A large domain common to sperm receptors (Zp2 and Zp3) and TGF-beta type III receptor. *FEBS Letters, 300*, 237–240.

Bork, P. (1993). A trefoil domain in the major rabbit zona pellucida protein. *Protein Science, 2*, 669–670.

Braga Emidio, N., Hoffmann, W., Brierley, S. M., & Muttenthaler, M. (2019). Trefoil factor family: Unresolved questions and clinical perspectives. *Trends in the Biochemical Sciences, 44*, 387–390.

Burkart, A. D., Xiong, B., Baibakov, B., Jimenez-Movilla, M., & Dean, J. (2012). Ovastacin, a cortical granule protease, cleaves ZP2 in the zona pellucida to prevent polyspermy. *Journal of Cell Biology, 197*, 37–44.

Byskov, A. G. (1982). *Primordial germ cells and regulation of meiosis. Reproduction in mammals 1, germ cells and fertilization*. Cambridge, UK: Cambridge University Press 1–16.

Callebaut, I., Mornon, J. P., & Monget, P. (2007). Isolated ZP-N domains constitute the N-terminal extensions of zona pellucida proteins. *Bioinformatics (Oxford, England), 23*, 1871–1874.

Cao, Q., Zhao, C., Zhang, X., Zhang, H., Lu, Q., Wang, C., & Huo, R. J. (2020). Heterozygous mutations in ZP1 and ZP3 cause formation disorder of ZP and female infertility in human. *Cellular and Molecular Medicine, 24*, 8557–8566.

Carabatsos, M. J., Sellitto, C., Goodenough, D. A., & Albertini, D. F. (2000). Oocyte-granulosa cell heterologous gap junctions are required for the coordination of nuclear and cytoplasmic meiotic competence. *Developmental Biology, 226*, 167–179.

Chamberlin, M. E., & Dean, J. (1989). Genomic organization of a sex specific gene: The primary sperm receptor of the mouse zona pellucida. *Developmental Biology, 131*, 207–214.

Chamberlin, M. E., & Dean, J. (1990). Human homolog of the mouse sperm receptor. *Proceedings of the National Academy of Sciences, USA, 87*, 6014–6018.

Chaudhuri, O., Cooper-White, J., Janmey, P. A., Mooney, D. J., & Shenoy, V. B. (2020). Effects of extracellular matrix viscoelasticity on cellular behaviour. *Nature, 584*, 535–546.

Clarke, H. J. (2018). Regulation of germ cell development by intercellular signalling in the mammalian ovarian follicle. *Wiley Interdisciplinary Reviews Developmental Biology, 7*. https://doi.org/10.1002/wdev.294.

Claw, K. G., & Swanson, W. J. (2012). Evolution of the egg: New findings and challenges. *Annual Review of Genomics and Human Genetics, 13*, 109–125.

Cole, F. (2023). Meiosis in development and disease. *Current Topics in Developmental Biology, 151*, 1–369.

Conner, S. J., & Hughes, D. C. (2003). Analysis of fish ZP1/ZPB homologous genes – Evidence for both genome duplication and species-specific amplification modesl of evolution. *Reproduction (Cambridge, England), 126*, 347–352.

Crozet, F., Letort, G., Bulteau, R., Da Silva, C., Eichmuller, A., Tortorelli, A. F., & Terret, M. E. (2023). Filipodia-like protrusions of adjacent somatic cells shape the developmental potential of oocytes. *Life Sciences Alliance, 6*, e202301963.

Dell, A., Chalabi, S., Easton, R. L., Haslam, S. M., Sutton-Smith, M., Patankar, M. S., & Clark, G. F. (2003). Murine and human zona pellucida 3 derived from mouse eggs express identical O-glycans. *Proceedings of the National Academy of Sciences, USA, 100*, 15631–15636.

Deshpande, P. S., & Gupta, A. S. P. (2019). Causes and prevalence of factors causing infertility in a public health facility. *Journal of Human Reproductive Sciences, 12*, 287–293.

Dickmann, Z. (1965). Sperm penetration into and through the zona pellucida of the mammalian egg. *Ciba Foundation Symposium*, 169–182.

Doherty, C. A., Amargant, F., Shvartsman, S. Y., Duncan, F. E., & Gavis, E. R. (2022). Bidirectional communication in oogenesis: A dynamic conversation in mice and Drosophila. *Trends in Cell Biology, 32*, 311–323.

Duarte, M. A., Rodriguez-Fernandes, C., Heckel, G., da Luz Mathias, M., & Bastos-Silveira, C. (2021). Variation and selection in the putative sperm-binding region of ZP3 in muroid rodents: A comparison between cricetids and murines. *Genes (Basel), 12*, 1450.

Egge, N., Muthusubramanian, A., & Cornwall, G. A. (2015). Amyloid properties of the mouse egg zona pellucida. *PLoS One, 10*, e0129907.

Eisenbach, M., & Giojalas, L. V. (2006). Sperm guidance in mammals – An unpaved road to the egg. *Nature Reviews. Molecular Cell Biology, 7*, 276–285.

El-Hayek, S., & Clarke, H. J. (2016). Control of oocyte growth and development by intercellular communication within the follicular niche. *Results and Problems in Cell Differentiation, 58*, 191–224.

El-Hayek, S., Yang, Q., Abbassi, L., FitzHarris, G., & Clarke, H. J. (2018). Mammalian oocytes locally remodel follicular architecture to provide the foundation for germline-soma communication. *Current Biology, 28*, 258–270.

El-Mestrah, M., Castle, P. E., Borossa, G., & Kan, F. W. K. (2002). Subcellular distribution of ZP1, ZP2, and ZP3 glycoproteins during folliculogenesis and demonstration of their topological disposition within the zona matrix of mouse ovarian oocytes. *Biology of Reproduction, 66*, 866–876.

Elofsson, A., Han, L., Bianchi, E., Wright, G. J., & Jovine, L. (2024). Deep learning insights into the architecture of the mammalian egg-sperm fusion synapse. *eLife, 13*, RP93131.

Epifano, O., Liang, L.-F., Familiari, M., Moos, M. C., & Dean, J. (1995a). Coordinate expression of the three zona pellucida genes during mouse oogenesis. *Development (Cambridge, England), 121*, 1947–1956.

Epifano, O., Liang, L.-F., & Dean, J. (1995b). Mouse ZP1 encodes a zona pellucida protein homologous to egg envelope proteins in mammals and fish. *Journal of Biological Chemistry, 270*, 27254–27258.

Eppig, J. J., Wigglesworth, K., & Pendola, F. L. (2002). The mammalian oocyte orchestrates the rate of ovarian follicular development. *Proceedings of the National Academy of Sciences, USA, 99*, 2890–2894.

Evans, J. P. (2020). Preventing polyspermy in mammalian eggs – Contributions of the membrane block and other mechanisms. *Molecular Reproduction and Development, 87*, 341–349.

Evans, J. P., & Florman, H. M. (2002). The state of the union: The cell biology of fertilization. *Nature Cell Biology, 4*, s57–s63.

Fahrenkamp, E., Algarra, B., & Jovine, L. (2020). Mammalian egg coat modifications and the block to polyspermy. *Molecular Reproduction and Development, 87*, 326–340.

Familiari, G., Relucenti, M., Heyn, R., Micara, G., & Correr, S. (2006). Three-dimensional structure of the zona pellucida at ovulation. *Microscopy Research and Technique, 69*, 415–426.

Feng, J.-M., Tian, H.-F., Hu, Q.-M., Meng, Y., & Xiao, H.-B. (2018). Evolution and multiple origins of zona pellucida genes in vertebrates. *Biology Open, 7*, 36137.

Findlay, J. K., Hutt, K. J., Hickey, M., & Anderson, R. A. (2015). How is the number of primordial follicles in the ovarian reserve established? *Biology of Reproduction, 93*, 1–7.

Florman, H. M., & Storey, B. T. (1982). Mouse gamete interactions: The zone pellucida is the site of the acrosome reaction leading to fertilization in vitro. *Developmental Biology, 91*, 121–130.

Florman, H. M., Bechtol, K. B., & Wassarman, P. M. (1984). Enzymatic dissection of the functions of the mouse egg's receptor for sperm. *Developmental Biology, 106*, 243–255.

Florman, H. M., & Wassarman, P. M. (1985). O-Linked oligosaccharides of mouse egg ZP3 account for its sperm receptor activity. *Cell, 41*, 313–324.

Florman, H. M., & Ducibella, T. (2006). *Fertilization in mammals. The physiology of reproduction, 1*, San Diego, CA: Academic Press/Elsevier149–196.

Florman, H. M., Jungnickel, M. K., & Sutton, K. A. (2008). Regulating the acrosome reaction. *International Journal of Developmental Biology, 52*, 503–510.

Florman, H. M., & Fissore, R. A. (2015). *Fertilization in mammals. Knobil and Neill's physiology of reproduction, 1*, San Diego, CA: Academic Press/Elsevier 55–112.

Frankenberg, S., & Renfree, M. B. (2018). Conceptus coats of marsupials and monotremes. *Current Topics in Developmental Biology, 130*, 357–378.

Gahlay, G., Gauthier, L., Baibakov, B., Epifano, O., & Dean, J. (2010). Gamete recognition in mice depends on the cleavage status of an egg's zona pellucida protein. *Science (New York, N. Y.), 329*, 216–219.

Gittens, J. E., & Kidder, G. M. (2005). Differential contributions of connexin 37 and connexin 43 to oogenesis revealed in chimeric reaggregated mouse ovaries. *Journal of Cell Science, 118*, 5071–5078.

Goodenough, D. A., & Paul, D. L. (2009). Gap junctions. *Cold Spring Harbor Perspectives in Biology, 1*, a002576.

Goudet, G., Mugnier, S., Callebaut, I., & Monget, P. (2008). Phylogenetic analysis and identification of pseudogenes reveal a progressive loss of zona pellucida genes during evolution of vertebrates. *Biology of Reproduction, 78*, 796–806.

Greve, J. M., & Wassarman, P. M. (1985). Mouse egg extracellular coat is a matrix of interconnected fibrils possessing a structural repeat. *Journal of Molecular Biology, 181*, 253–264.

Greve, J. M., Salzmann, G. S., Roller, R. J., & Wassarman, P. M. (1982). Biosynthesis of the major zona pellucida glycoprotein secreted by oocytes during mammalian oogenesis. *Cell, 31*, 749–759.

Gupta, S. K. (2018). The human egg's zona pellucida. *Current Topics in Developmental Biology, 130*, 379–412.

Gupta, S. K. (2023). Zona pellucida glycoproteins: Relevance in fertility and development of contraceptive vaccines. *American Journal of Reproductive Immunology, 89*, e13535.

Gwatkin, R. B. L. (1977). *Fertilization mechanisms in man and mammals*. New York, NY: Plenum.

Haddad, A., & Nagai, M. E. (1977). Radioautographic study of glycoprotein biosynthesis and renewal in the ovarian follicles of mice and the origin of the zona pellucida. *Cell and Tissue Research, 177*, 347–369.

Han, L., Monné, M., Okumura, H., Schwend, T., Cherry, A. L., Flot, D., & Jovine, L. (2010). Insights into egg-coat assembly and egg-sperm interaction from the X-ray structure of full-length ZP3. *Cell, 143*, 1–12.

Hanna, W. F., Kerr, C. L., Shaper, J. H., & Wright, W. W. (2004). Lewis X-containing neoglycoproteins mimic the intrinsic ability of zona pellucida glycoprotein ZP3 to induce the acrosome reaction in capacitated mouse sperm. *Biology of Reproduction, 71*, 778–789.

Hastings, R. A., Enders, A. C., & Schlafke, S. (1972). Permeability of the zona pellucida to protein tracers. *Biology of Reproduction, 7*, 288–296.

Hewetson, A., Do, H. Q., Myers, C., Muthusubramanian, A., Sutton, R. B., Wylie, B. J., & Cornwall, G. A. (2017). Functional amyloids in reproduction. *Biomolecules, 7*, 46.

Hoodbhoy, T., Aviles, Mk, Baibakov, B., Epifano, O., Jimenez-Movilla, M., Gauthier, L., & Dean, J. (2006). ZP2 and ZP3 traffic independently within oocytes prior to assembly into the extracellular zona pellucida. *Molecular and Cellular Biology, 26*, 7991–7998.

Huang, H. L., Lv, C., Zhao, Y. C., Li, W., He, X. M., Li, P., & Xiao, H. M. (2014). Mutant ZP1 in familial infertility. *New England Journal of Medicine, 370*, 1220–1226.

Hughes, D. C., & Barratt, C. L. R. (1999). Identification of the true orthologue of the mouse ZP1 gene: Evidence for greater complexity in the mammalian zona pellucida. *Biochimica et Biophysica Acta, 1447*, 303–306.

Hughes, D. C. (2007). ZP genes in avian species illustrate the dynamic evolution of the vertebrate egg envelope. *Cytogenetic and Genome Research, 117*, 86–91.

Hutchinson, E. G., & Thornton, J. M. (1993). The Greek key motif: Extraction, classification and analysis. *Protein Engineering, 6*, 233–245.

Inoue, N., Ikawa, M., Isotani, A., & Okabe, M. (2005). The immunoglobulin superfamily protein Izumo is required for sperm to fuse with eggs. *Nature, 434*, 234–238.

Izquierdo-Rico, M. J., Moros-Nicolás, C., Pérez-Crespo, M., Laguna-Barraza, R., Gutiérrez-Adán, A., Veryrunes, F., & Avilés, M. (2021). ZP4 is present in murine zona pellucida and is not responsible for the specific gamete interaction. *Frontiers in Cell and Developmental Biology, 8*, 626679.

Järva, M. A., Lingford, J. P., John, A., Soler, N. M., Scott, N. E., & Goddard-Borger, E. D. (2020). Trefoil factors share a lectin activity that defines their role in mucus. *Nature Communications, 11*, 2265.

Jia, W., Xi, Q., Zhu, L., Luo, Z., Li, M., Hou, M., Zhang, D., Yang, X., Hu, J., Jin, L., & Zhang, X. (2022). Novel mutations in ZP2 and ZP3 cause female infertility in three patients. *Journal of Assisted Reproduction and Genetics, 39*, 1205–1215.

Jimenez-Movilla, M., & Dean, J. (2011). ZP2 and ZP3 cytoplasmic tails prevent premature interactions and ensure incorporation into the zona pellucida. *Journal of Cell Science, 124*, 940–950.

John, B., & Lewis, K. R. (1973). The meiotic mechanism. In J. J. Head (Ed.). *Oxford biology readers*. London, UK: Oxford University Press.

Johnston, D. S., Wright, W. W., Shaper, J. H., Hokke, C. H., Van den Eijnden, D. H., & Joziasse, D. H. (1998). Murine sperm-zona binding, a fucosyl residue is required for a high affinity sperm-binding ligand. *Journal of Biological Chemistry, 273*, 1888–1895.

Jovine, L., Qi, H., Williams, Z., Litscher, E., & Wassarman, P. M. (2002). The ZP domain is a conserved module for protein polymerization. *Nature Cell Biology, 4*, 457–461.

Jovine, L., Qi, H., Williams, Z., Litscher, E. S., & Wassarman, P. M. (2004). A duplicated motif controls assembly of zona pellucida domain proteins. *Proceedings of the National Academy of Sciences, USA, 101*, 5922–5927.

Jovine, L., Darie, C. C., Litscher, E. S., & Wassarman, P. M. (2005). Zona pellucida domain proteins. *Annual Review of Biochemistry, 74*, 83–114.

Jovine, L., Janssen, W. G., Litscher, E. S., & Wassarman, P. M. (2006). The PLAC1-homology region of the ZP domain is sufficient for protein polymerization. *BMC Biochemistry, 7*, 11.

Jovine, L., Qi, H., Williams, Z., Litscher, E. S., & Wassarman, P. M. (2007). Features that affect secretion and assembly of zona pellucida glycoproteins during mammalian oogenesis. In S. K. Gupta, K. Koyama, & J. F. Murray (Eds.). *Gamete biology* (pp. 187–201)UK: Nottingham University Press.

Jungnickel, M. K., Sutton, K. A., & Florman, H. M. (2003). In the beginning: Lessons from fertilization in mice and worms. *Cell, 114*, 401–404.

Jungnickel, M. K., Sutton, K. A., Wang, Y., & Florman, H. M. (2007). Phosphoinositide-dependent pathways in mouse sperm are regulated by egg ZP3 and drive the acrosome reaction. *Developmental Biology, 304*, 116–126.

Keefe, D., Tran, P., Pellegrini, C., & Oldenberg, R. (1997). Polarized light microscopy and digital image processing identify a multilaminar structure of the hamster zona pellucida. *Human Reproduction, 12*, 1250–1252.

Kerr, C. L., Hanna, W. F., Shaper, J. H., & Wright, W. W. (2004). Lewis X-containing glycans are specific and potent competitive inhibitors of the binding of ZP3 to complementary sites on capacitated, acrosome-intact mouse sperm. *Biology of Reproduction, 71*, 770–777.

Kidder, J. M., & Mhawi, A. A. (2002). Gap junctions and ovarian folliculogenesis. *Reproduction (Cambridge, England), 123*, 613–620.

Killingbeck, E. E., & Swanson, W. J. (2018). Egg coat proteins across metazoan evolution. *Current Topics in Developmental Biology, 130*, 443–488.

Kim, J., & Kim, J. (2013). Viscoelastic characterization of mouse zona pellucida. *IEEE Transactions of Biomedical Engineering, 60*, 569–575.

Kinloch, R. A., Mortillo, S., Stewart, C. L., & Wassarman, P. M. (1991). Embryonal carcinoma cells transfected with ZP3 genes differentially glycosylate similar polypeptides and secrete active mouse sperm receptor. *Journal of Cell Biology, 115*, 655–664.

Kinloch, R. A., Sakai, Y., & Wassarman, P. M. (1995). Mapping the mouse ZP3 combining site for sperm by exon-swapping and site-directed mutagenesis. *Proceedings of the National Academy of Sciences, USA, 92*, 263–267.

Kirkpatrick, J. F., Rowan, A., Lamberski, N., Wallace, R., Frank, K., & Lyda, R. (2009). The practical side of immunocontraception: Zona proteins and wildlife. *Journal of Reproductive Immunology, 83*, 151–157.

Kong, N., Xu, Q., Shen, X., Zhu, X., & Cao, G. (2023). Case report: A novel homozygous varient in ZP3 is associated with human empty follicle syndrome. *Frontiers in Genetics, 14*, 1256549.

Krarup, T., Pedersen, T., & Faber, M. (1969). Regulation of oocyte growth in the mouse ovary. *Nature, 224*, 187–188.

Lambert, H. (1984). Role of sperm-surface glycoproteins in gamete recognition in two mouse species. *Journal of Reproduction and Fertilty, 70*, 281–284.

Lefievre, L., Conner, S. J., Salpekar, A., Olufowobi, O., Ashton, P., Pavlovic, B., & Barratt, C. L. R. (2004). Four zona pellucida glycoproteins are expressed in the human. *Human Reproduction, 19*, 1580–1586.

Legan, P. K., Rau, A., Keen, J. N., & Richardson, G. P. (1997). The mouse tectorins: modular matrix proteins of the inner ear homologous to components of the sperm-egg adhesion system. *Journal of Biological Chemistry, 272*, 8791–8801.

Legan, P. K., Lukashkina, V. A., Goodyear, R. J., Lukashkin, A. N., Verhoeven, K., Van Camp, G., & Richardson, G. P. (2005). A deafness mutation isolates a second role for the tectorial membrane in hearing. *Nature Neuroscience, 8*, 1035–1042.

Legge, M. (1995). Oocyte and zygote zona pellucida permeability to macromolecules. *Journal of Experimental Zoology, 271*, 145–150.

Lewandoski, M., Wassarman, K. M., & Martin, G. R. (1997). Zp3-cre, a transgenic mouse line for the activation or inactivation of loxP-flanked target genes specifically in the female germ line. *Current Biology, 7*, 148–151.

Li, R., & Albertini, D. F. (2013). The road to maturation: Somatic cell interaction and self-organization of the mammalian oocyte. *Nature Reviews of Molecular Cell Biology, 14*, 141–152.

Liang, L.-F., & Dean, J. (1993). Conservation of mammalian secondary sperm receptor genes enables the promoter of the human gene to function in mouse oocytes. *Developmental Biology, 156*, 399–408.

Liang, L.-F., Soyal, S. M., & Dean, J. (1997). FIGalpha, a germ cell specific transcription factor involved in the coordinate expression of the zona pellucida genes. *Development (Cambridge, England), 124*, 4939–4947.

Lira, S. A., Kinloch, R. A., Mortillo, S., & Wassarman, P. M. (1990). An upstream region of the mouse ZP3 gene directs expression of firefly luciferase specifically to growing oocytes in transgenic mice. *Proceedings of the National Academy of Sciences, USA, 87*, 7215–7219.

Lira, S. A., Schickler, M., & Wassarman, P. M. (1993). Cis-acting DNA elements involved in oocyte-specific expression of mouse sperm receptor gene mZP3 are located close to the gene's transcription start site. *Molecular Reproduction and Development, 36*, 494–499.

Litscher, E. S., Juntunen, K., Seppo, A., Penttila, L., Niemela, R., Renkonen, O., & Wassarman, P. M. (1995). Oligosaccharide constructs with defined structures that inhibit binding of mouse sperm to unfertilized eggs in vitro. *Biochemistry, 34*, 4662–4669.

Litscher, E. S., & Wassarman, P. M. (1996). Characterization of a mouse ZP3-derived glycopeptide, gp55, that exhibits sperm receptor and acrosome reaction-inducing activity in vitro. *Biochemistry, 35*, 3980–3986.

Litscher, E. S., Qi, H., & Wassarman, P. M. (1999). Mouse zona pellucida glycoproteins mZP2 and mZP3 undergo carboxy-terminal proteolytic processing in growing oocytes. *Biochemistry, 38*, 12280–12287.

Litscher, E. S., & Wassarman, P. M. (2007). Egg extracellular coat proteins: From fish to mammals. *Histology and Histopathology, 22*, 337–347.

Litscher, E. S., & Wassarman, P. M. (2014). Evolution, structure, and synthesis of vertebrate egg-coat proteins. *Trends in Developmental Biology, 8*, 65–76.

Litscher, E. S., & Wassarman, P. M. (2015). *A guide to zona pellucida domain proteins*. Hoboken, NJ: John Wiley & Sons.

Litscher, E. S., & Wassarman, P. M. (2020a). Zona pellucida proteins, fibrils, and matrix. *Annual Review of Biochemistry, 89*, 695–715.

Litscher, E. S., & Wassarman, P. M. (2020b). Zona pellucida genes and proteins and human fertility. *Trends in Developmental Biology, 13*, 21–33.

Liu, M., & Grigoriev, A. (2004). Protein domains correlate strongly with exons in mutiple eukaryotic genomes – Evidence of exon shuffling? *Trends in Genetics, 20*, 399–403.

Liu, W., Li, K., Bai, D., Yin, J., Tang, Y., Chi, F., & Gao, S. (2017). Dosage effects of ZP2 and ZP3 heterogeneous mutations cause human infertility. *Human Genetics, 136*, 1822–1827.

Liu, C., Litscher, E. S., Mortillo, S., Sakai, Y., Kinloch, R. A., Stewart, C. L., & Wassarman, P. M. (1996). Targeted disruption of the mZP3 gene results in production of eggs lacking a zona pellucida and infertility in female mice. *Proceedings of the National Academy of Sciences, USA, 93*, 5431–5436.

Liu, S.-L., Zuo, H.-Y., Zhao, B.-W., Guo, J.-N., Liu, W.-B., Lei, W.-L., & Wang, Z.-B. (2023). A heterozygous ZP2 mutation causes zona pellucida defects and female infertility in mouse and human. *iScience, 26*, 107828.

Louros, N. N., Iconomidou, V. A., Giannelou, P., & Hamodrakas, S. J. (2013). Structural analysis of peptide-analogues of human zona pellucida ZP1 protein with amyloidogenic properties: Insights into mammalian zona pellucida formation. *PLoS One, 8*, e73258.

Lunsford, R. D., Jenkins, N. A., Kozak, C. A., Liang, L. F., Silan, C. M., Copeland, N. G., & Dean, J. (1990). Genomic mapping of murine Zp-2 and Zp-3, two oocyte-specific loci encoding zona pellucida proteins. *Genomics, 6*, 184–187.

Männikkö, M., Törmälä, R. M., Tuuri, T., Haltia, A., Martikainen, H., Ala-Kokko, L., & Lakkakorpi, J. T. (2005). Association between sequence variations in genes encoding human zona pellucida glycoproteins and fertilization failure in IVF. *Human Reproduction, 20*, 1578–1585.

Margalit, M., Paz, G., Yavetz, H., Yogev, L., Amit, A., Hevlin-Schwartz, T., & Kleiman, S. E. (2012). Genetic and physiological study of morphologically abnormal human zona pellucida. *European Journal of Obstetrics, Gynecology, and Reproductive Biology, 165*, 70–76.

Marchais, M., Gilbert, I., Bastein, A., Macaulay, A., & Robert, C. (2022). Mammalian cumulus-oocyte complex communication: A dialog through long and short distance messaging. *Journal of Assisted Reproduction and Genetics, 39*, 1011–1025.

Marsh, J. A., & Teichmann, S. A. (2010). How do proteins gain new domains? *Genome Biology, 11*, 126.

Matassa, R., Gatti, M., Crociati, M., Brunelli, R., Battaglione, E., Papi, M., & Familiari, G. (2023). Self-assembly of glycoprotein nanostructured fibrils for modulating extracellular networks at long range. *Nanoscale*. https://doi.org/10.1039/d3nr02644b.

Matzuk, M. M., Burns, K. H., Viveiros, M. M., & Eppig, J. J. (2002). Intercellular communication in the mammalian ovary: Oocytes carry the conversation. *Science (New York, N. Y.), 296*, 2178–2180.

Millar, S. E., Lader, E. S., Liang, L.-F., & Dean, J. (1991). Oocyte-specific factors bind a conserved upstream sequence required for mouse zona pellucida promoter activity. *Molecular and Cellular Biology, 11*, 6197–6204.

Millar, S. E., Lader, E. S., & Dean, J. (1993). ZAP-1 DNA binding activity is first detected at the onset of zona pellucida gene expression in embryonic mnouse oocytes. *Developmental Biology, 158*, 410–413.

Miyado, K., Yamada, G., Yamada, S., Hasuwa, H., Nakamura, Y., Ryu, F., & Mekada, E. (2000). Requirement of CD9 on egg plasma membrane for fertilization. *Science (New York, N. Y.), 287*, 321–324.

Moller, C. C., & Wassarman, P. M. (1989). Characterization of a proteinase that cleaves zona pellucida glycoprotein ZP2 following activation of mouse eggs. *Developmental Biology, 132*, 103–112.

Monné, M., Han, L., Schwend, T., Burendahl, S., & Jovine, L. (2008). Crystal structure of the ZP-N domain of ZP3 reveals the core fold of animal egg coats. *Nature, 456*, 653–657.

Monné, M., & Jovine, L. (2011). A structural view of egg coat architecture and function in fertilization. *Biology of Reproduction, 85*, 661–669.

Morgan, A. A., & Rubenstein, E. (2013). Proline: The distribution, frequency, positioning, and common functional roles of proline and polyproline sequences in the human proteome. *PLoS One, 8*, e53785.

Moros-Nicolás, C., Chevret, P., Izquierdo-Rico, M. J., Holt, W. V., Esteban-Diaz, D., Lopez-Béjar, M., & Avilés, M. (2018). Composition of marsupial zona pellucida: A molecular and phylogenetic approach. *Reproduction, Fertility, and Development, 30*, 721–733.

Mortillo, S., & Wassarman, P. M. (1991). Differential binding of gold-labeled zona pellucida glycoproteins mZP2 and mZP3 to mouse sperm membrane compartments. *Development (Cambridge, England), 113*, 141–149.

Nishio, S., Okumura, H., & Matsuda, T. (2018). Egg-coat and zona pellucida proteins of chicken as a typical species of Aves. *Current Topics in Developmental Biology, 130*, 307–330.

Nishio, S., Emori, C., Wiseman, B., Fahrenkamp, D., Dioguardi, E., Zamora-Caballero, S., & Jovine, L. (2024). ZP2 cleavage blocks polyspermy by modulating the architecture of the egg coat. *Cell, 187*, 1440–1459.

Noda, T., Lu, Y., Fujihara, Y., Oura, S., Koyano, T., Kobayashi, S., & Ikawa, M. (2020). Sperm proteins SOF1, TMEM95, and SPACA6 are required for sperm-oocyte fusion in mice. *Proceedings of the National Academy of Sciences, USA, 117*, 11493–11502.

Notova, S., Bonnardel, F., Lisacek, F., Varrot, A., & Imberty, A. (2020). Structure and engineering of tandem repeat lectins. *Current Opinion in Structural Biology, 62*, 39–47.

Okabe, M. (2018). Sperm-egg interaction and fertilization: Past, present, and future. *Biology of Reproduction, 99*, 134–146.

Pang, P. C., Chiu, P. C., Lee, C. L., Chang, L. Y., Panico, M., Morris, H. R., & Dell, A. (2011). Human sperm binding is mediated by the sialyl-Lewis(x) oligosaccharide on the zona pellucida. *Science (New York, N. Y.), 333*, 1761–1764.

Patthy, L. (2021). Exon shuffling played a decisive role in the evolution of the genetic toolkit for the multicellular body plan of Metazoa. *Genes (Basel), 12*, 382.

Pelletier, C., Keefe, D. L., & Trimarchi, J. R. (2004). Noninvasive polarized light microscopy quantitatively distinguishes the multilaminar structure of the zona pellucida of living human eggs and embryos. *Fertility and Sterility, 81*, 850–856.

Peters, H., Byskov, S., Lintern-Moore, S., & Faber, M. (1973). Proceedings: Follicle growth initiation in the immature mouse ovary: Extraovarian or intraovarian control? *Journal of Reproduction and Fertility, 35*, 619–620.

Phillips, D. M., & Shalgi, R. M. (1980). Surface architecture of the mouse and hamster zona pellucida and oocyte. *Journal of Ultrastructural Research, 72*, 1–12.

Philpott, C. C., Ringuette, M. R., & Dean, J. (1987). Oocyte-specific expression and developmental regulation of ZP3, the sperm receptor of the mouse zona pellucida. *Developmental Biology, 121*, 568–575.

Plaza, S., Chanut-Delalande, H., Fernandes, I., Wassarman, P. M., & Payre, F. (2010). From A to Z: Apical structures and zona pellucida-domain proteins. *Trends in Cell Biology, 20* 534-532.

Pökkylä, R. M., Lakkakorpi, J. T., Noujua-Huttunen, S. H., & Tapanainen, J. S. (2011). Sequence variations in human ZP genes as potential modifiers of zona pellucida architecture. *Fertility and Sterility, 95*, 2669–2672.

Polshakov, V. I., Williams, M. A., Gargaro, A. R., Frenkiel, T. A., Westley, B. R., Chadwick, M. P., & Feeney, J. (1997). High-resolution solution structure of human pNR-2/pS2; a single trefoil motif protein. *Journal of Molecular Biology, 267*, 418–432.

Qi, H., Williams, Z., & Wassarman, P. M. (2002). Secretion and assembly of zona pellucida glycoproteins by growing mouse oocytes microinjected with epitope-tagged cDNAs for mZP2 and mZP3. *Molecular Biology of the Cell, 13*, 530–541.

Que, E. L., Duncan, F. E., Bayer, A. R., Philips, S. J., Roth, E. W., Bleher, R., & O'Halloran, T. V. (2017). Zinc sparks induce physiochemical changes in the egg zona pellucida that prevent polyspermy. *Integrative Biology (Cambridge), 9*, 135–144.

Quesada, V., Sánchez, L. M., Alvarez, J., & López-Otin, C. (2004). Identification and characterization of human and mouse ovastacin: A novel metalloproteinase similar to hatching enzymes from arthropods, birds, amphibians, and fish. *Journal of Biological Chemistry, 279*, 26627–26634.

Raj, I., Sadat Al Hosseini, H., Dioguardi, E., Nishimura, K., Han, L., Villa, A., & Jovine, L. (2017). Structural basis of egg coat-sperm recognition at fertilization. *Cell, 169*, 1315–1326.

Rankin, T., Familiari, M., Lee, E., Ginsburg, A., Dwyer, N., Blanchette-Mackie, J., & Dean, J. (1996). Mice homozygous for an insertional mutation in the ZP3 gene lack a zona pellucida and are infertile. *Development (Cambridge, England), 122*, 2903–2910.

Rankin, T., Talbot, P., Lee, E., & Dean, J. (1999). Abnormal zonae pellucidae in mice lacking ZP1 result in early embryonic loss. *Development (Cambridge, England), 126*, 3847–3855.

Rankin, T., O'Brien, M., Lee, E., Wigglesworth, K., Eppig, J., & Dean, J. (2001). Defective zonae pellucidae in Zp2-null mice disrupt folliculogenesis, fertility, & development. *Development (Cambridge, England), 128*, 1119–1126.

Rath, A., Davidson, A. R., & Deber, C. M. (2005). The structure of "unstructured" regions in peptides and proteins: Role of the polyproline II helix in protein folding and recognition. *Biopolymers, 80*, 179–185.

Ringuette, M. J., Sobieski, D. A., Chamow, S. M., & Dean, J. (1986). Oocyte-specific gene expression: Molecular characterization of a cDNA coding for ZP3, the sperm receptor of the mouse zona pellucida. *Proceedings of the National Academy of Sciences, USA, 83*, 4341–4345.

Rivera, A. M., Wilburn, D. B., & Swanson, W. J. (2022). Domain expansion and functional diversification in vertebrate reproductive proteins. *Molecular and Biological Evolution, 39, msac1*, 05.

Robert, C. (2021). Nurturing the egg: The essential connection between cumulus cells and the oocyte. *Reproduction, Fertility, and Development, 34*, 149–159.

Roller, R. J., Kinloch, R. A., Hiraoka, B. Y., Li, S. S.-L., & Wassarman, P. M. (1989). Gene expression during mammalian oogenesis and early embryogenesis: Quantification of three messenger-RNAs abundant in fully grown mouse oocytes. *Development (Cambridge, England), 106*, 251–261.

Rosiere, T. K., & Wassarman, P. M. (1992). Identification of a region of mouse zona pellucida glycoprotein mZP3 that possesses sperm receptor activity. *Developmental Biology, 154*, 309–317.

Saleh, A., Kashir, J., Thanassoulas, A., Sefieh-Garabedian, B., Lia, F. A., & Nomikos, M. (2020). Essential role of sperm factor PLC-zeta in egg activation and male factor infertility: An update. *Frontiers in Cell and Developmental Biology, 8* fcell.2020.00028.

Salzmann, G. S., Greve, J. M., Roller, R. J., & Wassarman, P. M. (1983). Biosynthesis of the sperm receptor during oogenesis in the mouse. *EMBO Journal, 2*, 1451–1456.

Schickler, M., Lira, S. A., Kinloch, R. A., & Wassarman, P. M. (1992). A mouse oocyte-specific protein that binds to a region of mZP3 promoter involved in regulating oocyte-specific expression of the mZP3 gene. *Molecular and Cellular Biology, 12*, 120–127.

Sexton, O. J., Bramble, J. E., Heisler, I. L., Phillips, C. A., & Cox, D. L. (2005). Eggshell composition of squamate reptiles: Relationship between eggshell permeabilty and amino acid distribution. *Journal of Chemical Ecology, 31*, 2391–2401.

Shalgi, R., Bleil, J. D., & Wassarman, P. M. (1990). Carbohydrate-mediated sperm-egg interactions in mammals. In S. Mashiach, (Ed.). *Advances in assisted reproductive technologies* (pp. 437–441). New York: Plenum Press (et al.).

Shimizu, S., Tsuji, M., & Dean, J. (1983). In vitro biosynthesis of three sulfated glycoproteins of murine zonae pellucidae by oocytes grown in follicle culture. *Journal of Biological Chemistry, 259*, 5858–5863.

Simon, A. M., Goodenough, D. A., Li, E., & Paul, D. I. (1997). Female infertility in mice lacking connexin-37. *Nature, 385*, 525–529.

Siu, K. K., Sarrao, V. H. B., Ziyyat, A., & Lee, J. E. (2021). The cell biology of fertilization: Gamete attachment and fusion. *Journal of Cell Biology, 220*, e202102146.

Smith, J., Paton, I. R., Hughes, D. C., & Burt, D. W. (2005). Isolation and mapping the chicken zona pellucida genes: An insight into the evolution of orthologous genes in different species. *Molecular and Reproductive Development, 70*, 133–145.

Sorensen, R. A., & Wassarman, P. M. (1976). Relationship between growth and meiotic maturation of the mouse oocyte. *Developmental Biology, 50*, 531–536.

Soyal, S. M., Amleh, A., & Dean, J. (2000). FIGalpha, a germ cell-specific transcription factor required for ovarian follicle formation. *Development (Cambridge, England), 127*, 4645–4654.

Spargo, S. C., & Hope, R. M. (2003). Evolution and nomenclature of the zona pellucida gene family. *Biology of Reproduction, 68*, 358–362.

Stsiapanava, A., Xu, C., Brunati, M., Zamora-Caballero, S., Schaeffer, C., Bokhove, M., & Jovine, L. (2020). Cryo-EM structure of native human uromodulin, a zona pellucida module polymer. *EMBO Journal, 39*, e106807.

Sun, L., Tong, K., Liu, W., Tian, Y., Liu, D., Huang, G., & Li, J. (2023). Novel variants in ZP1, ZP2, and ZP3 associated with empty follicle syndrome and abnormal zona pellucida. *Reproductive Biomedicine Online, 46*, 847–855.

Swann, K., Saunders, C. M., Rogers, N. T., & Lai, F. A. (2006). PLCzeta: A sperm protein that triggers Ca^{2+} oscillations and egg activation in mammals. *Seminars in Cell and Developmental Biology, 17*, 264–273.

Swanson, W. J., Yang, Z., Wolfner, M. F., & Aquadro, C. F. (2001). Positive Darwinian selection drives the evolution of several female reproductive proteins in mammals. *Proceedings of the National Academy of Sciences, USA, 98*, 2509–2514.

Swanson, W. J., Nielsen, R., & Yang, Q. (2003). Pervasive adaptive evolution in mammalian fertilization proteins. *Molecular Biology and Evolution, 20*, 18–20.

Telfer, E., Grosbois, J., Odey, Y. L., Rosario, R., & Anderson, R. A. (2023). Making a good egg: Human oocyte health, aging, and in vitro development. *Physiological Reviews, 103*, 2623–2677.

Thim, L., & May, F. E. (2005). Structure of mammalian trefoil factors and functional insights. *Cellular Molecular Life Sciences, 62*, 2956–2973.

Tokuhiro, K., & Dean, J. (2018). Glycan-independent gamete recognition triggers egg zinc sparks and ZP2 cleavage to prevent polyspermy. *Developmental Cell, 46*, 627–640.

Tong, Z.-B., Nelson, L. M., & Dean, J. (1995). Inhibition of zona pellucida gene expression by antisense oligonucleotides injected into mouse oocytes. *Journal of Biological Chemistry, 270*, 849–853.

Toshimori, K., & Eddy, E. M. (2015). *The spermatozoon. Knobil and Neill's physiology of reproduction, 1*, San Diego, CA: Academic Press/Elsevier 99–148.

Upadhyay, S., & Zamboni, L. (1982). Ectopic germ cells: Natural model for the study of germ cell sexual differentiation. *Proceedings of the National Academy of Sciences, USA, 79*, 6584–6588.

Vazquez, M. H., Phillips, D. M., & Wassarman, P. M. (1989). Interaction of mouse sperm with purified sperm receptors covalently linked to silica beads. *Journal of Cell Science, 92*, 713–722.

Visconti, P. E., & Florman, H. M. (2010). Mechanisms of sperm-egg interactions: Between sugars and broken bonds. *Science Signaling, 3* pe35.

Vogt, E. J., Tokuhiro, K., Guo, M., Dale, R., Yang, G., Shin, S. W., & Dean, J. (2019). Anchoring cortical granules in the cortex ensures trafficking to the plasma membrane for post-fertilization exocytosis. *Nature Communications, 10*, 2271.

Wassarman, P. M. (1987). The biology and chemistry of fertilization. *Science (New York, N. Y.), 235*, 553–560.

Wassarman, P. M. (1988a). Zona pellucida glycoproteins. *Annual Review of Biochemistry, 57*, 415–442.

Wassarman, P. M. (1999). Mammalian fertilization: Molecular aspects of gamete adhesion. *Cell, 96*, 175–183.

Wassarman, P. M. (2002). Channels of communication in the ovary. *Nature Cell Biology (Supplement)*, s7–9.

Wassarman, P. M. (2008). Zona pellucida glycoproteins. *Journal of Biological Chemistry, 283*, 24285–24289.

Wassarman, P. M., & Josefowicz, W. J. (1978). Oocyte development in the mouse: An ultrastructural comparison of oocytes isolated at various stages of growth and meiotic competence. *Journal of Morphology, 156*, 209–235.

Wassarman, P. M., Bleil, J. D., Florman, H. M., Greve, J. M., Roller, R. J., Salzmann, G. S., & Samuels, F. G. (1985). The mouse egg's sperm receptor: What is it and how does it work? *Cold Spring Harbor Symposium on Quantitative Biology, 50*, 11–18.

Wassarman, P. M., & Mortillo, S. (1991). Structure of the mouse egg extracellular coat, the zona pellucida. *International Review of Cytology, 130*, 85–110.

Wassarman, P. M., & Albertini, D. F. (1994). *The mammalian ovum. The physiology of reproduction, 1*, New York, NY: Raven Press79–122.

Wassarman, P. M., & Litscher, E. S. (1995). Sperm-egg recognition mechanisms in mammals. *Current Topics in Developmental Biology, 30*, 1–19.

Wassarman, P. M., Liu, C., & Litscher, E. S. (1996). Constructing the mouse egg zona pellucida: Some new pieces of an old puzzle. *Journal Cell Science, 109*, 2001–2004.

Wassarman, P. M., Qi, H., & Litscher, E. S. (1997). Mutant female mice carrying a single mZP3 allele produce eggs with a thin zona pellucida, but reproduce normally. *Proceedings Royal Society, Biological Sciences, 264*, 323–328.

Wassarman, P. M., & Litscher, E. S. (1998). Ovarian development in mice bearing homozygous or heterozygous null mutations in zona pellucida gene mZP3. *Histology and Histopathology, 13*, 293–300.

Wassarman, P. M., & Litscher, E. S. (2008). Mammalian fertilization is dependent on multiple membrane fusion events. In E. H. Chen (Ed.). *Cell fusion* (pp. 99–114). Totowa, NJ: Humana Press.

Wassarman, P. M., & Litscher, E. S. (2012). Influence of the zona pellucida of the mouse egg on folliculogenesis and fertility. *International Journal of Developmental Biology, 56*, 833–839.

Wassarman, P. M., & Litscher, E. S. (2018). The mouse egg's zona pellucida. *Current Topics in Developmental Biology, 130*, 331–356.

Wassarman, P. M., & Litscher, E. S. (2021). Zona pellucida genes and proteins: Essential players in mammalian oogenesis and fertility. *Genes (Basel), 12*, 1266.

Wassarman, P. M., & Litscher, E. S. (2022a). Female fertility and the zona pellucida. *eLife, 11*, e76106.
Wassarman, P. M., & Litscher, E. S. (2022b). Mouse zona pellucida proteins as receptors for binding of sperm to eggs. *Trends in Developmental Biology, 15*, 1–13.
Wilburn, D. B., & Swanson, W. J. (2017). The "ZP domain" is not one, but likely two independent domains. *Molecular Reproduction and Development, 84*, 284–285.
Williams, A. F., & Barclay, A. N. (1988). The immunoglobulin superfamily – Domains for cell surface recognition. *Annual Review of Immunology, 6*, 381–405.
Williams, Z., & Wassarman, P. M. (2001). Secretion of mouse ZP3, the sperm receptor, requires cleavage of its polypeptide at a consensus furin cleavage-site. *Biochemistry, 40*, 929–937.
Williams, Z., Litscher, E. S., Jovine, L., & Wassarman, P. M. (2006). Polypeptide encoded by mouse ZP3 exon-7 is necessary and sufficient for sperm binding in vitro. *Journal of Cellular Physiology, 207*, 30–39.
Xiong, B., Zhao, Y., Beall, S., Sadusky, A. B., & Dean, J. (2017). A unique egg cortical granule localization motif Is required for ovastacin sequestration to prevent premature ZP2 cleavage and ensure female fertility in mice. *PLoS Genetics, 13*, e1006580.
Yanagimachi, R. (1994). *Mammalian fertilization. The physiology of reproduction, 1*, New York, NY: Raven Press189–317.
Yoshihara, M., Wagner, M., Damdimopoulos, A., Zhao, C., Petropoulos, S., Katayama, S., & Damdimopoulou, P. (2023). The continued absence of functional germline stem cells in adult ovaries. *Stem Cells, 41*, 105–110.
Zhao, M., Gold, L., Ginsberg, A. M., Liang, L.-F., & Dean, J. (2002). Conserved furin cleavage-site not essential for secretion and integration of ZP3 into the extracellular coat of transgenic mice. *Molecular and Cellular Biology, 22*, 3111–3120.
Zhao, M., Gold, L., Dorward, H., Liang, L.-F., Hoodbhoy, T., Boja, E., & Dean, J. (2003). Mutation of a conserved hydrophobic patch prevents incorporation of ZP3 into the zona pellucida surrounding mouse eggs. *Molecular and Cellular Biology, 23*, 8982–8991.
Zhou, Z., Ni, C., Wu, L., Chen, B., Xu, Y., Zhang, S., & Wang, L. (2019). Novel mutations in ZP1, ZP2, and ZP3 cause female infertility due to abnormal zona pellucida formation. *Human Genetics, 138*, 327–337.

Further reading

Extracellular matrix and egg coats. In E. S. Litscher, & P. M. Wassarman (Vol. Eds.), *Current Topics in Developmental Biology: 130*, (pp. 1–488).
Fernandes, I., Chanut-Delalande, H., Ferrer, P., Latapie, Y., Waltzer, L., Affolter, M., Payre, F., & Plaza, S. (2010). Zona pellucida domain proteins remodel the apical compartment for localized cell shape changes. *Developmental Cell, 18*, 64–76.
Litscher, E. S., & Wassarman, P. M. (2018b). The fish egg's zona pellucida. *Current Topics in Developmental Biology, 130*, 275–306.
Wassarman, P. M. (1988b). Fertilization in mammals. *Scientific American, 255*, 78–84.

CHAPTER EIGHT

Luteinizing hormone-induced changes in the structure of mammalian preovulatory follicles

Corie M. Owen* **and Laurinda A. Jaffe***
Department of Cell Biology, University of Connecticut Health Center, Farmington, CT, United States
*Corresponding authors. e-mail address: coowen@uchc.edu; ljaffe@uchc.edu

Contents

Abstract

Ovulation of a mammalian oocyte from its follicle, which occurs in response to luteinizing hormone (LH), requires complex restructuring of the ~20 layers of surrounding somatic cells. This chapter describes the cellular architecture of preovulatory follicles, the localization of the receptors for LH, and the LH-induced changes in follicular structure, focusing on mice and other small mammals. The multiple interrelated processes that result in ovulation include breakdown of existing extracellular matrix, generation of new extracellular matrix, thinning of the follicular apex where the oocyte will be released, invagination of the follicular surface, and responses of the vascular system to support these dynamic changes. However, much remains unknown about how these events function together to release a fertilizable egg.

1. Introduction

Ovulation is a universal process by which animal eggs are released from the ovary such that they can participate in fertilization. The events of

Current Topics in Developmental Biology, Volume 162
ISSN 0070-2153, https://doi.org/10.1016/bs.ctdb.2024.10.011

mammalian ovulation are similar in many ways to those in other vertebrates, starfish, and even jellyfish. In each of these species, the ovulatory signal acts on a G-protein coupled receptor in the somatic cells of the follicle that surrounds the oocyte. In mammals and other vertebrates, the signal is luteinizing hormone (LH) from the pituitary (Hunzicker-Dunn & Mayo, 2015; Omony et al., 2022), named "luteinizing" because in addition to triggering ovulation, LH causes the follicle to transform into the corpus luteum that supports pregnancy in mammals. In starfish, the stimulus for ovulation is a relaxin-like peptide from nerves in the gonoduct (Feng et al., 2023; Mita, 2023), and in jellyfish, it is light (Quiroga Artigas et al., 2020). In all of these animals, activation of the receptor in the somatic cells of the follicle results in resumption of meiosis and cytoplasmic changes in the oocyte that prepare it for fertilization. Receptor activation also results in rupture of the follicle, releasing the oocyte.

This chapter reviews structural changes in mammalian follicles that are stimulated by luteinizing hormone, and discusses some of the unanswered questions about how these structural changes in the somatic cells act together to enable oocyte release. The ovulatory process in mammalian ovaries is especially complex because the follicular structure has ~20 cellular layers, in contrast to starfish where there is just a single layer (Feng et al., 2023), or frogs and fish where there are ~3 layers (Dumont & Brummett, 1978; Takahashi, Hagiwara, & Ogiwara, 2018). We focus on studies of mice, rats, hamsters, and rabbits – small mammals that have been studied extensively because they are accessible to experimental manipulation. We review some of the classic and current studies about the changes in cellular and tissue structure that are induced by LH in the follicle, and discuss how these structural changes may act together to release the oocyte from the follicle at ovulation. Other aspects of ovulation including gene regulation in the granulosa cells, oocyte maturation, and mechanisms of ovulation in humans and other large mammals, are reviewed elsewhere (Duffy, Ko, Jo, Brannstrom, & Curry, 2019; Hughes & Murphy, 2021; Hunzicker-Dunn & Mayo, 2015; Jaffe & Egbert, 2017; Mogessie, Scheffler, & Schuh, 2018; Richards & Ascoli, 2018; Robker, Hennebold, & Russell, 2018).

2. Structure of mammalian preovulatory follicles

In small mammals such as mice, an ovary at the preovulatory stage has multiple fully grown follicles located around its surface. The ovary also contains thousands of smaller follicles that will grow in subsequent reproductive cycles. At ovulation, oocytes will be released from the "apical" side of each preovulatory follicle, the side that faces the ovary surface and is adjacent to the ovarian surface epithelium. The non-apical side of each follicle is referred to as "basolateral".

The elaborate complex of somatic cells that comprises each preovulatory follicle includes 2 concentric regions separated by a basal lamina: granulosa cells and theca cells (Fig. 1A). The multiple layers of granulosa cells are connected to each other and to the oocyte by gap junctions (see Norris et al., 2008). The ~2–3 layers of granulosa cells closest to the oocyte are called cumulus cells; these send fine processes through the extracellular matrix around the oocyte (zona pellucida) to form contacts with oocyte

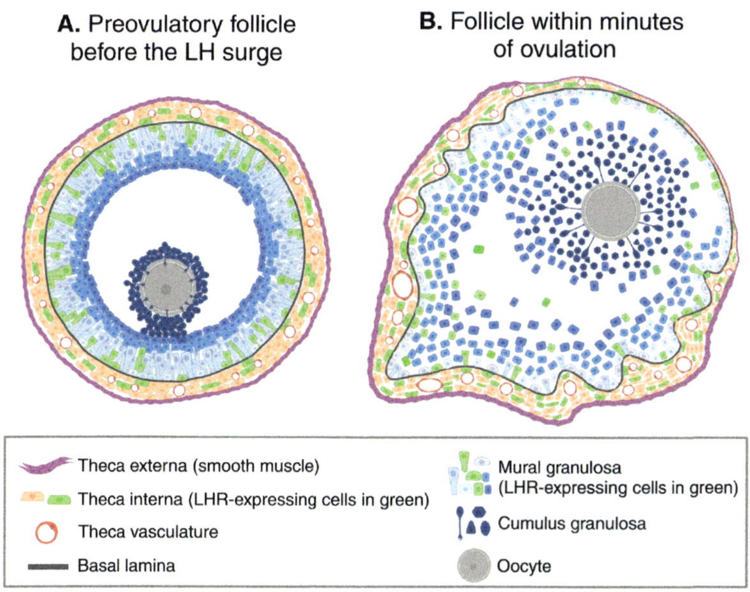

Fig. 1 Cellular structure of preovulatory follicles before and after the LH surge. (A) Before LH exposure. The oocyte is surrounded by several concentric layers of cells: cumulus granulosa, mural granulosa, theca interna (steroidogenic and non-steroidogenic cells with interspersed vasculature), and theca externa (smooth muscle). The graded shades of blue in the granulosa cells represent the gradient of signaling molecules emanating from the oocyte. Green indicates cells that express the receptor for LH (LHR) (see Fig. 3A). (B) After LH receptor activation, within minutes of ovulation. The mass of cumulus cells has expanded, and the follicular volume has increased. The apical region where ovulation will occur has thinned. Basolateral regions have invaginated. Blood vessels within the theca have dilated. Figure made in Biorender.com.

microvilli (Baena & Terasaki, 2019; Clarke, 2022). The outer ~5–15 granulosa layers are called mural granulosa. The ~3 outermost mural cell layers form a pseudostratified epithelium with cell bodies extending projections that contact the basal lamina (Baena & Terasaki, 2019; Baena et al., 2020; Lipner & Cross, 1968). The cumulus and mural cells are separated by a fluid-filled antral space, except at one region where the cumulus-oocyte complex and mural cells are connected, until they are separated during ovulation. The position of the cumulus-oocyte complex with respect to the apical side of the follicle is variable (for examples of the variability, see Owen & Jaffe, 2024, supplementary material).

Cells at different radial positions within the granulosa cell compartment show different patterns of gene expression, as a consequence of opposing signaling gradients within the follicle. The oocyte in the center of the follicle releases transforming growth factor beta signaling proteins that suppress mural transcripts, such as the LH receptor, and promote cumulus transcripts in the cells closest to the oocyte – in the cumulus cells as well as some of the mural cells closest to the antrum (Diaz, Wigglesworth, & Eppig, 2007; Eppig, Wigglesworth, Pendola, & Hirao, 1997; Salustri, Yanagishita, Underhill, Laurent, & Hascall, 1992; Su et al., 2008). Conversely, follicle stimulating hormone promotes mural transcripts and suppresses cumulus transcripts, and it has been proposed that due to its delivery to the follicle by the blood vessels in the theca, FSH may be present at a higher concentration in the outer mural granulosa cells vs the follicle interior (Diaz et al., 2007). In addition, a signal from the basal lamina acts on the outer granulosa cells that contact it to localize LH receptor expression to these cells (Eppig et al., 1997; Furman, Rotmensch, Kohen, Mashiach, & Amsterdam, 1986).

Outside of the basal lamina is a multi-layered theca region that is comprised of multiple types of cells (Fig. 1A). The theca interna contains interspersed steroidogenic and non-steroidogenic cells (Nicol, Rodriguez, & Yao, 2020) as well as endothelial cells that form a dense network of blood capillaries that support the avascular granulosa region and deliver hormones including LH (Macchiarelli, Jiang, Nottola, & Sato, 2006; McKey, Cameron, Lewis, Batchvarov, & Capel, 2020). Surrounding the theca interna is a layer of smooth muscle cells (theca externa) (Nicol et al., 2020). The theca on the apical side of each preovulatory follicle abuts against the ovarian surface epithelium, which surrounds the ovary.

The structure of the preovulatory follicle is similar in all mammals, except that its diameter is larger in larger species – ~0.4 mm in a mouse,

~2 cm in a woman, ~12 cm in an Indian rhinoceros (Bächler, Menshykau, De Geyter, & Iber, 2014). Oocyte diameters range from ~60 μm to ~130 μm, with a slight correlation with body weight (Gosden & Telfer, 1987). The numbers of mural cell layers and cumulus cell layers in the preovulatory follicle show only small differences among different species, but the antral space and total number of granulosa cells are much larger in the follicles of larger animals (Bächler et al., 2014; Gosden & Telfer, 1987). The larger number of granulosa cells is essential in larger animals because in addition to providing a nest for the oocyte, the granulosa cells of the follicle produce estrogen. Larger animals need the greater mass of granulosa cells to produce more estrogen.

3. Luteinizing hormone receptor localization and signaling

In preovulatory follicles of mice, rats, and hamsters, LH receptors are expressed in some of the mural granulosa cells and in some of the theca cells (Fig. 1A). Before the preovulatory stage, only the theca cells, not the granulosa cells, express receptors for LH. The localization of these receptors was originally determined by in situ hybridization of probes for mRNA and by binding of a radioactively labeled agonist (Bortolussi, Marini, & Dal Lago, 1977; Camp, Rahal, & Mayo, 1991; Oxberry & Greenwald, 1982; Richards et al., 1976). More recently, higher resolution localization was accomplished by use of mice with an HA epitope tag at the N-terminus of the endogenous receptor, such that HA antibodies could be used for immunolocalization (Baena et al., 2020; Owen & Jaffe, 2024) (see Fig. 3A below). The epitope tag was needed because native levels of the LH receptor have not been conclusively detectable with antibodies against the receptor protein itself. Within the mural granulosa cells, the LH receptors are present only in the outer few layers of cells that extend processes that contact the basal lamina (Baena et al., 2020). Among these cells, only some express the LH receptor (Baena et al., 2020). In the theca, LH receptor expression is also heterogeneous. Apical and basolateral regions show no differences in the distribution of LH receptors (see examples in the supplementary material of Owen & Jaffe, 2024).

LH receptors in the mural granulosa cells are well established to transduce the signal from the LH surge to cause ovulation, but while LH receptors in theca cells are essential for follicle growth (Pakarainen, Zhang,

Nurmi, Poutanen, & Huhtaniemi, 2005), the role of theca LH receptors in responding to the LH surge is uncertain. Which cell types in the theca express the LH receptor has also not been determined. Single cell RNA sequencing databases (Fan et al., 2019; Morris et al., 2022) indicate that some cells express both LH receptors and steroidogenic enzymes, but have not conclusively defined whether only the steroidogenic cells express LH receptors. Analysis of spatial transcriptomics databases (Huang et al., 2024; Mantri, Zhang, Spanos, Ren, & De Vlaminck, 2024) could be informative. Studies of primate endothelial cell lines have raised the question of whether LH receptors may also be expressed in cells of the theca vasculature (Lund, Pearson, Sage, & Duffy, 2023). mRNA encoding the LH receptor has been detected in these cell lines, and LH elevates their cAMP levels and elicits other responses. However, the lack of well validated LH receptor antibodies has limited conclusions regarding the presence of LH receptors in theca vasculature of intact follicles. Immunolocalization using thin sections of mouse ovaries expressing the HA-tagged LH receptor, together with immunolocalization of proteins marking other cell types in the theca, could address these questions.

Following LH release from the pituitary, plasma LH levels remain above baseline for several hours, in mice (Czieselsky et al., 2016; Kovacic & Parlow, 1972; Owen, Zhou, Bernard, & Jaffe, 2021), rats (Daane & Parlow, 1971), and hamsters (Norman & Greenwald, 1972). The ~30 kDa LH protein is small enough to permeate through the capillaries in the theca layers of the follicle (Hess, Chen, & Larsen, 1998). LH then crosses the basal lamina and activates its receptors in the mural granulosa cells. In mice (Kovacic & Parlow, 1972), rats (Daane & Parlow, 1971), and hamsters (Norman & Greenwald, 1972), ovulation occurs approximately 11–14 h after the start of the LH surge.

The LH receptor is coupled to multiple G-proteins, including G_s and $G_{q/11}$. Most of the granulosa cell responses are mediated by G_s activation of adenylyl cyclase and the resulting elevation of cAMP and activation of protein kinase A (PKA) (Hunzicker-Dunn & Mayo, 2015). cAMP diffuses through gap junctions to all of the granulosa cells (Lyga et al., 2016), such that the signal initiated in the outer mural granulosa cells results in rapid protein phosphorylation and ensuing signaling cascades throughout the entire granulosa compartment. Signals also spread inwards from the outer to the inner mural granulosa cells and to the cumulus cells by extracellular diffusion of EGF receptor ligands that are generated by LH receptor activation (Conti, Hsieh, Zamah, & Oh, 2012).

Elevating cAMP by applying inhibitors of cAMP phosphodiesterases is sufficient to trigger ovulation, and to increase synthesis of progesterone and progesterone receptors to levels comparable to those in response to LH (Vigone et al., 2018). Progesterone signaling is essential for ovulation (Kim, Bagchi, & Bagchi, 2009; Lydon et al., 1995). Many other transcriptional events that are also essential for ovulation, including synthesis of prostaglandin synthase (Sirois et al., 2004), are triggered at least in part by LH activation of PKA (Hunzicker-Dunn & Mayo, 2015).

However, mice with granulosa cell-specific deletion of $G\alpha_q$ and $G\alpha_{11}$, G-proteins that act to elevate Ca^{2+} rather than cAMP, ovulate only about half the number of oocytes as controls (Breen et al., 2013). Correspondingly, LH-induced increases in mRNA encoding several proteins required for ovulation (including the progesterone receptor, endothelin, and the protease ADAMTS1) are reduced in ovaries from these mice (Breen et al., 2013). Pharmacological inhibition of $G_{q/11}$ also delays ovulation (Egbert et al., 2019). These results indicate that signaling by both G_s and $G_{q/11}$ is required for optimal ovulation. LH induces elevation of Ca^{2+} in both mural granulosa and theca cells (Egbert et al., 2019), but the spatiotemporal dynamics of the Ca^{2+} elevation in the follicle during the later phases of ovulation have not been investigated.

4. Methods for inducing and observing ovulation

Studies of follicular structure during ovulation were initially conducted using ovaries from naturally cycling animals on the day of proestrous (for example, Norman & Greenwald, 1972). For mice (Czieselsky et al., 2016; Kovacic & Parlow, 1972), rats (Daane & Parlow, 1971), and hamsters (Norman & Greenwald, 1972), the LH surge begins a few hours before the circadian lights go off in the animal room, with some variability in timing for individual animals (Czieselsky et al., 2016). Because of this variability, most studies of ovulation are now performed using ovaries from animals stimulated by injection of LH or an LH analog, or by isolating follicles and exposing them to LH or an LH analog in vitro (Tsafriri, Lindner, Zor, & Lamprecht, 1972; Vigone et al., 2018; Thomas, Marx, Penir, & Schuh, 2024).

Many studies have used a different LH receptor agonist, human chorionic gonadotropin, (hCG) in place of LH. hCG is produced by trophoblast cells of the human placenta, and is readily available from urine of pregnant women. hCG acts on the corpus luteum to maintain early pregnancy (Stouffer & Hennebold, 2015). In women and other primates, LH and chorionic gonadotropin are both natural agonists of the same receptor, produced by the *LHCGR* gene. Mice, rats, hamsters, and rabbits do not produce chorionic gonadotropin, but hCG is an effective agonist for their LH receptors. The use of hCG in place of LH originated with early bioassays for human pregnancy, in which mice or rabbits were injected with urine from women; ovulation of the mouse or rabbit indicated pregnancy. Follicular growth and LH receptor expression are usually stimulated by injection of animals on the first day of their 4-day estrous cycle with equine chorionic gonadotropin (eCG, also known as PMSG), which mimics the action of follicle stimulating hormone. Subsequent hCG injection ~2 days later causes ovulation with a time course close to that following an endogenous LH surge. However, some aspects of LH receptor signaling, such as the kinetics of cAMP production, are not identical for LH and hCG (Casarini et al., 2012).

By 10 min after intraperitoneal injection, hCG reaches the ovarian follicles, as indicated by its rapid stimulation of an increase in permeability of theca blood vessels (Powers, Chen, Russell, & Larsen, 1995) (see below). However, the concentration of hCG at the follicle, and the kinetics of its concentration, are not known exactly, and some aspects of LH receptor signaling depend on agonist concentration (for example, Vigone et al., 2018). Alternatively, ovulation can be stimulated by injection of neuropeptides that induce an endogenous LH surge by signaling in the hypothalamus. During a natural LH surge, kisspeptin release stimulates gonadotropin releasing hormone release, which then stimulates LH release from the pituitary (Skorupskaite, George, & Anderson, 2014). Injection of kisspeptin results in a transient increase in serum LH with concentration and kinetics close to that in cycling mice (Owen et al., 2021) and also induces ovulation in mice (Owen & Jaffe, 2024; Owen et al., 2021) and women (Abbara et al., 2020). Likewise, GnRH injection stimulates ovulation in rats (Dekel, Sherizly, Tsafriri, & Naor, 1983). Injection of kisspeptin or GnRH to elicit an endogenous LH surge provides a more physiological method for stimulation of ovulation compared with injection of hCG or LH.

Many studies of ovulation use 3–4-week old prepubertal mice or rats. Injection of these animals with eCG causes a set of their follicles to grow to the preovulatory stage and induces expression of the LH receptor. One advantage of using prepubertal instead of adult mice is the ability to obtain ovaries at a precisely defined stage of development. When subsequently injected with hCG, LH, or kisspeptin, these hormonally "primed" prepubertal mice ovulate with normal kinetics (for example, see Owen & Jaffe, 2024).

Follicles isolated from prepubertal mice and incubated with follicle stimulating hormone to allow progression to the preovulatory stage have been another useful system for studies of ovulation (Egbert et al., 2019; Vigone et al., 2018; Thomas, Marx, Penir, & Schuh, 2024). These follicles are optically clearer than follicles obtained from eCG-primed mice, and when cultured on optically clear organotypic membranes (Millicell cell culture inserts) they flatten somewhat, facilitating live imaging of the ovulatory process. Ovulation in these isolated follicles occurs with normal or close to normal kinetics (Egbert et al., 2019; Thomas, Marx, Penir, & Schuh, 2024). Isolated follicles that have been cultured in a 3-dimensional alginate matrix can also ovulate, following removal from the alginate and transfer to a culture plate (Skory, Xu, Shea, & Woodruff, 2015). Use of isolated follicles has many advantages, particularly for imaging, but blood circulation to the follicle is not preserved, and the amount of theca retained after isolation is variable. A more physiological but more challenging approach is to image ovulation in ovaries that are still connected to the vasculature (Blandau, 1955; Migone et al., 2016).

Overall, these various approaches have provided consistent and complementary information about the ovulatory process. In the sections below, we discuss conclusions and questions that have emerged from these studies about several interrelated structural changes in preovulatory follicles that occur in response to LH (Fig. 1B).

5. Cumulus expansion and follicular enlargement, interrelated events

By 5 h after LH receptor stimulation, the extracellular spaces between the cumulus cells, and between the mural cells adjacent to the antrum increase (see Norman & Greenwald, 1972; Parr, 1974) (Fig. 1B). The spaces arise as a consequence of a dramatic increase in amount of extracellular matrix, due to the transcription of hyaluronic acid synthase (HAS2), the secretion of hyaluronic acid, and the association of several proteins with the hyaluronic acid (Salustri, Campagnolo, Klinger, & Camaioni, 2019). This process, known as "cumulus expansion" is accompanied by loss of gap junctional communication between the cumulus cells and the oocyte at 6–9 h after LH receptor stimulation (Eppig, 1982).

In mice lacking the hyaluronic acid-linking protein inter-α-trypsin inhibitor protein, cumulus expansion fails to occur, but oocytes are still ovulated, as assayed by their presence in the oviduct (Zhuo et al., 2001). However, the number of oocytes present in the oviduct is only 40 % of control (Zhuo et al., 2001). It is unknown if this reduction is due to fewer oocytes being ovulated or to less efficient capture of the oocytes by the ciliary action of the oviduct.

Accompanying cumulus expansion, the volume of the antral space in the follicle increases (Martin & Talbot, 1987; Norman & Greenwald, 1972; Thomas, Marx, Penir, & Schuh, 2024). Based on measurements of cross-sectional area, the antral volume of hamster follicles increases by almost 4-fold during the 11 h after hCG injection; it then decreases during the 3 h preceding ovulation, to about half of the peak value (Martin & Talbot, 1987). Follicular enlargement followed by contraction is also seen in isolated mouse follicles, although the magnitude of the changes is smaller (Thomas, Marx, Penir, & Schuh, 2024). The increase in antral volume has been proposed to be due to hyaluronic acid generating an osmotic gradient that draws fluid into the antrum (Martin & Talbot, 1987). This hypothesis was recently tested by applying an inhibitor of HAS2 (4-MU) to isolated mouse follicles (Thomas, Marx, Penir, & Schuh, 2024). 4-MU partially inhibits the increase in follicular volume, although 4-MU can have effects other than inhibiting HAS2 (Pibuel et al., 2023). The concept that an osmotic gradient causes the increase in follicular volume is also supported by inhibition of the volume increase in response to applying dextran outside of the follicle (Thomas, Marx, Penir, & Schuh, 2024).

6. Apical thinning, a multistep process

Release of the oocyte requires thinning of the cellular layers of the mural granulosa, theca and surface epithelium on the apical side of the follicle (Figs. 1B and 2A). The thinning of the apical mural granulosa layer begins within 6 h of the ovulatory stimulus (Fig. 2B); by 8 h, it decreases to about 30–70 % of its original thickness, reflecting a decrease in the number of cell layers (Martin & Talbot, 1987; Owen & Jaffe, 2024). The basolateral mural granulosa layer also thins in hamster, but not in mouse follicles. There is little or no change in the thickness of the theca interna during the initial 8 h (Martin & Talbot, 1987).

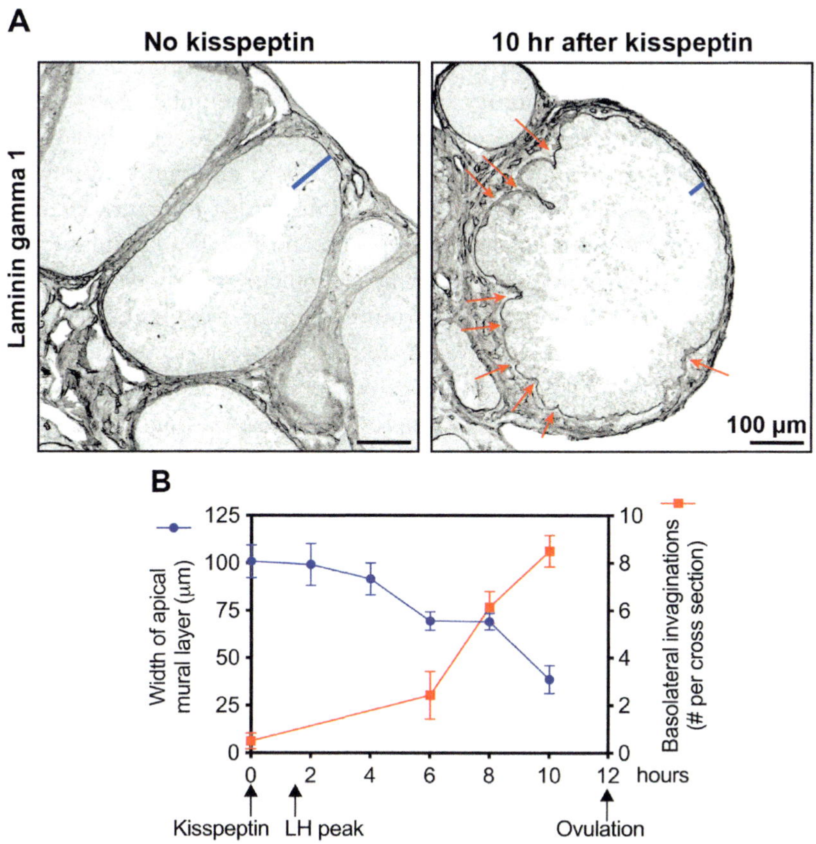

Fig. 2 Apical thinning and basolateral invaginations induced by LH in mouse preovulatory follicles. (A) Preovulatory follicles within ovaries from eCG-primed ~24-day-old mice, before or 10 h after kisspeptin injection to induce an LH surge. Cryosections were labeled with an antibody against laminin gamma 1 (Santa Cruz Biotechnology, sc-65643), to mark the basal lamina. Images are aligned with the apical surface on the right. The blue lines indicate thinning of the mural granulosa layer at the follicle apex, and the red arrows indicate basolateral invaginations. (B) Kinetics of apical thinning of the mural granulosa and formation of basolateral invaginations after the LH surge. *B is modified from Owen and Jaffe (2024), with permission.*

It is not well understood how LH signaling causes the initial thinning of the mural granulosa layer, although one factor appears to be the secretion of hyaluronic acid by the inner layers of the mural granulosa, resulting in these cells separating from the mural layer and adhering to the cumulus mass (Salustri et al., 1992) (Fig. 1B). These "mucified" mural cells are recovered with cumulus masses when follicles are punctured at 9 h after hCG injection. Correspondingly, the number of cells in the cumulus mass increases from

~1000 in an eCG-primed mouse before hCG injection to ~3000 at 15 h after hCG injection (Salustri et al., 1992). Cumulus cell proliferation is a possible alternative explanation of the increase in cell number (Salustri et al., 1992), although mitosis is infrequent after the LH surge, and the few cells that were classified as mitotic might be difficult to distinguish from atretic cells (Norman & Greenwald, 1972). The concept that the mural granulosa layer thins because some of its cells separate and join the cumulus mass is supported by images of hamster and mouse follicles that show mural granulosa cells sloughing off into the antrum during this time frame (Martin & Talbot, 1987; Norman & Greenwald, 1972; Zhuo et al., 2001). Loss of the gap junctions that hold the mural granulosa cells together, due to their internalization (Larsen, Wert, & Brunner, 1987), and possibly loss of other adhesion proteins due to internalization or proteolysis (see more about proteolysis below) might facilitate cells leaving the mural epithelium.

Additionally, within the first 8 h after the ovulatory stimulus, some of the LH receptor-expressing mural granulosa cells in the outer half of the mural layer migrate into the inner half of the mural layer, and some of these cells come into contact with the cumulus mass (Owen & Jaffe, 2024) (Fig. 3). Whether cells that do not express the LH receptor also migrate inwards, and whether cellular ingression contributes to apical thinning, is unknown. Possibly the cellular migration acts together with the several factors discussed above to cause relocation of cells from the mural layer to the cumulus-oocyte complex, resulting in the initial thinning of the apical mural granulosa layer. Another factor that might contribute to the thinning of the mural granulosa region is the enlargement of the antral volume, causing the cells of the mural epithelium to redistribute into a thinner layer with a larger surface area (Martin & Talbot, 1987).

In the final ~3 h preceding follicle rupture, the number of cell layers in the the apical mural granulosa layer, and in the apical theca layer, drops precipitously (Martin & Talbot, 1987). Proteolysis of collagen is required for ovulation (Curry & Smith, 2006; Duffy et al., 2019; Ichikawa, Ohta, Morioka, & Murao, 1983; Robker et al., 2018), suggesting that proteolysis of extracellular matrix could be a factor in apical thinning. In support of this concept, local protease activity in the apical region of the mural granulosa layer was detected by placing frozen sections of rat ovaries, 12 h after hCG injection, on a fluorescent gelatin substrate (Curry, Song, & Wheeler, 2001). The protease activity is metal-dependent, based on inhibition by EDTA. The time course of this local increase in metalloprotease activity has not been reported.

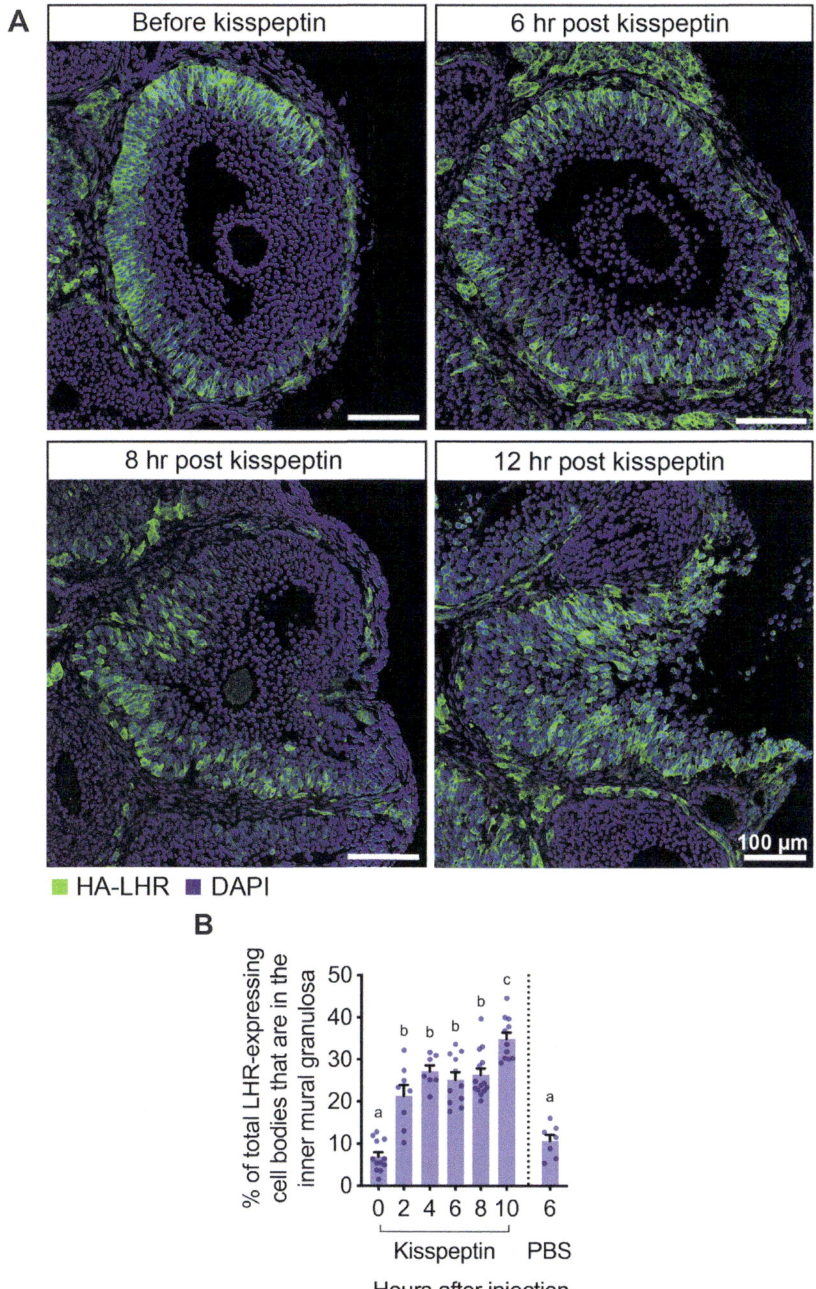

Fig. 3 Migration of LH receptor-expressing mural granulosa cells into the ovarian follicle interior in response to LH. (A) Confocal images of cryosections of follicles in ovaries before and at 6, 8, and 12 h after injection of HA-LHR-expressing

(Continued)

Several different extracellular matrix-targeting proteases are synthesized in response to an ovulatory stimulus, in both granulosa and theca cells (Ohnishi, Ohnishi, Shibuya, & Takahashi, 2005). ADAMTS1 is of particular interest because the mRNA encoding it increases > 100-fold fold by 8 h after injection of rats with hCG; the mRNA is localized in the mural granulosa cells (Espey et al., 2000). Mice lacking *Adamts1* show reduced ovulation, with ~20–30% of the control number of oocytes ovulated (Brown et al., 2010; Mittaz et al., 2004). In the *Adamts1*$^{-/-}$ mice, oocytes are seen trapped in unruptured follicles. However, the initial thinning of the apical mural granulosa layer still occurs, with only 1–2 layers of mural cells present at the apex in images at 12 h after hCG (Brown et al., 2010; Mittaz et al., 2004). In contrast, images of the theca in the *Adamts1*$^{-/-}$ mice did not show thinning, suggesting that the inhibition of ovulation in these mice might be primarily due to the theca remaining intact. Possibly the burst of ADAMTS1 synthesis in the mural granulosa cells acts to degrade the extracellular matrix of the basal lamina separating the granulosa and theca cells, and the matrix between the theca cells, causing the final thinning of the mural granulosa and theca layer that allows the follicle to rupture. Mice lacking various other individual proteases have shown little or no effect on ovulation, possibly due their redundant functions (Ohnishi et al., 2005).

7. Invagination and contraction of the follicular surface

While the follicular apex is thinning in preparation for rupture, invaginations appear in the basolateral surface, as seen in ovarian histological sections, (Brown et al., 2010; Szołtys, Galas, Jabłonka, & Tabarowski, 1994; Zhuo et al., 2001) (Figs. 1B and 2A). These invaginations start as early as 6 h before ovulation (Owen & Jaffe, 2024) (Fig. 2B) and become progressively larger, often forming a V-shape at the base of the follicle (Gothié, 1967; Martin & Talbot, 1981; Owen & Jaffe, 2024) (Figs. 1B and 3A). The apical

Fig. 3—Cont'd mice with kisspeptin to induce an endogenous LH surge. At 6 and 8 h, HA-LHR-expressing cells are migrating inwards. At 12 h, the follicle wall has ruptured, releasing the cumulus-oocyte complex. Sections were labeled for HA-LHR immunofluorescence and nuclei (DAPI). Images were obtained from ~24-day-old eCG-primed mice. Similar inward migration of mural granulosa cells occurs in naturally cycling adult mice (Owen & Jaffe, 2024). (B) Time course of the inward migration. *Modified from Owen and Jaffe (2024), with permission.*

side of the follicle sometimes collapses inwards as well (Owen & Jaffe, 2024; Zhuo et al., 2001) (Fig. 3A). Correlating with these indications of contractile activity, follicular volume decreases in the few hours preceding ovulation (Martin & Talbot, 1981; Thomas, Marx, Penir, & Schuh, 2024).

Contraction of smooth muscle cells in the outer theca has been proposed to be a factor in causing contraction of the follicular surface, based on electron micrographs of ovaries just before ovulation showing smooth muscle cells with a contracted shape, relative to what is seen 3 h earlier (Martin & Talbot, 1981). Another indicator of smooth muscle contraction is phosphorylation of the myosin regulatory light chain protein (MRLC, also known as myosin light chain 2 or MYL9) on its regulatory serine (Ito, Okamoto, Ito, Zhe, & Dohi, 2022). Measurements at 6 and 12 h after hCG application to isolated follicles or injection into mice have shown that MRLC is phosphorylated, but its phosphorylation state before LH exposure has not been examined (Thomas, Marx, Penir, & Schuh, 2024). Spatiotemporal imaging of Ca^{2+} in the theca smooth muscle cells might provide information as to whether or not the theca smooth muscle cells show an increase in contraction just preceding ovulation.

Measurements of contractile force generated by application of endothelin to rat or mouse ovary tissue strips indicate that ovarian smooth muscle cells have receptors for endothelin (Cacioppo et al., 2017; Ko et al., 2006), and endothelin-2 mRNA is synthesized in mural granulosa cells at ~11 h after hCG injection (Palanisamy et al., 2006). Thus, endothelin from the granulosa cells could cause theca smooth muscle cells to contract in the final minutes before the follicle ruptures. Mice lacking endothelin in their granulosa cells ovulate fewer oocytes, supporting the hypothesis that endothelin-induced smooth muscle contraction could help to expel the oocyte (Cacioppo et al., 2017). However, endothelin synthesis is too slow to explain the earlier surface invaginations. Contraction of theca smooth muscle could also result from activation of other $G_{q/11}$-linked receptors by agonists from the mural cells that are synthesized earlier, such as prostaglandins (Land, Lane, Fugate, & Hannon, 2021) and neurotensin (Shrestha et al., 2023).

Alternatively, the surface invaginations may result because as the antral volume expands, the wall of the follicle buckles inwards due to its confinement by the stiff ovarian tissue around it (see (Trushko et al., 2020)). The LH-induced invaginations are dependent on the ADAMTS1 protease (Brown et al., 2010), and protease action on the extracellular matrix of the follicle could weaken the theca and mural layers, allowing them to buckle.

Forces generated by the inward migration of mural granulosa cells that adhere to the basal lamina have also been proposed to contribute to pulling in of the follicular surface (Owen & Jaffe, 2024). In summary, the factors causing the LH-induced invaginations in the follicle surface are uncertain, and whether the invaginations and/or theca smooth muscle contraction enable the release of the oocyte remains to be determined.

8. Increases in vascular permeability, capillary diameter, and angiogenesis

In response to LH, the blood vessels in the theca rapidly increase their permeability to proteins. Before LH receptor stimulation, the permeability of blood capillaries is limited to proteins smaller than ~200 kDa and is also determined by protein charge (Hess et al., 1998; Powers et al., 1995). However, by 10 min after intraperitoneal injection of mice with hCG, a 220 kDa protein (inter-α-trypsin inhibitor protein) that was previously unable to pass from the serum into the follicular fluid, appears in the follicular fluid (Hess et al., 1998; Powers et al., 1995). Proteins larger than ~700 kDa are retained within the capillaries, even after the ovulatory stimulus (Hess et al., 1998). The permeability increase has been confirmed by injection of mice with fluorescently labeled Ficoll (Mitsube, Brännström, & Haraldsson, 2013).

As mentioned above (see Section 5), inter-α-trypsin inhibitor protein is critical for linking hyaluronic acid, allowing formation of the cumulus matrix, and in the absence of this protein, the number of oocytes found in the oviduct is reduced (Zhuo et al., 2001). In addition, after mating, very few eggs from mice lacking inter-α-trypsin inhibitor protein undergo cleavage, suggesting that they failed to fertilize (Zhuo et al., 2001). These observations indicate the LH-induced increase in vascular permeability serves the essential function of allowing transfer from the serum to the follicle of a protein needed for successful reproduction.

Whether LH acts to increase capillary permeability by way of LH receptors in the endothelial cell membranes, as suggested by studies of monkey ovarian endothelial cell lines (Lund et al., 2023), or by way of paracrine signaling from LH receptors in nearby theca or mural granulosa cells, is unknown. Nitric oxide may be a mediator of the LH-induced permeability increase, based on mimicking of the permeability increase by the nitric oxide donor sodium nitroprusside and inhibition of the increase by inhibitors of nitric oxide synthase (Powers et al., 1995).

LH receptor stimulation also regulates capillary diameter (Fig. 1B). Blood volume in the rat ovary starts to increase by 4 h after hCG injection (Tanaka, Espey, & Okamura, 1989), associated with an increase in diameter of the capillaries in the theca (Kranzfelder, Reich, Abisogun, & Tsafriri, 1992; Macchiarelli et al., 2006), which is detected by 3 h after hCG (Kranzfelder et al., 1992). The LH-induced increases in capillary permeability and diameter within the follicle could reflect an increased need for gas exchange, nutrient supply and waste removal during the dynamic process of ovulation. A recent study showed that just before ovulation, blood capillaries in the follicular apex constrict, such that blood flow is inhibited, and it was suggested that the decrease in blood flow could cause apical thinning and follicular rupture (Migone et al., 2016). However, isolated follicles that are disconnected from the blood circulation can also ovulate in response to LH (Egbert et al., 2019; Tsafriri, Lindner, Zor, & Lamprecht, 1972; Vigone et al., 2018; Thomas, Marx, Penir, & Schuh, 2024).

In addition to its regulation of capillary permeability and diameter, LH stimulates angiogenesis in the follicle. Angiogenic genes are upregulated in granulosa cells within 4 h of injecting mice with hCG, and the number of endothelial cells increases by 8 h (Guzmán, Hughes, & Murphy, 2022). The angiogenesis that begins before ovulation increases dramatically after ovulation, as blood vessels grow into the interior of the forming corpus luteum (Stocco, Telleria, & Gibori, 2007; Stouffer & Hennebold, 2015).

9. Conclusions

Despite intensive investigation over the past century, much remains mysterious about the structural changes in the ovarian follicle that result in ovulation in response to luteinizing hormone. Both the theca and mural granulosa layers of the follicle thin, and clearly this is essential for the follicle to rupture. However, the events causing apical thinning, such as the role of proteolysis of the extracellular matrix, are not well understood. Whether the signals initiating the thinning arise solely from LH receptors in the mural cells, or also from those in the theca, is uncertain. There are indications that contraction of smooth muscle cells in the basal region of the theca might provide pressure that breaks the thinned follicle wall. However, direct evidence as to whether this smooth muscle contraction is essential, when it occurs, and what controls it, is lacking. Thinning of the apical mural granulosa region and invaginations in the basolateral region begin many hours before the rupture of the follicle, accompanied by inward migration of mural granulosa cells. However, the functional significance of these early structural changes for ovulation, and for the subsequent formation of the corpus luteum, remains to be determined.

Acknowledgments

We thank John Eppig his insightful comments on the manuscript, and Jeremy Egbert, Lisa Mehlmann, Rachael Norris, Siu-Pok Yee, and Mark Terasaki for helpful discussions. Research in our laboratory is supported by the Eunice Kennedy Shriver National Institute of Child Health and Human Development (F31 HD107918 to C.M.O., and R37 HD014939 to L.A.J.).

References

Abbara, A., Eng, P. C., Phylactou, M., Clarke, S. A., Richardson, R., Sykes, C. M., ... Dhillo, W. S. (2020). Kisspeptin receptor agonist has therapeutic potential for female reproductive disorders. *The Journal of Clinical Investigation, 130*, 6739–6753. https://doi.org/10.1172/JCI139681.

Bächler, M., Menshykau, D., De Geyter, C., & Iber, D. (2014). Species-specific differences in follicular antral sizes result from diffusion-based limitations on the thickness of the granulosa cell layer. *Molecular Human Reproduction, 20*, 208–221. https://doi.org/10.1093/molehr/gat078.

Baena, V., Owen, C. M., Uliasz, T. F., Lowther, K. M., Yee, S.-P., Terasaki, M., ... Jaffe, L. A. (2020). Cellular heterogeneity of the luteinizing hormone receptor and its significance for cyclic GMP signaling in mouse preovulatory follicles. *Endocrinology, 161*, bqaa074. https://doi.org/10.1210/endocr/bqaa074.

Baena, V., & Terasaki, M. (2019). Three-dimensional organization of transzonal projections and other cytoplasmic extensions in the mouse ovarian follicle. *Scientific Reports, 9*, 1–13. https://doi.org/10.1038/s41598-018-37766-2.

Blandau, R. J. (1955). Ovulation in the living albino rat. *Fertility and Sterility, 6*, 391–404. https://doi.org/10.1016/S0015-0282(16)32100-8.

Bortolussi, M., Marini, G., & Dal Lago, A. (1977). Autoradiographic study of the distribution of LH(HCG) receptors in the ovary of untreated and gonadotrophin-primed immature rats. *Cell and Tissue Research, 183*, 329–342. https://doi.org/10.1007/BF00220640.

Breen, S. M., Andric, N., Ping, T., Xie, F., Offermans, S., Gossen, J. A., & Ascoli, M. (2013). Ovulation involves the luteinizing hormone-dependent activation of G(q/11) in granulosa cells. *Molecular Endocrinology (Baltimore, Md.), 27*, 1483–1491. https://doi.org/10.1210/me.2013-1130.

Brown, H. M., Dunning, K. R., Robker, R. L., Boerboom, D., Pritchard, M., Lane, M., & Russell, D. L. (2010). ADAMTS1 cleavage of versican mediates essential structural remodeling of the ovarian follicle and cumulus-oocyte matrix during ovulation in mice. *Biology of Reproduction, 83*, 549–557. https://doi.org/10.1095/biolreprod.110.084434.

Cacioppo, J. A., Lin, P.-C. P., Hannon, P. R., McDougle, D. R., Gal, A., & Ko, C. (2017). Granulosa cell endothelin-2 expression is fundamental for ovulatory follicle rupture. *Scientific Reports, 7*, 817. https://doi.org/10.1038/s41598-017-00943-w.

Camp, T. A., Rahal, J. O., & Mayo, K. E. (1991). Cellular localization and hormonal regulation of follicle-stimulating hormone and luteinizing hormone receptor messenger RNAs in the rat ovary. *Molecular Endocrinology (Baltimore, Md.), 5*, 1405–1417. https://doi.org/10.1210/mend-5-10-1405.

Casarini, L., Lispi, M., Longobardi, S., Milosa, F., La Marca, A., Tagliasacchi, D., ... Simoni, M. (2012). LH and hCG action on the same receptor results in quantitatively and qualitatively different intracellular signalling. *PLoS One, 7*, e46682. https://doi.org/10.1371/journal.pone.0046682.

Clarke, H. J. (2022). Transzonal projections: Essential structures mediating intercellular communication in the mammalian ovarian follicle. *Molecular Reproduction and Development, 89*, 509–525. https://doi.org/10.1002/mrd.23645.

Conti, M., Hsieh, M., Zamah, A. M., & Oh, J. S. (2012). Novel signaling mechanisms in the ovary during oocyte maturation and ovulation. *Molecular and Cellular Endocrinology, 356*, 65–73. https://doi.org/10.1016/j.mce.2011.11.002.

Curry, T., & Smith, M. (2006). Impact of extracellular matrix remodeling on ovulation and the folliculo-luteal transition. *Seminars in Reproductive Medicine, 24*, 228–241. https://doi.org/10.1055/s-2006-948552.

Curry, T. E., Song, L., & Wheeler, S. E. (2001). Cellular localization of gelatinases and tissue inhibitors of metalloproteinases during follicular growth, ovulation, and early luteal formation in the rat. *Biology of Reproduction, 65*, 855–865. https://doi.org/10.1095/biolreprod65.3.855.

Czieselsky, K., Prescott, M., Porteous, R., Campos, P., Clarkson, J., Steyn, F. J., ... Herbison, A. E. (2016). Pulse and surge profiles of luteinizing hormone secretion in the mouse. *Endocrinology, 157*, 4794–4802. https://doi.org/10.1210/en.2016-1351.

Daane, T. A., & Parlow, A. F. (1971). Periovulatory patterns of rat serum follicle stimulating hormone and luteinizing hormone during the normal estrous cycle as revealed by radioimmunoassays: Effects of pentobarbital. *Endocrinology, 88*, 653–663. https://doi.org/10.1210/endo-88-3-653.

Dekel, N., Sherizly, I., Tsafriri, A., & Naor, Z. (1983). A comparative study of the mechanism of action of luteinizing hormone and a gonadotropin releasing hormone analog on the ovary. *Biology of Reproduction, 28*, 161–166.

Diaz, F. J., Wigglesworth, K., & Eppig, J. J. (2007). Oocytes determine cumulus cell lineage in mouse ovarian follicles. *Journal of Cell Science, 120*, 1330–1340. https://doi.org/10.1242/jcs.000968.

Duffy, D. M., Ko, C., Jo, M., Brannstrom, M., & Curry, T. E. (2019). Ovulation: Parallels with inflammatory processes. *Endocrine Reviews, 40*, 369–416. https://doi.org/10.1210/er.2018-00075.

Dumont, J. N., & Brummett, A. R. (1978). Oogenesis in *Xenopus laevis* (Daudin). V. Relationships between developing oocytes and their investing follicular tissues. *Journal of Morphology, 155*, 73–97. https://doi.org/10.1002/jmor.1051550106.

Egbert, J. R., Fahey, P. G., Reimer, J., Owen, C. M., Evsikov, A. V., Nikolaev, V. O., ... Jaffe, L. A. (2019). Follicle-stimulating hormone and luteinizing hormone increase Ca^{2+} in the granulosa cells of mouse ovarian follicles. *Biology of Reproduction, 101*, 433–444. https://doi.org/10.1093/biolre/ioz085.

Eppig, J. J. (1982). The relationship between cumulus cell-oocyte coupling, oocyte meiotic maturation, and cumulus expansion. *Developmental Biology, 89*, 268–272. https://doi.org/10.1016/0012-1606(82)90314-1.

Eppig, J. J., Wigglesworth, K., Pendola, F., & Hirao, Y. (1997). Murine oocytes suppress expression of luteinizing hormone receptor messenger ribonucleic acid by granulosa cells. *Biology of Reproduction, 56*, 976–984. https://doi.org/10.1095/biolreprod56.4.976.

Espey, L. L., Yoshioka, S., Russell, D. L., Robker, R. L., Fujii, S., & Richards, J. S. (2000). Ovarian expression of a disintegrin and metalloproteinase with thrombospondin motifs during ovulation in the gonadotropin-primed immature rat. *Biology of Reproduction, 62*, 1090–1095. https://doi.org/10.1095/biolreprod62.4.1090.

Fan, X., Bialecka, M., Moustakas, I., Lam, E., Torrens-Juaneda, V., Borggreven, N. V., ... de, S. (2019). Single-cell reconstruction of follicular remodeling in the human adult ovary. *Nature Communications, 10*, 1–13. https://doi.org/10.1038/s41467-019-11036-9.

Feng, Y., Piñon Gonzalez, V. M., Lin, M., Egertová, M., Mita, M., & Elphick, M. R. (2023). Localization of relaxin-like gonad-stimulating peptide expression in starfish reveals the gonoducts as a source for its role as a regulator of spawning. *The Journal of Comparative Neurology, 531*, 1299–1316. https://doi.org/10.1002/cne.25496.

Furman, A., Rotmensch, S., Kohen, F., Mashiach, S., & Amsterdam, A. (1986). Regulation of rat granulosa cell differentiation by extracellular matrix produced by bovine corneal endothelial cells. *Endocrinology, 118*, 1878–1885. https://doi.org/10.1210/endo-118-5-1878.

Gosden, R. G., & Telfer, E. E. (1987). Scaling of follicular sizes in mammalian ovaries. *Journal of Zoology, 211*, 157–168. https://doi.org/10.1111/j.1469-7998.1987.tb07459.x.

Gothié, S. (1967). Contribution à l'etude cinétique de l'ovulatoin chez la Souris prépubère. *Comptes Rendus des Seances de la Societe de Biologie et de ses Filiales, 161*, 554–556.

Guzmán, A., Hughes, C. H. K., & Murphy, B. D. (2022). Liver receptor homolog 1 (LRH-1) regulates follicle vasculature during ovulation in mice. *Reproduction (Cambridge, England), 164*, 283–289. https://doi.org/10.1530/REP-22-0135.

Hess, K. A., Chen, L., & Larsen, W. J. (1998). The ovarian blood follicle barrier Is both charge- and size-selective in mice. *Biology of Reproduction, 58*, 705–711. https://doi.org/10.1095/biolreprod58.3.705.

Huang, R., Kratka, C. E., Pea, J., McCann, C., Nelson, J., Bryan, J. P., ... Goods, B. A. (2024). Single-cell and spatiotemporal profile of ovulation in the mouse ovary. *BioRxiv Prepr. Serv. Biol.*, 2024.05.20.594719. https://doi.org/10.1101/2024.05.20.594719.

Hughes, C. H. K., & Murphy, B. D. (2021). Nuclear receptors: Key regulators of somatic cell functions in the ovulatory process. *Molecular Aspects of Medicine, 78*, 100937. https://doi.org/10.1016/j.mam.2020.100937.

Hunzicker-Dunn, M., & Mayo, K. (2015). *Gonadotropin signaling in the ovary. Knobil and Neill's physiology of reproduction*. Elsevier, 895–945. https://doi.org/10.1016/B978-0-12-397175-3.00020-X.

Ichikawa, S., Ohta, M., Morioka, H., & Murao, S. (1983). Blockage of ovulation in the explanted hamster ovary by a collagenase inhibitor. *Reproduction (Cambridge, England), 68*, 17–19. https://doi.org/10.1530/jrf.0.0680017.

Ito, M., Okamoto, R., Ito, H., Zhe, Y., & Dohi, K. (2022). Regulation of myosin light-chain phosphorylation and its roles in cardiovascular physiology and pathophysiology. *Hypertension Research, Official Publication of the Japanese Society of Hypertension, 45*, 40–52. https://doi.org/10.1038/s41440-021-00733-y.

Jaffe, L. A., & Egbert, J. R. (2017). Regulation of mammalian oocyte meiosis by intercellular communication within the ovarian follicle. *Annual Review of Physiology, 79*, 237–260. https://doi.org/10.1146/annurev-physiol-022516-034102.

Kim, J., Bagchi, I. C., & Bagchi, M. K. (2009). Control of ovulation in mice by progesterone receptor-regulated gene networks. *Molecular Human Reproduction, 15*, 821–828. https://doi.org/10.1093/molehr/gap082.

Ko, C., Gieske, M. C., Al-Alem, L., Hahn, Y., Su, W., Gong, M. C., Iglarz, M., & Koo, Y. (2006). Endothelin-2 in ovarian follicle rupture. *Endocrinology, 147*, 1770–1779. https://doi.org/10.1210/en.2005-1228.

Kovacic, N., & Parlow, A. F. (1972). Alterations in serum FSH/LH ratios in relation to the estrous cycle, pseudopregnancy, and gonadectomy in the mouse. *Endocrinology, 91*, 910–915. https://doi.org/10.1210/endo-91-4-910.

Kranzfelder, D., Reich, R., Abisogun, A. O., & Tsafriri, A. (1992). Preovulatory changes in the perifollicular capillary network in the rat: role of eicosanoids. *Biology of Reproduction, 46*, 379–385. https://doi.org/10.1095/biolreprod46.3.379.

Land, K. L., Lane, M. E., Fugate, A. C., & Hannon, P. R. (2021). Ovulation is inhibited by an environmentally relevant phthalate mixture in mouse antral follicles in vitro. *Toxicological Sciences: Official Journal of the Society of Toxicology, 179*, 195–205. https://doi.org/10.1093/toxsci/kfaa170.

Larsen, W. J., Wert, S. E., & Brunner, G. D. (1987). Differential modulation of rat follicle cell gap junction populations at ovulation. *Developmental Biology, 122*, 61–71. https://doi.org/10.1016/0012-1606(87)90332-0.

Lipner, H., & Cross, N. L. (1968). Morphology of the membrana granulosa of the ovarian follicle. *Endocrinology, 82*, 638–641. https://doi.org/10.1210/endo-82-3-638.

Lund, M., Pearson, A. C., Sage, M. A. G., & Duffy, D. M. (2023). Luteinizing hormone receptor promotes angiogenesis in ovarian endothelial cells of Macaca fascicularis and Homo sapiens. *Biology of Reproduction, 108*, 258–268. https://doi.org/10.1093/biolre/ioac189.

Lydon, J. P., DeMayo, F. J., Funk, C. R., Mani, S. K., Hughes, A. R., Montgomery, C. A., ... O'Malley, B. W. (1995). Mice lacking progesterone receptor exhibit pleiotropic reproductive abnormalities. *Genes & Development, 9*, 2266–2278. https://doi.org/10.1101/gad.9.18.2266.

Lyga, S., Volpe, S., Werthmann, R. C., Götz, K., Sungkaworn, T., Lohse, M. J., & Calebiro, D. (2016). Persistent cAMP signaling by internalized LH receptors in ovarian follicles. *Endocrinology, 157*, 1613–1621. https://doi.org/10.1210/en.2015-1945.

Macchiarelli, G., Jiang, J., Nottola, S. A., & Sato, E. (2006). Morphological patterns of angiogenesis in ovarian follicle capillary networks. A scanning electron microscopy study of corrosion cast. *Microscopy Research and Technique, 69*, 459–468. https://doi.org/10.1002/jemt.20305.

Mantri, M., Zhang, H. H., Spanos, E., Ren, Y. A., & De Vlaminck, I. (2024). A spatiotemporal molecular atlas of the ovulating mouse ovary. *Proceedings of the National Academy of Sciences, 121*, e2317418121. https://doi.org/10.1073/pnas.2317418121.

Martin, G. G., & Talbot, P. (1981). The role of follicular smooth muscle cells in hamster ovulation. *The Journal of Experimental Zoology, 216*, 469–482. https://doi.org/10.1002/jez.1402160316.

Martin, G. G., & Talbot, P. (1987). Formation of the rupture site in preovulatory hamster follicles: Morphological and morphometric analysis of thinning of the granulosa and thecal layers. *Gamete Research, 17*, 303–320. https://doi.org/10.1002/mrd.1120170404.

McKey, J., Cameron, L. A., Lewis, D., Batchvarov, I. S., & Capel, B. (2020). Combined iDISCO and CUBIC tissue clearing and lightsheet microscopy for in toto analysis of the adult mouse ovary. *Biology of Reproduction. 102*, 1080–1089. https://doi.org/10.1093/biolre/ioaa012.

Migone, F. F., Cowan, R. G., Williams, R. M., Gorse, K. J., Zipfel, W. R., & Quirk, S. M. (2016). In vivo imaging reveals an essential role of vasoconstriction in rupture of the ovarian follicle at ovulation. *Proceedings of the National Academy of Sciences of the United States of America, 113*, 2294–2299. https://doi.org/10.1073/pnas.1512304113.

Mita, M. (2023). Relaxin-like gonad-stimulating peptides in asteroidea. *Biomolecules, 13*, 781. https://doi.org/10.3390/biom13050781.

Mitsube, K., Brännström, M., & Haraldsson, B. (2013). Modulation of microvascular permeability in the preovulatory rat ovary by an ovulatory gonadotropin stimulus. *Fertility and Sterility, 99*, 903–909. https://doi.org/10.1016/j.fertnstert.2012.11.017.

Mittaz, L., Russell, D. L., Wilson, T., Brasted, M., Tkalcevic, J., Salamonsen, L. A., ... Pritchard, M. A. (2004). Adamts-1 is essential for the development and function of the urogenital system. *Biology of Reproduction, 70*, 1096–1105. https://doi.org/10.1095/biolreprod.103.023911.

Mogessie, B., Scheffler, K., & Schuh, M. (2018). Assembly and positioning of the oocyte meiotic spindle. *Annual Review of Cell and Developmental Biology, 34*, 381–403. https://doi.org/10.1146/annurev-cellbio-100616-060553.

Morris, M. E., Meinsohn, M.-C., Chauvin, M., Saatcioglu, H. D., Kashiwagi, A., Sicher, N. A., ... Pépin, D. (2022). A single-cell atlas of the cycling murine ovary. *eLife, 11*, e77239. https://doi.org/10.7554/eLife.77239.

Nicol, B., Rodriguez, K., & Yao, H. H.-C. (2020). Aberrant and constitutive expression of FOXL2 impairs ovarian development and functions in mice. *Biology of Reproduction, 103*, 966–977. https://doi.org/10.1093/biolre/ioaa146.

Norman, R. L., & Greenwald, G. S. (1972). Follicular histology and physiological correlates in the preovulatory hamster. *The Anatomical Record, 173*, 95–107. https://doi.org/10.1002/ar.1091730108.

Norris, R. P., Freudzon, M., Mehlmann, L. M., Cowan, A. E., Simon, A. M., Paul, D. L., ... Jaffe, L. A. (2008). Luteinizing hormone causes MAP kinase-dependent phosphorylation and closure of connexin 43 gap junctions in mouse ovarian follicles: One of two paths to meiotic resumption. *Development (Cambridge, England), 135*, 3229–3238. https://doi.org/10.1242/dev.025494.

Ohnishi, J., Ohnishi, E., Shibuya, H., & Takahashi, T. (2005). Functions for proteinases in the ovulatory process. *Biochimica et Biophysica Acta, 1751*, 95–109. https://doi.org/10.1016/j.bbapap.2005.05.002.

Omony, J. B., Biran, J., Kahwa, D., Aizen, J., Golan, M., Nyatia, E., ... Rutaisire, J. (2022). Cloning of gonadotropin Gph-alpha, FSH-beta and LH-beta subunits and seasonal profiles of steroid hormones in wild-caught Nile perch, Lates niloticus. *General and Comparative Endocrinology*, 323–324. https://doi.org/10.1016/j.ygcen.2022.114035.

Owen, C. M., & Jaffe, L. A. (2024). Luteinizing hormone stimulates ingression of mural granulosa cells within the mouse preovulatory follicle. *Biology of Reproduction, 142*. https://doi.org/10.1093/biolre/ioad142.

Owen, C. M., Zhou, X., Bernard, D. J., & Jaffe, L. A. (2021). Kisspeptin-54 injection induces a physiological luteinizing hormone surge and ovulation in mice. *Biology of Reproduction, 104*, 1181–1183. https://doi.org/10.1093/biolre/ioab067.

Oxberry, B. A., & Greenwald, G. S. (1982). An autoradiographic study of the binding of 125 I-labeled follicle-stimulating hormone, human chorionic gonadotropin and prolactin to the hamster ovary throughout the estrous cycle. *Biology of Reproduction, 27*, 505–516. https://doi.org/10.1095/biolreprod27.2.505.

Pakarainen, T., Zhang, F.-P., Nurmi, L., Poutanen, M., & Huhtaniemi, I. (2005). Knockout of luteinizing hormone receptor abolishes the effects of follicle-stimulating hormone on preovulatory maturation and ovulation of mouse graafian follicles. *Molecular Endocrinology (Baltimore, Md.), 19*, 2591–2602. https://doi.org/10.1210/me.2005-0075.

Palanisamy, G. S., Cheon, Y.-P., Kim, J., Kannan, A., Li, Q., Sato, M., ... Bagchi, I. C. (2006). A novel pathway involving progesterone receptor, endothelin-2, and endothelin receptor B controls ovulation in mice. *Molecular Endocrinology (Baltimore, Md.), 20*, 2784–2795. https://doi.org/10.1210/me.2006-0093.

Parr, E. L. (1974). Histological examination of the rat ovarian follicle wall prior to ovulation. *Biology of Reproduction, 11*, 483–503. https://doi.org/10.1095/biolreprod11.5.483.

Pibuel, M. A., Poodts, D., Sias, S. A., Byrne, A., Hajos, S. E., Franco, P. G., & Lompardía, S. L. (2023). 4-Methylumbelliferone enhances the effects of chemotherapy on both temozolomide-sensitive and resistant glioblastoma cells. *Scientific Reports, 13*, 9356. https://doi.org/10.1038/s41598-023-35045-3.

Powers, R. W., Chen, L., Russell, P. T., & Larsen, W. J. (1995). Gonadotropin-stimulated regulation of blood-follicle barrier is mediated by nitric oxide. *American Journal of Physiology – Endocrinology and Metabolism, 269*, E290–E298. https://doi.org/10.1152/ajpendo.1995.269.2.E290.

Quiroga Artigas, G., Lapébie, P., Leclère, L., Bauknecht, P., Uveira, J., Chevalier, S., ... Houliston, E. (2020). A G protein–coupled receptor mediates neuropeptide-induced oocyte maturation in the jellyfish Clytia. *PLoS Biology, 18*, e3000614. https://doi.org/10.1371/journal.pbio.3000614.

Richards, J. S., & Ascoli, M. (2018). Endocrine, paracrine, and autocrine signaling pathways that regulate ovulation. *Trends in Endocrinology & Metabolism TEM, 29*, 313–325. https://doi.org/10.1016/j.tem.2018.02.012.

Richards, J. S., Ireland, J. J., Rao, M. C., Bernath, G. A., Midgley, A. R., & Reichert, L. E. (1976). Ovarian follicular development in the rat: Hormone receptor regulation by estradiol, follicle stimulating hormone and luteinizing hormone. *Endocrinology, 99*, 1562–1570. https://doi.org/10.1210/endo-99-6-1562.

Robker, R. L., Hennebold, J. D., & Russell, D. L. (2018). Coordination of ovulation and oocyte maturation: A good egg at the right time. *Endocrinology, 159*, 3209–3218. https://doi.org/10.1210/en.2018-00485.

Salustri, A., Campagnolo, L., Klinger, F. G., & Camaioni, A. (2019). Molecular organization and mechanical properties of the hyaluronan matrix surrounding the mammalian oocyte. *Matrix Biology: Journal of the International Society for Matrix Biology, 78–79*, 11–23. https://doi.org/10.1016/j.matbio.2018.02.002.

Salustri, A., Yanagishita, M., Underhill, C. B., Laurent, T. C., & Hascall, V. C. (1992). Localization and synthesis of hyaluronic acid in the cumulus cells and mural granulosa cells of the preovulatory follicle. *Developmental Biology, 151*, 541–551. https://doi.org/10.1016/0012-1606(92)90192-J.

Shrestha, K., Al-Alem, L., Garcia, P., Wynn, M. A. A., Hannon, P. R., Jo, M., Drnevich, J., ... Curry Jr, T. E. (2023). Neurotensin expression, regulation, and function during the ovulatory period in the mouse ovary. *Biology of Reproduction, 108*, 107–120. https://doi.org/10.1093/biolre/ioac191.

Sirois, J., Sayasith, K., Brown, K. A., Stock, A. E., Bouchard, N., & Doré, M. (2004). Cyclooxygenase-2 and its role in ovulation: A 2004 account. *Human Reproduction Update, 10*, 373–385. https://doi.org/10.1093/humupd/dmh032.

Skorupskaite, K., George, J. T., & Anderson, R. A. (2014). The kisspeptin-GnRH pathway in human reproductive health and disease. *Human Reproduction Update, 20*, 485–500. https://doi.org/10.1093/humupd/dmu009.

Skory, R. M., Xu, Y., Shea, L. D., & Woodruff, T. K. (2015). Engineering the ovarian cycle using in vitro follicle culture. *Human Reproduction (Oxford, England), 30*, 1386–1395. https://doi.org/10.1093/humrep/dev052.

Stocco, C., Telleria, C., & Gibori, G. (2007). The molecular control of corpus luteum formation, function, and regression. *Endocrine Reviews, 28*, 117–149. https://doi.org/10.1210/er.2006-0022.

Stouffer, R. L., & Hennebold, J. D. (2015). *Structure, function, and regulation of the corpus luteum. Knobil and Neill's physiology of reproduction*. Elsevier Inc, 1023–1076. https://doi.org/10.1016/B978-0-12-397175-3.00023-5.

Su, Y.-Q., Sugiura, K., Wigglesworth, K., O'Brien, M. J., Affourtit, J. P., Pangas, S. A., ... Eppig, J. J. (2008). Oocyte regulation of metabolic cooperativity between mouse cumulus cells and oocytes: BMP15 and GDF9 control cholesterol biosynthesis in cumulus cells. *Development (Cambridge, England), 135*, 111–121. https://doi.org/10.1242/dev.009068.

Szołtys, M., Galas, J., Jabłonka, A., & Tabarowski, Z. (1994). Some morphological and hormonal aspects of ovulation and superovulation in the rat. *The Journal of Endocrinology, 141*, 91–100. https://doi.org/10.1677/joe.0.1410091.

Takahashi, T., Hagiwara, A., & Ogiwara, K. (2018). Follicle rupture during ovulation with an emphasis on recent progress in fish models. *Reproduction (Cambridge, England)*. https://doi.org/10.1530/REP-18-0251.

Tanaka, N., Espey, L. L., & Okamura, H. (1989). Increase in ovarian blood volume during ovulation in the gonadotropin-primed immature rat. *Biology of Reproduction, 40*, 762–768. https://doi.org/10.1095/biolreprod40.4.762.

Thomas, C., Marx, T. L., Penir, S. M., & Schuh, M. (2024). Ex vivo imaging reveals the spatiotemporal control of ovulation. *Nature Cell Biology* (in press).

Trushko, A., Di Meglio, I., Merzouki, A., Blanch-Mercader, C., Abuhattum, S., Guck, J., ... Roux, A. (2020). Buckling of an epithelium growing under spherical confinement. *Developmental Cell, 54*, 655–668.e6. https://doi.org/10.1016/j.devcel.2020.07.019.

Tsafriri, A., Lindner, H. R., Zor, U., & Lamprecht, S. A. (1972). In vitro induction of meiotic division in follicle-enclosed rat oocytes by LH, cyclic AMP and prostaglandin E2. *Reproduction (Cambridge, England), 31*, 39–50. https://doi.org/10.1530/jrf.0.0310039.

Vigone, G., Shuhaibar, L. C., Egbert, J. R., Uliasz, T. F., Movsesian, M. A., & Jaffe, L. A. (2018). Multiple cAMP phosphodiesterases act together to prevent premature oocyte meiosis and ovulation. *Endocrinology, 159*, 2142–2152. https://doi.org/10.1210/en.2018-00017.

Zhuo, L., Yoneda, M., Zhao, M., Yingsung, W., Yoshida, N., Kitagawa, Y., ... Kimata, K. (2001). Defect in SHAP-hyaluronan complex causes severe female infertility: A study by inactivation of the bikunin gene in mice. *The Journal of Biological Chemistry, 276*, 7693–7696. https://doi.org/10.1074/jbc.C000899200.

CHAPTER NINE

Cellular and molecular regulations of oocyte selection and activation in mammals

Xuebing Yang, Yan Zhang, and Hua Zhang*
State Key Laboratory of Animal Biotech Breeding, College of Biological Sciences, China Agricultural University, Beijing, P.R. China
*Corresponding author. e-mail address: huazhang@cau.edu.cn

Contents

Abstract

Oocytes, a uniquely pivotal cell population, play a central role in species continuity. In mammals, oogenesis involves distinct processes characterized by sequential rounds of selection, arrest, and activation to produce a limited number of mature eggs, fitting their high-survival yet high-cost fertility. During the embryonic phase, oocytes undergo intensive selection via cytoplasmic and organelle enrichment, accompanied by the onset and arrest of meiosis, thereby establishing primordial follicles (PFs) as a finite reproductive reserve. Subsequently, the majority of primary oocytes enter a dormant state and are gradually recruited through a process termed follicle activation, essential for maintaining orderly fertility. Following activation, oocytes undergo rapid growth, experiencing cycles of arrest and activation regulated by endocrine and paracrine signals, ultimately forming fertilizable eggs. Over the past two decades, advancements in genetically modified animal models, high-resolution imaging, and omics technologies have significantly enhanced our understanding of the cellular and

Current Topics in Developmental Biology, Volume 162
ISSN 0070-2153, https://doi.org/10.1016/bs.ctdb.2024.11.003

molecular mechanisms that govern mammalian oogenesis. These advances offer profound insights into the regulatory mechanisms of mammalian reproduction and associated female infertility disorders. In this chapter, we provide an overview of current knowledge in mammalian oogenesis, with a particular emphasis on oocyte selection and activation in vivo.

1. Introduction

Reproduction is fundamental to the biological continuity of living organisms. In mammals, successful reproduction depends on the generation of specialized gametes by each sex, which merge to form a zygote, thereby initiating the development of a new organism. Males and females, distinguished by their roles in species perpetuation, have evolved distinct yet sophisticated patterns of gamete production. In males, the testes continuously produce a large number of sperms to ensure reproductive success at the critical moment of winning reproductive competition. Conversely, females generate a finite number of oocytes during the embryonic period, which are then meticulously recruited over time, thereby significantly reducing the energy consumed in gamete formation after reaching adulthood, while ensuring the continuous production of mature oocytes throughout their extended reproductive lifespan (Pedersen & Peters, 1968; Zhang et al., 2012; Zhang, Adhikari, Zheng, & Liu, 2013).

Given that all oocytes are formed during the embryonic stage, mammalian female reproduction faces the critical challenge of maintaining oocyte health over an extended lifespan. To address this, oogenesis in female mammals has evolved distinct phase-specific developmental characteristics. These features are evident not only during well-known ovulatory phases of the menstrual or estrus cycle but also throughout the entire lifecycle, involving the formation, growth, and maturation of oocytes (McGee & Hsueh, 2000). During this process, oocytes undergo multiple rounds of selection, arrest, and activation to balance the storage and consumption of oocyte reserves, which is essential for maintaining a high-quality of ovulated oocytes in an extended reproductive lifespan in females (Endo, Mikedis, Nicholls, Page, & de Rooij, 2019; Hsueh, Kawamura, Cheng, & Fauser, 2015; McGee & Hsueh, 2000).

Examining the developmental dynamics of oocytes over the lifespan within the ovaries, the quantity of oocytes peaks during embryonic development, after which the majority are sacrificed to nurture a select few that form primordial follicles (PFs) as the ovarian reserve (Kezele, Nilsson,

& Skinner, 2002; Pepling, 2006). These selected surviving oocytes in PFs are arrested at the prophase of the first meiotic division (meiosis I) and enter a dormant state, serving as a reproductive reserve for adulthood. Postpartum, dormant oocytes are gradually activated in coordination with follicle recruitment and growth, ultimately maturing into fertilizable eggs. As the ovarian reserve declines over time, women transition into reproductive aging and eventually reach menopause (Adhikari & Liu, 2009). Therefore, the processes of arrest and activation of oocytes directly determine the reproductive lifespan of individual females.

From the perspective of individual gametogenesis, in contrast to spermatogenesis, which is characterized by a "subtraction" process to form streamlined sperm, oocyte maturation follows an "addition" process through several rounds of selection and activation, equipping them with the necessary energy and materials required for early embryonic development (Fig. 1). Thus, from a cell biology perspective, the development and maturation of oocytes involve direct early intercellular competition to select oocytes that will survive long-term, whereas in adulthood, complex interactions between oocytes and surrounding somatic cells facilitate their maturation at both the cytoplasmic and nuclear levels. These processes ensure not only the reproductive lifespan but also the reproductive quality in females.

In this review, we summarize recent advancements in the understanding of oogenesis, emphasizing the crucial balance between selection, arrest, and activation of oocytes as a key factor in maintaining the female reproductive lifespan. By synthesizing current research, we aim to provide a comprehensive overview of how these processes interplay to affect reproductive health and longevity, offering insights that may inform future studies and potential therapeutic approaches.

2. Oocyte selection and arrest before the establishment of female reproductive reserve in fetal ovaries

The process of oogenesis begins with primordial germ cells (PGCs) transforming into oogonia within the differentiated gonads. Subsequently, these oogonia rapidly enter meiosis and are then referred to as oocytes. This transition marks the first, and perhaps the most important, event where the developmental pathways of male and female gametes diverge, occurring

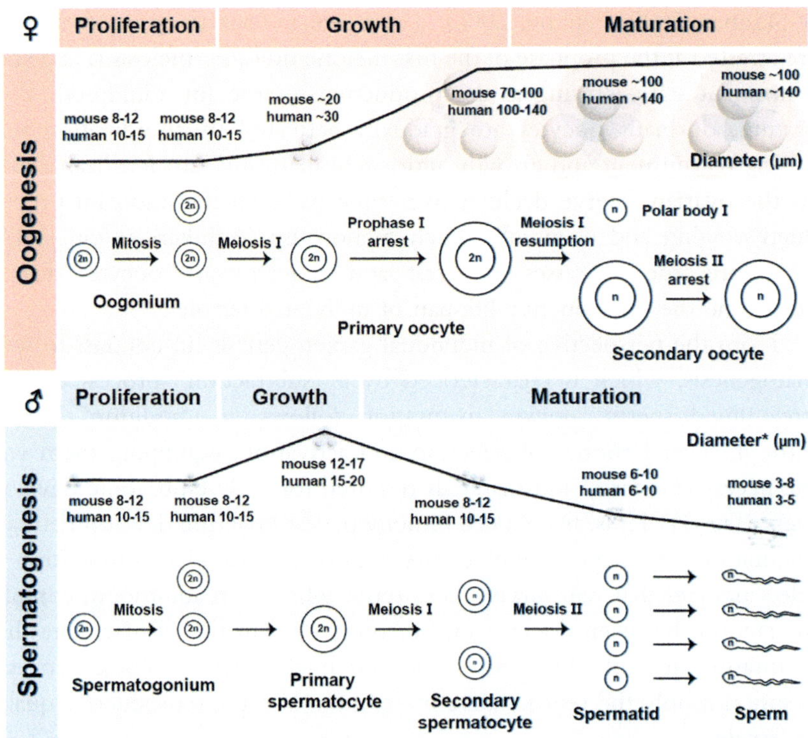

Fig. 1 Overview of mammalian gametogenesis, highlighting gender differences in germ cell development and size dynamics. This figure illustrates the developmental processes of oogenesis and spermatogenesis in mammals, highlighting contrasting changes in germ cell size between females and males. Oogenesis begins with oogonia in the ovaries, which proliferate through mitosis in the embryonic ovaries. These oogonia develop into primary oocytes that enter and arrest in meiosis I prophase until puberty. At puberty, each menstrual cycle prompts some primary oocytes to resume meiosis I, dividing to form a large secondary oocyte and a smaller first polar body. The secondary oocyte then initiates meiosis II but halts at metaphase II, becoming available for fertilization upon ovulation. Throughout oogenesis, female germ cells progressively increase in size, reaching their largest form as mature eggs. In contrast, spermatogenesis in males involves a marked reduction in germ cell size. Primary spermatocytes divide to produce two haploid secondary spermatocytes, which subsequently undergo a second meiotic division (meiosis II) to yield four spermatids. These spermatids then mature into sperm, which are notably smaller. The labels "n" and "2 n" denote the ploidy level (number of chromosome sets) at each stage. Additionally, "diameter* " refers specifically to the width of the sperm head rather than its overall length.

during a very short window in the embryonic period. Taking the well-studied model animal, the mouse, as an example, it is well-documented that this process occurs from 12.5 days post coitum (dpc) in fetus until 5 days postpartum (dpp), spanning an approximately two-week window (O'Connell & Pepling, 2021; Pepling, 2006). During this period, female germ cells undergo entry into meiosis and arrest, and oocytes are subjected to selection where about two-thirds are eliminated, eventually establishing a fixed ovarian reserve through a complex developmental process. For female individuals, the orderly occurrence of meiotic arrest and oocyte selection during this process directly determines the number and quality of oocytes available after reaching sexual maturity (Grive & Freiman, 2015).

2.1 The selection of oocytes during the PF formation

Selecting optimal germ cells by sacrificing others to form fertilizable eggs is a conserved process across sexual reproductive species, from worms to mammals. In mammals, about two-thirds of the oocytes undergo cell death during the selection process in the embryonic period, leaving only a small number to constitute the individual's lifetime reproductive reserve (Findlay, Hutt, Hickey, & Anderson, 2015; Lei & Spradling, 2013). In *Drosophila*, only one out of sixteen germ cells ultimately survive to become a mature oocyte. It is well-accepted that the selection process involves not merely the death of some oocytes and the survival of others but also the transfer of organelles and other cytoplasmic materials from the sacrificed oocytes to the surviving ones. Therefore, the "selection and nursing" mechanism ensures and enhances the initial quality of the surviving oocytes.

Studies on oogenesis using invertebrate models, particularly *Drosophila*, reveal that germline cysts—groups of germ cells derived from one progenitor cell and interconnected by intercellular bridges (IBs)—are believed to be the structure base for the crucial events during early oogenesis, including meiosis, the formation of the fusome, oocyte selection, and nursing (Bolivar et al., 2001; Cox & Spradling, 2003; Spradling, Niu, Yin, Pathak, & Maurya, 2022). Following four rounds of mitotic divisions to form a sixteen germ-cell cyst, one germ cell in the cyst is designated as the oocyte based on asymmetric cytoplasm transfer from the other 15 cyst cells, which become nurse cells (NCs). A simple selection model shows that microtubule minus ends of fusome, a microtubule-dependent organelle, orient to transport cargos from any germ cell toward the initial cell, which becomes the mature oocyte (Spradling et al., 2022). Recent study has discovered that a novel key protein, Patronin/CAMSAP, plays a critical role in determining the fate of *Drosophila* oocytes by regulating the formation of a non-centrosomal microtubule network (Nashchekin, Busby, Jakobs, Squires, & St, 2021). Moreover, NCs support the oocyte by producing piRNAs, promoting oocyte growth, and establishing oocyte polarity and development (Guzzardo, Muerdter, & Hannon, 2013; Milas & Telley, 2022; Spradling et al., 2022; Wang & Lin, 2021). Disruption of germline cyst structure or fusome function leads to the failure to produce normal oocytes and results in female sterility. Overall, the connection and communication of oocytes through IBs are essential for oocyte fate determination in *Drosophila* (Spradling et al., 2022).

In mammals, early oocyte selection is less understood. Lei and Spradling proposed a mechanism following their tracing of oocyte selection in mouse fetal ovaries: initial large cysts fragment into smaller ones, each selecting a dominant oocyte while others become NCs and later die (Ikami et al., 2023; Lei & Spradling, 2013; Lei & Spradling, 2016). Recently, Niu et al. investigated the molecular mechanisms involved in cyst breakdown and oocyte selection process using single-cell RNA sequencing (Niu & Spradling, 2022). Genes in both NCs and oocytes begin to express differently at 14.5 dpc when cyst structures just form and germ cells enter into meiosis. NCs from 14.5 dpc to postnatal day (PD) 1 have less mRNA content than other oocytes. Specific genes, such as the oocyte surviving related transcription factor *NOBOX oogenesis homeobox* (*Nobox*), are expressed specifically in one oocyte within cysts as early as 18.5 dpc. Additionally, *follistatin* (*Fst*) is upregulated, and the component of IBs *testis expressed gene 14* (*Tex14*) is downregulated in NCs, playing roles in oocyte selection process (Niu & Spradling, 2022). Beside the recent study using omics technology, previous studies have also identified various factors that are involved in the regulations of germline cyst breakdown in mice using an in vitro fetal ovarian culture system. Notably, all these studies suggested that disrupting the normal cyst breakdown and oocyte selection result in a failure of the construction of ovarian reserve, highlighting the early selection is indispensable for the female reproduction.

Although early oocyte selection is widely accepted as a conserved process across different species, the mechanism in mammals exhibits some notable and distinct characteristics. Firstly, the migration of mammalian PGCs into the gonads is not synchronized, resulting in the asynchronous formation of germline cyst or the entry into meiosis in different regions of the gonads (Lei & Spradling, 2013). Additionally, tracking studies using genetically modified reporter mouse model have shown that germ cells may undergo varying rounds of mitotic divisions, leading to an inconsistent number of germ cells within different cysts (Lei & Spradling, 2013). Meanwhile, the survival ratio of oocytes in mammals is much higher than it in the fly. Moreover, *Tex14* knockout female mice, despite lacking IBs, remain fertile, suggesting that IBs or even cyst structures may not be essential for oocyte selection in mammals as they are in flies (Greenbaum, Iwamori, Agno, & Matzuk, 2009; Ikami et al., 2021). This indicates that cyst-dependent mechanisms might not be the only ones at play in mammals. Given the extended lifespan of mammalian oocytes, it is likely that other specific strategies have evolved to enhance oocyte quality, aligning with the longer reproductive lifespan in mammals.

2.2 Onset and arrest of meiosis

During mammalian sexual differentiation, a crucial event is the timely entry of germ cells into meiosis within the fetal ovaries. After migrating to the gonads, PGCs undergo several rounds of mitosis to increase their numbers (Monk & McLaren, 1981). In females, these germ cells must enter meiosis, making oocytes a non-renewable resource in adulthood (Zhang et al., 2014). In contrast, male germ cells remain in a mitotic state, allowing continuous sperm production postnatally. Although entering meiosis is not mandatory for germ cells to develop into oocyte-like cells, failure to do so compromises their long-term survival in the ovaries (Dokshin, Baltus, Eppig, & Page, 2013). Therefore, the initiation of meiosis is crucial for establishing the foundation of oogenesis, including the subsequent stages of arrest, dormancy, and activation that define the female reproductive lifespan.

The initiation of meiosis in female germ cells is primarily regulated by retinoic acid (RA), which is produced by both the mesonephros and the ovaries (Bowles et al., 2006; Bowles et al., 2016; Chassot et al., 2020; Koubova et al., 2006; Mu et al., 2013). Around 12.5 dpc, RA enters the gonads and creates a concentration gradient along the ovarian axis. This gradient stimulates the expression of the "*stimulated by RA gene 8*" (*Stra8*) gene, which is essential for triggering the onset of meiosis (Bowles & Koopman, 2007). Consequently, germ cells in the anterior region of the ovary begin meiosis first, followed by those further along the ovarian axis, resulting in an anterior-to-posterior wave of meiotic initiation (Bullejos & Koopman, 2004). Interestingly, recent three-dimensional imaging studies have revealed an additional radial wave preceding this pattern (Soygur et al., 2021). This radial wave highlights the role of IBs in coordinating the transition of germ cells within the cyst from a pluripotent state to meiosis, ensuring a synchronized and orderly progression (Soygur et al., 2021).

This process of meiotic entry is tightly controlled by epigenetic factors, notably Polycomb repressive complex 1 (PRC1) (Hu et al., 2022; Yokobayashi et al., 2013). PRC1 serves as a crucial epigenetic regulator that prevents premature RA-induced meiosis in PGCs by silencing key meiosis-associated genes, including *Stra8*. When the RA signal reaches a sufficient threshold, the repression by PRC1 is lifted, allowing effective RA-mediated initiation of meiosis (Hu et al., 2022). During this transition, extensive epigenetic modifications, such as DNA methylation and histone modifications, occur to regulate chromatin state and gene expression, thereby ensuring proper developmental progression and synchronization of meiotic entry across the ovarian axis (Fu et al., 2017).

While RA is a key player in meiotic initiation, it is not the sole regulator. Studies have shown that meiosis can still proceed in ovaries lacking all RA receptors, suggesting the existence of additional regulatory pathways (Vernet et al., 2020). One such pathway involves bone morphogenetic protein (BMP) signaling (Miyauchi et al., 2017). The downstream effector of BMP, Zinc finger GATA like protein-1 (ZGLP1), is necessary for STRA8-mediated activation of meiotic genes (Nagaoka et al., 2020). In addition, the STRA8-interacting proteins meiosis initiator (MEIOSIN) and Retinoblastoma (RB) have also been identified as crucial for the timely entry of oocytes into meiosis (Ishiguro et al., 2020; Shimada et al., 2023). These findings indicate that successful initiation of meiosis requires a coordinated network of RA, BMP, and possibly other signaling pathways working together to ensure precise regulation.

Once oocytes enter meiosis, they become arrested at the dictyate stage of prophase I, where they remain until the oocyte grows and resumes meiosis prior to ovulation (Borum, 1961). During this prolonged arrest, PRC1 plays a critical role in establishing a repressive chromatin environment that suppresses the expression of genes required for meiotic progression (Hu et al., 2022). The oocytes maintain a low level of DNA methylation until they resume growth, which is then accompanied by increased histone modifications and global DNA methylation. This extended arrest is essential for preserving oocyte dormancy throughout the female reproductive lifespan, but it also presents challenges for maintaining oocyte quality over time. For example, protein phosphatase 6 (PP6) is known to protect genome integrity during this extended meiotic arrest, which is crucial for oocyte survival (Hu et al., 2016).

The maintenance of this prolonged meiotic arrest is predominantly regulated by high intracellular levels of cyclic adenosine monophosphate (cAMP) (Pan & Li, 2019). The Gs protein–coupled receptor 3 (GPR3)-Gs-adenylyl cyclase (ADCY) signaling pathway is responsible for generating and sustaining these elevated cAMP levels, which activate protein kinase A (PKA). Activated PKA then inactivates maturation-promoting factors (MPFs), ensuring that the oocytes remain arrested in meiosis I (Adhikari et al., 2016). In addition, the inhibition of phosphodiesterase 3 (PDE3), which degrades cAMP, is crucial for maintaining these elevated cAMP levels (Vaccari, Horner, Mehlmann, & Conti, 2008). This inhibition is facilitated by cyclic guanosine monophosphate (cGMP) produced by adjacent granulosa cells through the natriuretic peptide precursor type C (NPPC)/ natriuretic peptide receptor 2 (NPR2) signaling system (Zhang, Su, Sugiura, Xia, & Eppig, 2010).

While much of the research on meiotic arrest focuses on the final stages of oocyte maturation, the mechanisms regulating early meiotic arrest remain less understood. Studies on perinatal mice suggest that elevated cAMP levels are also involved in regulating early meiotic progression (Wang et al., 2015). Continued research into the mechanisms of long-term meiotic arrest is necessary to gain a deeper understanding of the unique biological processes underlying female reproduction.

3. Oocytes in PF: a balance of dormancy and activation

After the selection of oocytes, while the surviving oocytes become enriched with cytoplasm and organelles from NCs, they remain in an intermediate developmental state. At this state, in terms of cell size and critical organelles, these oocytes account for only about one percent of the volume and possess merely one one-thousandth of the mitochondria compared to mature eggs. Furthermore, these oocytes, arrested in the dictyate stage of meiosis I, are characterized by low transcriptional and translational activity, indicative of a dormant state. Notably, maintaining this intermediate dormant state solely through the oocytes themselves is a considerable challenge. The formation of PFs is a crucial prerequisite for oocyte dormancy.

3.1 Oocyte dormancy entering and PF formation

During the embryonic development of the ovaries, oocytes arrest at the dictyate stage of prophase I and begin to interact with pre-granulosa cells, initiating the formation of PFs (Childs, Cowan, Kinnell, Anderson, & Saunders, 2011; Le Bouffant et al., 2010). This process establishes the PF pool, which serves as the lifelong reproductive reserve. The timing of PF formation varies significantly across species. In mice, the process commences around embryonic day 17.5 (E17.5) and continues until PD3–5, by which time the PF pool is fully established (Mork et al., 2012; Pepling, 2012). In contrast, in humans, PF formation begins in utero as early as 16 weeks of gestation (Kurilo, 1981). However, the exact timing of the complete establishment of the PF pool remains a topic of debate. Recent studies in non-human primate models, such as cynomolgus monkeys, have shown that PF formation begins before E90 and is completed within two weeks post-birth (Tu et al., 2022). The encapsulation of oocytes within PFs shields them from the surrounding ovarian microenvironment, which is critical for maintaining oocyte dormancy.

Structurally, a PF consists of a single dormant oocyte surrounded by a monolayer of flattened pre-granulosa cells. The formation of PFs thus involves both the oocytes and the recruitment of pre-granulosa cells. Cell lineage tracing studies have revealed that pre-granulosa cells have multiple origins, primarily deriving from the mesonephros and the ovarian surface epithelium (Mork et al., 2012). Despite their varied origins, these cells uniformly express essential transcription factors such as *Forkhead box L2* (*Foxl2*), and function identically once differentiated into pre-granulosa cells (Mork et al., 2012; Uhlenhaut et al., 2009).

The formation of PFs is regulated by a complex network of signaling pathways, including wingless-related integration site (WNT), NOTCH, KIT, and c-Jun amino-terminal kinase (JNK), as well as various molecular factors (Grive & Freiman, 2015; Pepling, 2006, 2012; Wang, Zhou, & Xia, 2017). Among these, a series of transcription factors that are specifically expressed in oocytes or granulosa cells have been shown to play pivotal roles in the regulation of PF formation (Jagarlamudi & Rajkovic, 2012). Recent advancements in omics technologies have facilitated comprehensive analyses of transcriptional and epigenetic landscapes during this critical developmental process, providing deeper insights into the molecular mechanisms involved (O'Connell & Pepling, 2021). For instance, Wang et al. demonstrated that the transcription factors *Factor in the germline alpha* (*Figla*), *LIM homeobox protein 8* (*Lhx8*), and *spermatogenesis- and oogenesis-specific helix-loop-helix transcription factor 1* (*Sohlh1*) not only regulate each other but also form a nuclear complex within oocytes to coordinate PF assembly (Wang, Liu, Zhao, & Dean, 2020). Additionally, cis-regulome mapping in mouse oocytes has identified other key transcription factors, such as *transcription factor 3/12* (*Tcf3/12*), which interact with cis-regulatory elements and contribute significantly to the regulation of folliculogenesis (Liu et al., 2024). These transcription factors function in concert with *Figla* and *Lhx8*, highlighting a sophisticated and highly coordinated transcriptional network essential for the proper development of PFs. Together, these findings emphasize the critical roles of oocyte- and granulosa cell-specific transcriptional regulators in the successful formation and establishment of PFs.

The duration of oocyte dormancy within PFs varies greatly among species. In mice, oocyte dormancy can last for up to two years, whereas in humans, it extends for more than 50 years. Studies have shown that the maintenance of oocyte dormancy over such prolonged periods is closely associated with exceptional protein stability and homeostasis. Oocytes achieve this by sequestering protein aggregates in endolysosomal vesicular assemblies and inhibiting mitochondrial complex I to maintain a reactive oxygen species (ROS)-free environment (Rodriguez-Nuevo et al., 2022; Zaffagnini et al., 2024). These mechanisms not only preserve oocyte quality but are also vital for sustaining female fertility over extended reproductive lifespans. Recent research has further elucidated that extreme protein longevity in oocytes is a key factor in preserving their functional integrity and viability during long-term dormancy (Harasimov et al., 2024).

The strategy of forming a finite pool of PFs during early development and maintaining them in a dormant state throughout the reproductive lifespan is a fundamental aspect of mammalian reproduction. This strategy minimizes the need for continuous germ cell production in adulthood, thereby conserving metabolic energy—especially compared to the high energetic costs of gametogenesis observed in lower animals such as insects and fish. However, this approach also imposes a relatively short reproductive lifespan on females compared to their overall lifespan, potentially serving as an evolutionary mechanism to ensure offspring quality.

3.2 Molecular regulations of oocyte activation in PFs

After the establishment of the PF pool as reproductive reserve, a key challenge in maintaining female reproductive potential is the orderly recruitment of these dormant follicles to balance fertility and reproductive lifespan. This process, known as PF activation or initial recruitment, involves the transition of oocytes from dormancy to an active growth phase, accompanied by the differentiation and proliferation of pre-granulosa cells (Adhikari & Liu, 2009; McGee & Hsueh, 2000). During this activation, oocytes significantly increase in volume and enhance their transcriptional activity, while pre-granulosa cells differentiate into granulosa cells and undergo rapid proliferation, changing in morphology from flattened to cuboidal. Studies in both mice and humans have shown that the activation of PFs is regulated through a complex interplay between oocytes, pre-granulosa cells, and the ovarian microenvironment. Under physiological conditions, this process is primarily initiated by the mechanistic target of rapamycin complex 1 (mTORC1) signaling in pre-granulosa cells (Zhang et al., 2014). In dormant follicles, the mTORC1 pathway is inhibited; however, upon receiving activation signals, mTORC1 activity in pre-granulosa cells is stimulated, promoting their differentiation and proliferation. This activation also leads to the upregulation of KIT ligand (KITL) expression, which, upon binding to KIT receptors on oocyte surfaces, activates the phosphatidylinositol 3 kinase (PI3K)/ thymoma viral proto-oncogene 1 (AKT) signaling pathway within the oocyte (Zhang et al., 2014). Subsequently, activated AKT phosphorylates and sequesters the transcription factor Forkhead box O3 (FOXO3) in the cytoplasm, preventing it from initiating the cell cycle inhibitor $p27^{Kip1}$, thereby enabling the resumption of oocyte growth (Zhang & Liu, 2015). Thus, PF activation in vivo is predominantly regulated by the mTORC1-KITL-KIT-PI3K/AKT signaling cascade within both pre-granulosa cells and oocytes (Zhang & Liu, 2015). Dysregulation of this pathway can lead to aberrant follicle activation. For example, deletion of *Rptor*, a crucial component of the mTORC1 complex in pre-granulosa cells, or *3-phosphoinositide-dependent kinase-1* (*Pdk1*), a key activator of the PI3K pathway in oocytes, inhibits PF activation, resulting in a significant loss of follicles and primary ovarian failure (POF) (Gorre et al., 2014). Conversely, knockout of negative regulators of mTORC1, such as *tuberous sclerosis complex 1* (*Tsc1*) and *tuberous sclerosis complex 2* (*Tsc2*) in pre-granulosa cells, or *phosphatase and tensin homolog deleted on chromosome 10* (*Pten*) in oocytes, causes premature and global activation of the PF pool before puberty (John, Gallardo, Shirley, & Castrillon, 2008; Reddy et al., 2008). In humans, abnormalities in the TSC/mTOR pathway are also associated with excessive PF activation, contributing to premature ovarian insufficiency (POI) (Xu et al., 2022). Therefore, precise regulation of mTORC1 signaling in pre-granulosa cells and PI3K/AKT signaling in oocytes is crucial for maintaining the delicate balance between dormancy and activation of PFs.

In addition to the canonical mTORC1-KITL-KIT-PI3K/AKT pathway, several upstream regulators and alternative signaling pathways are involved in PF activation (Fig. 2). Notably, dormant PFs are located in the ovarian cortex and migrate to the medulla upon activation. Recent studies have highlighted the importance of the local microenvironment, including hypoxia and extracellular matrix (ECM)-mediated mechanical stress, in regulating PF dormancy and activation. Under hypoxic conditions, increased expression of hypoxia-inducible factors (HIFs) maintains the nuclear localization of FOXO3 in oocytes, thereby supporting the

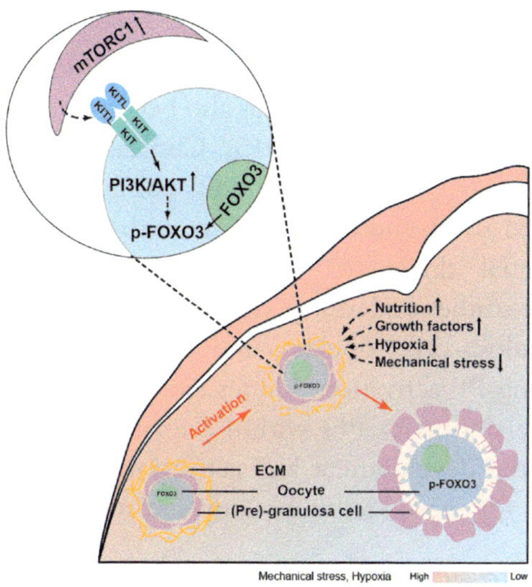

Factors involved in activation

cAMP	Zheng et al., 2023
CDC42	Yan et al., 2018
CDK6	Ataei-Nazari et al., 2022
E-cadherin	Yan et al., 2019
EGF	Zhang et al., 2020
FOXL2	Schmidt et al., 2004
Glycolysis	Zhang et al., 2022
MAPK3/1	Zhao et al., 2018
NGF	Zhang et al., 2024
NOBOX	Rajkovic et al., 2004
PARP1	Chen et al., 2023
SOHLH1/2	Choi et al., 2008
WNT signaling	Habara et al., 2021

Factors involved in inhibiting activation

AMH	Tanimoto et al., 2021
CLPP	Wang et al., 2018
HDAC6	Zhang et al, 2021
HIFs	Shimamoto et al., 2019
Hippo-YAP1	Hu et al., 2019
LHX8	Ren et al., 2015
LKB1-AMPK	Jiang et al., 2016
Mechanical stress	Nagamatsu et al., 2019
$p27^{Kip1}$-CDK	Rajareddy et al., 2007
PTEN	Reddy et al., 2008
TGFβ	Wang et al., 2014
TSC1	Zhang et al., 2014

Fig. 2 Key factors and signaling pathways involved in the regulation of oocyte activation in PFs. Left: PFs reside in the ovarian cortex, where their surrounding environment—including ECM-mediated high-intensity mechanical stress and hypoxia—helps maintain low mTORC1 signaling activity in pre-granulosa cells, preserving the dormancy of PFs. With increased nutrient availability, elevated growth factors, and reduced hypoxia and mechanical stress, the mTORC1 signaling pathway in pre-granulosa cells becomes activated. This activation promotes the differentiation and proliferation of pre-granulosa cells, leading to the secretion of KITL, which binds to KIT receptors on the oocyte surface. This interaction activates the PI3K signaling pathway in the dormant oocyte, ultimately triggering its activation. During this process, FOXO3 translocates from the nucleus to the cytoplasm in the oocyte, releasing it from dormancy. Right: Summary of key factors reported over the past two decades that either promote or inhibit oocyte activation. (Ataei-Nazari et al., 2022; Chen et al., 2023; Choi et al., 2008; Habara et al., 2021; Hu et al., 2019; Jiang et al., 2016; Rajareddy et al., 2007; Rajkovic et al., 2004; Ren et al., 2015; Schmidt et al. 2004; Tanimoto et al., 2021; Wang et al., 2018; Wang et al., 2014; Yan et al., 2019; Zhang et al., 2024; Zhang et al., 2022; Zhao et al., 2018; Zheng et al., 2023).

maintenance of dormancy (Shimamoto et al., 2019). Moreover, treatment of ovarian tissue with collagenase to digest the ECM has been shown to activate oocytes in the cortical region (Nagamatsu, Shimamoto, Hamazaki, Nishimura, & Hayashi, 2019). Interestingly, culturing collagenase-treated ovaries under high air pressure reversed the activation and restored the number of dormant oocytes with nuclear FOXO3 localization to untreated levels (Nagamatsu et al., 2019). These findings suggest that mechanical stress contributes to maintaining oocyte quiescence in PFs and influences the rotational orientation of oocyte nuclei within the ovarian cortex. However, the molecular mechanisms through which mechanical signals are transmitted remain to be elucidated.

The Hippo signaling pathway, which is modulated by the physical and mechanical properties of the cellular environment, has also been implicated in the regulation of PF activation (Hsueh et al., 2015). Clinically, disruption of Hippo signaling through ovarian fragmentation, combined with AKT stimulation, has been used to promote follicular growth in patients with POI, offering a potential therapeutic approach for infertility (Kawamura et al., 2013). These studies underscore the complex regulatory interactions between pre-granulosa cells, oocytes, and the local ovarian microenvironment in controlling PF activation. Understanding the spatial distribution and activation mechanisms of PFs within the ovarian cortex provides new perspectives on the development and activation of oocytes and has significant implications for reproductive biology and medicine.

3.3 Heterogeneity of oocyte activation in PFs

As previously discussed, all PFs are formed around the time of birth. Following this, most oocytes within these PFs enter a dormant phase, awaiting recruitment to contribute to fertility later in life. However, some oocytes are activated immediately after formation. Notably, in mammals, there is a prolonged interval between birth and sexual maturity, meaning that these neonatally recruited oocytes and follicles cannot fully develop until puberty. This leads to a marked difference in the developmental dynamics of oocytes and follicles before and after puberty onset. This developmental distinction between neonatal and adult follicles gave rise to the concept of two waves of follicular development, first introduced by British reproductive biologist Dr. Hirshfield in the 1990s (Hirshfield, 1992). Through histological observations and radio-labeled tracing, her work, along with contributions from other pioneers, proposed that follicles located in the medulla of neonatal ovaries represent the first wave of follicles (Feng et al., 2016; Hirshfield, 1992; Mork et al., 2012). In contrast, PFs located in the ovarian cortex form the second wave, which remains dormant until adulthood and contributes to fertility throughout reproductive lifespan.

Over the past decade, significant progress has been made in understanding the origin, developmental dynamics, and molecular regulation of these two waves of PFs in mouse models, largely through the application of cell lineage tracing and other advanced techniques. These studies have revealed important differences in the developmental trajectories and physiological roles of oocytes in the two waves of follicles, which are activated at distinct times. Research by Mork et al. and Niu et al. has shown that this heterogeneity in follicle activation is closely linked to the distinct cellular origins of the pre-granulosa cells in the two waves (Mork et al., 2012; Niu & Spradling, 2020). Specifically, Mork et al. identified that the first wave of follicles, activated shortly after birth, originates from pre-granulosa cells derived from embryonic mesonephric mesenchyme rather than the ovarian surface epithelium (Mork et al., 2012). These early-activated follicles play a critical role in reproductive functions, contributing to the onset of puberty and early postnatal fertility. In contrast, the second wave of PFs, or the adult wave, is derived from pre-granulosa cells from postnatal ovarian surface epithelium (Mork et al., 2012). These follicles enter dormancy in the PF pool and are gradually activated throughout the reproductive lifespan, providing oocytes for long-term fertility (Hirshfield & DeSanti, 1995).

Further supporting this model, the study by Zheng and Zhang using genetically modified mouse models provided critical insights into the distinct developmental dynamics of these two follicle classes (Zheng et al., 2014; Zheng, Zhang, & Liu, 2014). By labeling and tracing oocytes and pre-granulosa cells, they demonstrated that the first-wave follicles not only initiate puberty but also support early fertility. In contrast, the second wave of PFs is gradually and selectively activated throughout the female's reproductive lifespan, providing mature oocytes essential for long-term fertility. Unlike the first wave, which is depleted relatively early, the second wave ensures sustained production of oocytes, thereby supporting fertility throughout the reproductive lifespan (Zheng et al., 2014).

Building on these findings, recent work from the Zhang lab has further elucidated the mechanisms underlying the activation of the two waves of follicles (Dai et al., 2022). Tracing the development of oocytes in fetal ovaries revealed that the asynchronous development of embryonic oocytes generates heterogeneity in postnatal follicle activation, specifically affecting the two waves (Dai et al., 2022). Oocytes in the first wave are derived from rapidly developing embryonic germ cells that enter meiosis earlier and are activated immediately after birth. In contrast, the second wave of follicles is associated with slower-developing embryonic oocytes, which enter the PF pool in a dormant state after encapsulation by pre-granulosa cells. Mechanistically, this study demonstrated that the activation of first-wave follicles relies primarily on intrinsic oocyte signals and is less dependent on external signaling pathways, such as KITL-KIT signaling (Dai et al., 2022). Deleting the key molecule KITL, which mediates the synchronous activation of oocytes and pre-granulosa cells, showed that first-wave follicles are activated spontaneously without the need for pre-granulosa cell initiation. In contrast, the activation of second-wave follicles is strictly driven by classic signaling caspase, i.e. pre-granulosa cell mTORC1-KITL— oocyte KIT-PI3K signaling (Dai et al., 2022).

Collectively, these studies underscore that the two waves of PFs exhibit distinct developmental dynamics and physiological functions. The first wave is pivotal for the onset of puberty and early fertility, while the second wave supports long-term reproductive capacity. While these findings provide critical insights into ovarian function in mouse models, further research is needed to explore whether similar mechanisms are at play in humans. Understanding the heterogeneity of follicle activation in human ovaries will be essential for elucidating the broader implications of these findings, particularly with respect to fertility, reproductive aging, and the timing of puberty.

3.4 Oocyte and follicle growth after activation

Following successful activation, follicles progress through a series of developmental stages—primary, secondary, and antral—culminating in the acquisition of the ability to release a mature egg. Although the initial activation is triggered by somatic pre-granulosa cells, the subsequent growth and development of the follicle are predominantly governed by the oocytes themselves (Zhang & Liu, 2015). During this phase, oocytes undergo rapid enlargement, marked by substantial increases in RNA and protein synthesis, bringing their diameter close to that of fully mature oocytes (Hamazaki et al., 2021; Hirshfield, 1991; Lintern-Moore & Moore, 1979). As the oocytes grow, they secrete a range of oocyte-secreted factors (OSFs), such as growth differentiation factor 9 (GDF9), BMP15, and R-spondin2 (RSPO2) (De Cian et al., 2020; Matzuk & Burns, 2012). These OSFs are critical for stimulating the proliferation and differentiation of surrounding somatic cells, thereby promoting overall follicular growth and development (Dong et al., 1996; Liu, Peng, Matzuk, & Yao, 2015; Mottershead et al., 2015). Disruptions in these signaling pathways can lead to compromised fertility and ovarian dysfunction, underscoring the essential role of oocytes in orchestrating follicle development and reproductive health.

During the initial stages of oocyte growth, the oocytes synthesize and secrete zona pellucida proteins, forming the Zona Pellucida (ZP) (Gook, Edgar, Borg, & Martic, 2008; Greve & Wassarman, 1985; Wassarman & Litscher, 2018). While the ZP is vital for preventing polyspermy during fertilization, it also acts as a physical barrier, limiting direct communication between the oocyte and granulosa cells during follicular development (Wassarman & Litscher, 2016; Wassarman & Litscher, 2021; Wassarman & Mortillo, 1991). To overcome this barrier, granulosa cells extend numerous transzonal projections (TZPs) through the ZP, connecting their tips to the oocyte membrane (Clarke, 2022; Edson, Nagaraja, & Matzuk, 2009; Eppig, Wigglesworth, & Pendola, 2002). Gap junctions at the tips of these TZPs facilitate the exchange of ions, metabolites, and signaling molecules between oocytes and granulosa cells (Doherty, Amargant, Shvartsman, Duncan, & Gavis, 2022). Two key gap junction proteins—connexin 37 (CX37) and connexin 43 (CX43), are essential for these interactions. CX37 is predominantly found at the oocyte-granulosa cell interface, while CX43 is present among granulosa cells themselves (Gershon, Plaks, & Dekel, 2008). Studies using *Cx37* knockout mice have shown impaired oocyte development and infertility, while the restoration of CX43 expression reinstated communication between oocytes and granulosa cells, partially rescuing fertility (Li, Colley, Barr, Yee, & Kidder, 2007; Teilmann, 2005). These findings suggest that both CX37 and CX43 play complementary roles in supporting normal oocyte and follicle development.

Despite extensive research on TZPs and gap junctions, the precise mechanisms by which oocytes regulate communication and signaling during follicle development have remained elusive. Recent studies using high-resolution imaging and genetically modified mouse models have identified specialized structures known as oocyte microvilli (Oo-Mvi) on the surfaces of growing oocytes (Zhang et al., 2021). These mushroom-shaped microvilli are crucial for the regulated release of OSFs, which in turn direct the orderly growth and differentiation of granulosa cells, driving follicle development (Zhang et al., 2021). Structurally, Oo-Mvi consist of a basal stalk and an apical vesicle enriched in OSFs. Interestingly, the apical end of Oo-Mvi does not directly contact granulosa cells. Instead, live-cell imaging has revealed that the apical vesicles of Oo-Mvi periodically rupture, releasing OSFs in bursts (Zhang et al., 2021). This strategy ensures that granulosa cells receive high concentrations of OSFs rapidly and efficiently, triggering downstream signaling cascades that promote granulosa cell proliferation and follicular growth. Mechanistically, the formation of Oo-Mvi in mice is dependent on the high expression of the Radixin protein in oocytes. The absence of *Radixin* disrupts Oo-Mvi formation, leading to impaired follicle development and reproductive abnormalities (Zhang et al., 2021).

The identification of Oo-Mvi completes the understanding of the communication network between oocytes and granulosa cells (Fig. 3). This coordination between oocyte signaling and granulosa cell communication is essential for synchronized follicular development (Clarke, 2018; El-Hayek, Yang, Abbassi, FitzHarris, & Clarke, 2018; Zhang et al., 2021). Disruptions in any component—OSFs, TZPs, gap junctions, or Oo-Mvi—can lead to abnormal follicle development and fertility issues.

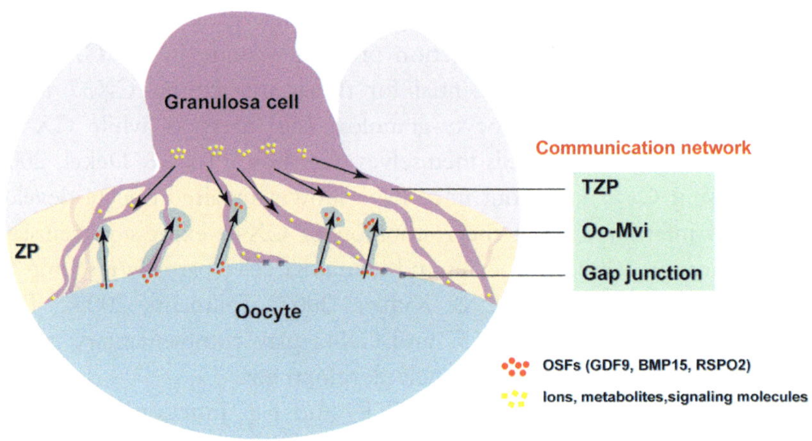

Fig. 3 Intricate crosstalk between the oocyte and granulosa cells across the ZP. Upon oocyte activation, the ZP, or egg coat, forms to support normal oocyte development while blocking direct communication between the oocyte and surrounding granulosa cells. To coordinate the synchronous development of the oocyte and granulosa cells, the oocyte develops mushroom-shaped structures, Oo-Mvi, to reduce the physical distance between the oocyte and granulosa cells, facilitating the efficient release of OSFs, such as GDF9 and BMP15. Simultaneously, each granulosa cell extends numerous TZPs that connect the membranes of the oocyte and granulosa cells, forming gap junctions to enable active exchange of materials and signals, including ions, metabolites, and signaling molecules.

4. Oocyte maturation: an orchestration between nucleus and cytoplasm

Following oocyte activation and growth, oocyte maturation stands as the pivotal final step for oocytes to attain the capacity for successful fertilization and early embryonic development (Eppig, 1996). This intricate process encompasses both nuclear and cytoplasmic maturation—the former involving the resumption and completion of meiosis to produce haploid mature eggs, and the latter encompassing the synthesis and accumulation of essential organelles and biomolecules (Eppig, 1996). Given its fundamental role in reproductive success, the maturation of oocytes is intricately tied to clinical outcomes in fertility treatments. As such, for nearly a century, this process has been rigorously studied and extensively reviewed. In this review, we have subsequently provided only a brief outline summary of the essential regulatory frameworks of oocyte maturation, attempting to facilitate a comprehensive understanding of the overall processes involved in oocyte activation and selection.

Nuclear maturation primarily revolves around the resumption of meiosis I, which is closely regulated by a surge in luteinizing hormone (LH) preceding ovulation (Mehlmann, 2005). This hormonal influx triggers a series of biochemical alterations in the surrounding granulosa cells, beginning with the dephosphorylation and inactivation of guanylate cyclase NPR2 (Jaffe & Egbert, 2017). As a result, there is a notable reduction in cGMP levels within granulosa cells and consequently within the oocyte itself. This reduction further prompts the activation of cAMP phosphodiesterase, leading to decreased cAMP levels in oocytes that activate MPFs essential for continuing meiotic progression (Egbert et al., 2014; Vaccari, Weeks, Hsieh, Menniti, & Conti, 2009). Additionally, LH stimulates the granulosa cell production of epidermal growth factor (EGF)-like growth factors in granulosa cells, enhancing these signal cascades via EGF receptor pathway, which further supports the resumption of meiosis (Jaffe & Egbert, 2017; Norris, Freudzon, Nikolaev, & Jaffe, 2010; Park et al., 2004). Recent studies also underscore the role of histone deacetylase 3 in modulating the expression of EGF-like growth factors like amphiregulin (AREG) post-LH surge, critical for meiosis resumption (Wang et al., 2019). Moreover, a variety of post-translational protein modifications - including ubiquitination (Yu, Ji, Sha, Sun, & Fan, 2015; Zhou et al., 2022), SUMOylation (Feitosa, Hwang, & Morris, 2018; Yuan et al., 2014), acetylation (Jiang et al., 2023), and glycosylation (Li et al., 2020)—have been identified as key regulators of this nuanced nuclear maturation process.

In contrast, cytoplasmic maturation, though less understood, is starkly imperative for the accumulation of maternal factors crucial for subsequent embryonic development. Investigations reveal that mature oocytes contain substantial amounts of maternal mRNA, influential in early embryonic cell differentiation and structural organization (Meng et al., 2023; Zhang et al., 2023). These mRNAs are preserved within a mitochondria-associated ribonucleoprotein domain, regulated by specific components like the Smith-like

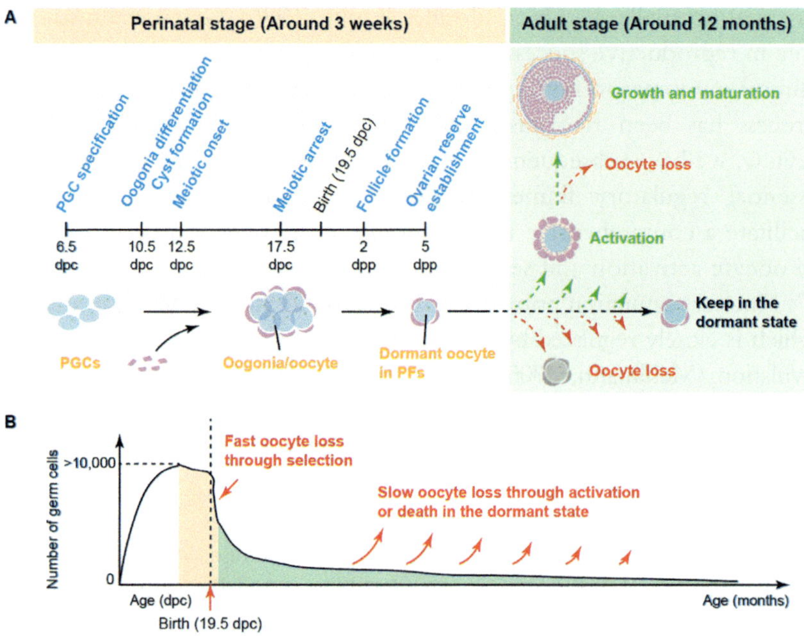

Fig. 4 Lifespan of female germ cells in mice: effects of oocyte selection and activation on female reproduction. (A) Key developmental events, timings, and fates from PGCs to oocyte activation and growth. During embryonic development, PGCs migrate into the ovaries and develop into oogonia, which proliferate to peak in number around 15.5 dpc in fetal ovaries. As oogonia enter meiosis, they transform into oocytes, with a dramatic decrease in number due to intense selection during the perinatal stage. Surviving oocytes enter a dormant state, becoming enclosed by pre-granulosa cells to form PFs, which constitute the ovarian reserve for each individual female (light yellow frame). After birth, a portion of these dormant oocytes gradually activate and grow, with some developing to the ovulation stage through strict selection, ultimately contributing to fertility. Meanwhile, a significant number of oocytes undergo cell death in the dormant state, leading to a continuous loss of oocytes in the ovaries (light green frame). (B) Curve graph showing the number of female germ cells throughout the lifespan in mice, highlighting the rapid oocyte loss during the perinatal stage and the gradual oocyte loss during adulthood.

family member 14 B (LSM14B) (Cheng et al., 2022; Shan et al., 2023). Furthermore, the identification of the subcortical maternal complex (SCMC) has been a significant advance, underscoring its necessity during early embryogenesis (Li, Baibakov, & Dean, 2008). The loss of any single SCMC component typically results in compromised fertility due to abnormal oocyte and embryo development (Jentoft et al., 2023; Lu, Gao, Qin, & Li, 2017). Recent findings also suggest that interactions between SCMC components and maternal effect genes such as *Zygote arrest-1* (*Zar1*) and *Zygote arrest-2* (*Zar2*) enhance the processes of meiotic maturation (Rong et al., 2019).

To conclude, the maturation of oocytes is a finely-tuned orchestration of both nuclear and cytoplasmic events, where each component plays a vital role in ensuring the viability and developmental potential of the oocyte. An enriched understanding of these processes not only illuminates fundamental biological phenomena but also holds promise for improving clinical interventions in human reproductive health.

5. Conclusions and future perspective

The selection and activation of oocytes in mammals are fundamental processes crucial for maintaining fertility (Fig. 4). In mammalian ovaries, a finite number of oocytes are selected during fetal development, which subsequently undergo dormancy and are recruited throughout the reproductive lifespan. These processes are finely regulated by complex molecular and cellular mechanisms, involving interactions between oocytes and surrounding somatic cells, particularly granulosa cells. Understanding the intricacies of these processes is vital because they determine not only the quality and quantity of oocytes available for fertilization but also influence female reproductive aging (Laisk et al., 2019).

While reproduction is one of life's most basic and instinctive functions, modern society faces increasing reproductive challenges. The global incidence of infertility is rising, and with it, a marked increase in ovarian diseases such as POI and polycystic ovary syndrome (PCOS) (Cox et al., 2022; Davis, Pinkerton, Santoro, & Simoncini, 2023; Qiao et al., 2021). These challenges make it more important than ever to delve deeply into the molecular mechanisms governing oocyte selection and activation. Additionally, societal trends such as delayed childbearing have introduced a biological conflict with the natural age of optimal fertility in women, highlighting the need for novel approaches to preserve and extend reproductive health (Rienzi et al., 2017).

From the perspective of ovarian biology, the ability to modulate oocyte selection and activation could have profound implications. If we can develop strategies to slow the depletion of ovarian reserves and prevent the premature activation of dormant follicles, it may be possible to delay reproductive aging. Such advances would not only increase fertility windows for women but also reduce the incidence of age-related reproductive diseases.

Recent advancements in understanding the regulatory mechanisms governing oocyte activation, including the role of the mTORC1-KITL-KIT-PI3K/AKT signaling pathway, have led to innovative fertility treatments. For instance, the "in vitro activation" (IVA) technique has emerged as a potential therapeutic approach for women with POI or those lacking growing follicles (Li et al., 2010; Sun et al., 2015; Yan et al., 2018; Zhang et al., 2020). However, despite its promise, IVA remains limited by its low success rates and the need for multiple invasive procedures (Lee & Chang, 2019). Enhancing the efficiency of IVA and developing less invasive alternatives will be critical future steps in addressing fertility challenges associated with ovarian reserve depletion.

In addition to assisted reproductive technology (ART) techniques like IVA, the importance of preserving the ovarian reserve through non-invasive methods has gained traction. Classical fertility preservation strategies, such as cryopreservation of oocytes and embryos, are well-established and effective, yet they require invasive procedures and do not protect the long-term function of the ovaries (Davis et al., 2023; Rienzi et al., 2017; Stoop, Cobo, & Silber, 2014; Suzuki et al., 2015). On the other hand, preserving ovarian function by regulating oocyte activation offers a more holistic approach to maintaining fertility, reducing the risk of ovarian dysfunction and extending the reproductive lifespan (Xu et al., 2022). The development of new treatments to modulate oocyte dormancy and prevent premature follicle activation is key to improving outcomes for women who face age-related fertility declines or ovarian diseases.

Looking ahead, future research should focus on several critical areas. First, exploring the molecular and cellular mechanisms that regulate oocyte selection and activation across the entire reproductive lifespan is essential. Understanding how these processes are coordinated at the organ level during different life stages will offer insights into the maintenance of ovarian reserves and overall female fertility. Additionally, investigating the communication between oocytes and their surrounding somatic cells, especially the oocyte specific communicating mechanisms, is crucial, as these interactions play a key role in follicle activation. A deeper understanding of the molecular pathways governing this communication, especially those involved in long-term oocyte storage and health, will provide valuable knowledge about how oocyte dormancy is maintained and follicles are activated in a regulated manner. Furthermore, oocytes remain dormant for extended periods, and understanding the mechanisms controlling dormancy and why some oocytes are lost during this phase is vital. Identifying the pathways responsible for oocyte loss and developing strategies to prevent it will be critical for extending reproductive lifespan and delaying ovarian aging. The use of advanced technologies, such as high-resolution imaging, multi-layered omics, and in vivo cell fate tracking, will be instrumental in mapping the spatial and temporal dynamics of follicular development, providing a comprehensive understanding of how ovarian reserves are managed over time. Lastly, translating these research findings into clinical applications is essential for improving women's reproductive health. Collaboration between researchers and clinicians will pave the way for new diagnostic tools and treatments for infertility, ovarian diseases, and reproductive aging. By applying these insights to human studies, novel therapies can be more effectively implemented in clinical practice.

In conclusion, while significant strides have been made in understanding the mechanisms of oocyte selection and activation, there remains much to uncover. The future holds exciting possibilities for extending female reproductive health through targeted therapies that regulate oocyte dormancy and activation. By focusing on the maintenance of the ovarian reserve and developing new fertility preservation techniques, we may be able to enhance women's reproductive health, delay reproductive aging, and improve outcomes for those facing infertility or ovarian disorders.

Acknowledgements

This study was supported by the Key Program of National Naturaal Science Foundation of China to H.Z. (82230051), the National Key Research and Development Program of China to H.Z. (2022YFC2703803), and the 2115 Talent Development Program of China Agricultural University to H.Z. (1021-00109035).

References

Adhikari, D., Busayavalasa, K., Zhang, J., Hu, M., Risal, S., Bayazit, M. B., et al. (2016). Inhibitory phosphorylation of Cdk1 mediates prolonged prophase I arrest in female germ cells and is essential for female reproductive lifespan. *Cell Research, 26*, 1212–1225. https://doi.org/10.1038/cr.2016.119.

Adhikari, D., & Liu, K. (2009). Molecular mechanisms underlying the activation of mammalian primordial follicles. *Endocrine Reviews, 30*, 438–464. https://doi.org/10.1210/er.2008-0048.

Ataei-Nazari, S., Amoushahi, M., Madsen, J. F., Jensen, J., Heuck, A., Mohammadi-Sangcheshmeh, A., et al. (2022). Cyclin-dependent kinase 6 (CDK6) as a potent regulator of the ovarian primordial-to-primary follicle transition. *Frontiers in Cell and Developmental Biology, 10*, 1036917. https://doi.org/10.3389/fcell.2022.1036917.

Bolivar, J., Huynh, J. R., Lopez-Schier, H., Gonzalez, C., St, J. D., & Gonzalez-Reyes, A. (2001). Centrosome migration into the *Drosophila* oocyte is independent of *BicD* and *egl*, and of the organisation of the microtubule cytoskeleton. *Development (Cambridge, England), 128*, 1889–1897. https://doi.org/10.1242/dev.128.10.1889.

Borum, K. (1961). Oogenesis in the mouse. A study of the meiotic prophase. *Experimental Cell Research, 24*, 495–507. https://doi.org/10.1016/0014-4827(61)90449-9.

Bowles, J., Feng, C. W., Miles, K., Ineson, J., Spiller, C., & Koopman, P. (2016). ALDH1A1 provides a source of meiosis-inducing retinoic acid in mouse fetal ovaries. *Nature Communications, 7*, 10845. https://doi.org/10.1038/ncomms10845.

Bowles, J., Knight, D., Smith, C., Wilhelm, D., Richman, J., Mamiya, S., et al. (2006). Retinoid signaling determines germ cell fate in mice. *Science (New York, N. Y.), 312*, 596–600. https://doi.org/10.1126/science.1125691.

Bowles, J., & Koopman, P. (2007). Retinoic acid, meiosis and germ cell fate in mammals. *Development (Cambridge, England), 134*, 3401–3411. https://doi.org/10.1242/dev.001107.

Bullejos, M., & Koopman, P. (2004). Germ cells enter meiosis in a rostro-caudal wave during development of the mouse ovary. *Molecular Reproduction and Development, 68*, 422–428. https://doi.org/10.1002/mrd.20105.

Chassot, A. A., Le Rolle, M., Jolivet, G., Stevant, I., Guigonis, J. M., Da, S. F., et al. (2020). Retinoic acid synthesis by ALDH1A proteins is dispensable for meiosis initiation in the mouse fetal ovary. *Science Advances, 6*, eaaz1261. https://doi.org/10.1126/sciadv.aaz1261.

Chen, W. E. Q., Sun, B., Zhang, P., Li, N., Fei, S., et al. (2023). PARP1-catalyzed PARylation of YY1 mediates endoplasmic reticulum stress in granulosa cells to determine primordial follicle activation. *Cell Death & Disease, 14*, 524. https://doi.org/10.1038/s41419-023-05984-w.

Cheng, S., Altmeppen, G., So, C., Welp, L. M., Penir, S., Ruhwedel, T., et al. (2022). Mammalian oocytes store mRNAs in a mitochondria-associated membraneless compartment. *Science (New York, N. Y.), 378*, eabq4835. https://doi.org/10.1126/science.abq4835.

Childs, A. J., Cowan, G., Kinnell, H. L., Anderson, R. A., & Saunders, P. T. (2011). Retinoic acid signalling and the control of meiotic entry in the human fetal gonad. *PLoS One, 6*, e20249. https://doi.org/10.1371/journal.pone.0020249.

Choi, Y., Ballow, D. J., Xin, Y., & Rajkovic, A. (2008). Lim homeobox gene, *Lhx8*, is essential for mouse oocyte differentiation and survival. *Biology of Reproduction, 79*, 442–449. https://doi.org/10.1095/biolreprod.108.069393.

Clarke, H. J. (2018). Regulation of germ cell development by intercellular signaling in the mammalian ovarian follicle. *Wiley Interdisciplinary Reviews: Developmental Biology, 7*. https://doi.org/10.1002/wdev.294.

Clarke, H. J. (2022). Transzonal projections: Essential structures mediating intercellular communication in the mammalian ovarian follicle. *Molecular Reproduction and Development, 89*, 509–525. https://doi.org/10.1002/mrd.23645.

Cox, R. T., & Spradling, A. C. (2003). A Balbiani body and the fusome mediate mitochondrial inheritance during *Drosophila* oogenesis. *Development (Cambridge, England), 130*, 1579–1590. https://doi.org/10.1242/dev.00365.

Cox, C. M., Thoma, M. E., Tchangalova, N., Mburu, G., Bornstein, M. J., Johnson, C. L., et al. (2022). Infertility prevalence and the methods of estimation from 1990 to 2021: A systematic review and meta-analysis. *Human Reproduction Open, 2022*. https://doi.org/10.1093/hropen/hoac051.

Dai, Y., Bo, Y., Wang, P., Xu, X., Singh, M., Jia, L., et al. (2022). Asynchronous embryonic germ cell development leads to a heterogeneity of postnatal ovarian follicle activation and may influence the timing of puberty onset in mice. *BMC Biology, 20*, 109. https://doi.org/10.1186/s12915-022-01318-y.

Davis, S. R., Pinkerton, J., Santoro, N., & Simoncini, T. (2023). Menopause-Biology, consequences, supportive care, and therapeutic options. *Cell, 186*, 4038–4058. https://doi.org/10.1016/j.cell.2023.08.016.

De Cian, M. C., Gregoire, E. P., Le Rolle, M., Lachambre, S., Mondin, M., Bell, S., et al. (2020). R-spondin2 signaling is required for oocyte-driven intercellular communication and follicular growth. *Cell Death and Differentiation, 27*, 2856–2871. https://doi.org/10.1038/s41418-020-0547-7.

Doherty, C. A., Amargant, F., Shvartsman, S. Y., Duncan, F. E., & Gavis, E. R. (2022). Bidirectional communication in oogenesis: A dynamic conversation in mice and *Drosophila*. *Trends in Cell Biology, 32*, 311–323. https://doi.org/10.1016/j.tcb.2021.11.005.

Dokshin, G. A., Baltus, A. E., Eppig, J. J., & Page, D. C. (2013). Oocyte differentiation is genetically dissociable from meiosis in mice. *Nature Genetics, 45*, 877–883. https://doi.org/10.1038/ng.2672.

Dong, J., Albertini, D. F., Nishimori, K., Kumar, T. R., Lu, N., & Matzuk, M. M. (1996). Growth differentiation factor-9 is required during early ovarian folliculogenesis. *Nature, 383*, 531–535. https://doi.org/10.1038/383531a0.

Edson, M. A., Nagaraja, A. K., & Matzuk, M. M. (2009). The mammalian ovary from genesis to revelation. *Endocrine Reviews, 30*, 624–712. https://doi.org/10.1210/er.2009-0012.

Egbert, J. R., Shuhaibar, L. C., Edmund, A. B., Van Helden, D. A., Robinson, J. W., Uliasz, T. F., et al. (2014). Dephosphorylation and inactivation of NPR2 guanylyl cyclase in granulosa cells contributes to the LH-induced decrease in cGMP that causes resumption of meiosis in rat oocytes. *Development (Cambridge, England), 141*, 3594–3604. https://doi.org/10.1242/dev.112219.

El-Hayek, S., Yang, Q., Abbassi, L., FitzHarris, G., & Clarke, H. J. (2018). Mammalian oocytes locally remodel follicular architecture to provide the foundation for germline-soma communication. *Current Biology, 28*, 1124–1131. https://doi.org/10.1016/j.cub.2018.02.039.

Endo, T., Mikedis, M. M., Nicholls, P. K., Page, D. C., & de Rooij, D. G. (2019). Retinoic acid and germ cell development in the ovary and testis. *Biomolecules, 9*, 775. https://doi.org/10.3390/biom9120775.

Eppig, J. J. (1996). Coordination of nuclear and cytoplasmic oocyte maturation in eutherian mammals. *Reproduction, Fertility, and Development, 8*, 485–489. https://doi.org/10.1071/rd9960485.

Eppig, J. J., Wigglesworth, K., & Pendola, F. L. (2002). The mammalian oocyte orchestrates the rate of ovarian follicular development. *Proceedings of the National Academy of Sciences of the United States of America, 99*, 2890–2894. https://doi.org/10.1073/pnas.052658699.

Feitosa, W. B., Hwang, K., & Morris, P. L. (2018). Temporal and SUMO-specific SUMOylation contribute to the dynamics of Polo-like kinase 1 (PLK1) and spindle integrity during mouse oocyte meiosis. *Developmental Biology, 434*, 278–291. https://doi.org/10.1016/j.ydbio.2017.12.011.

Feng, L., Wang, Y., Cai, H., Sun, G., Niu, W., Xin, Q., et al. (2016). ADAM10-Notch signaling governs the recruitment of ovarian pregranulosa cells and controls folliculogenesis in mice. *Journal of Cell Science, 129*, 2202–2212. https://doi.org/10.1242/jcs.184267.

Findlay, J. K., Hutt, K. J., Hickey, M., & Anderson, R. A. (2015). How is the number of primordial follicles in the ovarian reserve established? *Biology of Reproduction, 93*, 111. https://doi.org/10.1095/biolreprod.115.133652.

Fu, X. F., Yang, F., Cheng, S. F., Feng, Y. N., Li, L., Dyce, P. W., et al. (2017). The epigenetic modifications and the anterior to posterior characterization of meiotic entry during mouse oogenesis. *Histochemistry and Cell Biology, 148*, 61–72. https://doi.org/10.1007/s00418-017-1545-9.

Gershon, E., Plaks, V., & Dekel, N. (2008). Gap junctions in the ovary: Expression, localization and function. *Molecular and Cellular Endocrinology, 282*, 18–25. https://doi.org/10.1016/j.mce.2007.11.001.

Gook, D. A., Edgar, D. H., Borg, J., & Martic, M. (2008). Detection of zona pellucida proteins during human folliculogenesis. *Human Reproduction, 23*, 394–402. https://doi.org/10.1093/humrep/dem373.

Gorre, N., Adhikari, D., Lindkvist, R., Brannstrom, M., Liu, K., & Shen, Y. (2014). mTORC1 Signaling in oocytes is dispensable for the survival of primordial follicles and for female fertility. *PLoS One, 9*, e110491. https://doi.org/10.1371/journal.pone.0110491.

Greenbaum, M. P., Iwamori, N., Agno, J. E., & Matzuk, M. M. (2009). Mouse TEX14 is required for embryonic germ cell intercellular bridges but not female fertility. *Biology of Reproduction, 80*, 449–457. https://doi.org/10.1095/biolreprod.108.070649.

Greve, J. M., & Wassarman, P. M. (1985). Mouse egg extracellular coat is a matrix of interconnected filaments possessing a structural repeat. *Journal of Molecular Biology, 181*, 253–264. https://doi.org/10.1016/0022-2836(85)90089-0.

Grive, K. J., & Freiman, R. N. (2015). The developmental origins of the mammalian ovarian reserve. *Development (Cambridge, England), 142*, 2554–2563. https://doi.org/10.1242/dev.125211.

Guzzardo, P. M., Muerdter, F., & Hannon, G. J. (2013). The piRNA pathway in flies: Highlights and future directions. *Current Opinion in Genetics & Development, 23*, 44–52. https://doi.org/10.1016/j.gde.2012.12.003.

Habara, O., Logan, C. Y., Kanai-Azuma, M., Nusse, R., & Takase, H. M. (2021). WNT signaling in pre-granulosa cells is required for ovarian folliculogenesis and female fertility. dev198846 *Development (Cambridge, England), 148*. https://doi.org/10.1242/dev.198846.

Hamazaki, N., Kyogoku, H., Araki, H., Miura, F., Horikawa, C., Hamada, N., et al. (2021). Reconstitution of the oocyte transcriptional network with transcription factors. *Nature, 589*, 264–269. https://doi.org/10.1038/s41586-020-3027-9.

Harasimov, K., Gorry, R. L., Welp, L. M., Penir, S. M., Horokhovskyi, Y., Cheng, S., et al. (2024). The maintenance of oocytes in the mammalian ovary involves extreme protein longevity. *Nature Cell Biology, 26*, 1124–1138. https://doi.org/10.1038/s41556-024-01442-7.

Hirshfield, A. N. (1991). Development of follicles in the mammalian ovary. *International Review of Cytology, 124*, 43–101. https://doi.org/10.1016/s0074-7696(08)61524-7.

Hirshfield, A. N. (1992). Heterogeneity of cell populations that contribute to the formation of primordial follicles in rats. *Biology of Reproduction, 47*, 466–472. https://doi.org/10.1095/biolreprod47.3.466.

Hirshfield, A. N., & DeSanti, A. M. (1995). Patterns of ovarian cell proliferation in rats during the embryonic period and the first three weeks postpartum. *Biology of Reproduction, 53*, 1208–1221. https://doi.org/10.1095/biolreprod53.5.1208.

Hsueh, A. J., Kawamura, K., Cheng, Y., & Fauser, B. C. (2015). Intraovarian control of early folliculogenesis. *Endocrine Reviews, 36*, 1–24. https://doi.org/10.1210/er.2014-1020.

Hu, M. W., Meng, T. G., Jiang, Z. Z., Dong, M. Z., Schatten, H., Xu, X., et al. (2016). Protein phosphatase 6 protects prophase I-arrested oocytes by safeguarding genomic integrity. *PLoS Genetics, 12*, e1006513. https://doi.org/10.1371/journal.pgen.1006513.

Hu, L. L., Su, T., Luo, R. C., Zheng, Y. H., Huang, J., Zhong, Z. S., et al. (2019). Hippo pathway functions as a downstream effector of AKT signaling to regulate the activation of primordial follicles in mice. *Journal of Cellular Physiology, 234*, 1578–1587. https://doi.org/10.1002/jcp.27024.

Hu, M., Yeh, Y. H., Munakata, Y., Abe, H., Sakashita, A., Maezawa, S., et al. (2022). PRC1-mediated epigenetic programming is required to generate the ovarian reserve. *Nature Communications, 13*, 4510. https://doi.org/10.1038/s41467-022-31759-6.

Ikami, K., Nuzhat, N., Abbott, H., Pandoy, R., Haky, L., Spradling, A. C., et al. (2021). Altered germline cyst formation and oogenesis in Tex14 mutant mice. *Biology Open, 10*, bio058807. https://doi.org/10.1242/bio.058807.

Ikami, K., Shoffner-Beck, S., Tyczynska, W. M., Schnell, S., Yoshida, S., Diaz, M. E., et al. (2023). Branched germline cysts and female-specific cyst fragmentation facilitate oocyte determination in mice. *Proceedings of the National Academy of Sciences of the United States of America, 120*, e2075284176. https://doi.org/10.1073/pnas.2219683120.

Ishiguro, K. I., Matsuura, K., Tani, N., Takeda, N., Usuki, S., Yamane, M., et al. (2020). MEIOSIN directs the switch from mitosis to meiosis in mammalian germ cells. *Developmental Cell, 52*, 429–445. https://doi.org/10.1016/j.devcel.2020.01.010.

Jaffe, L. A., & Egbert, J. R. (2017). Regulation of mammalian oocyte meiosis by intercellular communication within the ovarian follicle. *Annual Review of Physiology, 79*, 237–260. https://doi.org/10.1146/annurev-physiol-022516-034102.

Jagarlamudi, K., & Rajkovic, A. (2012). Oogenesis: Transcriptional regulators and mouse models. *Molecular and Cellular Endocrinology, 356*, 31–39. https://doi.org/10.1016/j.mce.2011.07.049.

Jentoft, I., Bauerlein, F., Welp, L. M., Cooper, B. H., Petrovic, A., So, C., et al. (2023). Mammalian oocytes store proteins for the early embryo on cytoplasmic lattices. *Cell, 186*, 5308–5327. https://doi.org/10.1016/j.cell.2023.10.003.

Jiang, X., Cheng, Y., Zhu, Y., Xu, C., Li, Q., Xing, X., et al. (2023). Maternal NAT10 orchestrates oocyte meiotic cell-cycle progression and maturation in mice. *Nature Communications, 14*, 3729. https://doi.org/10.1038/s41467-023-39256-0.

Jiang, Z. Z., Hu, M. W., Ma, X. S., Schatten, H., Fan, H. Y., Wang, Z. B., et al. (2016). LKB1 acts as a critical gatekeeper of ovarian primordial follicle pool. *Oncotarget, 7*, 5738–5753. https://doi.org/10.18632/oncotarget.6792.

John, G. B., Gallardo, T. D., Shirley, L. J., & Castrillon, D. H. (2008). Foxo3 is a PI3K-dependent molecular switch controlling the initiation of oocyte growth. *Developmental Biology, 321*, 197–204. https://doi.org/10.1016/j.ydbio.2008.06.017.

Kawamura, K., Cheng, Y., Suzuki, N., Deguchi, M., Sato, Y., Takae, S., et al. (2013). Hippo signaling disruption and Akt stimulation of ovarian follicles for infertility treatment. *Proceedings of the National Academy of Sciences of the United States of America, 110*, 17474–17479. https://doi.org/10.1073/pnas.1312830110.

Kezele, P., Nilsson, E., & Skinner, M. K. (2002). Cell-cell interactions in primordial follicle assembly and development. *Frontiers in Bioscience: A Journal and Virtual Library, 7*, d1990–d1996. https://doi.org/10.2741/kezele.

Koubova, J., Menke, D. B., Zhou, Q., Capel, B., Griswold, M. D., & Page, D. C. (2006). Retinoic acid regulates sex-specific timing of meiotic initiation in mice. *Proceedings of the National Academy of Sciences of the United States of America, 103*, 2474–2479. https://doi.org/10.1073/pnas.0510813103.

Kurilo, L. F. (1981). Oogenesis in antenatal development in man. *Human Genetics, 57*, 86–92. https://doi.org/10.1007/BF00271175.

Laisk, T., Tsuiko, O., Jatsenko, T., Horak, P., Otala, M., Lahdenpera, M., et al. (2019). Demographic and evolutionary trends in ovarian function and aging. *Human Reproduction Update, 25*, 34–50. https://doi.org/10.1093/humupd/dmy031.

Le Bouffant, R., Guerquin, M. J., Duquenne, C., Frydman, N., Coffigny, H., Rouiller-Fabre, V., et al. (2010). Meiosis initiation in the human ovary requires intrinsic retinoic acid synthesis. *Human Reproduction, 25*, 2579–2590. https://doi.org/10.1093/humrep/deq195.

Lee, H. N., & Chang, E. M. (2019). Primordial follicle activation as new treatment for primary ovarian insufficiency. *Clinical and Experimental Reproductive Medicine-CERM, 46*, 43–49. https://doi.org/10.5653/cerm.2019.46.2.43.

Lei, L., & Spradling, A. C. (2013). Mouse primordial germ cells produce cysts that partially fragment prior to meiosis. *Development (Cambridge, England), 140*, 2075–2081. https://doi.org/10.1242/dev.093864.

Lei, L., & Spradling, A. C. (2016). Mouse oocytes differentiate through organelle enrichment from sister cyst germ cells. *Science (New York, N. Y.), 352*, 95–99. https://doi.org/10.1126/science.aad2156.

Li, L., Baibakov, B., & Dean, J. (2008). A subcortical maternal complex essential for preimplantation mouse embryogenesis. *Developmental Cell, 15*, 416–425. https://doi.org/10.1016/j.devcel.2008.07.010.

Li, T. Y., Colley, D., Barr, K. J., Yee, S. P., & Kidder, G. M. (2007). Rescue of oogenesis in Cx37-null mutant mice by oocyte-specific replacement with Cx43. *Journal of Cell Science, 120*, 4117–4125. https://doi.org/10.1242/jcs.03488.

Li, J., Kawamura, K., Cheng, Y., Liu, S., Klein, C., Liu, S., et al. (2010). Activation of dormant ovarian follicles to generate mature eggs. *Proceedings of the National Academy of Sciences of the United States of America, 107*, 10280–10284. https://doi.org/10.1073/pnas.1001198107.

Li, H., You, L., Tian, Y., Guo, J., Fang, X., Zhou, C., et al. (2020). DPAGT1-mediated protein N-glycosylation is indispensable for oocyte and follicle development in mice. *Advanced Science, 7*, 2000531. https://doi.org/10.1002/advs.202000531.

Lintern-Moore, S., & Moore, G. P. (1979). The initiation of follicle and oocyte growth in the mouse ovary. *Biology of Reproduction, 20*, 773–778. https://doi.org/10.1095/biolreprod20.4.773.

Liu, B., He, Y., Wu, X., Lin, Z., Ma, J., Qiu, Y., et al. (2024). Mapping putative enhancers in mouse oocytes and early embryos reveals TCF3/12 as key folliculogenesis regulators. *Nature Cell Biology, 26*, 962–974. https://doi.org/10.1038/s41556-024-01422-x.

Liu, C., Peng, J., Matzuk, M. M., & Yao, H. H. (2015). Lineage specification of ovarian theca cells requires multicellular interactions via oocyte and granulosa cells. *Nature Communications, 6*, 6934. https://doi.org/10.1038/ncomms7934.

Lu, X., Gao, Z., Qin, D., & Li, L. (2017). A maternal functional module in the mammalian oocyte-to-embryo transition. *Trends in Molecular Medicine, 23*, 1014–1023. https://doi.org/10.1016/j.molmed.2017.09.004.

Matzuk, M. M., & Burns, K. H. (2012). Genetics of mammalian reproduction: Modeling the end of the germline. *Annual Review of Physiology, 74*, 503–528. https://doi.org/10.1146/annurev-physiol-020911-153248.

McGee, E. A., & Hsueh, A. J. (2000). Initial and cyclic recruitment of ovarian follicles. *Endocrine Reviews, 21*, 200–214. https://doi.org/10.1210/edrv.21.2.0394.

Mehlmann, L. M. (2005). Stops and starts in mammalian oocytes: Recent advances in understanding the regulation of meiotic arrest and oocyte maturation. *Reproduction (Cambridge, England), 130*, 791–799. https://doi.org/10.1530/rep.1.00793.

Meng, T. G., Guo, J. N., Zhu, L., Yin, Y., Wang, F., Han, Z. M., et al. (2023). NLRP14 safeguards calcium homeostasis via regulating the K27 ubiquitination of Nclx in oocyte-to-embryo transition. *Advanced Science, 10*, e2301940. https://doi.org/10.1002/advs.202301940.

Milas, A., & Telley, I. A. (2022). Polarity events in the *Drosophila melanogaster* oocyte. *Frontiers in Cell and Developmental Biology, 10*, 895876. https://doi.org/10.3389/fcell.2022.895876.

Miyauchi, H., Ohta, H., Nagaoka, S., Nakaki, F., Sasaki, K., Hayashi, K., et al. (2017). Bone morphogenetic protein and retinoic acid synergistically specify female germ-cell fate in mice. *Embo Journal, 36*, 3100–3119. https://doi.org/10.15252/embj.201796875.

Monk, M., & McLaren, A. (1981). X-chromosome activity in foetal germ cells of the mouse. *Journal of Embryology and Experimental Morphology, 63*, 75–84.
Mork, L., Maatouk, D. M., McMahon, J. A., Guo, J. J., Zhang, P., McMahon, A. P., et al. (2012). Temporal differences in granulosa cell specification in the ovary reflect distinct follicle fates in mice. *Biology of Reproduction, 86*, 37. https://doi.org/10.1095/biolreprod.111.095208.
Mottershead, D. G., Sugimura, S., Al-Musawi, S. L., Li, J. J., Richani, D., White, M. A., et al. (2015). Cumulin, an oocyte-secreted heterodimer of the transforming growth factor-beta family, is a potent activator of granulosa cells and improves oocyte quality. *Journal of Biological Chemistry, 290*, 24007–24020. https://doi.org/10.1074/jbc.M115.671487.
Mu, X., Wen, J., Guo, M., Wang, J., Li, G., Wang, Z., et al. (2013). Retinoic acid derived from the fetal ovary initiates meiosis in mouse germ cells. *Journal of Cellular Physiology, 228*, 627–639. https://doi.org/10.1002/jcp.24172.
Nagamatsu, G., Shimamoto, S., Hamazaki, N., Nishimura, Y., & Hayashi, K. (2019). Mechanical stress accompanied with nuclear rotation is involved in the dormant state of mouse oocytes. *Science Advances, 5*, eaav9960. https://doi.org/10.1126/sciadv.aav9960.
Nagaoka, S. I., Nakaki, F., Miyauchi, H., Nosaka, Y., Ohta, H., Yabuta, Y., et al. (2020). ZGLP1 is a determinant for the oogenic fate in mice. *Science (New York, N. Y.), 367*, eaaw4115. https://doi.org/10.1126/science.aaw4115.
Nashchekin, D., Busby, L., Jakobs, M., Squires, I., & St, J. D. (2021). Symmetry breaking in the female germline cyst. *Science (New York, N. Y.), 374*, 874–879. https://doi.org/10.1126/science.abj3125.
Niu, W., & Spradling, A. C. (2020). Two distinct pathways of pregranulosa cell differentiation support follicle formation in the mouse ovary. *Proceedings of the National Academy of Sciences of the United States of America, 117*, 20015–20026. https://doi.org/10.1073/pnas.2005570117.
Niu, W., & Spradling, A. C. (2022). Mouse oocytes develop in cysts with the help of nurse cells. *Cell, 185*, 2576–2590. https://doi.org/10.1016/j.cell.2022.05.001.
Norris, R. P., Freudzon, M., Nikolaev, V. O., & Jaffe, L. A. (2010). Epidermal growth factor receptor kinase activity is required for gap junction closure and for part of the decrease in ovarian follicle cGMP in response to LH. *Reproduction (Cambridge, England), 140*, 655–662. https://doi.org/10.1530/REP-10-0288.
O'Connell, J. M., & Pepling, M. E. (2021). Primordial follicle formation – Some assembly required. *Curr Opin Endocr Metab Res, 18*, 118–127. https://doi.org/10.1016/j.coemr.2021.03.005.
Pan, B., & Li, J. (2019). The art of oocyte meiotic arrest regulation. *Reproductive Biology and Endocrinology, 17*, 8. https://doi.org/10.1186/s12958-018-0445-8.
Park, J. Y., Su, Y. Q., Ariga, M., Law, E., Jin, S. L., & Conti, M. (2004). EGF-like growth factors as mediators of LH action in the ovulatory follicle. *Science (New York, N. Y.), 303*, 682–684. https://doi.org/10.1126/science.1092463.
Pedersen, T., & Peters, H. (1968). Proposal for a classification of oocytes and follicles in the mouse ovary. *Journal of Reproduction and Fertility, 17*, 555–557. https://doi.org/10.1530/jrf.0.0170555.
Pepling, M. E. (2006). From primordial germ cell to primordial follicle: mammalian female germ cell development. *Genesis (New York, N. Y.: 2000), 44*, 622–632. https://doi.org/10.1002/dvg.20258.
Pepling, M. E. (2012). Follicular assembly: Mechanisms of action. *Reproduction (Cambridge, England), 143*, 139–149. https://doi.org/10.1530/REP-11-0299.
Qiao, J., Wang, Y., Li, X., Jiang, F., Zhang, Y., Ma, J., et al. (2021). A Lancet Commission on 70 years of women's reproductive, maternal, newborn, child, and adolescent health in China. *Lancet, 397*, 2497–2536. https://doi.org/10.1016/S0140-6736(20)32708-2.

Rajareddy, S., Reddy, P., Du, C., Liu, L., Jagarlamudi, K., Tang, W., et al. (2007). p27kip1 (cyclin-dependent kinase inhibitor 1B) controls ovarian development by suppressing follicle endowment and activation and promoting follicle atresia in mice. *Molecular Endocrinology (Baltimore, Md.), 21*, 2189–2202. https://doi.org/10.1210/me.2007-0172.

Rajkovic, A., Pangas, S. A., Ballow, D., Suzumori, N., & Matzuk, M. M. (2004). NOBOX deficiency disrupts early folliculogenesis and oocyte-specific gene expression. *Science (New York, N. Y.), 305*, 1157–1159. https://doi.org/10.1126/science.1099755.

Reddy, P., Liu, L., Adhikari, D., Jagarlamudi, K., Rajareddy, S., Shen, Y., et al. (2008). Oocyte-specific deletion of *Pten* causes premature activation of the primordial follicle pool. *Science (New York, N. Y.), 319*, 611–613. https://doi.org/10.1126/science.1152257.

Ren, Y., Suzuki, H., Jagarlamudi, K., Golnoski, K., McGuire, M., Lopes, R., et al. (2015). *Lhx8* regulates primordial follicle activation and postnatal folliculogenesis. *BMC Biology, 13*, 39. https://doi.org/10.1186/s12915-015-0151-3.

Rienzi, L., Gracia, C., Maggiulli, R., LaBarbera, A. R., Kaser, D. J., Ubaldi, F. M., et al. (2017). Oocyte, embryo and blastocyst cryopreservation in ART: Systematic review and meta-analysis comparing slow-freezing versus vitrification to produce evidence for the development of global guidance. *Human Reproduction Update, 23*, 139–155. https://doi.org/10.1093/humupd/dmw038.

Rodriguez-Nuevo, A., Torres-Sanchez, A., Duran, J. M., De Guirior, C., Martinez-Zamora, M. A., & Boke, E. (2022). Oocytes maintain ROS-free mitochondrial metabolism by suppressing complex I. *Nature, 607*, 756–761. https://doi.org/10.1038/s41586-022-04979-5.

Rong, Y., Ji, S. Y., Zhu, Y. Z., Wu, Y. W., Shen, L., & Fan, H. Y. (2019). ZAR1 and ZAR2 are required for oocyte meiotic maturation by regulating the maternal transcriptome and mRNA translational activation. *Nucleic Acids Research, 47*, 11387–11402. https://doi.org/10.1093/nar/gkz863.

Schmidt, D., Ovitt, C. E., Anlag, K., Fehsenfeld, S., Gredsted, L., Treier, A. C., et al. (2004). The murine winged-helix transcription factor Foxl2 is required for granulosa cell differentiation and ovary maintenance. *Development (Cambridge, England), 131*, 933–942. https://doi.org/10.1242/dev.00969.

Shan, L. Y., Tian, Y., Liu, W. X., Fan, H. T., Li, F. G., Liu, W. J., et al. (2023). LSM14B controls oocyte mRNA storage and stability to ensure female fertility. *Cellular and Molecular Life Sciences, 80*, 247. https://doi.org/10.1007/s00018-023-04898-2.

Shimada, R., Kato, Y., Takeda, N., Fujimura, S., Yasunaga, K. I., Usuki, S., et al. (2023). STRA8-RB interaction is required for timely entry of meiosis in mouse female germ cells. *Nature Communications, 14*, 6443. https://doi.org/10.1038/s41467-023-42259-6.

Shimamoto, S., Nishimura, Y., Nagamatsu, G., Hamada, N., Kita, H., Hikabe, O., et al. (2019). Hypoxia induces the dormant state in oocytes through expression of *Foxo3*. *Proceedings of the National Academy of Sciences of the United States of America, 116*, 12321–12326. https://doi.org/10.1073/pnas.1817223116.

Soygur, B., Jaszczak, R. G., Fries, A., Nguyen, D. H., Malki, S., Hu, G., et al. (2021). Intercellular bridges coordinate the transition from pluripotency to meiosis in mouse fetal oocytes. *Science Advances, 7*, eabc6747. https://doi.org/10.1126/sciadv.abc6747.

Spradling, A. C., Niu, W., Yin, Q., Pathak, M., & Maurya, B. (2022). Conservation of oocyte development in germline cysts from *Drosophila* to mouse. *ELife, 11*, e83230. https://doi.org/10.7554/eLife.83230.

Stoop, D., Cobo, A., & Silber, S. (2014). Fertility preservation for age-related fertility decline. *Lancet, 384*, 1311–1319. https://doi.org/10.1016/S0140-6736(14)61261-7.

Sun, X., Su, Y., He, Y., Zhang, J., Liu, W., Zhang, H., et al. (2015). New strategy for in vitro activation of primordial follicles with mTOR and PI3K stimulators. *Cell Cycle (Georgetown, Tex.), 14*, 721–731. https://doi.org/10.1080/15384101.2014.995496.

Suzuki, N., Yoshioka, N., Takae, S., Sugishita, Y., Tamura, M., Hashimoto, S., et al. (2015). Successful fertility preservation following ovarian tissue vitrification in patients with primary ovarian insufficiency. *Human Reproduction, 30*, 608–615. https://doi.org/10.1093/humrep/deu353.

Tanimoto, R., Sekii, K., Morohaku, K., Li, J., Pepin, D., & Obata, Y. (2021). Blocking estrogen-induced AMH expression is crucial for normal follicle formation. *Development (Cambridge, England), 148*, dev197459. https://doi.org/10.1242/dev.197459.

Teilmann, S. C. (2005). Differential expression and localisation of connexin-37 and connexin-43 in follicles of different stages in the 4-week-old mouse ovary. *Molecular and Cellular Endocrinology, 234*, 27–35. https://doi.org/10.1016/j.mce.2004.10.014.

Tu, W., Ni, D., Yang, H., Zhao, F., Yang, C., Zhao, X., et al. (2022). Deciphering the dynamics of the ovarian reserve in cynomolgus monkey through a quantitative morphometric study. *Science Bulletin, 67*, 1854–1859. https://doi.org/10.1016/j.scib.2022.08.027.

Uhlenhaut, N. H., Jakob, S., Anlag, K., Eisenberger, T., Sekido, R., Kress, J., et al. (2009). Somatic sex reprogramming of adult ovaries to testes by FOXL2 ablation. *Cell, 139*, 1130–1142. https://doi.org/10.1016/j.cell.2009.11.021.

Vaccari, S., Horner, K., Mehlmann, L. M., & Conti, M. (2008). Generation of mouse oocytes defective in cAMP synthesis and degradation: endogenous cyclic AMP is essential for meiotic arrest. *Developmental Biology, 316*, 124–134. https://doi.org/10.1016/j.ydbio.2008.01.018.

Vaccari, S., Weeks, J. N., Hsieh, M., Menniti, F. S., & Conti, M. (2009). Cyclic GMP signaling is involved in the luteinizing hormone-dependent meiotic maturation of mouse oocytes. *Biology of Reproduction, 81*, 595–604. https://doi.org/10.1095/biolreprod.109.077768.

Vernet, N., Condrea, D., Mayere, C., Feret, B., Klopfenstein, M., Magnant, W., et al. (2020). Meiosis occurs normally in the fetal ovary of mice lacking all retinoic acid receptors. *Science Advances, 6*, eaaz1139. https://doi.org/10.1126/sciadv.aaz1139.

Wang, T., Babayev, E., Jiang, Z., Li, G., Zhang, M., Esencan, E., et al. (2018). Mitochondrial unfolded protein response gene *Clpp* is required to maintain ovarian follicular reserve during aging, for oocyte competence, and development of pre-implantation embryos. *Aging Cell, 17*, e12784. https://doi.org/10.1111/acel.12784.

Wang, H., Cai, H., Wang, X., Zhang, M., Liu, B., Chen, Z., et al. (2019). HDAC3 maintains oocyte meiosis arrest by repressing amphiregulin expression before the LH surge. *Nature Communications, 10*, 5719. https://doi.org/10.1038/s41467-019-13671-8.

Wang, C., & Lin, H. (2021). Roles of piRNAs in transposon and pseudogene regulation of germline mRNAs and lncRNAs. *Genome Biology, 22*, 27. https://doi.org/10.1186/s13059-020-02221-x.

Wang, Z., Liu, C. Y., Zhao, Y., & Dean, J. (2020). FIGLA, LHX8 and SOHLH1 transcription factor networks regulate mouse oocyte growth and differentiation. *Nucleic Acids Research, 48*, 3525–3541. https://doi.org/10.1093/nar/gkaa101.

Wang, Z. P., Mu, X. Y., Guo, M., Wang, Y. J., Teng, Z., Mao, G. P., et al. (2014). Transforming growth factor-beta signaling participates in the maintenance of the primordial follicle pool in the mouse ovary. *Journal of Biological Chemistry, 289*, 8299–8311. https://doi.org/10.1074/jbc.M113.532952.

Wang, Y., Teng, Z., Li, G., Mu, X., Wang, Z., Feng, L., et al. (2015). Cyclic AMP in oocytes controls meiotic prophase I and primordial folliculogenesis in the perinatal mouse ovary. *Development (Cambridge, England), 142*, 343–351. https://doi.org/10.1242/dev.112755.

Wang, C., Zhou, B., & Xia, G. (2017). Mechanisms controlling germline cyst breakdown and primordial follicle formation. *Cellular and Molecular Life Sciences, 74*, 2547–2566. https://doi.org/10.1007/s00018-017-2480-6.

Wassarman, P. M., & Litscher, E. S. (2016). A bespoke coat for eggs: Getting ready for fertilization. *Current Topics in Developmental Biology, 117*, 539–552. https://doi.org/10.1016/bs.ctdb.2015.10.018.

Wassarman, P. M., & Litscher, E. S. (2018). The mouse egg's zona pellucida. *Current Topics in Developmental Biology, 130*, 331–356. https://doi.org/10.1016/bs.ctdb.2018.01.003.

Wassarman, P. M., & Litscher, E. S. (2021). Zona pellucida genes and proteins: Essential players in mammalian oogenesis and fertility. *Genes, 12*, 1266. https://doi.org/10.3390/genes12081266.

Wassarman, P. M., & Mortillo, S. (1991). Structure of the mouse egg extracellular coat, the zona pellucida. *International Review of Cytology, 130*, 85–110. https://doi.org/10.1016/s0074-7696(08)61502-8.

Xu, B., Li, Z., Li, S., Ke, H., Zhang, Q., Qin, Y., et al. (2022). Pathogenic variants in TSC2 might cause premature ovarian insufficiency through activated mTOR induced hyperactivation of primordial follicles. *Fertility and Sterility, 118*, 1139–1149. https://doi.org/10.1016/j.fertnstert.2022.08.853.

Xu, X., Mu, L., Li, L., Liang, J., Zhang, S., Jia, L., et al. (2022). Imaging and tracing the pattern of adult ovarian angiogenesis implies a strategy against female reproductive aging. eabi8683 *Science Advances, 8*. https://doi.org/10.1126/sciadv.abi8683.

Yan, H., Wen, J., Zhang, T., Zheng, W., He, M., Huang, K., et al. (2019). Oocyte-derived E-cadherin acts as a multiple functional factor maintaining the primordial follicle pool in mice. *Cell Death & Disease, 10*, 160. https://doi.org/10.1038/s41419-018-1208-3.

Yan, H., Zhang, J., Wen, J., Wang, Y., Niu, W., Teng, Z., et al. (2018). CDC42 controls the activation of primordial follicles by regulating PI3K signaling in mouse oocytes. *BMC Biology, 16*, 73. https://doi.org/10.1186/s12915-018-0541-4.

Yokobayashi, S., Liang, C. Y., Kohler, H., Nestorov, P., Liu, Z., Vidal, M., et al. (2013). PRC1 coordinates timing of sexual differentiation of female primordial germ cells. *Nature, 495*, 236–240. https://doi.org/10.1038/nature11918.

Yu, C., Ji, S. Y., Sha, Q. Q., Sun, Q. Y., & Fan, H. Y. (2015). CRL4-DCAF1 ubiquitin E3 ligase directs protein phosphatase 2A degradation to control oocyte meiotic maturation. *Nature Communications, 6*, 8017. https://doi.org/10.1038/ncomms9017.

Yuan, Y. F., Zhai, R., Liu, X. M., Khan, H. A., Zhen, Y. H., & Huo, L. J. (2014). SUMO-1 plays crucial roles for spindle organization, chromosome congression, and chromosome segregation during mouse oocyte meiotic maturation. *Molecular Reproduction and Development, 81*, 712–724. https://doi.org/10.1002/mrd.22339.

Zaffagnini, G., Cheng, S., Salzer, M. C., Pernaute, B., Duran, J. M., Irimia, M., et al. (2024). Mouse oocytes sequester aggregated proteins in degradative super-organelles. *Cell, 187*, 1109–1126. https://doi.org/10.1016/j.cell.2024.01.031.

Zhang, H., Adhikari, D., Zheng, W., & Liu, K. (2013). Combating ovarian aging depends on the use of existing ovarian follicles, not on putative oogonial stem cells. *Reproduction (Cambridge, England), 146*, R229–R233. https://doi.org/10.1530/REP-13-0202.

Zhang, T., He, M., Zhao, L., Qin, S., Zhu, Z., Du, X., et al. (2021). HDAC6 regulates primordial follicle activation through mTOR signaling pathway. *Cell Death & Disease, 12*, 559. https://doi.org/10.1038/s41419-021-03842-1.

Zhang, H., Liu, L., Li, X., Busayavalasa, K., Shen, Y., Hovatta, O., et al. (2014). Life-long in vivo cell-lineage tracing shows that no oogenesis originates from putative germline stem cells in adult mice. *Proceedings of the National Academy of Sciences of the United States of America, 111*, 17983–17988. https://doi.org/10.1073/pnas.1421047111.

Zhang, H., & Liu, K. (2015). Cellular and molecular regulation of the activation of mammalian primordial follicles: Somatic cells initiate follicle activation in adulthood. *Human Reproduction Update, 21*, 779–786. https://doi.org/10.1093/humupd/dmv037.

Zhang, H., Risal, S., Gorre, N., Busayavalasa, K., Li, X., Shen, Y., et al. (2014). Somatic cells initiate primordial follicle activation and govern the development of dormant oocytes in mice. *Current Biology, 24*, 2501–2508. https://doi.org/10.1016/j.cub.2014.09.023.

Zhang, M., Su, Y. Q., Sugiura, K., Xia, G., & Eppig, J. J. (2010). Granulosa cell ligand NPPC and its receptor NPR2 maintain meiotic arrest in mouse oocytes. *Science (New York, N. Y.), 330*, 366–369. https://doi.org/10.1126/science.1193573.

Zhang, T., Tong, Y., Zhu, R., Liang, Y., Zhang, J., Hu, C., et al. (2024). HDAC6-dependent deacetylation of NGF dictates its ubiquitination and maintains primordial follicle dormancy. *Theranostics, 14*, 2345–2366. https://doi.org/10.7150/thno.95164.

Zhang, Y., Wang, Y., Feng, X., Zhang, S., Xu, X., Li, L., et al. (2021). Oocyte-derived microvilli control female fertility by optimizing ovarian follicle selection in mice. *Nature Communications, 12*, 2523. https://doi.org/10.1038/s41467-021-22829-2.

Zhang, J., Yan, L., Wang, Y., Zhang, S., Xu, X., Dai, Y., et al. (2020). In vivo and in vitro activation of dormant primordial follicles by EGF treatment in mouse and human. *Clinical and Translational Medicine, 10*, e182. https://doi.org/10.1002/ctm2.182.

Zhang, X., Zhang, W., Wang, Z., Zheng, N., Yuan, F., Li, B., et al. (2022). Enhanced glycolysis in granulosa cells promotes the activation of primordial follicles through mTOR signaling. *Cell Death & Disease, 13*, 87. https://doi.org/10.1038/s41419-022-04541-1.

Zhang, W., Zhang, R., Wu, L., Zhu, C., Zhang, C., Xu, C., et al. (2023). NLRP14 deficiency causes female infertility with oocyte maturation defects and early embryonic arrest by impairing cytoplasmic UHRF1 abundance. *Cell Reports, 42*, 113531. https://doi.org/10.1016/j.celrep.2023.113531.

Zhang, H., Zheng, W., Shen, Y., Adhikari, D., Ueno, H., & Liu, K. (2012). Experimental evidence showing that no mitotically active female germline progenitors exist in post-natal mouse ovaries. *Proceedings of the National Academy of Sciences of the United States of America, 109*, 12580–12585. https://doi.org/10.1073/pnas.1206600109.

Zhao, Y., Zhang, Y., Li, J., Zheng, N., Xu, X., Yang, J., et al. (2018). MAPK3/1 participates in the activation of primordial follicles through mTORC1-KITL signaling. *Journal of Cellular Physiology, 233*, 226–237. https://doi.org/10.1002/jcp.25868.

Zheng, W., Zhang, H., & Liu, K. (2014). The two classes of primordial follicles in the mouse ovary: Their development, physiological functions and implications for future research. *Molecular Human Reproduction, 20*, 286–292. https://doi.org/10.1093/molehr/gau007.

Zheng, W., Zhang, H., Gorre, N., Risal, S., Shen, Y., & Liu, K. (2014). Two classes of ovarian primordial follicles exhibit distinct developmental dynamics and physiological functions. *Human Molecular Genetics, 23*, 920–928. https://doi.org/10.1093/hmg/ddt486.

Zheng, W., Zhang, T., Zhao, T., Zhu, Z., Qin, S., Yan, H., et al. (2023). cAMP controls the balance between dormancy and activation of primordial follicles in mouse ovaries. *PNAS Nexus, 2*, pgad055. https://doi.org/10.1093/pnasnexus/pgad055.

Zhou, C. J., Wang, X. Y., Dong, Y. H., Wang, D. H., Han, Z., Zhang, X. J., et al. (2022). CENP-F-dependent DRP1 function regulates APC/C activity during oocyte meiosis I. *Nature Communications, 13*, 7732. https://doi.org/10.1038/s41467-022-35461-5.

CHAPTER TEN

The characteristics of the calcium signals that activate mammalian eggs at fertilization

Karl Swann*
School of Biosciences, Cardiff University, Sir Martin Evans Building, Museum Avenue, Cardiff, Wales, United Kingdom
*Corresponding author. e-mail address: Swannk1@cardiff.ac.uk

Contents

Abstract

Gamete membrane fusion in mammals brings the paternal genome into the cytoplasm of the egg. It also enables signals to pass from the sperm into the egg to trigger the completion of meiosis and the start of embryo development. The essential signal to activate development in all mammals studied, consists of a series of transient increases in the cytosolic Ca^{2+} concentration driven by cycles of $InsP_3$ production. This review focusses on the characteristics of these sperm-induced Ca^{2+} signals. I consider how some specific features of sperm-derived phospholipase C-zeta (PLCζ), along with the known properties of the type 1 $InsP_3$ receptor, provide a basis for understanding the mechanisms of the dynamic changes in Ca^{2+} observed in fertilizing eggs. I describe how the PLCζ targeting of cytoplasmic vesicles in the egg cytoplasm, that contain $PI(4,5)P_2$, is necessary to explain the rapid waves associated with the rising phase of each Ca^{2+} transient. I also discuss the importance of the repetitive Ca^{2+} rises for egg activation and the way mitochondrial ATP production may modulate Ca^{2+} release in eggs. Finally, I consider the role that a sperm-induced ATP increase may play in the egg activation process.

Current Topics in Developmental Biology, Volume 162
ISSN 0070-2153, https://doi.org/10.1016/bs.ctdb.2024.12.002

1. The sperm-induced Ca^{2+} signals that cause egg activation

The Ca^{2+} signals that are initiated by the sperm at fertilization in mammalian eggs consists of a series of oscillations in the cytosolic Ca^{2+} concentration (Miyazaki, 2007). In all mammalian species studied, a series of repetitive increases in Ca^{2+} have been shown to persist for several hours after gamete fusion (Cuthbertson & Cobbold, 1985; Igusa & Miyazaki, 1986; Kline & Kline, 1992; Miyazaki & Igusa, 1981; Taylor, Lawrence, Kingsland, Biljan, & Cuthbertson, 1993). These Ca^{2+} oscillations are mediated by increased production of inositol (1,4,5)-trisphosphate ($InsP_3$) and Ca^{2+} release from the endoplasmic reticulum via the $InsP_3$ receptor (IP3R), which also acts as the release channel (Fissore, Longo, Anderson, Parys, & Ducibella, 1999; Kline & Kline, 1994; Malcuit et al. 2005; Miyazaki, 1988; Miyazaki et al. 1992; Swann, 1992). In mouse eggs the Ca^{2+} oscillations start within about a minute of gamete fusion and consist of 10–20 separate transient Ca^{2+} increases that each last for a few minutes (Cuthbertson & Cobbold, 1985; Igusa & Miyazaki, 1986; Kline & Kline, 1992; Lawrence, Whitaker, & Swann, 1997; Mehlmann & Kline, 1994; Miyazaki et al. 1986). The spike-like increases in Ca^{2+} occur about 10–20 min apart. Fig. 1 shows a typical series of Ca^{2+} oscillations that we measure using Oregon Green BAPTA dextran. This is a fluorescent Ca^{2+} sensitive dye that we find gives a very good signal to noise. We typically

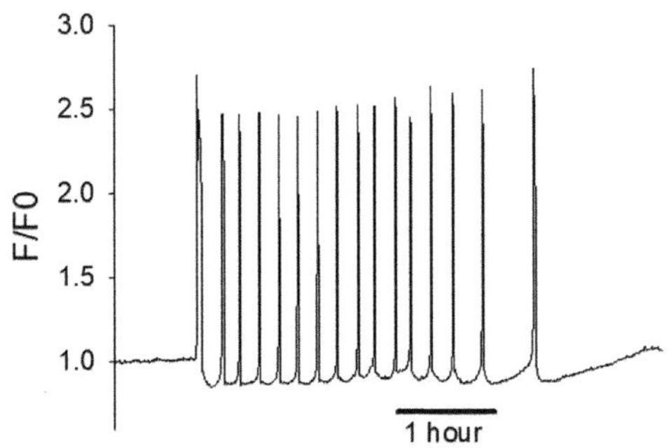

Fig. 1 Ca^{2+} oscillations at fertilization in mouse eggs. This trace shows a recording of cytosolic Ca^{2+} from a single mouse egg that was injected with the Ca^{2+} sensitive dye Oregon Green BAPTA dextran. The fluorescence level from the whole egg was measured every 10 s during in vitro fertilization. Full experimental details are given elsewhere (Ikie-Eshalomi et al. 2023). The fluorescence intensity is plotted as the fluorescence (F) divided by the initial fluorescence value (F0), hence F/F0. Ca^{2+} oscillations persisted for several hours.

observe 10 to 20 large Ca^{2+} spikes, that occur from a baseline Ca^{2+} level between spikes that is slightly lower than that in unfertilized eggs. These Ca^{2+} oscillations occur as the egg completes meiosis, forming a second polar body and two pronuclei (Larman, Saunders, Carroll, Lai, & Swann, 2004; Marangos, FitzHarris, & Carroll, 2003). The list of species where sperm-induced Ca^{2+} oscillations in eggs have been demonstrated includes mice, rats, hamsters, humans, pigs, cows and goats (Ben-Yosef, Oron, & Shalgi, 1995; Cuthbertson & Cobbold, 1985; Igusa & Miyazaki, 1986; Jellerette et al. 2006; Sun, Hoyland, Huang, Mason, & Moor, 1992; Sun, Bradshaw, Galli, & Moor, 1994; Taylor et al. 1993). A series of cytosolic Ca^{2+} oscillations are also seen in mouse and human eggs after intracytoplasmic sperm injection (ICSI) that is widely used as part of IVF treatment (Nakano, Shirakawa, Mitsuhashi, Kuwabara, & Miyazaki, 1997; Tesarik, Sousa, & Testart, 1994; Yanagida et al. 2001).

In mice and hamsters, the pattern of Ca^{2+} increases in fertilizing eggs have been measured with temporal resolution of seconds and spatial resolution of a few microns. The initial Ca^{2+} increase spreads in a wave-like manner from the region of sperm egg fusion to the antipode in about 5–10 s (Deguchi, Shirakawa, Oda, Mohri, & Miyazaki, 2000; Miyazaki et al. 1986). Once it has spread across the egg the Ca^{2+} levels stay high for several minutes after which Ca^{2+} decreases over a few seconds in all regions of the egg at the same time. Over a period of 10–20 min following gamete fusion, the Ca^{2+} waves associated with each transient increase in speed and eventually (~30 min) they cross the egg much more rapidly, in about 1 s (Deguchi et al. 2000; Miyazaki et al. 1986). This change in wave speed is also manifest as an increase in the rate of rise of each Ca^{2+} transient (Swann & Lai, 2013). The series of oscillations then stops just before the time of pronuclear formation. Studies in other species, including human eggs, also suggest that fertilization induced Ca^{2+} oscillations can last for several hours (Taylor et al. 1993). In mouse, hamster, and pig eggs it has been shown that Ca^{2+} oscillations are dependent upon Ca^{2+} influx (Cheek et al. 1993; Igusa & Miyazaki, 1983; Zhang, Chao, Jaeger, Papp, & Machaty, 2018). The Ca^{2+} channels have been identified in mouse eggs and their roles in fertilization is discussed elsewhere (Stein, Savy, Williams, & Williams, 2020; Wakai, Mehregan, & Fissore, 2019). I will concentrate here on other specific aspects of Ca^{2+} release via IP3Rs.

There are numerous publications showing measurements of Ca^{2+} oscillations in fertilizing mammalian eggs, but the amplitude of the Ca^{2+} transients is typically not calibrated. This is also the case in Fig. 1 where the y-axis is plotted as the ratio of fluorescence divided by the fluorescence before the start of Ca^{2+} oscillations (F/F0). The absolute concentration of the peak of each Ca^{2+} transient is worth considering because it is significant for understanding the mechanisms of oscillations that will be considered later. The most common dye used to measure Ca^{2+} in eggs is fura2. Some studies using Ca^{2+} sensitive fluorescent dyes like fura2 have suggested that the Ca^{2+} peaks are above 1 μM (Mehlmann & Kline, 1994), whilst other studies have suggested they might be in the range of 300 to 500 nM (Cheek et al. 1993; Miyazaki et al. 1992). These differences reflect the difficulties in calibrating Ca^{2+} sensitive fluorescence dyes in the cytosol of egg. Attempts to calibrate fluorescent dyes in vitro is also compromised by the fact that many dyes have a different affinity for Ca^{2+} in cytosol compared with calibration solutions, and this may require correction factors. Fura2 dye is also problematic because it has a high affinity for Ca^{2+} and is poor at resolving the amplitude of Ca^{2+} transients above 1 μM. Another issue is that many researchers use acetoxymethyl ester versions of dyes for loading them into eggs. This leads to dye compartmentalizing into organelles in eggs, so that much of the fluorescent signal is not emanating from the cytosol (Carroll, Swann, Whittingham, & Whitaker, 1994). We microinject dextran-linked Ca^{2+} dyes into eggs which provides a prolonged signal that is entirely derived from the cytosolic compartment, but even with these dextran-linked dyes it is difficult to obtain a fluorescence maximum and minimum which are need for the in situ calibration. Hence, despite their widespread use, fluorescent dyes have not provided an unequivocal determination of either the resting or the peak Ca^{2+} concentrations in fertilizing eggs. It is worth noting that if we assume the resting Ca^{2+} concentration in eggs is 100 nM, which is typical of mammalian cells, then the peak Ca^{2+} transients in mouse eggs in the experiment shown in Fig. 1 are generally > 2 μM. This is consistent with some older studies on Ca^{2+} oscillations in mammalian eggs.

The first studies that measured and calibrated Ca^{2+} oscillations in fertilizing mouse and human eggs used aequorin to measure cytosolic Ca^{2+} (Cuthbertson & Cobbold, 1985; Taylor et al. 1993). The luminescence from aequorin can be calibrated in drops of Ca^{2+} EGTA buffers and is less prone to problems in extrapolating from such buffers to conditions in the cytosol because aequorin is a protein. Furthermore, the calibration involves egg lysis at the end of an experiment which simply involves integrating all the light emitted. Aequorin is not ideal for measuring resting Ca^{2+}, but it is effective at higher Ca^{2+} concentrations and this study showed that the peak of each sperm-induced Ca^{2+} transient is close to 2 μM. This is consistent with another study in mouse eggs that used a low affinity Ca^{2+} dye (MagFura2) to measure the peaks of initial Ca^{2+} transients at fertilization (Halet, Tunwell, Parkinson, & Carroll, 2004). MagFura2 is not effective for measuring resting Ca^{2+}, but it can faithfully record the peak of each Ca^{2+} transient. The data with MagFura2 suggests that sperm induced Ca^{2+} transients are of an amplitude of between 1 and 3 μM, and this lasts for at least 1 min (Halet et al. 2004). Increases in cytosolic free Ca^{2+} around 2 μM will allow for the maximum stimulation of calmodulin which transduces the signal downstream. The calmodulin dependent protein kinase II (CAMKII) has been shown to be stimulated at fertilization by each rise in Ca^{2+} (Markoulaki, Matson, Abbott, & Ducibella, 2003; Markoulaki, Matson, & Ducibella, 2004), and fluorescent calmodulin indicators also show full stimulation of CAM kinases in response to each Ca^{2+} rise (Ardestani, West, Maresca, Fissore, & Stratton, 2019). The most important processes that are targets for Ca^{2+} and calmodulin dependent kinases are meiotic resumption and cortical granule exocytosis. The mechanism of exit from meiotic metaphase arrest is mediated by CAMKII (Knott et al. 2006). The exocytosis of cortical granules plays a key role in modifying the zona pellucida and blocking polyspermy (Evans, 2020). This is triggered by myosin light chain kinase that is also stimulated by Ca^{2+} binding to calmodulin (Matson, Markoulaki, & Ducibella, 2006).

We know that Ca^{2+} is the physiological stimulus for the calmodulin dependent kinases because it has been shown that a rise in cytosolic Ca^{2+} is essential for mouse egg activation at fertilization. Loading mouse eggs with the Ca^{2+} chelator BAPTA can be done in a way to prevent all Ca^{2+} increases at fertilization (Kline & Kline, 1992). BAPTA loading does not prevent sperm-egg fusion, but it blocks the two key markers of egg activation namely meiotic resumption and cortical granule exocytosis. Hence, if there are no Ca^{2+} increases there is no egg activation or development. Lower concentrations of BAPTA-AM can be used to stop most oscillations, but a single small Ca^{2+} transient can still occur (Kline & Kline, 1992). In this case pronuclear formation can still be prevented. However, the use of BAPTA-AM to block Ca^{2+} oscillations can also inhibit protein synthesis, and the inhibition of protein synthesis can in itself promote egg activation because it leads to loss of cyclin B (Lawrence, Ozil, & Swann, 1998; Moses & Kline, 1995). Hence, interpreting the effects of adding BAPTA-AM to oscillating eggs can be problematic. BAPTA loading of eggs can show that a Ca^{2+} change is necessary for initial meiotic resumption and for exocytosis, but the more subtle question of whether mammalian eggs require multiple Ca^{2+} transients to undergo full activation and development is more difficult to address.

2. The importance of multiple Ca^{2+} transients in egg activation

The question of how many Ca^{2+} transients, of whatever amplitude or duration, are needed to activate mammalian eggs was first considered over 20 years ago. It has returned as a point of discussion because of the clinical use of Ca^{2+} ionophores in oocyte activation after fertilization failure that occasionally occurs in ICSI. To investigate the role of multiple Ca^{2+} transients in egg activation, one can stimulate eggs with different patterns of Ca^{2+} oscillations. A single large Ca^{2+} increase can be triggered in mouse eggs by incubation in 7% ethanol, and this method was commonly used to induce parthenogenetic activation in mouse eggs in the 1980s (Cuthbertson, Whittingham, & Cobbold, 1981). This causes a large Ca^{2+} rise that persists for as long as ethanol is present, which is typically ~7 min (Swann & Ozil, 1994). Incubation in Ca^{2+} ionophores, such as A23187 or ionomycin, also causes one Ca^{2+} transient and can activate mouse or human eggs. These data imply that a single Ca^{2+} increase can activate eggs. However, both ethanol and ionophore generally cause a larger and longer lasting Ca^{2+} increase than the initial one seen at fertilization (Rinaudo, Pepperell, Buradgunta, Massobrio, & Keefe, 1997; Swann & Ozil, 1994). In mouse eggs treated with Ca^{2+} ionophore the activation rates, as judged by pronuclear formation, are often low (Moses, Kline, & Masui, 1995). Furthermore, mouse eggs are generally only activated when the eggs are aged in vitro before stimulation, and freshly ovulated mouse eggs generally do not activate in response to Ca^{2+} ionophore treatment. In human eggs, the first study using Ca^{2+} ionophores reported rates of ~60% egg activation (Winston, Johnson, Pickering, & Braude, 1991), but then two subsequent studies found that either no eggs, or only 16% of eggs, activated using Ca^{2+} ionophores (Balakier & Casper, 1993; Rinaudo et al. 1997; Winston et al. 1991). There are numerous clinical studies suggesting that A23187 can be effective in activating eggs after ICSI, when previous cycles of treatment had been ineffective (Murugesu et al. 2017). In most cases where ionophores are found to be effective, the eggs from the previous ICSI cycle were used as the control group. More recent studies on human eggs, using a parallel control group, reported that a single application of A23187 was ineffective in activating unfertilized (aged) human eggs (Esbert, Carmody, Ballesteros, Seli, & Scott, 2022). A single large Ca^{2+} rise is probably insufficient to induce human egg activation in most conditions.

In many cases in domesticated animal studies, egg activation protocols involve incubation in a Ca^{2+} ionophore, in combination with agents that cause a decrease in MPF activity such as protein kinase inhibitors, or protein synthesis inhibitors (Chung, Keefer, & Downey, 2000; Prather, Tao, & Macháty, 1999; Ross et al. 2008). This double treatment is used because the single Ca^{2+} increase with ionophore does not activate eggs effectively. A single Ca^{2+} increase can cause a decrease in MPF activity which may initiate meiotic resumption, but then a later return of MPF activity leads to the recondensation of chromosomes in a state referred to as MIII (Ducibella et al. 2002; Moses et al. 1995). In both mouse and bovine eggs, it was found that multiple Ca^{2+} increases can cause a decrease in MPF activity that is sustained and allows the complete exit from meiosis II and the formation of pronuclei (Collas, Fissore, Robl, Sullivan, & Barnes, 1993a; Collas, Sullivan, & Barnes, 1993b; Ducibella et al. 2002; Ozil & Swann, 1995). In mouse eggs the standard activation protocol involves incubation in Sr^{2+} containing medium. Medium containing millimolar Sr^{2+}, but no added Ca^{2+}, can cause long lasting Ca^{2+} oscillations in mouse and rat eggs (Kline & Kline, 1992; Tomashov-Matar et al. 2005). These oscillations often involve long duration Ca^{2+} increases but are otherwise similar in amplitude to those seen at fertilization. Unlike A23187, Sr^{2+} medium can activate freshly ovulated mouse eggs and is the most effective artificial stimulus for parthenogenetic activation up to the blastocyst stage (Bos-Mikich, Swann, & Whittingham, 1995; Ferrer-Buitrago et al. 2020). Sr^{2+} medium is very effective as a stimulus after ICSI in mouse eggs when they are injected with sperm that are either heat treated or lacking PLCζ (Ferrer-Buitrago et al. 2020; Sanusi, Yu, Nomikos, Lai, & Swann, 2015). Sr^{2+} medium is also routinely used to activate mouse eggs after nuclear transfer protocols designed for reproductive cloning. However, Sr^{2+} medium is not generally used in egg activation protocols for other species such as in humans, because it does not cause Ca^{2+} oscillations in human eggs (Lu et al. 2018; Storey, Elgmati, Wang, Knaggs, & Swann, 2021). The reasons for this lack of effect of Sr^{2+} in some species is discussed later. Regardless, the data obtained with Ca^{2+} ionophores and Sr^{2+} suggest that multiple Ca^{2+} transients are the most effective way to trigger egg activation. The most informative studies on Ca^{2+} and egg activation come from ways of mimicking Ca^{2+} transients in a more controllable manner.

The most effective way of artificially causing Ca^{2+} transients in eggs in a controlled manner is to use electroporation in medium containing Ca^{2+}. This causes Ca^{2+} influx and a large cytosolic Ca^{2+} increase (Ozil & Swann, 1995). Electroporation is applied to mammalian eggs in medium of low ionic strength with low concentrations of Ca^{2+}. The formation of pores due to the electrical field leads to an influx of Ca^{2+} from the extracellular medium. If appropriate culture medium is very rapidly perfused onto eggs after electroporation (within 200msec), the pores close and the eggs recover (Ozil & Swann, 1995). By using repeated perfusions and pulses, multiple Ca^{2+} transients can be generated with a variety of amplitudes and durations that depend on controllable variables such as the applied voltage. Using this approach Ozil and colleagues found that mouse eggs can be activated by a single large Ca^{2+} rise, but when a single increase is used, the amplitude and duration of such a rise needs to be higher for longer periods than that seen at fertilization (Ducibella et al. 2002; Ozil et al. 2005; Swann & Ozil, 1994; Tóth, Huneau, Banrezes, & Ozil, 2006). When physiologically sized Ca^{2+} transients are applied to mouse eggs, then 4 to 8 Ca^{2+} transients are only sufficient to initiate exocytosis and trigger metaphase to anaphase transition. To obtain full completion of meiosis and pronuclear formation, 24 Ca^{2+} transients were required (Ducibella et al. 2002). In addition, the number or frequency of Ca^{2+} transients can determine the timing of pronuclear formation, with higher frequency, or more total Ca^{2+} transients, resulting in more rapid pronuclear formation. These data show that in some cases a single large and rather prolonged Ca^{2+} increase can activate mouse eggs, but that a single physiologically sized Ca^{2+} increase is ineffective. Multiple physiological Ca^{2+} transients generally stimulate the greatest percentage of eggs to activate. Unfortunately, the repetitive electroporation technique has not been applied to human eggs, probably because the equipment required for the rapid perfusion is very complex and not commercially available.

3. PLCζ induced Ca^{2+} oscillations in eggs

In mammals the sperm stimulates Ca^{2+} release several seconds or minutes after sperm-egg membrane fusion, and it has been shown that fusion is necessary for the initiation of Ca^{2+} oscillations (Lawrence et al. 1997). The fusion allows the introduction of a soluble factor into the egg that can initiate Ca^{2+} release. PLCζ is the only sperm derived soluble protein that has been reproducibly shown to trigger Ca^{2+} oscillations in mammalian eggs. Microinjecting PLCζ RNA, or recombinant protein, has been found to cause Ca^{2+} oscillations in mouse, pig, cow, horse, and human eggs (Bedford-Guaus, Yoon, Fissore, Choi, & Hinrichs, 2008; Malcuit et al. 2005; Rogers et al. 2004; Saunders et al. 2002; Yoneda et al. 2006). PLCζ induced Ca^{2+} oscillations require a catalytically active site, and they cause downregulation of $InsP_3$ receptors (IP3Rs) (Malcuit et al. 2005; Saunders et al. 2002). The downregulation of IP3Rs only occurs in response to $InsP_3$ increases, so we know that PLCζ is mediating its affects via hydrolysis of phosphatidyl -inositol (4,5)-trisphosphate (PI(4,5)P2), and increased $InsP_3$ production, then leading to the opening of the $InsP_3$ receptor (IP_3R). The amount of PLCζ in single sperm is sufficient to cause a series of Ca^{2+} oscillations similar to those seen at fertilization (Saunders et al. 2002). PLCζ is remarkably effective in causing Ca^{2+} release in eggs and can trigger oscillations when present in eggs at ~10 nM. Given that mammalian eggs have considerable autofluorescence, this concentration is well below the detection limit for fluorescent proteins, and the only effective way to monitor the expression of PLCζ after injection of RNA is to use a luciferase tag. We have used this technique for many years to quantify the threshold for PLCζ induced Ca^{2+} release (Nomikos et al. 2005; Nomikos et al. 2011). PLCζ consists of four EF hand domains, X and Y catalytic domains and a C2 domain, and each domain is essential for its ability to cause Ca^{2+} oscillations in eggs (Kouchi et al. 2005; Nomikos et al. 2005). These findings and other aspects of the structure and function of PLCζ activity in eggs is covered in several other reviews (Miyazaki & Ito, 2006; Thanassoulas, Swann, Lai, & Nomikos, 2022). My interest here is to concentrate on specific aspects of PLCζ that have been linked to the characteristics of Ca^{2+} signals in eggs.

One of the key aspects of PLCζ enzyme activity studied in vitro is that it is very sensitive to Ca^{2+} (Kouchi et al. 2004; Nomikos et al. 2005) All mammalian PLC enzymes are sensitive to Ca^{2+} but PLCζ is unusual in that it is stimulated by resting Ca^{2+} concentrations in eggs. As with peak Ca^{2+} signals the resting Ca^{2+} concentration in mammalian eggs is not entirely clear, but it is likely to be around 100 nM. The resting level of Ca^{2+} will result in PLCζ being as much as 50 % active. Furthermore, when Ca^{2+} starts to increase during $InsP_3$ induced Ca^{2+} release, the production of $InsP_3$ will be enhanced by the rise in Ca^{2+} leading to a positive feedback loop of $InsP_3$ production. It's sensitivity to the physiological Ca^{2+} concentration is particularly significant when it is combined with another feature of PLCζ that is perhaps the most unusual compared to other mammalian PLC subtypes.

One of the most unusual features of PLCζ is that it does not localize to the plasma membrane (Larman et al. 2004; Yoda et al. 2004; Yu et al. 2012). Even, overexpressed GFP tagged versions of PLCζ show no localization in the plasma membrane (Yoda et al. 2004; Yu et al. 2012). Interestingly the fluorescently tagged versions PLCζ do localize to pronuclei after they form, and this localization appears to play a key role in the termination of Ca^{2+} oscillations during fertilization in the mouse (Yoda et al. 2004). However, as noted above, once one can detect the expression of PLCζ with a fluorescently tagged protein, then the expression level is an order of magnitude above the physiological concentrations. Luciferase tagged probes do not help with localization because they emit too little light, and they do not provide confocal images needed for subcellular resolution. The localization is best visualized with techniques that contain some signal amplification step, such as indirect immunofluorescence. Using a myc-tagged version of PLCζ, that was expressed at physiological levels, the main localization of PLCζ has been shown to be on small vesicles that are distributed throughout the egg cytoplasm (Yu et al. 2012). The identity of these vesicles is unclear, but they are not the endoplasmic reticulum or mitochondria. Significantly, a similar staining pattern on small cytoplasmic vesicles, was also seen with antibodies against PI(4,5)P2, and with a fluorescently tagged PI(4,5)P2 binding peptide called PBP10 (Sanders, Ashley, Moon, Woolley, & Swann, 2018; Yu et al. 2012). The data suggests that the majority of the PI(4,5)P2 in mature eggs is found in cytoplasmic vesicles and that PLCζ binds to, and presumably hydrolyzes the PI (4,5)P2, in these vesicles. The importance of these vesicles in fertilization is suggested by experiments using a simple drug intervention. We found that the drug propranolol can block sperm and PLCζ induced Ca^{2+} oscillations in mouse eggs (Sanders et al. 2018). This effect requires high concentrations and is probably related to the ability of propranolol to inhibit Golgi recycling to the endoplasmic reticulum. Whilst there may be concerns about the high concentrations used in these experiments it is important to note that propranolol does not block Sr^{2+}-induced, or $InsP_3$-induced Ca^{2+} oscillations (Sanders et al. 2018). The effects of propranolol are also fully reversible. We found that propranolol completely abolished the staining of vesicles with PBP10, which implies that the PI(4,5)P2 in vesicles is significant in generating PLCζ-induced Ca^{2+} oscillations. In addition, it suggests that the vesicles might be Golgi derived because propranolol acts upon recycling Golgi vesicles. It is worth noting that the Golgi is broken up into small vesicles in MII eggs, in a manner that resembles the Golgi 'haze' of somatic cells during mitosis. It is not clear how PLCζ might bind to the Golgi vesicles, nor how the PI(4,5)P2 is generated in the Golgi which is otherwise rich in PI(4)P.

Whilst most studies assay the effects of PLCζ induced production of $InsP_3$ and Ca^{2+} release, PLCζ it will also generate diacyclglycerol (DAG) which will be expected to stimulate protein kinase C (PKC). There is evidence that PKC is stimulated at fertilization in rat eggs (Eliyahu & Shalgi, 2002). It has been proposed to play a role in exocytosis, but as discussed above, myosin light chain kinase may well be the main target for Ca^{2+} stimulated exocytosis (Matson et al. 2006). As noted above, PLCζ is localized on cytoplasmic vesicles and the DAG generated by PLCζ should also be confined to the membrane phase of these vesicles. The consequences of this are unclear. What is known is that if one measures the production of DAG in mouse eggs using a GFP probe tagged to a C2 domain, then there is no indication of DAG production in the plasma membrane, either at fertilization or after PLCζ injection (Yu, Halet, Lai, & Swann, 2008a). This result is not due to a lack of sensitivity of the GFP-C2 domain indicator, because this probe shows clear plasma membrane localization in response to Ca^{2+} ionophores or phorbol esters (Yu et al. 2008a). Instead, the data imply that DAG is not generated in the plasma membrane of eggs during fertilization. Studies that have monitored PKC translocation, or measured PKC activity at fertilization in the plasma membrane, suggest that the conventional PKCs are stimulated by a rise in Ca^{2+} rather than by DAG (Halet, 2004).

4. The mechanism of PLCζ induced Ca^{2+} oscillations

The identification of PLCζ as the sperm factor that initiates Ca^{2+} oscillations enable an understanding and explanation of some of the characteristic features of the Ca^{2+} signals at fertilization. The two distinctive features of PLCζ that matter most are that PLCζ is stimulated by Ca^{2+} and that PLCζ is localized in vesicles throughout the cytoplasm. These two features come together in explaining the fast Ca^{2+} waves seen in fertilizing mouse and hamster eggs.

One of the characteristics of sperm-induced Ca^{2+} oscillations in mouse and hamster eggs is that, whilst the first few increases occur as slow waves of $\sim$10 μm/sec from the point of sperm-egg fusion, the later Ca^{2+} waves originate from different parts of the egg cortex and are much more rapid at over 50 μm/s (Deguchi et al. 2000; Miyazaki et al. 1986). The time of ~6 s for the initial slow Ca^{2+} wave to cross the mouse egg is consistent with the time for a molecule the size of $InsP_3$ to diffuse across an egg of ~70 μm (Swann, 2023). However, the later fast waves that cross the egg in less than 1 sec are not easily explained unless one considers the specific mechanism of Ca^{2+} oscillations.

There are two classes of models for explaining cytosolic based Ca^{2+} oscillations in cells that involve IP3Rs (Politi, Gaspers, Thomas, & Höfer, 2006; Sneyd et al. 2006). The first class is based upon the behavior of IP3Rs and was evidenced by experiments measuring single channel activity, either with isolated IP3Rs or with IP3Rs within the endoplasmic reticulum membrane from cells such as Xenopus oocytes. It was shown that Ca^{2+} has a biphasic effect on the IP3R; it opens the channel at low Ca^{2+} and then closes the channel at higher Ca^{2+} concentrations (Bezprozvanny, Watras, & Ehrlich, 1991; Finch, Turner, & Goldin, 1991). The stimulation, and then inhibition of IP3R is mediated by two different binding sites on the IP3R channel (see Fig. 2) (Foskett, White, Cheung, & Mak, 2007). The biphasic effect of Ca^{2+} provides a basis for a class of models of Ca^{2+} oscillations that involve stimulation and then inhibition of IP3R induced Ca^{2+} release (Atri, Amundson, Clapham, & Sneyd, 1993). These are referred to as IP3R based models since the IP3R provides the key element of the positive and

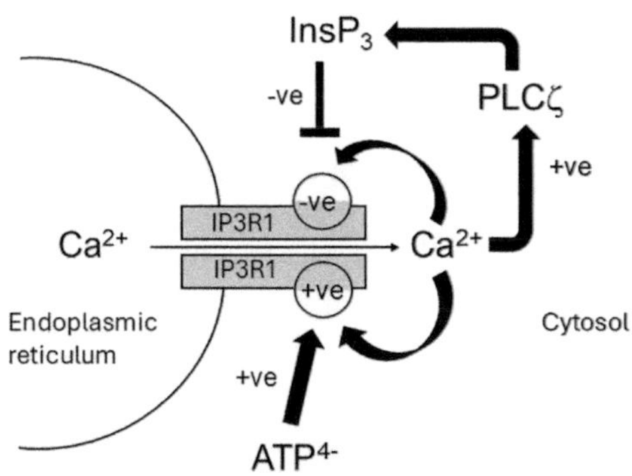

Fig. 2 A model of how $InsP_3$ and Ca^{2+} modulate the IP3R in eggs during fertilization. The IP3R1 is a Ca^{2+} channel and represented by the two blocks in the middle. IP3R1s have a Ca^{2+} binding site that opens the channel (+ve) at low Ca^{2+} (<μM) and one that closes the channel (–ve) at higher Ca^{2+} concentrations. Ca^{2+} released into the cytosol stimulates PLCζ to make $InsP_3$. The $InsP_3$ then opens the IP3R1 by inhibiting the inhibitory (–ve) Ca^{2+} binding site. ATP^{4-} promotes opening of the IP3R1 by enhancing the Ca^{2+} sensitivity of the stimulatory Ca^{2+} (+ve) binding site (Foskett et al. 2007).

negative feedback needed for oscillations. These models suggest that at fertilization the sperm causes a prolonged and constant increase in $InsP_3$ concentration. This model is supported by the finding that sustained microinjection of $InsP_3$ into hamster or mouse eggs causes Ca^{2+} oscillations (Galione, Swann, Georgiou, & Whitaker, 1994; Swann, 1992; Swann, Igusa, & Miyazaki, 1989). Also, microinjecting the potent IP3R agonist, adenophostin, causes Ca^{2+} oscillations in mouse and bovine eggs (Malcuit et al. 2005; Sato et al. 1998). Despite these observations, the effects of $InsP_3$ in these studies is not compelling because the oscillations do not effectively mimic the Ca^{2+} response at fertilization. For example, the $InsP_3$-induced oscillations in hamster eggs, are of higher frequency and lower amplitude than those seen at fertilization. Regardless, this model of Ca^{2+} oscillations is still used many schematics of $InsP_3$ signaling at fertilization.

The second class of models involves a different positive feedback loop in which the Ca^{2+} released via IP3Rs stimulates PLCζ to make more $InsP_3$, and hence increase Ca^{2+} release (Politi et al. 2006; Sneyd et al. 2006). These can be referred to as PLC based models since they rely on Ca^{2+} stimulated PLC activity. There is very good evidence for Ca^{2+} stimulated PLCζ as discussed above. Furthermore, measurements of $InsP_3$ in mouse eggs using fluorescent indicators show that there are oscillations in $InsP_3$ that accompany the Ca^{2+} oscillations, as expected by with PLC positive feedback loop models (Matsu-Ura et al. 2019; Shirakawa, Ito, Sato, Umezawa, & Miyazaki, 2006). The detection of $InsP_3$ oscillations does not, however, prove that such oscillations themselves are essential for Ca^{2+} oscillations. There is, however, a way one can determine which model of oscillations is operating in eggs.

The two different mechanisms of Ca^{2+} oscillations can be distinguished by examining the effects of sudden pulses of $InsP_3$, that can be delivered into the egg cytoplasm by uncaging $InsP_3$ (Sneyd et al. 2006) It has been shown that in all mathematical models based upon IP3Rs, a sudden pulse of $InsP_3$ temporarily increases the frequency of oscillations (Sneyd et al. 2006). In contrast, with all models based upon Ca^{2+} stimulated $InsP_3$ production, a sudden pulse of $InsP_3$ causes an immediate Ca^{2+} transient but this does not increase the frequency of oscillations. Instead, there is a resetting of the phase of Ca^{2+} oscillations. We have shown that when a pulse of $InsP_3$ is triggered in mouse eggs, that are oscillating in response to PLCζ, there is always a resetting of the phase of Ca^{2+} oscillations (Sanders et al. 2018). This is only consistent with the Ca^{2+} stimulated PLC models. This result is unequivocal and implies that during each Ca^{2+} rise there is an increase in

$InsP_3$ that plays an active role in a positive feedback loop. Therefore, $InsP_3$ acts as a dynamic variable and $InsP_3$ oscillations are absolutely necessary for PLCζ induced Ca^{2+} oscillations.

The Ca^{2+} waves that are generated by Ca^{2+} and $InsP_3$ production are not only rapid, but they travel through the center of the egg (Deguchi et al. 2000). It is very difficult to explain this type of Ca^{2+} wave if the $InsP_3$ is being produced at the plasma membrane, because the diffusion of $InsP_3$ from the membrane to the center of the egg will take several seconds (Swann, 2023). The only way that a Ca^{2+} wavefront can propagate through the egg, at the speeds that are observed, is if $InsP_3$ is generated within the cytoplasm (Sanders et al. 2018). This idea has been suggested from previous models of Ca^{2+} waves in fertilizing ascidian oocytes (Dupont & Dumollard, 2004). In mouse eggs the vesicles that we found to contain PI(4,5)P2 appear to be about 2–3 μm apart, which is sufficiently small to allow for the diffusion of $InsP_3$ to play its role in propagating the Ca^{2+} waves (Sanders et al. 2018). Hence, the mechanism of the Ca^{2+} oscillations requires a cytoplasmic source of $InsP_3$, and the vesicular distribution of PI(4,5)P2 is not an odd epiphonema, but an absolutely necessary feature of a mature fertilizable egg.

There are two specific new issues to resolve with the PLC based $InsP_3$ positive feedback model of Ca^{2+} oscillations in eggs. First, there is the one of how the Ca^{2+} reaches $> 1\ \mu M$. InsP3 opens all IP3Rs, and the Ca^{2+} plays a role in both opening and closing IP3Rs (see Fig. 2). It is notable that studies of IP3Rs imply that Ca^{2+} cannot rise particularly high because the IP3R will be mostly closed above 2 μM Ca^{2+} , in the same way that they are mostly closed at 100 nM Ca^{2+} (Bezprozvanny et al. 1991). This is problematic for eggs because we know the Ca^{2+} increase in fertilizing eggs is around 2 μM and this state lasts for at least 1 min. Mathematical models of Ca^{2+} oscillations in somatic cells, circumnavigate this problem by introducing a time delay so that the IP3R does not close in response to Ca^{2+} until after Ca^{2+} release is initiated (Atri et al. 1993; Politi et al. 2006; Sneyd et al. 2006). This delayed IP3R closing mechanism allows the models to generate Ca^{2+} oscillations. However, the evidence for this mechanism for IP3Rs undergoing delayed closure suggests it works on a time scale of about 1 s (Finch et al. 1991), which does not allow Ca^{2+} to stay elevated for a minute or more as seen in mammalian eggs.

An alternative take on the problem of high Ca^{2+} for all IP3R models is to recognize that the concentration at which Ca^{2+} inhibits IP3Rs can be very dependent upon the concentration of $InsP_3$. The predominant form of

IP3R in mammalian eggs is the type 1 receptor (IP3R1) and a version of this type is found in *Xenopus* oocytes (Mak, McBride, & Foskett, 1998). In studies of *Xenopus* oocytes, Mak and Fosket showed, that if the $InsP_3$ concentration is increased above 100 nM, then the IP3R1 is no longer inhibited by Ca^{2+} in the physiological range, and the Ca^{2+} inhibitory site (Fig. 2) is unable to close the IP3R1 (Mak et al. 1998). Hence, if there is a substantial increase in both Ca^{2+} and $InsP_3$ at the same time, then the IP3R1s do not close until Ca^{2+} reaches levels well above 10 μM. A large and simultaneous increase in $InsP_3$ and Ca^{2+} is exactly what occurs with PLCζ induced Ca^{2+} transients. Hence, the positive feedback of Ca^{2+} and $InsP_3$ production means that $InsP_3$ overrides the inhibitory effect of Ca^{2+} on the IP3Rs. This can explain why sperm and PLCζ cause larger Ca^{2+} transients than seen after $InsP_3$ injection. However, the release system now becomes locked into a state of high Ca^{2+}, high $InsP_3$, and open IP3R1s. This brings in the second problem of how each Ca^{2+} rise is terminated. To resolve this type of problem, some models of Ca^{2+} oscillations invoke the emptying of Ca^{2+} stores to terminate a cytosolic increase (Dupont & Goldbeter, 1993). However, it is known that Ca^{2+} stores do not run out of Ca^{2+} during each Ca^{2+} peak (Wakai, Zhang, Vangheluwe, & Fissore, 2013). Instead, I suggest it is possible that the IP3R1s desensitize to $InsP_3$ itself, and hence Ca^{2+} release decreases. This is plausible because IP3R mediated Ca^{2+} release shows clear and specific desensitization to $InsP_3$ in sea urchin egg homogenates (Jones, Cruttwell, Parrington, & Swann, 1998). This idea has not been fully investigated in mathematical models of Ca^{2+} oscillations.

5. Other soluble sperm factors and Ca^{2+} release in eggs

The key experiments to establish a role of PLCζ in fertilization have involved generating male mice that produce sperm with no functional PLCζ. Several groups have now made PLCζ knockout mouse lines in which sperm are produced that do not express PLCζ, or else contain a highly truncated and non-functional version of the enzyme (Hachem et al. 2017; Kashir et al. 2024; Nozawa, Satouh, Fujimoto, Oji, & Ikawa, 2018). Sperm that lack functional PLCζ fail to cause Ca^{2+} oscillations or egg activation after ICSI (Hachem et al. 2017; Nozawa et al. 2018). However, after normal in vitro fertilization, with the sperm fusing with the egg, there are 3 to 4 Ca^{2+} spikes that appear to be delayed by ~40 mins from the time of sperm egg fusion (Nozawa et al. 2018). These delayed Ca^{2+} spikes can activate eggs and can account for a small number of live young being born after mating female mice with PLCζ deficient males (Hachem et al. 2017; Nozawa et al. 2018). The delayed nature of the Ca^{2+} spikes means that many PLCζ knockout sperm fertilized eggs do not develop an effective block to extra sperm entry and most eggs become polyspermic (Nozawa et al. 2018). These data clearly prove that PLCζ accounts for Ca^{2+} oscillations and egg activation after ICSI, but the results during normal fertilization are more complex to interpret. The data suggest that PLCζ initiates Ca^{2+} release and causes the majority of Ca^{2+} oscillations, but that some other factor, or signaling system from the sperm also causes Ca^{2+} release in eggs. It is unclear why this other mechanism, or sperm factor, is not active in ICSI. Nor why it should work with a distinct delay. It is also unclear how much this applies to human fertilization. There are now cases of men, attending IVF clinics, who have been identified as having homologous mutations in PLCζ that lead to a loss of function (Abdulsamad et al. 2023; Escoffier et al. 2016). These men and their female partners came to attention of researchers because of repeated total fertilization failure during ICSI treatment, and presumably because of failed natural conception. So, in humans a lack of functional PLCζ appears to lead to a complete loss of fertility. It is not clear if there is a genuine difference between mice and men in these findings, and if so, why this would arise.

The fact that PLCζ deficient sperm lead to delayed Ca^{2+} oscillations and cause some mouse eggs to activate, implies that another sperm factor could be involved in Ca^{2+} release. This factor has been variously referred to as the primitive, cryptic, or secondary sperm factor (Jones, 2018; Nozawa et al. 2018; Swann, 2020). It seems likely that this secondary factor would also be active at fertilization. This is based upon the observations that spermatogenesis and sperm motility seem to be completely unaffected by the lack of a functional PLCζ (Hachem et al. 2017). This implies that PLCζ cannot affect spermatogenesis and that the first time that a lack of PLCζ becomes evident is after the sperm fuses with the egg. In such case it is difficult to see how any secondary factor could be 'upregulated' to compensate for a lack of PLCζ. Hence the secondary factor is expected to enter the egg in normal fertilization. In a sense it is only a 'secondary' sperm factor by virtue of the fact that its effects are delayed about an hour after PLCζ initiates Ca^{2+} oscillations. Any secondary factor is constrained in how it might act to cause Ca^{2+} oscillations, because we have a clear idea of what factors can cause Ca^{2+} oscillations in mouse eggs. For example, increasing Ca^{2+} influx by itself does not cause Ca^{2+} oscillations in mouse eggs so that is not a viable mechanism to use (Swann & Ozil, 1994). All the evidence suggests that fertilization induced Ca^{2+} oscillations are mediated by the IP3R, which means that a secondary factor needs to stimulate the IP3R.

When considering other non-PLCζ sperm derived factors, it is notable that there have already been other proteins that have been suggested to cause Ca^{2+} oscillations in mammalian eggs. These proteins include tr-kit, PAWP and citrate synthase (Aarabi et al. 2014; Kang et al. 2020; Sette et al. 1997; Wu et al. 2007). One of the general issues with these candidate sperm factors is the reproducibility of the original findings. The original paper suggesting the tr-kit could activate mouse eggs was published in 1997 and yet no one has replicated the data, nor has anyone ever shown that tr-kit causes Ca^{2+} oscillations. The paper suggesting PAWP protein could cause Ca^{2+} oscillations in mouse and human eggs was published in 2012. We tried to replicate the results in this study but could not find any evidence that PAWP causes Ca^{2+} release in eggs (Nomikos et al. 2014; Nomikos et al. 2015). More recently it was suggested that citrate synthase might be a Ca^{2+} releasing sperm factor in mouse eggs (Kang et al. 2020). Citrate synthase is commercially available, and it is simple to test if it activates eggs but there have been no positive reports from other labs so far. We have injected citrate synthase into mouse eggs and failed to detect any Ca^{2+} oscillations. In contrast with the lack of replications in such studies, it took just two years for two independent laboratories to reproduce our key finding that PLCζ causes Ca^{2+} oscillations and egg activation in the mouse. This was followed by numerous other replication studies in other species.

6. The sensitivity of Ca^{2+} release to different stimuli

Whilst all mammalian eggs tested show Ca^{2+} release at fertilization, and in response to the microinjection of $InsP_3$ or PLCζ, there are differences in the sensitivity of eggs to various stimuli. These differences were evident in studies from the 1980s on hamster eggs. A remarkable feature of hamster eggs is that, when the zona pellucida is removed, they can be fertilized by sperm from other mammalian species and even become polyspermic, with multiple sperm entering the egg. When mouse sperm fertilize hamster eggs, they trigger Ca^{2+} oscillations and egg activation. However, it requires between 5 and 10 mouse sperm to fuse before the hamster egg generates any Ca^{2+} transients and becomes activated (Igusa, Miyazaki, & Yamashita, 1983).

Obviously one mouse sperm will cause a series of Ca^{2+} oscillations in a mouse egg, so we can deduce that the hamster egg is somehow much less sensitive than the mouse egg to the mouse factor - now known to be PLCζ. This difference in sensitivity of eggs from different species is also seen with experiments injecting $InsP_3$. $InsP_3$ can be injected continuously into eggs by inserting a sharp micropipette containing a high concentration of $InsP_3$ (Galione et al. 1994; Swann, 1994). In mouse eggs this causes sustained Ca^{2+} oscillations with pipette concentrations as low as 10 μM $InsP_3$ (Swann, 1994). In hamster eggs the same technique causes Ca^{2+} oscillations but one needs to use 1 mM $InsP_3$ in the micropipette, and the oscillations are damped and generally do not persist beyond 10 mins (Galione et al. 1994). This suggests that mouse eggs are more sensitive to $InsP_3$ than hamster eggs, and that this can explain their greater sensitivity to a fertilizing sperm.

A significant difference in sensitivity is also seen between mouse eggs and human eggs. Human PLCζ is very potent at causing Ca^{2+} oscillations in mouse eggs. It requires as little as 30 times less human PLCζ to trigger Ca^{2+} oscillations in mouse eggs than mouse PLCζ (Yu, Saunders, Lai, & Swann, 2008b). Since human PLCζ appears to cause a similar series of Ca^{2+} oscillations in human eggs to what is seen in fertilizing mouse eggs then we can deduce that the greater potency of human PLCζ must be because the human egg requires a more potent stimulus. Some of this requirement may be because the human egg is larger than the mouse egg, but the volume difference is only about 5-fold, so it appears that the human egg is somehow less sensitive to PLCζ induced signaling. As with the hamster egg, this difference is potentially explained by IP3R sensitivity. We used UV pulses to un-cage and release $InsP_3$ to stimulate IP3R mediated Ca^{2+} transients in mouse eggs and human eggs (Storey et al. 2021). The amount of UV light needed to stimulate Ca^{2+} release in human eggs is about 10 times greater than that required to cause Ca^{2+} release in a mouse egg, implying that human eggs are many times less sensitive to $InsP_3$ than mouse eggs (Storey et al. 2021).

The differences in the sensitivity of Ca^{2+} release between different species of mammalian eggs is also evident from the way they respond to Sr^{2+} medium. As noted above both mouse and rat eggs can be stimulated to undergo prolonged Ca^{2+} oscillations in Sr^{2+} medium (Kline & Kline, 1992; Lu et al. 2018; Storey et al. 2021; Tomashov-Matar et al. 2005). This is consistent with the findings that Sr^{2+} stimulates Ca^{2+} release via the IP3R (Marshall & Taylor, 1994; Sanders et al. 2018). However, human eggs do not undergo Ca^{2+} oscillations in Sr^{2+} medium (Lu et al. 2018; Storey et al. 2021).

Sr^{2+} probably enters eggs via TRPV3 channels, but both human eggs and mouse eggs have TRPV3 channels so the inability of human eggs to respond cannot be explained by a lack of Sr^{2+} influx (Lu et al. 2018). Its notable that there are no reports of pig, cow, or hamster eggs undergoing Ca^{2+} oscillations in Sr^{2+} medium, despite many studies of Ca^{2+} release in these types of eggs. The implication is that mouse and rat eggs have a much more sensitive Ca^{2+} release system than eggs from other species such as humans and hamsters. It's not obvious why there is such a marked difference in sensitivity because all mammalian eggs appear to predominantly have the same type 1 IP3R (Fissore et al. 1999; Mann, Lowther, & Mehlmann, 2010). We recently found evidence that the difference is due to the modulation of IP3R1.

Single channel recordings of IP3R1s have identified three ligands that are likely to be physiological: Ca^{2+}, $InsP_3$ and ATP (Foskett et al. 2007). The interacting roles of Ca^{2+} and $InsP_3$ are widely acknowledged, but the potential role of ATP is generally overlooked. When referring to 'ATP', it should be understood that I am referring to the total ATP in a cell or in the cytosol. However, in all cells ATP exists in two forms: $MgATP^{2-}$ and ATP^{4-}. Most of the cellular ATP is bound to Mg^{2+}, but some of the ATP is Mg^{2+} free in the form of ATP^{4-} (Grubbs, 2002). The studies of IP3R1s in *Xenopus* oocytes show that the Mg^{2+} free form ATP^{4-} can open the IP3R1 channel, and in fact that ATP^{4-} acts as an allosteric stimulator of the IP3R1 at the Ca^{2+} stimulatory site (see Fig. 2) (Mak, McBride, & Foskett, 1999). The half maximal concentration of ATP^{4-} to open IP3R1s is around 270 μM which is physiologically relevant (Mak et al. 1999). Most mammalian cells, including eggs, maintain a concentration of 1 to 4 mM total ATP. With this level of total ATP there will be about 100–200 μM of unbound ATP^{4-}. Hence ATP^{4-} is normally within the range where it could modulate IP3R1 activity and increases in ATP^{4-} could potentiate, or else help trigger, Ca^{2+} release. Its notable that ATP^{4-} is the immediate product of mitochondrial ATP production because it is exported from mitochondria via the adenine nucleotide exchanger (Foskett et al. 2007). This is important because mammalian eggs are unusual in that they generate nearly all their ATP from mitochondria (Dumollard, Campbell, Halet, Carroll, & Swann, 2008). The type 1 $InsP_3$ receptor is one of three main subtypes in mammalian cells and is the one most sensitive to ATP^{4-}, and most likely to respond in the physiological range (Tu, Wang, Nosyreva, De Smedt, & Bezprozvanny, 2005). This makes eggs particularly likely to be sensitive to an effect of ATP on Ca^{2+} release.

We recently showed that the concentration of ATP in mouse eggs affects their ability to generate Ca^{2+} oscillations in response to Sr^{2+}. Mouse eggs invariably generate Ca^{2+} oscillations when incubated in Sr^{2+} medium, but if all the carbohydrate metabolites are removed from the incubation medium, mouse eggs fail to respond to Sr^{2+} (Storey et al. 2021). The lack of metabolites causes the egg to have a low ATP level. When pyruvate is added back to such mouse eggs the ATP concentration rises back to normal levels as mitochondrial activity increases. At the same time as ATP increases, Ca^{2+} oscillations are generated. These data suggest that the concentration of ATP in mouse eggs critically affects their ability to respond to Sr^{2+}, and that with low ATP levels, Ca^{2+} oscillations are not generated. This is potentially significant because we also found that the concentration of ATP (as total ATP) in human eggs is consistently lower than that in mouse eggs, and in fact the level of ATP in a human egg is similar to that of a pyruvate deprived mouse egg (Storey et al. 2021). This implies that the relatively low ATP concentration in human eggs can explain why they do not activate in response to Sr^{2+}. This may also be the case for eggs from domestic animals. It certainly appears to be the case in hamster eggs. As discussed above, hamster eggs appear to be significantly less sensitive to $InsP_3$-induced Ca^{2+} release than mouse eggs. One study that involved measuring ATP in both mouse and hamster eggs (to look at the effects of ageing) found that the mouse egg ATP level was about six times higher than that in hamster eggs (Simsek-Duran et al. 2013). This is consistent with the hypothesis that ATP levels set the sensitivity of Ca^{2+} release in eggs. It should be noted that these measurements of ATP, with cross-species comparisons, involve egg lysis and use kits that measure the total ATP. However, if we assume that intracellular free Mg^{2+} concentrations are similar in eggs from different species, then the concentration of ATP^{4-} will be proportional to the total ATP. Hence, ATP^{4-} is expected to be significantly higher in mouse eggs compared to human or hamster eggs. Unfortunately, there are no probes to measure ATP^{4-} since firefly luciferase, and all the available fluorescent ATP probes, respond to $MgATP^{2-}$, which in most cases, is almost the same as the total ATP.

7. Changes in ATP concentration at fertilization

The oscillatory changes in cytosolic Ca^{2+} at fertilization cause an increase in mitochondrial activity. In mouse eggs it was found that each

Ca^{2+} transient leads to a distinct increase NADH and FADH levels (Dumollard et al. 2004). The increases in NADH and FADH outlast each Ca^{2+} transient. Measurements using Ca^{2+} probes inside the mitochondrial matrix show that matrix Ca^{2+} increases occur with each cytosolic Ca^{2+} increase (Dumollard et al. 2008). This suggests that each NADH or FADH change is a result of Ca^{2+} influx into mitochondria which can then stimulate mitochondrial dehydrogenases. Hence Ca^{2+} oscillations stimulate mitochondria during the process of fertilization. Stimulating mitochondrial activity may be a way of avoiding a decrease in ATP levels, because the pumping of Ca^{2+} into the endoplasmic reticulum is likely to consume a significant fraction of the egg's ATP. We know that mitochondrial ATP production is critical in maintaining Ca^{2+} oscillations because if mouse eggs are fertilized in the absence of pyruvate, to reduce ATP production, then Ca^{2+} oscillations are interrupted by a persistent high level of Ca^{2+} (Dumollard et al. 2008; Liu, Hammar, Smith, Inoue, & Keefe, 2001). Hence, the Ca^{2+} oscillations appear to act to simultaneously increase the egg's supply and demand of ATP.

The idea of Ca^{2+} mediated increased ATP supply and demand at fertilization, is however, more complex. The changes in ATP in mouse eggs can be measured by microinjecting firefly luciferase and imaging the luminescence of luciferase alongside the fluorescence from Ca^{2+} sensitive dyes. During the sperm-induced Ca^{2+} oscillations there is a substantial increase in the concentration of ATP in eggs (Campbell & Swann, 2006; Ikie-Eshalomi, Aliyev, Hoehn, Jurkowski, & Swann, 2023). Fig. 3 shows the increase in ATP alongside Ca^{2+} oscillations in a fertilizing mouse egg. The ATP increase occurs in two phases, the first starts during the initial Ca^{2+} transient, and then a secondary ATP increase occurs about 1 h later into the Ca^{2+} oscillations. The two-step change in ATP can also be seen when luciferase is targeted to the egg's mitochondria (Dumollard et al. 2008). These data imply that a substantial 'overproduction' of ATP occurs in mitochondria during fertilization.

Some of this increase in ATP concentration in fertilizing eggs can be explained by Ca^{2+} entering mitochondria and stimulating matrix dehydrogenases, as discussed above. This is shown by the fact that PLCζ can also cause an increase in ATP (Ikie-Eshalomi et al. 2023). However, the ATP increase with PLCζ is only about half of that seen at fertilization (Ikie-Eshalomi et al. 2023). Furthermore, with PLCζ there is no secondary increase in ATP. Interestingly, the secondary increase in ATP still occurs at around 1 h after sperm egg fusion even when eggs are loaded with BAPTA

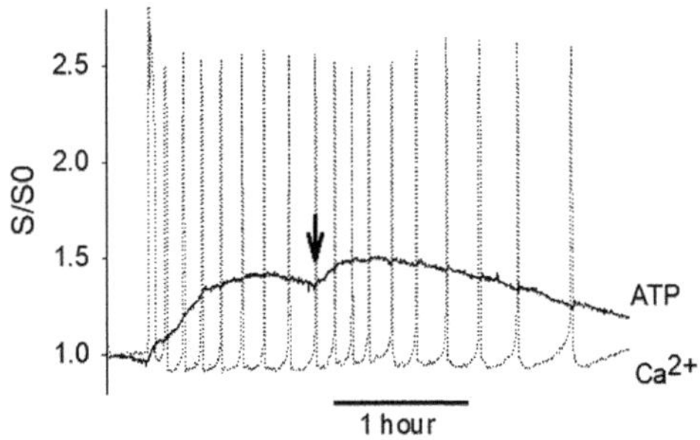

Fig. 3 Changes in ATP and Ca^{2+} at fertilization in mouse eggs. Mouse eggs were injected with firefly luciferase, to monitor ATP (solid trace), and with Oregon Green BAPTA dextran, to measure Ca^{2+} (dotted trace) as described previously (Ikie-Eshalomi et al. 2023). During fertilization in this mouse egg there were a series of Ca^{2+} oscillations and at the same time a two phased increase in ATP. The second phase of increase is only seen with sperm-induced Ca^{2+} oscillations and starts at the point indicated by the arrow. The fluorescence signals or luminescence signals (S) are plotted as intensity values divided by starting values (S0).

to prevent all but the initial Ca^{2+} transient (Ikie-Eshalomi et al. 2023). This shows that the secondary increase in ATP is sperm mediated, and not driven by PLCζ, or by Ca^{2+}. The simplest way to interpret these data is to propose that the sperm provides a stimulus, separate to PLCζ, that increases mitochondrial activity after sperm–egg fusion. It is not clear what this factor might be, but it is possible that a sperm protein, or set of proteins, targets egg mitochondria. This idea is similar to one previously suggested by Sakkas and colleagues for glucose metabolism at fertilization (Urner & Sakkas, 1999). It was found that glucose metabolism increases through both glycolysis and the pentose phosphate pathway during fertilization in mouse eggs (Urner & Sakkas, 1999). The increase in glucose metabolism is not mimicked by parthenogenetic activation and so it appears to be sperm specific. Together with our observations, the data imply that sperm stimulates increases in both glycolysis and oxidative phosphorylation in eggs in a way that does not involve PLCζ.

It is not clear what role, if any, the increase in ATP plays at fertilization in mouse eggs. It is possible that an increase in ATP concentration just reflects an imbalance between supply and demand. However, I consider this unlikely because adding excess pyruvate to eggs, does not increase the ATP concentration which suggests that ATP levels are specifically regulated (Dumollard, Carroll, Duchen, Campbell, & Swann, 2009). Overproduction of ATP is also unlikely because the increase in ATP at fertilization lasts for more than 2 h, and yet the turnover time for ATP in eggs is likely to be about 1 min (Leese, 1991). Hence, the homeostatic system controlling ATP levels has ample time to readjust. Overproduction is also literally a waste of energy for the cell. Furthermore, as judged by the changes in luciferase signal, the increase in ATP at fertilization in mouse eggs appears to be larger than the increase in ATP detected in pancreatic beta cells responding to glucose (Campbell & Swann, 2006; Ikie-Eshalomi et al. 2023; Kennedy et al. 1999). The ATP increase in beta cells plays a vital role in insulin secretion and depends upon the specific rewiring of metabolism (Rutter, Georgiadou, Martinez-Sanchez, & Pullen, 2020). It seems unusual to suppose that an even larger change of ATP in eggs during fertilization were just an accident, a type of epiphenomenon. I suggest it is more likely that the sperm induced ATP increase has a consequence, and a role to play in the process of egg activation.

One possible function of the increase is that the extra ATP promotes energy using steps in fertilization. The energy requiring usage of ATP in cells is dominated by protein turnover, and Na^{+} and Ca^{2+} pumping (Rolfe & Brown, 1997). Protein synthesis rates increase after fertilization in mouse eggs by about 40 % and it is possible that increases in ATP drives some of this change (Schultz, Letourneau, & Wassarman, 1979). Another major use of ATP, as mentioned above, will be associated with pumping Ca^{2+} back into stores, or out of the egg. A doubling of ATP levels in a cell will make little difference to the free energy available for pumping Ca^{2+}. This is because the free energy available depends upon the logarithm of the ATP concentration, and the reaction ATP $\leftrightarrow$ ADP + Pi is already ~10 orders of magnitude away from equilibrium (Wilson, 2015). However, an increase in the concentration of ATP will speed up the kinetics of Ca^{2+} pumping and this could increase the rate of Ca^{2+} uptake into stores (Verjovski-Almeida & Inesi, 1979). Increases in the activity of Ca^{2+} ATPases will lower cytosolic Ca^{2+} and hence the increase in ATP could explain why the 'inter-spike' Ca^{2+} concentration at fertilization is lower than the resting Ca^{2+} concentration before fertilization (see Fig. 1).

Another possible effect of the ATP increase could be to promote Ca^{2+} release in eggs. This is related to the stimulation of the IP3R by ATP^{4-} discussed above. It is notable that the secondary increase in ATP starts about an hour after the initial Ca^{2+} spike at fertilization. This secondary ATP increase occurs about 10 min before a small increase in the frequency of Ca^{2+} oscillations, that is seen in many eggs. It is notable that the transient increase in frequency of Ca^{2+} oscillations occur during the rising phase of ATP (technically $MgATP^{2-}$). This is the time when mitochondrial ATP^{4-}

production is likely to be at a maximum because the increase in total ATP is driven by mitochondria. The mitochondrially derived ATP^{4-} may act as a co-agonist with $InsP_3$ in promoting Ca^{2+} release and increasing the number of Ca^{2+} oscillations at fertilization. Experiments to investigate these potential effects of ATP^{4-} are, however, non-trivial because any interference in mitochondrial ATP production, for example by removing pyruvate or adding mitochondrial poisons, causes a rapid rise in Ca^{2+} due to the inhibition of Ca^{2+} pumps (Dumollard et al. 2004; Dumollard et al. 2008). It is difficult to separate out an effect of ATP^{4-}, as an allosteric activator of the IP3R1, from the role of $MgATP^{2-}$ as an energy source for Ca^{2+} pumping.

8. Unresolved problems

There are various issues to resolve about sperm induced Ca^{2+} signals in mammalian eggs. We still do not know how PLCζ binds to vesicles in eggs and whether this involves a specific egg protein. It's also unclear how the vesicles containing PI(4,5)P2 generate and recycle their inositol containing lipids. There is also the problem of the nature of the putative secondary sperm factor that can trigger Ca^{2+} oscillations in the absence of PLCζ. This issue is clearly important to understand for mouse egg activation, but it is important to establish the nature of the factor in order to know whether it has implications for egg activation in humans, or even non mammalian eggs. I would add to this specific problem another one of establishing the nature of potential sperm factors that increase metabolism in eggs at fertilization, and specifically how the sperm causes the secondary increase in ATP.

Another issue that would benefit from further investigation is the mechanism by which Ca^{2+} signals during egg activation could influence events much later in development. The connection has been difficult to address because the main phase of embryonic genome expression occurs days after the Ca^{2+} oscillations at fertilization have stopped. However, recent studies suggest that mammalian embryonic gene expression may be initiated during the first cell cycle and a possible link to the pattern of Ca^{2+} spikes may need to be revisited (Perry, Asami, Lam, & Yeo, 2023).

Finally, it is worth noting that there is a significant clinical problem with regards to Ca^{2+} signals in eggs. ICSI is used in most IVF treatment cycles around the world (Haddad et al. 2021). Total fertilization failure, where all ovulated eggs fail to fertilize, occurs in about 1–4 % of ICSI cases and a significant cause is failed egg activation (Bonte et al. 2019; Ferrer-Buitrago et al. 2018; Ferrer-Buitrago et al. 2019). However, low fertilization rates occur in about one third of all ICSI cycles, so the potental for improving the success rates of ICSI treatment is probably much larger (Scaravelli et al. 2021). In many cases of failed fertilization, it is now becoming common to offer artificial egg activation as a therapy. The main treatment approved for clinical use to induce artificial egg activation is Ca^{2+} ionophore. The Ca^{2+} ionophore A23187 was first shown to be capable of causing egg activation in the 1970s and it can only generate a single large Ca^{2+} increase. There is evidence that multiple Ca^{2+} transients would be a more effective stimulus for egg activation but it's not clear how this can be achieved in clinical practice. The obvious solution would be to microinject PLCζ, but it is a very unstable enzyme, and it has been difficult to develop a commercially available form of the recombinant protein. It would be invaluable to find a way to improve PLCζ's stability or else provide an alternative way of activating human eggs that invokes multiple Ca^{2+} transients.

Acknowledgments

I thank Zizhen Huang and Maryam Al Shaikh for discussion of the manuscript.

References

Aarabi, M., Balakier, H., Bashar, S., Moskovtsev, S. I., Sutovsky, P., Librach, C. L., & Oko, R. (2014). Sperm-derived WW domain-binding protein, PAWP, elicits calcium oscillations and oocyte activation in humans and mice. *The FASEB Journal, 28*(10), 4434–4440. https://doi.org/10.1096/fj.14-256495.

Abdulsamad, H. M. R., et al. (2023). The therapeutic and diagnostic potential of phospholipase C zeta, oocyte activation, and calcium in treating human infertility. *Pharmaceuticals (Basel), 16*(3), https://doi.org/10.3390/ph16030441.

Ardestani, G., West, M. C., Maresca, T. J., Fissore, R. A., & Stratton, M. M. (2019). FRET-based sensor for CaMKII activity (FRESCA): A useful tool for assessing CaMKII activity in response to Ca(2+) oscillations in live cells. *The Journal of Biological Chemistry, 294*(31), 11876–11891. https://doi.org/10.1074/jbc.RA119.009235.

Atri, A., Amundson, J., Clapham, D., & Sneyd, J. (1993). A single-pool model for intracellular calcium oscillations and waves in the Xenopus laevis oocyte. *Biophysical Journal, 65*(4), 1727–1739. https://doi.org/10.1016/s0006-3495(93)81191-3.

Balakier, H., & Casper, R. F. (1993). Experimentally induced parthenogenetic activation of human oocytes. *Human Reproduction (Oxford, England), 8*(5), 740–743. https://doi.org/10.1093/oxfordjournals.humrep.a138132.

Bedford-Guaus, S. J., Yoon, S. Y., Fissore, R. A., Choi, Y. H., & Hinrichs, K. (2008). Microinjection of mouse phospholipase C zeta complementary RNA into mare oocytes induces long-lasting intracellular calcium oscillations and embryonic development. *Reproduction, Fertility, and Development, 20*(8), 875–883. https://doi.org/10.1071/rd08115.

Ben-Yosef, D., Oron, Y., & Shalgi, R. (1995). Low temperature and fertilization-induced Ca2+ changes in rat eggs. *Molecular Reproduction and Development, 42*(1), 122–129. https://doi.org/10.1002/mrd.1080420116.

Bezprozvanny, I., Watras, J., & Ehrlich, B. E. (1991). Bell-shaped calcium-response curves of Ins(1,4,5)P3- and calcium-gated channels from endoplasmic reticulum of cerebellum. *Nature, 351*(6329), 751–754. https://doi.org/10.1038/351751a0.

Bonte, D., et al. (2019). Assisted oocyte activation significantly increases fertilization and pregnancy outcome in patients with low and total failed fertilization after intracytoplasmic sperm injection: A 17-year retrospective study. *Fertility and Sterility, 112*(2), 266–274. https://doi.org/10.1016/j.fertnstert.2019.04.006.

Bos-Mikich, A., Swann, K., & Whittingham, D. G. (1995). Calcium oscillations and protein synthesis inhibition synergistically activate mouse oocytes. *Molecular Reproduction and Development, 41*(1), 84–90. https://doi.org/10.1002/mrd.1080410113.

Campbell, K., & Swann, K. (2006). Ca2+ oscillations stimulate an ATP increase during fertilization of mouse eggs. *Developmental Biology, 298*(1), 225–233. https://doi.org/10.1016/j.ydbio.2006.06.032.

Carroll, J., Swann, K., Whittingham, D., & Whitaker, M. (1994). Spatiotemporal dynamics of intracellular [Ca2+]i oscillations during the growth and meiotic maturation of mouse oocytes. *Development (Cambridge, England), 120*(12), 3507–3517. https://doi.org/10.1242/dev.120.12.3507.

Cheek, T. R., McGuinness, O. M., Vincent, C., Moreton, R. B., Berridge, M. J., & Johnson, M. H. (1993). Fertilisation and thimerosal stimulate similar calcium spiking patterns in mouse oocytes but by separate mechanisms. *Development (Cambridge, England), 119*(1), 179–189. https://doi.org/10.1242/dev.119.1.179.

Chung, J. T., Keefer, C. L., & Downey, B. R. (2000). Activation of bovine oocytes following intracytoplasmic sperm injection (ICSI). *Theriogenology, 53*(6), 1273–1284. https://doi.org/10.1016/s0093-691x(00)00271-5.

Collas, P., Fissore, R., Robl, J. M., Sullivan, E. J., & Barnes, F. L. (1993a). Electrically induced calcium elevation, activation, and parthenogenetic development of bovine oocytes. *Molecular Reproduction and Development, 34*(2), 212–223. https://doi.org/10.1002/mrd.1080340214.

Collas, P., Sullivan, E. J., & Barnes, F. L. (1993b). Histone H1 kinase activity in bovine oocytes following calcium stimulation. *Molecular Reproduction and Development, 34*(2), 224–231. https://doi.org/10.1002/mrd.1080340215.

Cuthbertson, K. S., & Cobbold, P. H. (1985). Phorbol ester and sperm activate mouse oocytes by inducing sustained oscillations in cell Ca2+. *Nature, 316*(6028), 541–542. https://doi.org/10.1038/316541a0.

Cuthbertson, K. S., Whittingham, D. G., & Cobbold, P. H. (1981). Free Ca2+ increases in exponential phases during mouse oocyte activation. *Nature, 294*(5843), 754–757. https://doi.org/10.1038/294754a0.

Deguchi, R., Shirakawa, H., Oda, S., Mohri, T., & Miyazaki, S. (2000). Spatiotemporal analysis of Ca(2+) waves in relation to the sperm entry site and animal-vegetal axis during Ca(2+) oscillations in fertilized mouse eggs. *Developmental Biology, 218*(2), 299–313. https://doi.org/10.1006/dbio.1999.9573.

Ducibella, T., et al. (2002). Egg-to-embryo transition is driven by differential responses to Ca(2+) oscillation number. *Developmental Biology, 250*(2), 280–291.

Dumollard, R., Campbell, K., Halet, G., Carroll, J., & Swann, K. (2008). Regulation of cytosolic and mitochondrial ATP levels in mouse eggs and zygotes. *Developmental Biology, 316*(2), 431–440. https://doi.org/10.1016/j.ydbio.2008.02.004.

Dumollard, R., Carroll, J., Duchen, M. R., Campbell, K., & Swann, K. (2009). Mitochondrial function and redox state in mammalian embryos. *Seminars in Cell & Developmental Biology, 20*(3), 346–353. https://doi.org/10.1016/j.semcdb.2008.12.013.

Dumollard, R., Marangos, P., Fitzharris, G., Swann, K., Duchen, M., & Carroll, J. (2004). Sperm-triggered [Ca2+] oscillations and Ca2+ homeostasis in the mouse egg have an absolute requirement for mitochondrial ATP production. *Development (Cambridge, England), 131*(13), 3057–3067. https://doi.org/10.1242/dev.01181.

Dupont, G., & Dumollard, R. (2004). Simulation of calcium waves in ascidian eggs: Insights into the origin of the pacemaker sites and the possible nature of the sperm factor. *Journal of Cell Science, 117*(Pt 18), 4313–4323. https://doi.org/10.1242/jcs.01278.

Dupont, G., & Goldbeter, A. (1993). One-pool model for Ca2+ oscillations involving Ca2+ and inositol 1,4,5-trisphosphate as co-agonists for Ca2+ release. *Cell Calcium, 14*(4), 311–322. https://doi.org/10.1016/0143-4160(93)90052-8.

Eliyahu, E., & Shalgi, R. (2002). A role for protein kinase C during rat egg activation. *Biology of Reproduction, 67*(1), 189–195. https://doi.org/10.1095/biolreprod67.1.189.

Esbert, M., Carmody, A., Ballesteros, A., Seli, E., & Scott, R. T. (2022). Calcium Ionophore A23187 treatment to rescue unfertilized oocytes: A prospective randomized analysis of sibling oocytes. *Reproductive Biomedicine Online, 45*(5), 878–883. https://doi.org/10.1016/j.rbmo.2022.06.021.

Escoffier, J., et al. (2016). Homozygous mutation of PLCZ1 leads to defective human oocyte activation and infertility that is not rescued by the WW-binding protein PAWP. *Human Molecular Genetics, 25*(5), 878–891. https://doi.org/10.1093/hmg/ddv617.

Evans, J. P. (2020). Preventing polyspermy in mammalian eggs-Contributions of the membrane block and other mechanisms. *Molecular Reproduction and Development, 87*(3), 341–349. https://doi.org/10.1002/mrd.23331.

Ferrer-Buitrago, M., Bonte, D., Dhaenens, L., Vermorgen, S., Lu, Y., De Sutter, P., & Heindryckx, B. (2019). Assessment of the calcium releasing machinery in oocytes that failed to fertilize after conventional ICSI and assisted oocyte activation. *Reproductive Biomedicine Online, 38*(4), 497–507. https://doi.org/10.1016/j.rbmo.2018.12.035.

Ferrer-Buitrago, M., et al. (2018). Human oocyte calcium analysis predicts the response to assisted oocyte activation in patients experiencing fertilization failure after ICSI. *Human Reproduction (Oxford, England), 33*(3), 416–425. https://doi.org/10.1093/humrep/dex376.

Ferrer-Buitrago, M., et al. (2020). Comparative study of preimplantation development following distinct assisted oocyte activation protocols in a PLC-zeta knockout mouse model. *Molecular Human Reproduction, 26*(11), 801–815. https://doi.org/10.1093/molehr/gaaa060.

Finch, E. A., Turner, T. J., & Goldin, S. M. (1991). Calcium as a coagonist of inositol 1,4,5-trisphosphate-induced calcium release. *Science (New York, N. Y.), 252*(5004), 443–446. https://doi.org/10.1126/science.2017683.

Fissore, R. A., Longo, F. J., Anderson, E., Parys, J. B., & Ducibella, T. (1999). Differential distribution of inositol trisphosphate receptor isoforms in mouse oocytes. *Biology of Reproduction, 60*(1), 49–57. https://doi.org/10.1095/biolreprod60.1.49.

Foskett, J. K., White, C., Cheung, K. H., & Mak, D. O. (2007). Inositol trisphosphate receptor Ca2+ release channels. *Physiological Reviews, 87*(2), 593–658. https://doi.org/10.1152/physrev.00035.2006.

Galione, A., Swann, K., Georgiou, P., & Whitaker, M. (1994). Regenerative and non-regenerative calcium transients in hamster eggs triggered by inositol 1,4,5-trisphosphate. *The Journal of Physiology, 480*(Pt 3), 465–474. https://doi.org/10.1113/jphysiol.1994.sp020375.

Grubbs, R. D. (2002). Intracellular magnesium and magnesium buffering. *Biometals: An International Journal on the Role of Metal Ions in Biology, Biochemistry, and Medicine, 15*(3), 251–259. https://doi.org/10.1023/a:1016026831789.

Hachem, A., et al. (2017). PLCζ is the physiological trigger of the Ca(2+) oscillations that induce embryogenesis in mammals but conception can occur in its absence. *Development (Cambridge, England), 144*(16), 2914–2924. https://doi.org/10.1242/dev.150227.

Haddad, M., et al. (2021). Thoughts on the popularity of ICSI. *Journal of Assisted Reproduction and Genetics, 38*(1), 101–123. https://doi.org/10.1007/s10815-020-01987-0.

Halet, G. (2004). PKC signaling at fertilization in mammalian eggs. *Biochimica et Biophysica Acta, 1742*(1-3), 185–189. https://doi.org/10.1016/j.bbamcr.2004.09.012.

Halet, G., Tunwell, R., Parkinson, S. J., & Carroll, J. (2004). Conventional PKCs regulate the temporal pattern of Ca2+ oscillations at fertilization in mouse eggs. *The Journal of Cell Biology, 164*(7), 1033–1044. https://doi.org/10.1083/jcb.200311023.

Igusa, Y., & Miyazaki, S. (1983). Effects of altered extracellular and intracellular calcium concentration on hyperpolarizing responses of the hamster egg. *The Journal of Physiology, 340*, 611–632. https://doi.org/10.1113/jphysiol.1983.sp014783.

Igusa, Y., & Miyazaki, S. (1986). Periodic increase of cytoplasmic free calcium in fertilized hamster eggs measured with calcium-sensitive electrodes. *The Journal of Physiology, 377*, 193–205. https://doi.org/10.1113/jphysiol.1986.sp016181.

Igusa, Y., Miyazaki, S., & Yamashita, N. (1983). Periodic hyperpolarizing responses in hamster and mouse eggs fertilized with mouse sperm. *The Journal of Physiology, 340*, 633–647. https://doi.org/10.1113/jphysiol.1983.sp014784.

Ikie-Eshalomi, C., Aliyev, E., Hoehn, S., Jurkowski, T. P., & Swann, K. (2023). Sperm induce a secondary increase in ATP levels in mouse eggs that is independent of Ca2+ oscillations. *The Biochemical Journal, 480*(24), 2023–2035. https://doi.org/10.1042/bcj20230065.

Jellerette, T., Melican, D., Butler, R., Nims, S., Ziomek, C., Fissore, R., & Gavin, W. (2006). Characterization of calcium oscillation patterns in caprine oocytes induced by IVF or an activation technique used in nuclear transfer. *Theriogenology, 65*(8), 1575–1586. https://doi.org/10.1016/j.theriogenology.2005.08.025.

Jones, K. T. (2018). Mammalian sperm contain two factors for calcium release and egg activation: Phospholipase C zeta and a cryptic activating factor. *Molecular Human Reproduction, 24*(10), 465–468. https://doi.org/10.1093/molehr/gay038.

Jones, K. T., Cruttwell, C., Parrington, J., & Swann, K. (1998). A mammalian sperm cytosolic phospholipase C activity generates inositol trisphosphate and causes Ca^{2+} release in sea urchin egg homogenates. *FEBS Letters, 437*(3), 297–300. Available at: https://www.ncbi.nlm.nih.gov/pubmed/9824311; https://doi.org/10.1016/s0014-5793(98)01254-x.

Kang, W., et al. (2020). Extra-mitochondrial citrate synthase initiates calcium oscillation and suppresses age-dependent sperm dysfunction. *Laboratory Investigation; A Journal of Technical Methods and Pathology, 100*(4), 583–595. https://doi.org/10.1038/s41374-019-0353-3.

Kashir, J., et al. (2024). The mammalian sperm factor phospholipase C zeta is critical for early embryo division and pregnancy in humans and mice. *Human Reproduction (Oxford, England), 39*(6), 1256–1274. https://doi.org/10.1093/humrep/deae078.

Kennedy, H. J., Pouli, A. E., Ainscow, E. K., Jouaville, L. S., Rizzuto, R., & Rutter, G. A. (1999). Glucose generates sub-plasma membrane ATP microdomains in single islet beta-cells. Potential role for strategically located mitochondria. *The Journal of Biological Chemistry, 274*(19), 13281–13291. https://doi.org/10.1074/jbc.274.19.13281.

Kline, D., & Kline, J. T. (1992). Repetitive calcium transients and the role of calcium in exocytosis and cell cycle activation in the mouse egg. *Developmental Biology, 149*(1), 80–89. https://doi.org/10.1016/0012-1606(92)90265-i.

Kline, J. T., & Kline, D. (1994). Regulation of intracellular calcium in the mouse egg: Evidence for inositol trisphosphate-induced calcium release, but not calcium-induced calcium release. *Biology of Reproduction, 50*(1), 193–203. https://doi.org/10.1095/biolreprod50.1.193.

Knott, J. G., Gardner, A. J., Madgwick, S., Jones, K. T., Williams, C. J., & Schultz, R. M. (2006). Calmodulin-dependent protein kinase II triggers mouse egg activation and embryo development in the absence of Ca2+ oscillations. *Developmental Biology, 296*(2), 388–395. https://doi.org/10.1016/j.ydbio.2006.06.004.

Kouchi, Z., Fukami, K., Shikano, T., Oda, S., Nakamura, Y., Takenawa, T., & Miyazaki, S. (2004). Recombinant phospholipase Czeta has high Ca2+ sensitivity and induces Ca2+ oscillations in mouse eggs. *The Journal of Biological Chemistry, 279*(11), 10408–10412. https://doi.org/10.1074/jbc.M313801200.

Kouchi, Z., Shikano, T., Nakamura, Y., Shirakawa, H., Fukami, K., & Miyazaki, S. (2005). The role of EF-hand domains and C2 domain in regulation of enzymatic activity of phospholipase Czeta. *The Journal of Biological Chemistry, 280*(22), 21015–21021. https://doi.org/10.1074/jbc.M412123200.

Larman, M. G., Saunders, C. M., Carroll, J., Lai, F. A., & Swann, K. (2004). Cell cycle-dependent Ca2+ oscillations in mouse embryos are regulated by nuclear targeting of PLCzeta. *Journal of Cell Science, 117*(Pt 12), 2513–2521. https://doi.org/10.1242/jcs.01109.

Lawrence, Y., Ozil, J. P., & Swann, K. (1998). The effects of a Ca2+ chelator and heavy-metal-ion chelators upon Ca2+ oscillations and activation at fertilization in mouse eggs suggest a role for repetitive Ca2+ increases. *The Biochemical Journal, 335*(Pt 2), 335–342. https://doi.org/10.1042/bj3350335.

Lawrence, Y., Whitaker, M., & Swann, K. (1997). Sperm-egg fusion is the prelude to the initial Ca2+ increase at fertilization in the mouse. *Development (Cambridge, England), 124*(1), 233–241. Available at: https://www.ncbi.nlm.nih.gov/pubmed/9006083.

Leese, H. J. (1991). Metabolism of the preimplantation mammalian embryo. *Oxford Reviews of Reproductive Biology, 13*, 35–72.

Liu, L., Hammar, K., Smith, P. J., Inoue, S., & Keefe, D. L. (2001). Mitochondrial modulation of calcium signaling at the initiation of development. *Cell Calcium, 30*(6), 423–433. https://doi.org/10.1054/ceca.2001.0251.

Lu, Y., et al. (2018). Strontium fails to induce Ca(2+) release and activation in human oocytes despite the presence of functional TRPV3 channels. *Human Reproduction Open, 2018*(3), hoy005. https://doi.org/10.1093/hropen/hoy005.

Mak, D. O., McBride, S., & Foskett, J. K. (1998). Inositol 1,4,5-trisphosphate [correction of tris-phosphate] activation of inositol trisphosphate [correction of tris-phosphate] receptor Ca2+ channel by ligand tuning of Ca2+ inhibition. *Proceedings of the National Academy of Sciences of the United States of America, 95*(26), 15821–15825. https://doi.org/10.1073/pnas.95.26.15821.

Mak, D. O., McBride, S., & Foskett, J. K. (1999). ATP regulation of type 1 inositol 1,4,5-trisphosphate receptor channel gating by allosteric tuning of Ca(2+) activation. *The Journal of Biological Chemistry, 274*(32), 22231–22237. https://doi.org/10.1074/jbc.274.32.22231.

Malcuit, C., Knott, J. G., He, C., Wainwright, T., Parys, J. B., Robl, J. M., & Fissore, R. A. (2005). Fertilization and inositol 1,4,5-trisphosphate (IP3)-induced calcium release in type-1 inositol 1,4,5-trisphosphate receptor down-regulated bovine eggs. *Biology of Reproduction, 73*(1), 2–13. https://doi.org/10.1095/biolreprod.104.037333.

Mann, J. S., Lowther, K. M., & Mehlmann, L. M. (2010). Reorganization of the endoplasmic reticulum and development of Ca2+ release mechanisms during meiotic maturation of human oocytes. *Biology of Reproduction, 83*(4), 578–583. https://doi.org/10.1095/biolreprod.110.085985.

Marangos, P., FitzHarris, G., & Carroll, J. (2003). Ca2+ oscillations at fertilization in mammals are regulated by the formation of pronuclei. *Development (Cambridge, England), 130*(7), 1461–1472. https://doi.org/10.1242/dev.00340.

Markoulaki, S., Matson, S., Abbott, A. L., & Ducibella, T. (2003). Oscillatory CaMKII activity in mouse egg activation. *Developmental Biology, 258*(2), 464–474. https://doi.org/10.1016/s0012-1606(03)00133-7.

Markoulaki, S., Matson, S., & Ducibella, T. (2004). Fertilization stimulates long-lasting oscillations of CaMKII activity in mouse eggs. *Developmental Biology, 272*(1), 15–25. https://doi.org/10.1016/j.ydbio.2004.04.008.

Marshall, I. C., & Taylor, C. W. (1994). Two calcium-binding sites mediate the interconversion of liver inositol 1,4,5-trisphosphate receptors between three conformational states. *The Biochemical Journal, 301*(Pt 2), 591–598. https://doi.org/10.1042/bj3010591.

Matson, S., Markoulaki, S., & Ducibella, T. (2006). Antagonists of myosin light chain kinase and of myosin II inhibit specific events of egg activation in fertilized mouse eggs. *Biology of Reproduction, 74*(1), 169–176. https://doi.org/10.1095/biolreprod.105.046409.

Matsu-Ura, T., et al. (2019). Dual-FRET imaging of IP(3) and Ca(2+) revealed Ca(2+)-induced IP(3) production maintains long lasting Ca(2+) oscillations in fertilized mouse eggs. *Scientific Reports, 9*(1), 4829. https://doi.org/10.1038/s41598-019-40931-w.

Mehlmann, L. M., & Kline, D. (1994). Regulation of intracellular calcium in the mouse egg: Calcium release in response to sperm or inositol trisphosphate is enhanced after meiotic maturation. *Biology of Reproduction, 51*(6), 1088–1098. https://doi.org/10.1095/biolreprod51.6.1088.

Miyazaki, S. (1988). Inositol 1,4,5-trisphosphate-induced calcium release and guanine nucleotide-binding protein-mediated periodic calcium rises in golden hamster eggs. *The Journal of Cell Biology, 106*(2), 345–353. https://doi.org/10.1083/jcb.106.2.345.

Miyazaki, S. (2007). Intracellular calcium oscillations in mammalian eggs at fertilization. *The Journal of Physiology, 584*(Pt 3), 713–714. https://doi.org/10.1113/jphysiol.2007.144238.

Miyazaki, S., Hashimoto, N., Yoshimoto, Y., Kishimoto, T., Igusa, Y., & Hiramoto, Y. (1986). Temporal and spatial dynamics of the periodic increase in intracellular free calcium at fertilization of golden hamster eggs. *Developmental Biology, 118*(1), 259–267. https://doi.org/10.1016/0012-1606(86)90093-x.

Miyazaki, S., & Igusa, Y. (1981). Fertilization potential in golden hamster eggs consists of recurring hyperpolarizations. *Nature, 290*(5808), 702–704. https://doi.org/10.1038/290702a0.

Miyazaki, S., & Ito, M. (2006). Calcium signals for egg activation in mammals. *Journal of Pharmacological Sciences, 100*(5), 545–552. https://doi.org/10.1254/jphs.cpj06003x.

Miyazaki, S., Yuzaki, M., Nakada, K., Shirakawa, H., Nakanishi, S., Nakade, S., & Mikoshiba, K. (1992). Block of Ca2+ wave and Ca2+ oscillation by antibody to the inositol 1,4,5-trisphosphate receptor in fertilized hamster eggs. *Science (New York, N. Y.), 257*(5067), 251–255. https://doi.org/10.1126/science.1321497.

Moses, R. M., & Kline, D. (1995). Release of mouse eggs from metaphase arrest by protein synthesis inhibition in the absence of a calcium signal or microtubule assembly. *Molecular Reproduction and Development, 41*(2), 264–273. https://doi.org/10.1002/mrd.1080410218.

Moses, R. M., Kline, D., & Masui, Y. (1995). Maintenance of metaphase in colcemid-treated mouse eggs by distinct calcium- and 6-dimethylaminopurine (6-DMAP)-sensitive mechanisms. *Developmental Biology, 167*(1), 329–337. https://doi.org/10.1006/dbio.1995.1027.

Murugesu, S., et al. (2017). Does the use of calcium ionophore during artificial oocyte activation demonstrate an effect on pregnancy rate? A meta-analysis. e463 *Fertility and Sterility, 108*(3), 468–482. https://doi.org/10.1016/j.fertnstert.2017.06.029.

Nakano, Y., Shirakawa, H., Mitsuhashi, N., Kuwabara, Y., & Miyazaki, S. (1997). Spatiotemporal dynamics of intracellular calcium in the mouse egg injected with a spermatozoon. *Molecular Human Reproduction, 3*(12), 1087–1093. https://doi.org/10.1093/molehr/3.12.1087.

Nomikos, M., et al. (2005). Role of phospholipase C-zeta domains in Ca2+-dependent phosphatidylinositol 4,5-bisphosphate hydrolysis and cytoplasmic Ca2+ oscillations. *The Journal of Biological Chemistry, 280*(35), 31011–31018. https://doi.org/10.1074/jbc.M500629200.

Nomikos, M., et al. (2011). Male infertility-linked point mutation disrupts the Ca2+ oscillation-inducing and PIP(2) hydrolysis activity of sperm PLCζ. *The Biochemical Journal, 434*(2), 211–217. https://doi.org/10.1042/bj20101772.

Nomikos, M., et al. (2015). Functional disparity between human PAWP and PLCζ in the generation of Ca2+ oscillations for oocyte activation. *Molecular Human Reproduction, 21*(9), 702–710. https://doi.org/10.1093/molehr/gav034.

Nomikos, M., et al. (2014). Sperm-specific post-acrosomal WW-domain binding protein (PAWP) does not cause Ca2+ release in mouse oocytes. *Molecular Human Reproduction, 20*(10), 938–947. https://doi.org/10.1093/molehr/gau056.

Nozawa, K., Satouh, Y., Fujimoto, T., Oji, A., & Ikawa, M. (2018). Sperm-borne phospholipase C zeta-1 ensures monospermic fertilization in mice. *Scientific Reports, 8*(1), 1315. https://doi.org/10.1038/s41598-018-19497-6.

Ozil, J. P., et al. (2005). Egg activation events are regulated by the duration of a sustained [Ca2+]cyt signal in the mouse. *Developmental Biology, 282*(1), 39–54. https://doi.org/10.1016/j.ydbio.2005.02.035.

Ozil, J. P., & Swann, K. (1995). Stimulation of repetitive calcium transients in mouse eggs. *The Journal of Physiology, 483*(Pt 2), 331–346. https://doi.org/10.1113/jphysiol.1995.sp020589.

Perry, A. C. F., Asami, M., Lam, B. Y. H., & Yeo, G. S. H. (2023). The initiation of mammalian embryonic transcription: To begin at the beginning. *Trends in Cell Biology, 33*(5), 365–373. https://doi.org/10.1016/j.tcb.2022.08.008.

Politi, A., Gaspers, L. D., Thomas, A. P., & Höfer, T. (2006). Models of IP3 and Ca2+ oscillations: Frequency encoding and identification of underlying feedbacks. *Biophysical Journal, 90*(9), 3120–3133. https://doi.org/10.1529/biophysj.105.072249.

Prather, R. S., Tao, T., & Macháty, Z. (1999). Development of the techniques for nuclear transfer in pigs. *Theriogenology, 51*(2), 487–498. https://doi.org/10.1016/s0093-691x(98)00242-8.

Rinaudo, P., Pepperell, J. R., Buradgunta, S., Massobrio, M., & Keefe, D. L. (1997). Dissociation between intracellular calcium elevation and development of human oocytes treated with calcium ionophore. *Fertility and Sterility, 68*(6), 1086–1092. https://doi.org/10.1016/s0015-0282(97)00406-8.

Rogers, N. T., Hobson, E., Pickering, S., Lai, F. A., Braude, P., & Swann, K. (2004). Phospholipase Czeta causes Ca2+ oscillations and parthenogenetic activation of human oocytes. *Reproduction (Cambridge, England), 128*(6), 697–702. https://doi.org/10.1530/rep.1.00484.

Rolfe, D. F., & Brown, G. C. (1997). Cellular energy utilization and molecular origin of standard metabolic rate in mammals. *Physiological Reviews,* 77(3), 731–758. https://doi.org/10.1152/physrev.1997.77.3.731.

Ross, P. J., et al. (2008). Parthenogenetic activation of bovine oocytes using bovine and murine phospholipase C zeta. *BMC Developmental Biology, 8*, 16. https://doi.org/10.1186/1471-213x-8-16.

Rutter, G. A., Georgiadou, E., Martinez-Sanchez, A., & Pullen, T. J. (2020). Metabolic and functional specialisations of the pancreatic beta cell: Gene disallowance, mitochondrial metabolism and intercellular connectivity. *Diabetologia, 63*(10), 1990–1998. https://doi.org/10.1007/s00125-020-05205-5.

Sanders, J. R., Ashley, B., Moon, A., Woolley, T. E., & Swann, K. (2018). PLCζ induced Ca(2+) oscillations in mouse eggs involve a positive feedback cycle of Ca(2+) induced InsP(3) formation from cytoplasmic PIP(2). *Frontiers in Cell and Developmental Biology, 6*, 36. https://doi.org/10.3389/fcell.2018.00036.

Sanusi, R., Yu, Y., Nomikos, M., Lai, F. A., & Swann, K. (2015). Rescue of failed oocyte activation after ICSI in a mouse model of male factor infertility by recombinant phospholipase Cζ. *Molecular Human Reproduction, 21*(10), 783–791. https://doi.org/10.1093/molehr/gav042.

Sato, Y., Miyazaki, S., Shikano, T., Mitsuhashi, N., Takeuchi, H., Mikoshiba, K., & Kuwabara, Y. (1998). Adenophostin, a potent agonist of the inositol 1,4,5-trisphosphate receptor, is useful for fertilization of mouse oocytes injected with round spermatids leading to normal offspring. *Biology of Reproduction, 58*(3), 867–873. https://doi.org/10.1095/biolreprod58.3.867.

Saunders, C. M., et al. (2002). PLC zeta: A sperm-specific trigger of Ca(2+) oscillations in eggs and embryo development. *Development (Cambridge, England), 129*(15), 3533–3544. https://doi.org/10.1242/dev.129.15.3533.

Scaravelli, G., et al. (2021). Fertilization rate as a novel indicator for cumulative live birth rate: A multicenter retrospective cohort study of 9,394 complete in vitro fertilization cycles. *Fertility and Sterility, 116*(3), 766–773. https://doi.org/10.1016/j.fertnstert.2021.04.006.

Schultz, R. M., Letourneau, G. E., & Wassarman, P. M. (1979). Program of early development in the mammal: Changes in patterns and absolute rates of tubulin and total protein synthesis during oogenesis and early embryogenesis in the mouse. *Developmental Biology, 68*(2), 341–359. https://doi.org/10.1016/0012-1606(79)90209-4.

Sette, C., Bevilacqua, A., Bianchini, A., Mangia, F., Geremia, R., & Rossi, P. (1997). Parthenogenetic activation of mouse eggs by microinjection of a truncated c-kit tyrosine kinase present in spermatozoa. *Development (Cambridge, England), 124*(11), 2267–2274. https://doi.org/10.1242/dev.124.11.2267.

Shirakawa, H., Ito, M., Sato, M., Umezawa, Y., & Miyazaki, S. (2006). Measurement of intracellular IP3 during Ca2+ oscillations in mouse eggs with GFP-based FRET probe. *Biochemical and Biophysical Research Communications, 345*(2), 781–788. https://doi.org/10.1016/j.bbrc.2006.04.133.

Simsek-Duran, F., Li, F., Ford, W., Swanson, R. J., Jones, H. W., Jr., & Castora, F. J. (2013). Age-associated metabolic and morphologic changes in mitochondria of individual mouse and hamster oocytes. *PLoS One, 8*(5), e64955. https://doi.org/10.1371/journal.pone.0064955.

Sneyd, J., Tsaneva-Atanasova, K., Reznikov, V., Bai, Y., Sanderson, M. J., & Yule, D. I. (2006). A method for determining the dependence of calcium oscillations on inositol trisphosphate oscillations. *Proceedings of the National Academy of Sciences of the United States of America, 103*(6), 1675–1680. https://doi.org/10.1073/pnas.0506135103.

Stein, P., Savy, V., Williams, A. M., & Williams, C. J. (2020). Modulators of calcium signalling at fertilization. *Open Biology, 10*(7), 200118. https://doi.org/10.1098/rsob.200118.

Storey, A., Elgmati, K., Wang, Y., Knaggs, P., & Swann, K. (2021). The role of ATP in the differential ability of Sr2+ to trigger Ca2+ oscillations in mouse and human eggs. *Molecular Human Reproduction, 27*(1), https://doi.org/10.1093/molehr/gaaa086.

Sun, F. Z., Bradshaw, J. P., Galli, C., & Moor, R. M. (1994). Changes in intracellular calcium concentration in bovine oocytes following penetration by spermatozoa. *Journal of Reproduction and Fertility, 101*(3), 713–719. https://doi.org/10.1530/jrf.0.1010713.

Sun, F. Z., Hoyland, J., Huang, X., Mason, W., & Moor, R. M. (1992). A comparison of intracellular changes in porcine eggs after fertilization and electroactivation. *Development (Cambridge, England), 115*(4), 947–956. https://doi.org/10.1242/dev.115.4.947.

Swann, K. (1992). Different triggers for calcium oscillations in mouse eggs involve a ryanodine-sensitive calcium store. *The Biochemical Journal, 287*(Pt 1), 79–84. https://doi.org/10.1042/bj2870079.

Swann, K. (1994). Ca2+ oscillations and sensitization of Ca2+ release in unfertilized mouse eggs injected with a sperm factor. *Cell Calcium, 15*(4), 331–339. https://doi.org/10.1016/0143-4160(94)90072-8.

Swann, K. (2020). The soluble sperm factor that activates the egg: PLCzeta and beyond. *Reproduction (Cambridge, England), 160*(1), V9–v11. https://doi.org/10.1530/rep-20-0079.

Swann, K. (2023). Sperm-induced Ca(2+) release in mammalian eggs: The roles of PLCζ, InsP(3), and ATP. *Cells, 12*(24), https://doi.org/10.3390/cells12242809.

Swann, K., Igusa, Y., & Miyazaki, S. (1989). Evidence for an inhibitory effect of protein kinase C on G-protein-mediated repetitive calcium transients in hamster eggs. *The EMBO Journal, 8*(12), 3711–3718. https://doi.org/10.1002/j.1460-2075.1989.tb08546.x.

Swann, K., & Lai, F. A. (2013). PLCζ and the initiation of Ca(2+) oscillations in fertilizing mammalian eggs. *Cell Calcium, 53*(1), 55–62. https://doi.org/10.1016/j.ceca.2012.11.001.

Swann, K., & Ozil, J. P. (1994). Dynamics of the calcium signal that triggers mammalian egg activation. *International Review of Cytology, 152*, 183–222. Available at: https://www.ncbi.nlm.nih.gov/pubmed/8206704; https://doi.org/10.1016/s0074-7696(08)62557-7.

Taylor, C. T., Lawrence, Y. M., Kingsland, C. R., Biljan, M. M., & Cuthbertson, K. S. (1993). Oscillations in intracellular free calcium induced by spermatozoa in human oocytes at fertilization. *Human Reproduction (Oxford, England), 8*(12), 2174–2179. https://doi.org/10.1093/oxfordjournals.humrep.a137999.

Tesarik, J., Sousa, M., & Testart, J. (1994). Human oocyte activation after intracytoplasmic sperm injection. *Human Reproduction (Oxford, England), 9*(3), 511–518. https://doi.org/10.1093/oxfordjournals.humrep.a138537.

Thanassoulas, A., Swann, K., Lai, F. A., & Nomikos, M. (2022). SPERM FACTORS AND EGG ACTIVATION: The structure and function relationship of sperm PLCZ1. *Reproduction (Cambridge, England), 164*(1), F1–f8. https://doi.org/10.1530/rep-21-0477.

Tomashov-Matar, R., Tchetchik, D., Eldar, A., Kaplan-Kraicer, R., Oron, Y., & Shalgi, R. (2005). Strontium-induced rat egg activation. *Reproduction (Cambridge, England), 130*(4), 467–474. https://doi.org/10.1530/rep.1.00746.

Tóth, S., Huneau, D., Banrezes, B., & Ozil, J. P. (2006). Egg activation is the result of calcium signal summation in the mouse. *Reproduction (Cambridge, England), 131*(1), 27–34. https://doi.org/10.1530/rep.1.00764.

Tu, H., Wang, Z., Nosyreva, E., De Smedt, H., & Bezprozvanny, I. (2005). Functional characterization of mammalian inositol 1,4,5-trisphosphate receptor isoforms. *Biophysical Journal, 88*(2), 1046–1055. https://doi.org/10.1529/biophysj.104.049593.

Urner, F., & Sakkas, D. (1999). Characterization of glycolysis and pentose phosphate pathway activity during sperm entry into the mouse oocyte. *Biology of Reproduction, 60*(4), 973–978. https://doi.org/10.1095/biolreprod60.4.973.

Verjovski-Almeida, S., & Inesi, G. (1979). Fast-kinetic evidence for an activating effect of ATP on the Ca2+ transport of sarcoplasmic reticulum ATPase. *The Journal of Biological Chemistry, 254*(1), 18–21.

Wakai, T., Mehregan, A., & Fissore, R. A. (2019). Ca(2+) signaling and homeostasis in mammalian oocytes and eggs. *Cold Spring Harbor Perspectives in Biology, 11*(12), https://doi.org/10.1101/cshperspect.a035162.

Wakai, T., Zhang, N., Vangheluwe, P., & Fissore, R. A. (2013). Regulation of endoplasmic reticulum Ca(2+) oscillations in mammalian eggs. *Journal of Cell Science, 126*(Pt 24), 5714–5724. https://doi.org/10.1242/jcs.136549.

Wilson, D. F. (2015). Programming and regulation of metabolic homeostasis. *American Journal of Physiology. Endocrinology and Metabolism, 308*(6), E506–E517. https://doi.org/10.1152/ajpendo.00544.2014.

Winston, N., Johnson, M., Pickering, S., & Braude, P. (1991). Parthenogenetic activation and development of fresh and aged human oocytes. *Fertility and Sterility, 56*(5), 904–912.

Wu, A. T., et al. (2007). PAWP, a sperm-specific WW domain-binding protein, promotes meiotic resumption and pronuclear development during fertilization. *The Journal of Biological Chemistry, 282*(16), 12164–12175. https://doi.org/10.1074/jbc.M609132200.

Yanagida, K., Katayose, H., Hirata, S., Yazawa, H., Hayashi, S., & Sato, A. (2001). Influence of sperm immobilization on onset of Ca(2+) oscillations after ICSI. *Human Reproduction (Oxford, England), 16*(1), 148–152. https://doi.org/10.1093/humrep/16.1.148.

Yoda, A., et al. (2004). Ca2+ oscillation-inducing phospholipase C zeta expressed in mouse eggs is accumulated to the pronucleus during egg activation. *Developmental Biology, 268*(2), 245–257. https://doi.org/10.1016/j.ydbio.2003.12.028.

Yoneda, A., et al. (2006). Molecular cloning, testicular postnatal expression, and oocyte-activating potential of porcine phospholipase Czeta. *Reproduction (Cambridge, England), 132*(3), 393–401. https://doi.org/10.1530/rep.1.01018.

Yu, Y., Halet, G., Lai, F. A., & Swann, K. (2008a). Regulation of diacylglycerol production and protein kinase C stimulation during sperm- and PLCzeta-mediated mouse egg activation. *Biology of the Cell/Under the Auspices of the European Cell Biology Organization, 100*(11), 633–643. https://doi.org/10.1042/BC20080033; https://www.ncbi.nlm.nih.gov/pubmed/18471090.

Yu, Y., Nomikos, M., Theodoridou, M., Nounesis, G., Lai, F. A., & Swann, K. (2012). PLCζ causes Ca(2+) oscillations in mouse eggs by targeting intracellular and not plasma membrane PI(4,5)P(2). *Molecular Biology of the Cell, 23*(2), 371–380. https://doi.org/10.1091/mbc.E11-08-0687.

Yu, Y., Saunders, C. M., Lai, F. A., & Swann, K. (2008b). Preimplantation development of mouse oocytes activated by different levels of human phospholipase C zeta. *Human Reproduction (Oxford, England), 23*(2), 365–373. https://doi.org/10.1093/humrep/dem350.

Zhang, L., Chao, C. H., Jaeger, L. A., Papp, A. B., & Machaty, Z. (2018). Calcium oscillations in fertilized pig oocytes are associated with repetitive interactions between STIM1 and ORAI1. *Biology of Reproduction, 98*(4), 510–519. https://doi.org/10.1093/biolre/ioy016.

CHAPTER ELEVEN

Ion channels and transporters involved in calcium flux regulation in mammalian sperm

Claudia Sánchez-Cárdenas[a,*,1], Enrique I. Oliver[b], Julio C. Chávez[a], Guillermina M. Luque[c], Arturo Hernández-Cruz[d], Mariano G. Buffone[c], Alberto Darszon[a], Pablo E. Visconti[e,*], and Ana Romarowski[c,*]

[a]Departamento de Genética del Desarrollo y Fisiología Molecular, Instituto de Biotecnología, UNAM, Cuernavaca, Mexico
[b]Department of Physiology and Biophysics, Miller School of Medicine, University of Miami, Miami, FL, United States
[c]Instituto de Biología y Medicina Experimental (IBYME), National Research Council of Argentina (CONICET), Buenos Aires, Argentina
[d]Departamento de Neuropatología Molecular y Laboratorio Nacional de Canalopatías, Instituto de Fisiología Celular UNAM, Ciudad Universitaria, Ciudad de México, Mexico
[e]Department of Veterinary and Animal Sciences, University of Massachusetts, Amherst, MA, United States
*Corresponding authors. e-mail address: claudia.sanchez@ibt.unam.mx; pvisconti@vasci.umass.edu; anaromarowski@gmail.com

Contents

[1] Dr. Sanchez-Cardenas, Dr. Oliver, Dr. Visconti and Dr. Romarowski contributed equally to the writing of this manuscript.

Current Topics in Developmental Biology, Volume 162
ISSN 0070-2153, https://doi.org/10.1016/bs.ctdb.2025.01.006

Abstract

After ejaculation, mammalian spermatozoa are not capable of fertilizing a metaphase II-arrested egg. They require to undergo a series of biochemical and physiological processes collectively known as capacitation. In all these processes, the regulation of calcium ions fluxes plays essential roles and involves participation of many channels and transporters localized in the plasma membrane as well as in the membrane of intracellular organelles. In mammalian sperm, a fraction of these molecules has been proposed to contribute to mature sperm function. However, in many cases, the evidence for the presence of a given protein is based on the use of agonists and antagonists with more than one target. In this review, we will critically analyze the published evidence supporting the presence of these molecules in mammalian sperm with special emphasis to methods involving tandem mass spectrometry identification, electrophysiological evidence and controlled immunoassays.

1. Introduction

After leaving the testis, spermatozoa require two additional post-testicular maturational processes, one in the male and the other in the female. The first process, known as epididymal maturation, occurs during their transit from the caput to the cauda epididymis. Once sperm are matured, they can undergo a second process known as sperm capacitation. Physiologically, capacitation occurs upon ejaculation in the female tract; however, this process can also be mimicked in vitro using either ejaculated or cauda epididymal sperm incubated in a simple media supplemented with serum albumin, energy substrates, calcium (Ca^{2+}) and bicarbonate (HCO_3^-) (Gervasi & Visconti, 2016). Both epididymal maturation and sperm capacitation activate complex signaling pathways which regulate ion homeostasis. Capacitation can be defined as the process that prepares mature sperm to fertilize. This definition encompasses those sperm obtained from cauda epididymis, from ejaculated semen, as well as those obtained upon ejaculation from the female tract, usually the uterus (e.g., mostly mice work) (Ferreira, Lybaert, Puga-Molina, & Santi, 2021). Physiologically, capacitation is associated with changes in the motility pattern, known as hyperactivation (HA) (Olson, Fauci, & Suarez, 2011) and for the preparation for acrosome exocytosis, also known as acrosome reaction (AR). At the molecular level, these physiological endpoints include but are not limited to a fast HCO_3^- -dependent increase in cAMP concentrations, activation of phosphorylation pathways, alkalinization of the acrosomal and the intracellular pH (pH_i), and in the sperm plasma membrane potential (Em) (De La Vega-Beltran et al., 2012; Visconti, 2009). Notably, one of the main signaling nodes in all these events is the exquisite regulation of intracellular calcium concentration ($[Ca^{2+}]_i$). In the last years, a significant number of reviews have covered the molecular basis of epididymal maturation (Breton, Nair, & Battistone, 2019; Dey, Brothag, & Vijayaraghavan, 2019; Gervasi & Visconti, 2017) and capacitation (Gervasi & Visconti, 2016; Lyon et al., 2023; Puga Molina et al., 2018; Takei, 2024). This study aims to present a critical update of molecular entities involved in regulating the Ca^{2+} fluxes in mammalian sperm.

2. Evaluating the evidence for the role of an ion transporter in calcium homeostasis

In every field, defining the presence and role of given molecules and signaling pathways is not straightforward. Conclusions rely on three types of strategies: (1) identification of a molecule in the cell of interest; (2) measurements of the molecule activity; (3) functional studies involving loss- and gain-of-function experiments. The challenges in spermatozoa are further exacerbated because these cells are transcriptionally and translationally inactive, precluding the direct use of molecular biology techniques. In addition, it is only at the end of sperm differentiation that the gamete can fertilize. Therefore, knock-out experiments are often inconclusive because loss of a particular property (e.g. fertilizing capacity) might reflect a problem in spermatogenesis or epididymal maturation. On the other hand, inhibitors and activators of given pathways can be misinterpreted due to off-target effects. Importantly, although a few molecules have collected all the necessary evidence to conclude that they are present and relevant in mammalian sperm, a lack of conclusive evidence should not be taken as evidence of the absence of a given molecule in sperm. Following is a series of strategic criteria that we used to evaluate the information presented in this review.

2.1 Identification criteria

Different Ca^{2+} channels and transporters have been proposed to be present in sperm. However, only a few of them have been fully confirmed. The various experimental approaches used for identification are assays based on: RNA (e.g. RT-PCR, RNA-seq), immunology (e.g. Western blot; immuno-fluorescence), enzymatic or electrophysiology activity and, on direct identification using tandem mass spectrometry. These methods have advantages and disadvantages. Although RNA-based assays have high sensitivity, results are difficult to interpret since sperm are transcriptionally and translationally silent; therefore, in many cases, RT-PCR experiments have been done using RNA from testicular germ cells (Treviño et al., 1998). As a second approach, immunological assays have been used to evaluate the presence and localization of given molecules. To be reliable, these assays require excellent antibodies validated in spermatozoa with conclusive negative controls. To be considered present in sperm, a protein present in sperm, the protein detected in a Western blot should have the expected MW and should not be detected by the antibody when the protein of interest is knocked out. An alternative reasonable control is the use of pre-absorbed antibodies with the respective antigen (or antigenic peptide). On the other hand, evidence obtained using proteomic approaches is very robust as it is based on direct sequencing of sperm peptides. However, considering extraction, proteolysis difficulties, and low expression levels typical of ion channels, the lack of sequencing data does not constitute proof of the absence of a given molecule.

2.2 Activity criteria

In addition to the presence and identity of a transporter, the second step is to show its activity in sperm. For channels, activity assays can involve direct electrophysiological measurements or indirect evaluation of the ions transported using fluorescent probes. In sperm, electrophysiological approaches using whole-cell patch-clamp have been used to characterize plasma membrane channels. Several sperm ion channels have been described with this methodology; in some cases, their activity was recorded before identifying the molecule, such as K-sper (Navarro, Kirichok, & Clapham, 2007) which was identified as Slo3 years later (Santi et al., 2010; Zeng, Yang, Kim, Lingle, & Xia, 2011), with some still pending questions, particularly in human sperm (Brenker et al., 2014; Mannowetz et al., 2013; Sánchez-Carranza, Torres-Rodríguez, Darszon, Treviño, & López-González, 2015). Although sperm electrophysiology has been reliably conducted in sperm as well as in isolated tails and heads (Navarro et al., 2007), this methodology has limitations for studying channels of intracellular organelles and these investigations have not yet been conducted.

Most relevant for the study of changes in $[Ca^{2+}]_i$ are those experiments using Ca^{2+} sensors, which can be divided into genetically-encoded and permeable fluorescent reagents (Rodriguez et al., 2017). Genetically-encoded sensors have the advantage of being less invasive than small molecules. However, in sperm, these sensors, can only be expressed by generating transgenic models (Cohen et al., 2022, 2023). On the other hand, small fluorescent molecules have several advantages, among them: (1) they are easy to use in mature sperm suspension or individual cells; (2) some of them are ratiometric and can give good quantitative estimates of the $[Ca^{2+}]_i$; (3) there is a variety of these sensors with different Ca^{2+} dissociation constants (Kd) and spectral wavelengths properties (Paredes, Etzler, Watts, Zheng, & Lechleiter, 2008), allowing to monitor different concentration ranges; (4) finally, because of their different chemical properties, while some of them distribute preferentially in the cytosol, others accumulate in intracellular organelles. In addition to single-cell real-time experiments, Ca^{2+} sensors can be used in population analyses employing fluorescence spectrometry (Lee & Storey, 1989) or flow-cytometry (Luque et al., 2018; Matamoros Volante et al., 2023). Importantly, fluorescent Ca^{2+} detection is essential for conducting real-time measurements in single sperm. These methods can be used to evaluate the extent to which pharmacologic agonists or antagonists can induce or block changes in $[Ca^{2+}]_i$. However, when using fluorescent sensors, it is important to consider the distribution of the dye, the different affinity for Ca^{2+}, and the fact that the Ca^{2+} binding motif changes depending on the environmental pH.

2.3 Functional criteria

Identification and activity strategies would reveal that proteins of interest are present and active. However, these assays are silent regarding their function in a cell. The classical approach to assigning a function to a given molecule is to either eliminate the molecule (loss-of-function), or to incorporate it into cells that lack the protein of interest (gain-of-function). While loss-of-function approaches can give strong evidence for the necessity of a molecule in a process, gain-of-function experiments are needed to show the sufficiency of a given molecule or pathway. There are multiple types of loss-of-function and gain-of-function strategies. Here, we will mention only those relevant for the study of Ca^{2+} in sperm: (1) Ca^{2+} chelators such as BAPTA, EGTA or EDTA are used to maintain extracellular Ca^{2+} below a certain concentration; (2) when chemically esterified (acetoxy-methyl ester, AM), these chelators can permeate cells and become tools to keep $[Ca^{2+}]_i$ to very low levels, providing evidence of the role of intracellular Ca^{2+} in a given process. However, as all chelators also have affinity for other ions such as Mg^{2+} and Mn^{2+}, these potential drawbacks must be carefully considered and the proper controls designed; (3) agonists and antagonists for channels and enzymes are routinely used to block or enhance their activities. Small molecules have advantages as well as disadvantages. They are ideal for analyzing kinetics and establishing signaling pathways, though specificity can become essential to arrive at conclusions. Some inhibitors have exquisite selectivity, but others have secondary targets that preclude correctly identifying a given protein. Off-targets effects become relevant when using pharmacological approaches to identify a given molecule without considering other possible targets; (4) knock-out and knock-in strategies have also advantages and disadvantages. There is always strong evidence that a protein is involved if a process is blocked when the protein under study is eliminated. However, these approaches are slow, laborious, and expensive, they may involve gene networks that influence cell differentiation, and the compensatory effects of other proteins must be considered (Nelson, 1997). Because spermatozoa are transcriptionally inactive, knock-down approaches using siRNAs are discarded. In the case of knock-out strategies, it is not possible to eliminate a gene only in mature sperm. Similarly to methods using genetically encoded sensors, conditional or total knock-out models are required. When the goal is to evaluate mature spermatozoa's function, it is essential to consider that genes of interest are eliminated in the testis long before spermatozoa are ready to mature in the epididymis and to be capacitated in the female tract. Therefore, sterility or subfertility phenotypes are, in most cases, silent regarding the role of a given protein in mature sperm. Determining that a Ca^{2+} signaling pathway is active, how it is turned on, and which physiological function is regulated, usually requires using of more than one strategy. As new methodologies are rapidly being developed, our conceptual framework regarding Ca^{2+} signaling in fertility adjusts and changes. This review attempts to update what is known about sperm Ca^{2+} channels and transporters responsible for regulating Ca^{2+} fluxes and homeostasis.

3. Calcium regulation in cells and sperm

As a ubiquitous second messenger, Ca^{2+} regulates numerous cellular functions (Berridge, Lipp, & Bootman, 2000; Berridge, Bootman, & Roderick, 2003). Examples of these are: cytoskeletal movement (Lopez-Guerrero et al., 2017), exocytosis (Pang & Südhof, 2010), and the regulation of transcription (Crabtree, 2001; West et al., 2001). Sperm are no exception to this, as regulated $[Ca^{2+}]_i$ changes are needed for capacitation, HA (Ishijima, 2015; Lesich et al., 2012) and the occurrence of AR (Mayorga, Tomes, & Belmonte, 2007; Yanagimachi, 1994). While extracellular Ca^{2+} concentrations ($[Ca^{2+}]_e$) are typically in the millimolar range, the concentration of Ca^{2+} inside healthy cells is maintained at nanomolar levels. Preserving this steep gradient is crucial to allow cells to dramatically increase $[Ca^{2+}]_i$ in response to the opening of Ca^{2+} channels (from ~100 nM to ~10 μM). This rise in $[Ca^{2+}]_i$ serves as a signal to trigger various cellular responses mediated by Ca^{2+}-binding proteins. Proper regulation of Ca^{2+} levels in both time and space is necessary for the precise execution of specific cellular functions (Berridge et al., 2003). During signaling activation, such as in response to an agonist, cells experience a rise in cytoplasmic Ca^{2+} levels. However, once the stimulus ends, cells must restore Ca^{2+} to basal levels to prevent continuous activation and prepare for future stimuli. Cells manage cytoplasmic Ca^{2+} levels through coordinated mechanisms that balance Ca^{2+} entry, removal, and storage.

These mechanisms can be divided into three main categories: (1) **Ca^{2+} entry mechanisms**: Proteins involved in regulating Ca^{2+} entry are located primarily in the plasma membrane and allow the influx of extracellular Ca^{2+} into the cytoplasm. (2) **Ca^{2+} extrusion mechanisms**: To decrease elevated $[Ca^{2+}]_i$ levels in the cytoplasm, cells employ Ca^{2+} pumps and transporters that extrude Ca^{2+} either to the extracellular space or into intracellular compartments. (3) **Intracellular Ca^{2+} storage**: Organelles such as the endoplasmic reticulum (ER), mitochondria, and lysosomes serve as Ca^{2+} reservoirs. These organelles store high concentrations of Ca^{2+}, which can be released into the cytoplasm when needed for signaling. Each of these groups is composed of a large number of molecules and reviewing them in detail is beyond the scope of this article. Instead, in the following sections we will focus on those molecules for which there is evidence of their presence in mature sperm.

4. Plasma membrane calcium entry mechanisms

Ca^{2+} can enter the cell following its electrochemical gradient. As mentioned above, $[Ca^{2+}]_e$ is at least 1000 times higher than $[Ca^{2+}]_i$. In addition, most cells have a negative resting Em, further promoting inward Ca^{2+} movement. Consequently, Ca^{2+} entry is favored thermodynamically and does not require intracellular adenosine triphosphate (ATP) hydrolysis. Also, Ca^{2+} enters cells directly through various cation channels or coupled to Na^+ extrusion using the Na^+/Ca^{2+} exchangers operating in reverse mode (Song, Chen, & Yu, 2014). Ca^{2+} channels can be classified, depending on their activation mechanisms, into voltage-operated channels (VOCs); intracellular second messenger-operated channels including those regulated by ions (SMOCs); extracellular ligand-dependent or receptor-operated channels (ROCs), transient receptor potential channels (TRPs), and store-operated channels (SOCs). Sometimes, Ca^{2+} can also enter cells using plasma membrane Na^+/Ca^{2+} exchangers (NCX) operating in reverse mode, driven by the relatively elevated intracellular Na^+ concentrations (Song et al., 2014). Below, we summarize each of these families and review the evidence of their presence and localization in mammalian spermatozoa (Fig. 1).

4.1 VOCs

VOCs are activated by changes in Em and are highly expressed in excitable cells such as neurons and cardiac muscle cells (Catterall, 2011; Simms & Zamponi, 2014). Among VOCs, two major types are those activated by high voltage (HVA) and those activated by low voltage (LVA). Although voltage-dependent Ca^{2+} channels have been proposed to be present in mature sperm, the evidence relies primarily on RT-PCR data or immunofluorescence of sperm or spermatogenic cells (for review see (Darszon, Nishigaki, Beltran, & Treviño, 2011)). However, results were inconclusive in most cases because of the lack of knock-out controls. Two exceptions are Cav2.3 and Cav3.2, which were detected by immunofluorescence using Western blot-validated antibodies. Both Cavs are present at the base of the acrosome in wild-type sperm but not in sperm from mice devoid of Cav2.3 and Cav3.2 (Cohen et al., 2014; Darszon & Hernández-Cruz, 2014; Wennemuth, Westenbroek, Xu, Hille, & Babcock, 2000). Interestingly, immunostaining of Cav3.2 is lost in acrosome-reacted sperm (Escoffier et al., 2007).

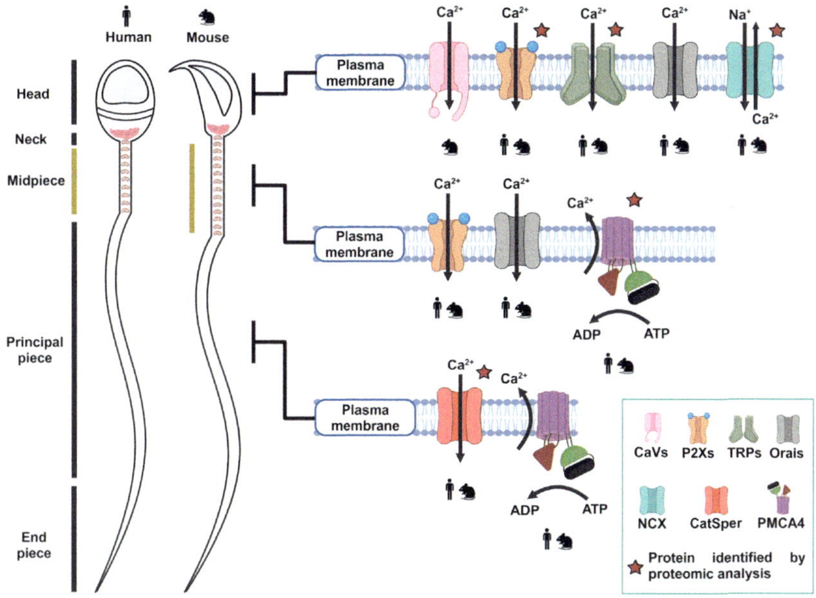

Fig. 1 Plasma membrane Ca^{2+} channels and transporters in sperm. Ca^{2+} channels and transporters hypothesized to be present in mouse and human sperm and their putative localization following evidence discussed in the main body of this review. Stars represent those molecules that have been confirmed by proteomic tandem mass spectrometry. The mouse and human symbol represent the species in which its presence has been demonstrated. This figure was done using Biorender.

4.2 SMOCs

SMOCs are Ca^{2+} channels regulated by second messengers. The presence of cyclic nucleotide-gated channels (CNG) has been proposed in bull sperm (Weyand et al., 1994). Evidence for this protein comes from Western blot and functional studies indicating an increase in $[Ca^{2+}]_i$ upon adding either cAMP or cGMP. More recently, the same group showed that the cyclic-nucleotide mediated $[Ca^{2+}]_i$ increase was due to CatSper and not CNG in human sperm (Brenker et al., 2012) and there is no proteomic evidence yet on the presence of CNG in human or mouse sperm. Although the role of CNG in sea urchin sperm is well-established, their presence and function in mammalian sperm need to be established and validated.

4.3 ROCs

Regarding receptor-gated channels or ROCs, proteomic data indicate the presence in sperm of the ionotropic purinergic receptors, P2RX2 (aka P2X2) and P2RX4 (aka P2X4) (Grahn et al., 2023; Skerrett-Byrne et al., 2022). Interestingly, although P2X4 can localize to the plasma membrane, this purinergic receptor is found preferentially in lysosomes (Suurväli, Boudinot, Kanellopoulos, & Rüütel Boudinot, 2017) where it gets activated by intra-lysosomal ATP only when the pH is increased after fusion. From these channels, P2X2 has been thoroughly studied with a combination of methods, including validated Western blots, electrophysiology and mice knock-out models. Navarro *et al.* conclusively showed that P2X2 is present and functional in mouse sperm. This receptor is responsible for a non-selective cationic ATP-induced inwardly rectifying current (I_{ATP}) (Navarro, Miki, & Clapham, 2011), abrogated in P2rx2$^{-/-}$ mice. Interestingly, although P2rx2$^{-/-}$ mice are fertile, their fertility declines over time after frequent mating (Navarro et al., 2011). This observation prompted the authors to suggest that the P2X2 expression can confer the sperm selection advantages. In addition, P2X7 has been reported in rat sperm by Western blots validated with their respective antigen-blocking controls.

4.4 Cation/calcium channels of sperm (CatSper) complex

A special case of sperm Ca^{2+} entry channel is CatSper, the best characterized Ca^{2+} channel in sperm. This channel has displayed several categorizations, including a mild voltage dependency, a strong pH_i dependency, and promiscuous activation or inhibition by multiple components. In human sperm progesterone is a potent agonist of CatSper by a mechanism involving lipids interaction (Kaupp & Strünker, 2016; Lishko & Mannowetz, 2018; Lishko, Botchkina, & Kirichok, 2011; Lishko et al. 2012). Evidence indicates that cAMP also regulates CatSper and that this effect is mediated by protein kinase A (PKA) (Orta et al., 2018). However, whether the cAMP/PKA action on CatSper is direct or indirect is still unclear. Reviewing the extensive literature regarding CatSper channel complex is beyond the scope of the present article; to learn about all the research done on this channel complex, we refer to an outstanding recent review by Hwang and Chung on this topic (Hwang & Chung, 2023).

4.5 TRPs

TRPs are cation channels described initially in *Drosophila melanogaster* (Clapham, 2003). These channels play key roles in various physiological processes, including sensory perception (such as vision, taste, and temperature sensing), cellular signaling, and homeostasis. TRP channels, besides Ca^{2+}, can conduct different cations, including Na^{+} and Mg^{2+}, depending on the channel subtype. In mammals, the TRP subfamily is comprised of 28 genes divided into seven subfamilies based on sequence homology and functional properties: TRPC (so-called canonical) is associated with receptor-operated and store-operated Ca^{2+} entry; TRPV (vanilloid) is involved in temperature sensation and pain signaling, TRPM (melastatin) thermoregulation and Mg^{2+} homeostasis; TRPA (ankyrin) plays a role in mechanosensation and pain; TRPML (mucolipin) is involved in lysosomal function and membrane trafficking; TRPP (polycistin) is implicated in ciliary function and associated with polycystic kidney disease; and TRPN (NOMPC-like) found in non-mammalian species, and is involved in mechanosensation.

For these channels, proteomic evidence in mice and humans (Skerrett-Byrne et al., 2022; Trötschel et al., 2020) confirms the presence of TRPM8 and TRPM4. In addition, menthol, a TRPM8 agonist, increased $[Ca^{2+}]_i$ and induced the AR in mouse and human sperm (De Blas et al., 2009; Gibbs et al., 2011). Since menthol is not specific (Nguyen, Itoh, Okumura, & Tominaga, 2021; Oz, El Nebrisi, Yang, Howarth, & Al Kury, 2017), it is unclear whether TRPM8 is responsible for these changes. In this regard, several potent and selective TRPM8 antagonists have been recently discovered but have not yet been tested in sperm (Weyer & Lehto, 2017). Regarding other members of the TRP family, there is evidence for TRPV1 (Gervasi et al., 2009), and TRPC1–3 (Sutton, Jungnickel, & Florman, 2003) expression by controlled immunoassays. Nonetheless, these TRPs are not found in the proteomic tables used for this review (Grahn et al., 2023; Greither, Dejung, Behre, Butter, & Herlyn, 2023; Hwang et al., 2019; Skerrett-Byrne, Teperino, & Nixon, 2024). In addition, capsaicin, a TRPV1 agonist induced Ca^{2+} responses in mice and human sperm (Francavilla et al., 2009; Gervasi et al., 2009, 2016), suggesting its presence. TRPC2, another member of the TRP family, has been hypothesized to be present in sperm and to be involved in Ca^{2+} mobilization in mouse sperm upon exposure to solubilized zona pellucida from the oocyte (Jungnickel, Marrero, Birnbaumer, Lémos, & Florman, 2001). Regarding TRPCs, enkurin, a TRPC-interacting protein, is highly expressed in the testis and is present in mouse sperm (Sutton et al., 2004). Despite all these data, except for TRPM8 and TRPM4, none of the other channels are found in proteomic databases (Grahn et al., 2023; Greither et al., 2023; Hwang et al., 2019; Skerrett-Byrne et al., 2024). In addition, it is crucial to consider that TRP mice knock-out models for genes encoding TRPC1–7, TRPV1–4, TRPA1, TRPM1–4, and TRPM8 are fertile (Lishko et al., 2012), suggesting that, even if the respective protein is present in mature mouse sperm, its absence can be compensated by other members of the family. For more information detailing most work on sperm TRP channels, we recommend other excellent reviews (Darszon et al., 2011; Darszon et al., 2012; Lishko et al., 2012).

4.6 SOCs

Conceptually, SOCs have been proposed as those Ca^{2+} entry channels activated by Ca^{2+} efflux from intracellular stores (Putney, 1986). The first direct evidence of the presence of SOCs was obtained using electrophysiological approaches, which showed that Ca^{2+} depletion from intracellular stores activated a non-voltage dependent, Ca^{2+} selective inward rectifying current (Hoth & Penner, 1992). The identity of the SOC channels has been controversial until recent years. TRP channels were the first Ca^{2+} entry channels to be considered as SOCs candidates. Initial evidence for SOCs was obtained in cells from patients suffering from a hereditary disease, known as severe combined immune deficiency (SCID) syndrome (Darszon et al., 2012; Feske et al., 2006). The SCID patients have a point mutation in the *ORAI1* ORF; expression of the wild-type *ORAI1* gene in their T cells restored extracellular Ca^{2+} influx with all the characteristics of the Calcium Release-Activated Current. There are three members of the Orai family (Orai1-3) with a similar structure encompassing four transmembrane-spanning domains (Tiffner & Derler, 2021). All of them form highly Ca^{2+} selective channels responsive to different members of the stromal interaction molecule (STIM) family following Ca^{2+} depletion from intracellular stores. Although these molecules belong to intracellular stores, we will discuss them in this section because their function is directly related to their interaction with Orai. The STIM family is encoded by two genes, *STIM1* and *STIM2* with a MW of ~77 and 85 kDa, respectively. In somatic cells, these molecules are localized to the endoplasmic reticulum lumen, where they sense intraorganellar Ca^{2+} concentration. When Ca^{2+} drops below a given threshold, Stim1 and Stim2 N-terminal calcium sensing domain changes its conformation and promotes their translocation connections between the endoplasmic reticulum and the plasma membrane, where they interact and activate the Orai family of Ca^{2+} channels (Emrich, Yoast, & Trebak, 2022).

In sperm, functional studies indicate that Ca^{2+} depletion induces transmembrane Ca^{2+} entry (O'Toole, Arnoult, Darszon, Steinhardt, & Florman, 2000) suggesting that STIM and ORAI are present in these cells. Western blot analyses validated using pre-absorbed antibodies with the respective antigenic peptides are consistent with this possibility (Darszon et al., 2012; Lefièvre et al., 2012). Proteomic data revealed the presence of Stim1 but not Orai family members in mice and human sperm (Greither et al., 2023; Hwang et al., 2019; Skerrett-Byrne et al., 2022; Trötschel et al., 2020). Interestingly, mice lacking Orai1 are sterile due to spermatogenesis defects with deficiencies in spermatid development (Davis et al., 2016). Because of their spermiogenesis defects, these knock-out mice are silent regarding the role of Orai1 in sperm. A pharmacological approach using penetrating peptides that bind and activate Orai members revealed that these peptides elicit a Ca^{2+} response in human sperm. Conditional knock-out models for *Orai1*, *Stim1* and *Stim2* have been generated (Oh-Hora & Lu, 2018) and warrant their use to further investigate their role in male reproduction.

5. Plasma membrane calcium extrusion mechanisms

Once Ca^{2+}-dependent signal transduction pathways are activated, cells must go back to resting $[Ca^{2+}]_i$, a process mediated by the transport of this ion outside of the cell or by re-sequestration to intracellular stores (e.g. endoplasmic reticulum; mitochondria). A decrease in cytosolic Ca^{2+} is essential to stopping Ca^{2+}-associated signaling pathways. Because even in conditions of high $[Ca^{2+}]_i$, extrusion of this ion is unfavorable, and removal of Ca^{2+} from the cytosol requires energy. The necessary energy is obtained either by ATP hydrolysis by Ca^{2+} pumps such as plasma membrane Ca^{2+} transporting ATPase (*ATP2B*, aka PMCA) family and the sarcoplasmic/endoplasmic reticulum Ca^{2+} ATPase (*ATP2A*, aka SERCA) (for intracellular stores, see below), or by coupling Ca^{2+} secretion to the inwardly favorable electrochemical gradient of Na^+ in the case of the Na^+/Ca^{2+} exchanger (*SLC8A* aka NCX). PMCA and NCX molecules play complementary roles; while NCX does not bind Ca^{2+} tightly, it can transport up 5000 Ca^{2+} ions per second. On the other hand, PMCA pumps have a high affinity for Ca^{2+} but lower capacity and are used primarily to maintain the low basal $[Ca^{2+}]_i$ found generally in cells. Proteomic data from mouse and human sperm revealed the presence of PMCA4. Interestingly, although PMCA4 is ubiquitously expressed in almost all cell types, when this gene is eliminated in mice, one of the few phenotypes observed is sterility due to failure of sperm to become hyperactive (Prasad, Okunade, Liu, Paul, & Shull, 2007). Regarding NCX, only Slc8a2 is found in mice and humans in the proteomic data consulted (Hwang et al., 2019). This exchanger contributes to Ca^{2+} clearance when $[Ca^{2+}]_i$ is elevated in mouse sperm (Wennemuth, Babcock, & Hille, 2003). In boar sperm, Western blots showed bands for the three isoforms (NCX1, 2 and 3); however, only anti NCX2 antibodies recognized a couple of bands of the correct MW (Yeste et al., 2024). The evidence for these molecules to be present in mature sperm is summarized in Fig. 1.

6. Intracellular calcium storage

In addition to Ca^{2+} channels and transporters in the plasma membrane, intracellular Ca^{2+} fluxes are coordinated by regulating organellar Ca^{2+} stores (aka Ca^{2+} reservoirs). In somatic cells, Ca^{2+} is stored in the endoplasmic reticulum (Makio, Chen, & Simmen, 2024), in the nucleus (Oliveira, Guimarães, Andrade, Menezes, & Leite, 2014), in the mitochondria (Oliveira et al., 2014; Rizzuto, De Stefani, Raffaello, & Mammucari, 2012), in peroxisomes (Sargsyan et al., 2021), and in endolysosomes (Alharbi & Parrington, 2021) among others (for review see (Zajac, Modi, & Krishnan, 2022)). These organelles have the dual role of removing excess Ca^{2+} and releasing it to the cytosol upon cell stimulation. Therefore, for an organelle to be considered a Ca^{2+} reservoir, it must possess at least two types of transporters, some that fulfill the function of Ca^{2+} uptake and others that are responsible for Ca^{2+} release. Spermatozoa are no exception; however, besides mitochondria, Ca^{2+} stores are not the traditional ones found in other cell types. Sperm are formed in the testis in a process known as spermatogenesis, which is organized into: (1) proliferative or spermatogonial phase in which spermatogonia undergo mitosis and generate spermatocytes; (2) meiotic phase generating haploid spermatids; and (3) spermiogenesis, whereby round spermatids differentiate into spermatozoa (Sassone-Corsi, 1997). In mammals, the last step of spermiogenesis is characterized by dramatic morphological changes in the haploid spermatid. These changes include chromatin condensation, flagellum assembly, and reorganization of most intracellular organelles, which differentiate into the acrosome and other less understood organelles, such as the redundant nuclear envelope (Ho, 2010; Teves et al., 2020). Together with the mitochondria, these unique organelles have been proposed to act as major intracellular Ca^{2+} stores in mature sperm (Herrick et al., 2005; Khawar, Gao, & Li, 2019; Treviño et al., 1998; Walensky & Snyder, 1995). These sperm Ca^{2+} stores have their own repertoire of channels and transporters. Below, we discussed each of these proteins as well as the evidence of their presence in spermatozoa. Fig. 2 summarizes this information.

7. Organellar calcium uptake in mammalian sperm

Generally, organellar Ca^{2+} uptake occurs against the electrochemical gradient for the ion and is associated with ATP hydrolysis (Olesen et al., 2007). In many cell types including mammalian sperm, ATPase pumps like the sarcoplasmic/endoplasmic reticulum Ca^{2+}-ATPases (SERCAs), or the secretory pathway Ca^{2+}-ATPases (SPCAs), expressed in the Golgi apparatus, can contribute to organellar Ca^{2+} uptake (Dyla, Kjærgaard, Poulsen, & Nissen, 2020; Vangheluwe et al., 2009). SERCAs and SPCAs are in the endoplasmic reticulum and Golgi apparatus, respectively. However, when studying the role of the intracellular ATPases in spermatozoa, we should consider that, in these cells, these classical organelles constitute other sperm-specific structures such as the acrosome and the redundant nuclear envelope. In addition, as in somatic cells, Ca^{2+} enters the mitochondrial

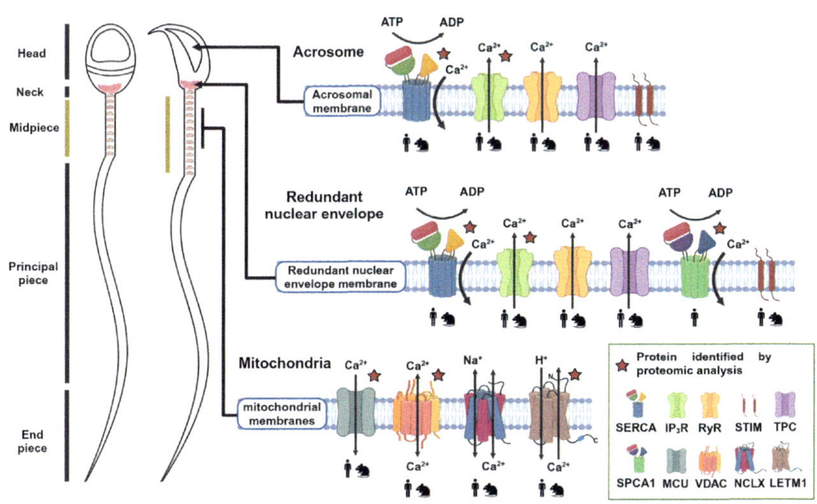

Fig. 2 Ca^{2+} channels and transporters in the membrane of sperm intracellular stores. Molecules in the membrane of intracellular stores of mouse and human sperm postulated to be present and have a functional role in transporting Ca^{2+} ions. Stars represent those molecules that have been confirmed by proteomic tandem mass spectrometry. The mouse and human symbol represent the species in which its presence has been demonstrated. This figure was done using Biorender.

matrix through voltage-dependent anion channels (VDACs) located in the outer mitochondrial membrane (OMM) and the mitochondrial Ca^{2+} uniporter (MCU), located in the inner mitochondrial membrane (IMM) (Pathak & Trebak, 2018). Although tightly regulated, Ca^{2+} transport into the mitochondrial matrix follows the Ca^{2+} electrochemical gradient and is not ATP-dependent.

7.1 SERCAs

The SERCA family is encoded by three different genes: *ATP2A1* (SERCAs 1), *ATP2A2* (SERCAs 2), and *ATP2A3* (SERCAs 3), which produce seven different isoforms with MW of ~110 kDa (SERCAs 1a/1b, SERCAs 2a/2b, and SERCAs 3a/3b/3c) (Periasamy & Kalyanasundaram, 2007). SERCAs are Ca^{2+} ATPases involved in maintaining low levels of cytosolic Ca^{2+} by pumping this ion from the cytosol into the endoplasmic reticulum lumen (Clapham, 2007; Sacchetto et al., 2012; Xu & Van Remmen, 2021). Ca^{2+} movement through these ATPases requires cyclical changes during which ATP phosphorylates the protein at a conserved aspartic acid side chain and then dephosphorylates it (Olesen et al., 2007). As a result, each ATP hydrolysis cycle transports two Ca^{2+} ions through SERCAs into the organelle (Wu et al., 2023).

In sperm, Atp2a1 and Atp2a2 are present in all sperm proteomic tables reviewed for this manuscript (Greither et al., 2023; Hwang et al., 2019; Skerrett-Byrne et al., 2022; Trötschel et al., 2020). In addition, immune-based assays suggest the presence of SERCA in the sperm acrosome from different mammalian species (Harayama, Okada, & Miyake, 2003; Lawson, Dorval, Goupil, & Leclerc, 2007). SERCA immunodetection in the acrosome is consistent with the hypothesis that this organelle is a major sperm Ca^{2+} store. The presence of SERCA in sperm is also in agreement with the effect of SERCA inhibitors on $[Ca^{2+}]_i$, hyperactivation and the acrosome reaction. Pharmacologically, two chemical compounds have been used in mammalian sperm. Thapsigargin (Tg) is a sesquiterpene lactone extracted from the plant *Thapsia garganica,* which binds to SERCA with subnanomolar affinity, and it is the most specific inhibitor of SERCA ATPases. Cyclopiazonic acid (CPA) is an indole tetratomic acid fungal toxin; this molecule binds to SERCAs with significantly lower affinity than Tg. Both compounds induce Ca^{2+} leakage from the endoplasmic reticulum in somatic cells (Treiman, Caspersen, & Christensen, 1998; Vangheluwe et al., 2009).

7.2 SPCAs

SPCA is a family of ATPases encoded by two different genes, *ATP2C1 (aka SPCA1)* and *ATP2C2 (aka SPCA2)* (Zong & Yue, 2023); different splicing variants represent both families with MW of ~100 kDa. The SPCAs carry Ca^{2+} and Mn^{2+} into the lumen of the Golgi apparatus; SPCA2 can also directly activate ORAIs independently of STIM 1, resulting in constitutive store-independent extracellular Ca^{2+} entry (Feng et al., 2010; Wootton, Argent, Wheatley, & Michelangeli, 2004). SPCA1 was identified earlier than SPCA2, as a result, the physiological function of this Ca^{2+} pump is better understood. This Ca^{2+}/Mn^{2+} pump transports these ions through a series of intermediate states (Dyla et al., 2020). Whereas SERCAs contain two high-affinity Ca^{2+}-binding sites, the SPCA1 pump transports only one divalent Ca^{2+} per cycle. Tg and CPA, both SERCA antagonists, have weaker inhibitory effects on SPCA1 (Wu et al., 2023).

In sperm, SPCA1 has only been detected in the neck/midpiece of human sperm using Western blotting and immunostaining, respectively (Harper et al., 2005). Additionally, the expression of these Ca^{2+} pumps has been confirmed using sperm proteomics (Greither et al., 2023; Wang et al., 2013). Pharmacologically, bis(2-hydroxy-3tert-butyl-5-methyl-phenyl)-methane (aka bis-phenol) and 2-aminoethoxydiphenylborate (aka 2-APB) have been reported as potent SPCA1 inhibitors with IC50 at the 100 nM level. Both compounds are more selective for SPCA than for SERCAs (60- and 8.3-fold greater potency, respectively). Both are capable of mobilizing Ca^{2+} from intracellular Ca^{2+} stores even when SERCA pumps are blocked by Tg (Lai & Michelangeli, 2012).

7.3 Inward mitochondrial calcium transport

Mitochondria, one of eukaryotic cells' most relevant Ca^{2+} storage organelles, regulates inward and outward Ca^{2+} transport. At the structural level, mitochondria comprise an outer (OMM) and an inner mitochondrial membrane (IMM) surrounding the mitochondrial matrix where most enzymatic processes occur, including the Krebs cycle. For inward transport, Ca^{2+} requires crossing both the OMM and the IMM. Transport through the OMM is mediated by voltage-dependent anion channels (VDAC) (Pathak & Trebak, 2018). Ca^{2+} transport through VDAC is Ca^{2+}-activated and depends on cytoplasmic $[Ca^{2+}]_i$ levels outside (Báthori, Csordás, Garcia-Perez, Davies, & Hajnóczky, 2006).

VDACs selectivity towards anions and cations is regulated by interaction with other proteins and by the electrochemical potential of the respective ion. On the other hand, permeability through the IMM is controlled by a highly Ca^{2+} selective channel known as the mitochondrial Ca^{2+} uniporter (MCU) (Baughman et al., 2011; Kirichok, Krapivinsky, & Clapham, 2004). Briefly, MCU Ca^{2+} transport is electrogenic, and is driven by the negative mitochondrial membrane potential generated by the respiratory electron transport chain (Szabadkai & Duchen, 2008). The MCU is a complex composed of the pore forming MCU, its dominant-negative isoform MCUb, three essential regulators and other non-essential modulatory proteins. In this holo-complex formed by different subunits, some are essential and found in all cell types, while others have regulatory roles and are tissue-specific. In addition to MCU, other essential proteins are Micu1, Micu2 and Smdt1 (aka Emre) (Yoo, 2022). MCU has a MW of ~40 kDa and forms a holo-complex with the C-terminal and N-terminal domain facing the mitochondrial matrix. In the middle of the structure resides a selective Ca^{2+} pore within the transmembrane domain containing an evolutionarily conserved sequence critical for Ca^{2+} binding known as DIME (for glutamic, isoleucine, methionine, and aspartic). When the Ca^{2+} concentration in the mitochondrial intermembrane space is high, it binds to the regulatory protein, mitochondrial calcium uptake 1 (MCU1), activating MCU and causing Ca^{2+} uptake into the mitochondrial matrix (Pathak & Trebak, 2018). Good MCU inhibitors and activators are not well-developed (Mishra et al., 2017). They lack specificity and, in many cases, they are not permeable. The most widely used MCU inhibitors are ruthenium red and its analogue Ru360. However, the results obtained with these inhibitors must be taken cautiously since they also block other plasma membrane and intracellular Ca^{2+} and K^+ channels (Mishra et al., 2017).

In sperm, the presence of VDAC1, VDAC2 and VDAC3, MCU, MICU1, MICU2, SMDT1 and MCUR1 have been confirmed by tandem mass spectrometry in mice and humans (Greither et al., 2023; Skerrett-Byrne et al., 2022; Trötschel et al., 2020) (Fig. 3). In addition, different works have shown the effects of ruthenium red in mammalian spermatozoa (Bravo et al., 2015; Ferreira, Cassina, et al., 2021; Rodriguez, Satorre, & Beconi, 2012). However, considering the lack of specificity of the inhibitor, the interpretation of these experiments requires further investigation.

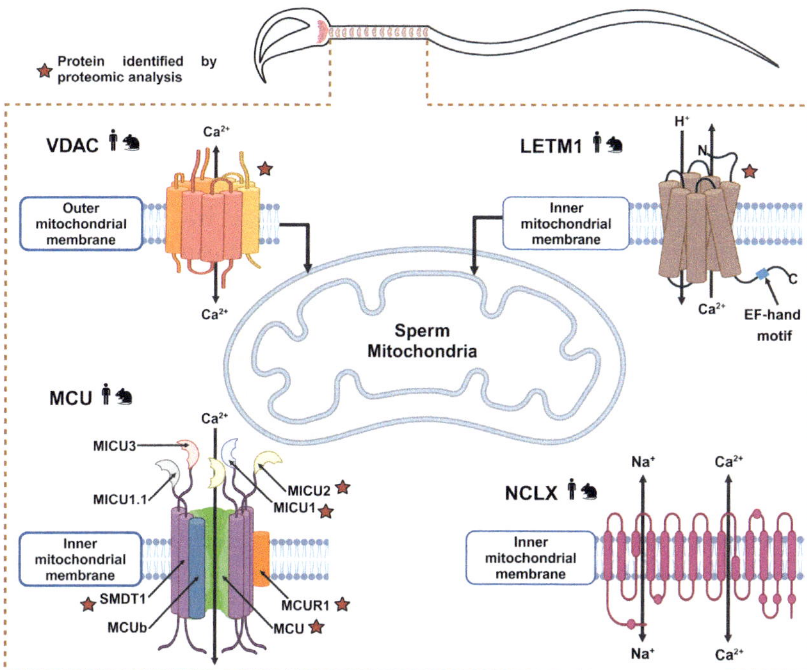

Fig. 3 Mitochondrial Ca^{2+} channels and transporters in mammalian sperm. Proteins in the sperm IMM and OMM that have supporting evidence of their presence in mouse and human sperm. Stars represent those molecules that have been confirmed by proteomic tandem mass spectrometry. The mouse and human symbol represent the species in which its presence has been demonstrated. This figure was done using Biorender.

8. Molecular basis of organellar calcium release in mammalian sperm

In addition to having an essential role in Ca^{2+} removal from the cytosol to avoid Ca^{2+} toxicity, intracellular organelles are also involved in storing this cation and releasing it under the right stimulus to activate Ca^{2+} signaling pathways. At the molecular level, outward Ca^{2+} transport from organelles requires different families of Ca^{2+} channels and exchangers, each activated by specific signaling pathways. Among the best characterized are: (1) the inositol 1,4,5-triphosphate (IP_3) receptors (IP_3Rs); (2) the ryanodine receptors (RyRs), (3) the two-pore channels family (TPCs 1/2); and (4) the mitochondrial Na^+ and H^+ exchangers in charge of releasing Ca^{2+} from these organelles.

8.1 Inositol 3-phosphate receptors (IP_3Rs)

IP_3Rs are homo- or heterotetramers composed of four subunits localized in the membrane of the sarco/endoplasmic reticulum (S/ER) (Santulli, Nakashima, Yuan, & Marks, 2017). The function of IP_3Rs is regulated by IP_3, a water-soluble second messenger which is produced primarily by hydrolysis of phosphatidylinositol 4,5-bisphosphate by phospholipase C upon activation of G-protein-coupled receptors (Luan & Wang, 2021; Santulli et al., 2017). Three different genes encode IP_3R subtypes: *ITPR1* (IP_3Rs 1), *ITPR2* (IP_3Rs 2), and *ITPR3* (IP_3Rs 3) (Gambardella et al., 2021). The multimodal regulation of the IP_3R highlights its functional relevance. Besides IP_3, these Ca^{2+} channels are also regulated by Ca^{2+}, phosphorylation, and interaction with multiple proteins inside and outside the endoplasmic reticulum (for review see: (Baker, Fan, Arige, Yule, & Serysheva, 2023; Ivanova & Atakpa-Adaji, 2023)).

Walensky and Snyder initially showed the presence of IP3 receptors in sperm in 1995 (Walensky & Snyder, 1995). In this manuscript, the authors used a battery of robust functional and immunological assays as evidence of the presence of IP_3Rs in the mouse sperm acrosome. First they documented high-affinity IP_3 binding sites using tritium labeled (3H)IP_3 and Scatchard plots; second, they used Western blots and antibodies against IP_3Rs with their respective antigen-blocked controls, immunofluorescence and immune-electron microscopy approaches; finally, they also showed that the antibody signal was lost in acrosome-reacted sperm. Other authors later confirmed these results (Gugssa et al., 2010; Ho & Suarez, 2001, 2003; Naaby-Hansen et al., 2001; Treviño et al., 1998). Interestingly, in bovine sperm IP_3 receptors were found in the base of the flagellum in a sperm-specialized structure known as the redundant nuclear envelope (Ho & Suarez, 2001, 2003). Similarly, in human sperm IP_3Rs have been localized to the acrosome and the RNE (Naaby-Hansen et al., 2001). Notably, the three IP3R gene products, ITPR1, ITPR2 and ITPR3 have been found in proteomic analyses of both mouse and human mature sperm (Grahn et al., 2023; Greither et al., 2023; Hwang et al., 2019; Skerrett-Byrne et al., 2024).

Pharmacologically, the most used IP_3R blockers are Xestospongin C and 2-APB. They have been used to evaluate the role of this Ca^{2+} transporter in mammalian sperm physiology. Xestospongin is a macrocyclic bis-1-oxa-quinolizidine obtained from the Australian marine sponge *Xestospongia exigua*, and it is regarded as the most specific IP_3R inhibitor (Gambardella et al., 2021). In contrast, 2-APB, obtained via the esterification of diphenyl boronic acid with aminoethanol, has a complex pharmacology and is used as an activator or inhibitor for different Ca^{2+} channels and transporters (Maruyama, Kanaji, Nakade, Kanno, & Mikoshiba, 1997; Zhou et al., 2007). On the other hand, adenosphostin A, a compound isolated from the culture broth of Penicillium brevicompactum is a potent and specific agonist of the three IP_3R isoforms (Saleem et al., 2013). Adenophostin A induces a rapid acrosomal Ca^{2+} release in capacitated streptolysin O-permeabilized human sperm (Ruete, Lucchesi, Bustos, & Tomes, 2014).

8.2 Ryanodine receptors

Ryanodine receptors (RyRs) are encoded by three different genes: *RyR1*, *RyRs2*, and *RyR3* (Santulli et al., 2017) encompassing tissue-specific splicing variants. They are large molecules with MW of ~560 kDa that forms homotetramers that bind four small FK506 binding proteins (FKBP) with a total MW of ~2200 kDa. RyRs are cation-selective channels that localize to the sarco/endoplasmic reticulum from somatic cells. RyRs activity is regulated by multiple signaling pathways including Ca^{2+}, ATP, calmodulin, phosphorylation, dephosphorylation, and redox-active species, and via direct protein-protein interactions with L-type voltage-dependent Ca^{2+} channels. Cyclic Adenosine diphosphate Ribose (cADPR) evokes Ca^{2+} transients in HEK293 cells stably expressing RyR1 or RyR3 (Ogunbayo et al., 2011). Each RyR subunit has Ca^{2+} activation sites in the luminal and cytoplasmic domains. These sites synergistically activate the gating mechanism for channel opening by Ca^{2+}. RyRs also possess two inhibitory sites in their cytoplasmic domains with Ca^{2+} affinities of 1 μM and 1 mM (Laver, 2018). Because of the large size of RyRs, their structure has been elusive; however, new developments in cryo-electron microscopy have started to reveal details at near-atomic resolution (Dulhunty, Beard, & Casarotto, 2018). Although many compounds can activate or inhibit RyRs in vitro, overall, the pharmacology is complex, and it is beyond the scope of this review to critically evaluate RyR pharmacology. We will only mention agents used to assess the presence of RyRs in mammalian sperm. One of them is ryanodine itself, an alkaloid plant toxin which gives the name to this family of cation channels (Willegems & Efremov, 2017). Ryanodine binds with high affinity and specificity to RyRs in vitro and can modulate their activity. It increases RyR open probability at sub-micromolar concentrations and completely blocks the channel at high micromolar concentrations. Another molecule used to modulate RyRs is dantrolene, a hydantoin derivative that can block Ca^{2+} release from the endoplasmic reticulum at micromolar concentrations (Santonastasi & Wehrens, 2007). Finally, like the mitochondrial MCU, RyRs can be inhibited by ruthenium red (Ma, 1993). RyRs were not find in mouse and human proteomic data used in this review; the evidence of RyRs expression in sperm is restricted to the effect of ryanodine and dantrolene as well as to the binding of ryanodine probes to sperm compartments. Contrary to permeable NAADP (NAADP-AM), the addition of permeable cADPR (cADPR-AM) did not elicit a Ca^{2+} response in human sperm (Sánchez-Tusie, Vasudevan, Churchill, Nishigaki, & Treviño, 2014). Although NAADP has been reported to activate RyRs, this activation is likely to be indirect through stimulation of two-pore domain channels (TPCs) (Laver, 2018). Clearly, more work is needed to elucidate the presence and putative role of these receptors in mammalian sperm.

8.3 Two-pore channels (TPC)

TPCs are key endo lysosomal cation channels responsive to NAADP. In vertebrates, three TPC paralogue genes exist (*Tpcn1*, *Tpcn2*, and *Tpcn3*); however, the *Tpcn3* gene has been lost in primates and rodents (Shah, Guan, & Yan, 2022). TPCs form homodimers and each of the subunits has a MW of ~94 kDa. TPCs are Na^{+}-selective with relatively low permeability for Ca^{2+}. This relatively low Ca^{2+} permeability is physiologically relevant given that TPCs mediate lysosomal Ca^{2+} release. TPCs 1 and 2 respond to second messengers: nicotinic acid adenine dinucleotide phosphate (NAADP), a cytosolic soluble second messenger, and the phosphatidylinositol 3,5-bisphosphate ($PI(3,5)P_2$), a lysosome -specific lipid (Morgan, Martucci, Davis, & Galione, 2022; Riederer, Cang, & Ren, 2023). Although initially, NAADP was thought to activate RyRs, it is now clear that this second messenger is primarily involved in TPC activation (Calcraft et al., 2009). However, NAADP action is mediated by TPC-interacting proteins such as Lsm12 (Zhang, Guan, Shah, & Yan, 2021) and Jupiter microtubule-associated homolog 2 (JPT2) (Gunaratne et al., 2021). In human sperm, permeable NAADP-AM increases $[Ca^{2+}]_i$ (Sánchez-Tusie et al., 2014). In another work, Western blots showed a band in TPC1 in mouse sperm that is not found in sperm from Tpc1 null mice models (Arndt et al., 2014). In the same work, TPC1 was localized to the acrosome of wild type mice using immunoelectron microscopy, a signal absent in sperm from knock out mice. Lack of TPC1 only has minor effects on reproduction, although this result could be due to compensation by TPC2. Nevertheless, proteomics detected the NAADP-binding and

TCCP-interacting proteins Lsm12 in human databases and JPT2 in human and mouse databases. As with other channels, TPC pharmacology is complex. As mentioned, NAADP and its permeable analog NAADP-AM have been used in sperm. In sperm, Ned19, a selective membrane-permeant non-competitive NAADP antagonist with IC50 on the nM range (Naylor et al., 2009) partially blocked the solubilized zona pellucida-induced acrosome reaction (Arndt et al., 2014) as well as Ca^{2+} release by compounds that increase intra-acrosomal pH (Oliver, Jabloñski, Buffone, & Darszon, 2023).

8.4 Mitochondrial calcium extrusion

As described in previous sections, a combination of VDAC in the OMM and MCU in the IMM allows mitochondrial uptake of high quantities of Ca^{2+}. However, Ca^{2+} overload of this organelle can trigger irreversible damage that leads to cell death (Rodrigues et al., 2022). To counterbalance Ca^{2+} uptake, mitochondria can use the electrochemical gradient to exchange Ca^{2+} for counter ions such as Na^+ or H^+. Although Na^+/Ca^{2+} exchangers are best known to participate in the plasma membrane, different lines of evidence support their presence in the IMM and OMM. They have been attributed relevant roles in controlling mitochondrial Ca^{2+} homeostasis. Among them, the Li^+-permeable Na^+/Ca^{2+} exchanger (NCLX) encoded by the *SLC8B1* gene is enriched in mitochondria (Palty et al., 2010). On the other hand, a Na^+ independent mechanism of mitochondrial Ca^{2+} extrusion is a Ca^{2+}/H^+ exchanger known as Leucine Zipper and EF- Hand Containing Transmembrane protein 1 (aka LETM1) (Shao et al., 2016). In sperm, proteomic data revealed the presence of LETM1 but not NCLX (Greither et al., 2023; Hwang et al., 2019; Skerrett-Byrne et al., 2022; Trötschel et al., 2020) (Fig. 3). However, as mentioned above NCX2 is found in mammalian sperm, and its participation in the sperm mitochondria cannot be discarded (Fig. 1).

9. Final remarks

In this article, we summarized and analyzed current evidence on the presence of channels and transporters involved in regulating Ca^{2+} fluxes in mammalian sperm. Changes in $[Ca^{2+}]_i$ are a central signaling pathway node in all cell types and spermatozoa are not an exception. Ca^{2+} plays essential roles in capacitation-associated pathways such as hyperactivation and the preparation for the acrosome reaction. In addition, a regulated Ca^{2+} increase is required for all events related to acrosomal exocytosis. Although not discussed here, at the molecular level, the role of Ca^{2+} is mediated by many Ca^{2+}-binding proteins, including but not limited to calmodulin (Naaby-Hansen et al., 2010). The relevance of the Ca^{2+} role in sperm fertility is highlighted by work from our group indicating that transient exposure to the Ca^{2+} ionophore A_{23187} bypasses other capacitation-associated pathways such as PRKACA (PKA) activation (Tateno et al., 2013). Moreover, transient exposure to Ca^{2+} ionophores also rescued sperm from sterile mice models and enable them able to fertilize metaphase-II arrested eggs in vitro (Navarrete et al., 2016). The best studied Ca^{2+} entry pathway in sperm is the Ca^{2+} channel complex CatSper, whose essential role in sperm hyperactivation has been conclusively shown. However, the restricted localization of this channel in the principal piece (Chung et al., 2014; Ren et al., 2001) raises question about its role in other sperm compartments and the acrosome reaction. In addition, it is clear from proteomic, immunological, pharmacological and, in some cases, knock-out studies, that other Ca^{2+} transport systems are present in mammalian sperm. As an example, studies from our group using Ca^{2+} fluorescent sensors indicate that, in sperm from CatSper1 null mice, it is still possible to detect capacitation and agonist-induced changes in $[Ca^{2+}]_i$ (Luque et al., 2018; Sánchez-Cárdenas et al., 2021). Together, these data warrant more investigations on the spatial and temporal regulation of Ca^{2+} homeostasis in mature sperm.

Acknowledgments

This study was supported by: NIH grants HD-038082 (to PEV and AD) and HD088571 (to JB, LL and PEV), and HD106968 (to Diego Krapf and MB); Male contraceptive initiative (to PEV), CONACHyT (314839 to AH-C and CF-2023-I-291 to AD) and PAPIIT-DGAPA (IN201124 to AH-C, IN218123 to JCC, and IN200919 to AD); *Agencia Nacional de Promoción Científica y Tecnológica* (PICT 2020-00098 to MGB and 2021-00031 to GML); and by the Chan Zuckerberg Initiative (CZI 2021–240504 to MGB).

References

Alharbi, A. F., & Parrington, J. (2021). Deciphering the role of endolysosomal Ca2+ channels in immunity. *Frontiers in Immunology,* 12. https://doi.org/10.3389/FIMMU.2021.656965.

Arndt, L., Castonguay, J., Arlt, E., Meyer, D., Hassan, S., Borth, H., ... Boekhoff, I. (2014). NAADP and the two-pore channel protein 1 participate in the acrosome reaction in mammalian spermatozoa. *Molecular Biology of the Cell, 25*(6), 949–962. https://doi.org/10.1091/MBC.E13-09-0523.

Baker, M. R., Fan, G., Arige, V., Yule, D. I., & Serysheva, I. I. (2023). Understanding IP3R channels: From structural underpinnings to ligand-dependent conformational landscape. *Cell Calcium, 114*. https://doi.org/10.1016/J.CECA.2023.102770.

Báthori, G., Csordás, G., Garcia-Perez, C., Davies, E., & Hajnóczky, G. (2006). Ca2+-dependent control of the permeability properties of the mitochondrial outer membrane and voltage-dependent anion-selective channel (VDAC. *The Journal of Biological Chemistry, 281*(25), 17347–17358. https://doi.org/10.1074/JBC.M600906200.

Baughman, J. M., Perocchi, F., Girgis, H. S., Plovanich, M., Belcher-Timme, C. A., Sancak, Y., ... Mootha, V. K. (2011). Integrative genomics identifies MCU as an essential component of the mitochondrial calcium uniporter. *Nature, 476*(7360), 341–345. https://doi.org/10.1038/NATURE10234.

Berridge, M. J., Bootman, M. D., & Roderick, H. L. (2003). Calcium signalling: Dynamics, homeostasis and remodelling. *Nature Reviews. Molecular Cell Biology, 4*(7), 517–529. https://doi.org/10.1038/NRM1155.

Berridge, M. J., Lipp, P., & Bootman, M. D. (2000). The versatility and universality of calcium signalling. *Nature Reviews. Molecular Cell Biology, 1*(1), 11–21. https://doi.org/10.1038/35036035.

Bravo, A., Treulen, F., Uribe, P., Boguen, R., Felmer, R., & Villegas, J. V. (2015). Effect of mitochondrial calcium uniporter blocking on human spermatozoa. *Andrologia, 47*(6), 662–668. https://doi.org/10.1111/AND.12314.

Brenker, C., Goodwin, N., Weyand, I., Kashikar, N. D., Naruse, M., Krähling, M., ... Strünker, T. (2012). The CatSper channel: A polymodal chemosensor in human sperm. *The EMBO Journal, 31*(7), 1654–1665. https://doi.org/10.1038/emboj.2012.30.

Brenker, C., Zhou, Y., Müller, A., Echeverry, F. A., Trötschel, C., Poetsch, A., ... Strünker, T. (2014). The Ca2+-activated K+ current of human sperm is mediated by Slo3. –e01438 *ELife, 3*, e01438. https://doi.org/10.7554/eLife.01438.

Breton, S., Nair, A. V., & Battistone, M. A. (2019). Epithelial dynamics in the epididymis: Role in the maturation, protection, and storage of spermatozoa. *Andrology,* 7(5), 631–643. https://doi.org/10.1111/ANDR.12632.

Calcraft, P. J., Ruas, M., Pan, Z., Cheng, X., Arredouani, A., Hao, X., ... Zhu, M. X. (2009). NAADP mobilizes calcium from acidic organelles through two-pore channels. *Nature, 459*(7246), 596–600. https://doi.org/10.1038/NATURE08030.

Catterall, W. A. (2011). Voltage-gated calcium channels. *Cold Spring Harbor Perspectives in Biology, 3*(8), 1–23. https://doi.org/10.1101/CSHPERSPECT.A003947.

Chung, J. J., Shim, S. H., Everley, R. A., Gygi, S. P., Zhuang, X., & Clapham, D. E. (2014). Structurally distinct Ca2+ signaling domains of sperm flagella orchestrate tyrosine phosphorylation and motility. *Cell, 157*(4), 808–822. https://doi.org/10.1016/j.cell.2014.02.056.

Clapham, D. E. (2003). TRP channels as cellular sensors. *Nature, 426*(6966), 517–524. https://doi.org/10.1038/nature02196.

Clapham, D. E. (2007). Calcium signaling. *Cell, 131*(6), 1047–1058. https://doi.org/10.1016/j.cell.2007.11.028.

Cohen, R., Buttke, D. E., Asano, A., Mukai, C., Nelson, J. L., Ren, D., ... Travis, A. J. (2014). Lipid modulation of calcium flux through CaV2.3 regulates acrosome exocytosis and fertilization. *Developmental Cell, 28*(3), 310–321. https://doi.org/10.1016/j.devcel.2014.01.005.

Cohen, R., Mukai, C., Nelson, J. L., Zenilman, S. S., Sosnicki, D. M., & Travis, A. J. (2022). A genetically targeted sensor reveals spatial and temporal dynamics of acrosomal calcium and sperm acrosome exocytosis. *Journal of Biological Chemistry, 298*(5), 101868–101869. https://doi.org/10.1016/j.jbc.2022.101868.

Cohen, R., Sosnicki, D. M., White, M. A., Nelson, J. L., Mukai, C., & Travis, A. J. (2023). Real-time imaging of acrosomal calcium dynamics and exocytosis in live mouse sperm. *Journal of Visualized Experiments: JOVE, 2023*(200), https://doi.org/10.3791/65962.

Crabtree, G. R. (2001). Calcium, calcineurin, and the control of transcription. *The Journal of Biological Chemistry, 276*(4), 2313–2316. https://doi.org/10.1074/JBC.R000024200.

Darszon, A., & Hernández-Cruz, A. (2014). T-type Ca2+ channels in spermatogenic cells and sperm. *Pflugers Archiv: European Journal of Physiology, 466*(4), 819–831. https://doi.org/10.1007/S00424-014-1478-2.

Darszon, A., Nishigaki, T., Beltran, C., & Treviño, C. L. (2011). Calcium channels in the development, maturation, and function of spermatozoa. *Physiological Reviews, 91*(4), 1305–1355. https://doi.org/10.1152/physrev.00028.2010.

Darszon, A., Sánchez-Cárdenas, C., Orta, G., Sánchez-Tusie, A. A., Beltrán, C., López-González, I., ... Treviño, C. L. (2012). Are TRP channels involved in sperm development and function? *Cell and Tissue Research, 349*(3), 749–764. https://doi.org/10.1007/s00441-012-1397-5.

Davis, F. M., Goulding, E. H., D'Agostin, D. M., Janardhan, K. S., Cummings, C. A., Bird, G. S., ... Putney, J. W. (2016). Male infertility in mice lacking the store-operated Ca2+ channel Orai1. *Cell Calcium, 59*(4), 189–197. https://doi.org/10.1016/j.ceca.2016.02.007.

De Blas, G. A., Darszon, A., Ocampo, A. Y., Serrano, C. J., Castellano, L. E., Hernández-Gonzá, E. O., ... Treviño, C. L. (2009). TRPM8, a versatile channel in human sperm. *PLoS One, 4*(6), https://doi.org/10.1371/JOURNAL.PONE.0006095.

De La Vega-Beltran, J. L., Sánchez-Cárdenas, C., Krapf, D., Hernandez-González, E. O., Wertheimer, E., Treviño, C. L., ... Darszon, A. (2012). Mouse sperm membrane potential hyperpolarization is necessary and sufficient to prepare sperm for the acrosome reaction. *Journal of Biological Chemistry, 287*(53), 44384–44393. https://doi.org/10.1074/jbc.M112.393488.

Dey, S., Brothag, C., & Vijayaraghavan, S. (2019). Signaling enzymes required for sperm maturation and fertilization in mammals. *Frontiers in Cell and Developmental Biology, 7*. https://doi.org/10.3389/FCELL.2019.00341.

Dulhunty, A. F., Beard, N. A., & Casarotto, M. G. (2018). Recent advances in understanding the ryanodine receptor calcium release channels and their role in calcium signalling. *F1000Research, 7*. https://doi.org/10.12688/F1000RESEARCH.16434.1/DOI.

Dyla, M., Kjærgaard, M., Poulsen, H., & Nissen, P. (2020). Structure and mechanism of P-Type ATPase ion pumps. *Annual Review of Biochemistry, 89*, 583–603. https://doi.org/10.1146/ANNUREV-BIOCHEM-010611-112801.

Emrich, S. M., Yoast, R. E., & Trebak, M. (2022). Physiological functions of CRAC channels. *Annual Review of Physiology, 84*, 355–379. https://doi.org/10.1146/ANNUREV-PHYSIOL-052521-013426.

Escoffier, J., Boisseau, S., Serres, C., Chen, C. C., Kim, D., Stamboulian, S., ... Arnoult, C. (2007). Expression, localization and functions in acrosome reaction and sperm motility of Ca(V)3.1 and Ca(V)3.2 channels in sperm cells: An evaluation from Ca(V)3.1 and Ca (V)3.2 deficient mice. *Journal of Cellular Physiology, 212*(3), 753–763. https://doi.org/10.1002/JCP.21075.

Feng, M., Grice, D. M., Faddy, H. M., Nguyen, N., Leitch, S., Wang, Y., ... Rao, R. (2010). Store-independent activation of Orai1 by SPCA2 in mammary tumors. *Cell, 143*(1), 84–98. https://doi.org/10.1016/J.CELL.2010.08.040.

Ferreira, J. J., Cassina, A., Irigoyen, P., Ford, M., Pietroroia, S., Peramsetty, N., ... Sapiro, R. (2021). Increased mitochondrial activity upon CatSper channel activation is required for mouse sperm capacitation. *Redox Biology, 48*. https://doi.org/10.1016/J.REDOX.2021.102176.

Ferreira, J. J., Lybaert, P., Puga-Molina, L. C., & Santi, C. M. (2021). Conserved mechanism of bicarbonate-induced sensitization of CatSper channels in human and mouse sperm. *Frontiers in Cell and Developmental Biology, 9*. https://doi.org/10.3389/FCELL.2021.733653.

Feske, S., Gwack, Y., Prakriya, M., Srikanth, S., Puppel, S. H., Tanasa, B., ... Rao, A. (2006). A mutation in Orai1 causes immune deficiency by abrogating CRAC channel function. *Nature, 441*(7090), 179–185. https://doi.org/10.1038/NATURE04702.

Francavilla, F., Battista, N., Barbonetti, A., Vassallo, M. R. C., Rapino, C., Antonangelo, C., ... Maccarrone, M. (2009). Characterization of the endocannabinoid system in human spermatozoa and involvement of transient receptor potential vanilloid 1 receptor in their fertilizing ability. *Endocrinology, 150*(10), 4692–4700. https://doi.org/10.1210/EN.2009-0057.

Gambardella, J., Morelli, M. B., Wang, X., Castellanos, V., Mone, P., & Santulli, G. (2021). The discovery and development of IP3 receptor modulators: An update. *Expert Opinion on Drug Discovery, 16*(6), 709–718. https://doi.org/10.1080/17460441.2021.1858792.

Gervasi, M. G., Osycka-Salut, C., Sanchez, T., Alonso, C. A. I., Llados, C., Castellano, L., ... Perez-Martinez, S. (2016). Sperm release from the oviductal epithelium depends on Ca2+ influx upon activation of CB1 and TRPV1 by anandamide. *Journal of Cellular Biochemistry, 117*(2), 320–333. https://doi.org/10.1002/JCB.25273.

Gervasi, M. G., Rapanelli, M., Ribeiro, M. L., Farina, M., Billi, S., Franchi, A. M., & Martinez, S. P. (2009). The endocannabinoid system in bull sperm and bovine oviductal epithelium: Role of anandamide in sperm-oviduct interaction. *Reproduction (Cambridge, England), 137*(3), 403–414. https://doi.org/10.1530/REP-08-0204.

Gervasi, M. G., & Visconti, P. E. (2016). Chang's meaning of capacitation: A molecular perspective. *Molecular Reproduction and Development, 83*(2016), 860–874. https://doi.org/10.1002/mrd.22663.

Gervasi, M. G., & Visconti, P. E. (2017). *Molecular changes and signaling events occurring in spermatozoa during epididymal maturation. Andrology, 5*, Blackwell Publishing Ltd, 204–218. https://doi.org/10.1111/andr.12320.

Gibbs, G. M., Orta, G., Reddy, T., Koppers, A. J., Martínez-López, P., De La Vega-Beltràn, J. L., ... O'Bryan, M. K. (2011). Cysteine-rich secretory protein 4 is an inhibitor of transient receptor potential M8 with a role in establishing sperm function. *Proceedings of the National Academy of Sciences of the United States of America, 108*(17), 7034–7039. https://doi.org/10.1073/PNAS.1015935108.

Grahn, E., Kaufmann, S. V., Askarova, M., Ninov, M., Welp, L. M., Berger, T. K., ... Kaupp, U. B. (2023). Control of intracellular pH and bicarbonate by CO_2 diffusion into human sperm. *Nature Communications, 14*(1), https://doi.org/10.1038/S41467-023-40855-0.

Greither, T., Dejung, M., Behre, H. M., Butter, F., & Herlyn, H. (2023). The human sperm proteome-toward a panel for male fertility testing. *Andrology, 11*(7), 1418–1436. https://doi.org/10.1111/ANDR.13431.

Gugssa, A., Williams, A., Tilahun, R., Eckberg, W., Anderson, W., Ahluwalia, B., ... Broomfield, D. (2010). Localization of guanylate cyclase receptor, inositol trisphosphate receptor, and calmodulin in boar spermatozoa. *International Journal of Biology, 2*(1), https://doi.org/10.5539/ijb.v2n1p13.

Gunaratne, G. S., Brailoiu, E., He, S., Unterwald, E. M., Patel, S., Slama, J. T., ... Marchant, J. S. (2021). Essential requirement for JPT2 in NAADP-evoked Ca2+ signaling. *Science Signaling, 14*(675), https://doi.org/10.1126/SCISIGNAL.ABD5605.

Harayama, H., Okada, K., & Miyake, M. (2003). Involvement of cytoplasmic free calcium in boar sperm: Head-to-head agglutination induced by a cell-permeable cyclic adenosine monophosphate analog. *Journal of Andrology, 24*(1), 91–99. https://doi.org/10.1002/J.1939-4640.2003.TB02646.X.

Harper, C., Wootton, L., Michelangeli, F., Lefièvre, L., Barratt, C., Publicover, S., ... Publicover, S. (2005). Secretory pathway Ca2+-ATPase (SPCA1) Ca2+ pumps, not SERCAs, regulate complex [Ca2+]i signals in human spermatozoa. *Journal of Cell Science, 118*(Pt 8), 1673–1685. https://doi.org/10.1242/jcs.02297.

Herrick, S. B., Schweissinger, D. L., Kim, S.-W., Bayan, K. R., Mann, S., & Cardullo, R. A. (2005). The acrosomal vesicle of mouse sperm is a calcium store. *Journal of Cellular Physiology, 202*(3), 663–671. https://doi.org/10.1002/jcp.20172.

Ho, H. C. (2010). Redistribution of nuclear pores during formation of the redundant nuclear envelope in mouse spermatids. *Journal of Anatomy, 216*(4), 525–532. https://doi.org/10.1111/J.1469-7580.2009.01204.X.

Ho, H. C., & Suarez, S. S. (2001). An inositol 1,4,5-trisphosphate receptor-gated intracellular Ca2+ store is involved in regulating sperm hyperactivated motility. *Biology of Reproduction, 65*(5), 1606–1615. https://doi.org/10.1095/biolreprod65.5.1606.

Ho, H. C., & Suarez, S. S. (2003). Characterization of the intracellular calcium store at the base of the sperm flagellum that regulates hyperactivated motility. *Biology of Reproduction, 68*(5), 1590–1596. https://doi.org/10.1095/biolreprod.102.011320.

Hoth, M., & Penner, R. (1992). Depletion of intracellular calcium stores activates a calcium current in mast cells. *Nature, 355*(6358), 353–356. https://doi.org/10.1038/355353A0.

Hwang, J. Y., & Chung, J. J. (2023). CatSper calcium channels: 20 Years on. *Physiology (Bethesda, Md.), 38*(3), 125–140. https://doi.org/10.1152/PHYSIOL.00028.2022.

Hwang, J. Y., Mannowetz, N., Zhang, Y., Everley, R. A., Gygi, S. P., Bewersdorf, J., ... Chung, J.-J. (2019). Dual sensing of physiologic pH and calcium by EFCAB9 regulates sperm motility. *Cell, 177*(6), 1480–1494.e19. https://doi.org/10.1016/j.cell.2019.03.047.

Ishijima, S. (2015). Ca2+ and cAMP regulations of microtubule sliding in hyperactivated motility of bull spermatozoa. *Proceedings of the Japan Academy. Series B, Physical and Biological Sciences, 91*(3), 99–108. https://doi.org/10.2183/PJAB.91.99.

Ivanova, A., & Atakpa-Adaji, P. (2023). Phosphatidylinositol 4,5-bisphosphate and calcium at ER-PM junctions - complex interplay of simple messengers. *Biochimica et Biophysica Acta. Molecular Cell Research, 1870*(6), https://doi.org/10.1016/J.BBAMCR.2023.119475.

Jungnickel, M. K., Marrero, H., Birnbaumer, L., Lémos, J. R., & Florman, H. M. (2001). Trp2 regulates entry of Ca2+ into mouse sperm triggered by egg ZP3. *Nature Cell Biology, 3*(5), 499–502. https://doi.org/10.1038/35074570.

Kaupp, U. B., & Strünker, T. (2016). Signaling in sperm: More different than similar. *Trends in Cell Biology, 27*(2), 101–109. https://doi.org/10.1016/j.tcb.2016.10.002.

Khawar, M. B., Gao, H., & Li, W. (2019). Mechanism of acrosome biogenesis in mammals. *Frontiers in Cell and Developmental Biology, 7*. https://doi.org/10.3389/FCELL.2019.00195.

Kirichok, Y., Krapivinsky, G., & Clapham, D. E. (2004). The mitochondrial calcium uniporter is a highly selective ion channel. *Nature, 427*(6972), 360–364. https://doi.org/10.1038/NATURE02246.

Lai, P., & Michelangeli, F. (2012). Bis(2-hydroxy-3-tert-butyl-5-methyl-phenyl)-methane (bis-phenol) is a potent and selective inhibitor of the secretory pathway Ca^{2+} ATPase (SPCA1. *Biochemical and Biophysical Research Communications, 424*(3), 616–619. https://doi.org/10.1016/J.BBRC.2012.07.004.

Laver, D. R. (2018). Regulation of the RyR channel gating by Ca2+ and Mg2. *Biophysical Reviews, 10*(4), 1087–1095. https://doi.org/10.1007/S12551-018-0433-4.

Lawson, C., Dorval, V., Goupil, S., & Leclerc, P. (2007). Identification and localisation of SERCA 2 isoforms in mammalian sperm. *Molecular Human Reproduction, 13*(5), 307–316. https://doi.org/10.1093/MOLEHR/GAM012.

Lee, M. A., & Storey, B. T. (1989). Endpoint of first stage of zona pellucida-induced acrosome reaction in mouse spermatozoa characterized by acrosomal H+ and Ca2+ permeability: population and single cell kinetics. *Gamete Research, 24*(3), 303–326. https://doi.org/10.1002/MRD.1120240307.

Lefièvre, L., Nash, K., Mansell, S., Costello, S., Punt, E., Correia, J., ... Publicover, S. (2012). 2-APB-potentiated channels amplify CatSper-induced Ca2+ signals in human sperm. *The Biochemical Journal, 448*(2), 189–200. https://doi.org/10.1042/BJ20120339.

Lesich, K. A., Kelsch, C. B., Ponichter, K. L., Dionne, B. J., Dang, L., & Lindemann, C. B. (2012). The calcium response of mouse sperm flagella: role of calcium ions in the regulation of dynein activity. *Biology of Reproduction, 86*(4), https://doi.org/10.1095/BIOLREPROD.111.094953.

Lishko, P. V., Botchkina, I. L., & Kirichok, Y. (2011). Progesterone activates the principal Ca2+ channel of human sperm. *Nature, 471*(7338), 387–391. https://doi.org/10.1038/nature09767.

Lishko, P. V., Kirichok, Y., Ren, D., Navarro, B., Chung, J.-J., Clapham, D. E., & Clapham, D. E. (2012). The control of male fertility by spermatozoan ion channels. *Annual Review of Physiology, 74*(18), 453–457. https://doi.org/10.1016/j.biotechadv.2011.08.021.Secreted.

Lishko, P. V., & Mannowetz, N. (2018). CatSper: A unique calcium channel of the sperm flagellum. *Current Opinion in Physiology, 2*, 109–113. https://doi.org/10.1016/j.cophys.2018.02.004.

Lopez-Guerrero, A. M., Tomas-Martin, P., Pascual-Caro, C., MacArtney, T., Rojas-Fernandez, A., Ball, G., ... Martin-Romero, F. J. (2017). Regulation of membrane ruffling by polarized STIM1 and ORAI1 in cortactin-rich domains. *Scientific Reports, 7*(1), https://doi.org/10.1038/S41598-017-00331-4.

Luan, S., & Wang, C. (2021). Calcium signaling mechanisms across kingdoms. *Annual Review of Cell and Developmental Biology, 37*, 311–340. https://doi.org/10.1146/ANNUREV-CELLBIO-120219-035210.

Luque, G. M., Dalotto-Moreno, T., Martín-Hidalgo, D., Ritagliati, C., Puga Molina, L. C., Romarowski, A., ... Buffone, M. G. (2018). Only a subpopulation of mouse sperm displays a rapid increase in intracellular calcium during capacitation. *Journal of Cellular Physiology, 233*(12), 9685–9700. https://doi.org/10.1002/jcp.26883.

Lyon, M. D., Ferreira, J. J., Li, P., Bhagwat, S., Butler, A., Anderson, K., ... Santi, C. M. (2023). SLO3: A conserved regulator of sperm membrane potential. *International Journal of Molecular Sciences, 24*(13), https://doi.org/10.3390/IJMS241311205.

Ma, J. (1993). Block by ruthenium red of the ryanodine-activated calcium release channel of skeletal muscle. *The Journal of General Physiology, 102*(6), 1031–1056. https://doi.org/10.1085/JGP.102.6.1031.

Makio, T., Chen, J., & Simmen, T. (2024). ER stress as a sentinel mechanism for ER Ca2+ homeostasis. *Cell Calcium, 124*. https://doi.org/10.1016/J.CECA.2024.102961.

Mannowetz, N., Naidoo, N. M., Choo, S.-A. A. S. S., Smith, J. F., Lishko, P. V., Sara Choo, S.-A., ... Lishko, P. V. (2013). Slo1 is the principal potassium channel of human spermatozoa. *ELife, 2*, e01009. https://doi.org/10.7554/eLife.01009.001.

Maruyama, T., Kanaji, T., Nakade, S., Kanno, T., & Mikoshiba, K. (1997). 2APB, 2-aminoethoxydiphenyl borate, a membrane-penetrable modulator of Ins(1,4,5)P3-induced Ca2+ release. *Journal of Biochemistry, 122*(3), 498–505. https://doi.org/10.1093/OXFORDJOURNALS.JBCHEM.A021780.

Matamoros Volante, A., Torres Rodriguez, P., Ramírez-Dominguez, L. B., Soriano, M., Carrasquel-Martinez, G., Flores, D., ... Agarwal, A. (2023). P-096 Prediction of fertilization success by voltage-dependent intracellular Ca2+ dynamics in non-normozoospermic human semen samples undergoing in vitro fertilization: Preliminary results. *Human Reproduction, 38*(Supplement_1). https://doi.org/10.1093/HUMREP/DEAD093.460.

Mayorga, L. S., Tomes, C. N., & Belmonte, S. A. (2007). Acrosomal exocytosis, a special type of regulated secretion. *IUBMB Life, 59*(4–5), 286–292. https://doi.org/10.1080/15216540701222872.

Mishra, J., Jhun, B. S., Hurst, S., O-Uchi, J., Csordás, G., & Sheu, S. S. (2017). The mitochondrial Ca2+ uniporter: Structure, function, and pharmacology. *Handbook of Experimental Pharmacology, 240*. https://doi.org/10.1007/164_2017_1.

Morgan, A. J., Martucci, L. L., Davis, L. C., & Galione, A. (2022). Two-pore channels: Going with the flows. *Biochemical Society Transactions, 50*(4), 1143–1155. https://doi.org/10.1042/BST20220229.

Naaby-Hansen, S., Diekman, A., Shetty, J., Flickinger, C. J., Westbrook, A., & Herr, J. C. (2010). Identification of calcium-binding proteins associated with the human sperm plasma membrane. *Reproductive Biology and Endocrinology: RB&E, 8*. https://doi.org/10.1186/1477-7827-8-6.

Naaby-Hansen, S., Wolkowicz, M. J., Klotz, K., Bush, L. A., Westbrook, V. A., Shibahara, H., ... Herr, J. C. (2001). Co-localization of the inositol 1,4,5-trisphosphate receptor and calreticulin in the equatorial segment and in membrane bounded vesicles in the cytoplasmic droplet of human spermatozoa. *Molecular Human Reproduction*, 7(10), 923–933. https://doi.org/10.1093/MOLEHR/7.10.923.

Navarrete, F. A., Alvau, A., Lee, H. C., Levin, L. R., Buck, J., Leon, P. M.-D., ... Visconti, P. E. (2016). Transient exposure to calcium ionophore enables in vitro fertilization in sterile mouse models. *Scientific Reports, 6*, 33589. https://doi.org/10.1038/srep33589.

Navarro, B., Kirichok, Y., & Clapham, D. E. (2007). KSper, a pH-sensitive K+ current that controls sperm membrane potential. *Proceedings of the National Academy of Sciences of the United States of America, 104*(18), 7688–7692. https://doi.org/10.1073/pnas.0702018104.

Navarro, B., Miki, K., & Clapham, D. E. (2011). ATP-activated P2X2 current in mouse spermatozoa. *Proceedings of the National Academy of Sciences of the United States of America*. https://doi.org/10.1073/pnas.1111695108.

Naylor, E., Arredouani, A., Vasudevan, S. R., Lewis, A. M., Parkesh, R., Mizote, A., ... Churchill, G. C. (2009). Identification of a chemical probe for NAADP by virtual screening. *Nature Chemical Biology, 5*(4), 220–226. https://doi.org/10.1038/NCHEMBIO.150.

Nelson, R. J. (1997). The use of genetic "knockout" mice in behavioral endocrinology research. *Hormones and Behavior, 31*(3), 188–196. https://doi.org/10.1006/HBEH.1997.1381.

Nguyen, T. H. D., Itoh, S. G., Okumura, H., & Tominaga, M. (2021). Structural basis for promiscuous action of monoterpenes on TRP channels. *Communications Biology, 4*(1), https://doi.org/10.1038/S42003-021-01776-0.

Ogunbayo, O. A., Zhu, Y., Rossi, D., Sorrentino, V., Ma, J., Zhu, M. X., & Evans, A. M. (2011). Cyclic adenosine diphosphate ribose activates ryanodine receptors, whereas NAADP activates two-pore domain channels. *The Journal of Biological Chemistry, 286*(11), 9136–9140. https://doi.org/10.1074/JBC.M110.202002.

Oh-Hora, M., & Lu, X. (2018). Function of orai/stim proteins studied in transgenic animal models. *Calcium Entry Channels in Non-Excitable Cells*, 107–126. https://doi.org/10.1201/9781315152592-6.

Olesen, C., Picard, M., Winther, A. M. L., Gyrup, C., Morth, J. P., Oxvig, C., ... Nissen, P. (2007). The structural basis of calcium transport by the calcium pump. *Nature, 450*(7172), 1036–1042. https://doi.org/10.1038/NATURE06418.

Oliveira, A. G., Guimarães, E. S., Andrade, L. M., Menezes, G. B., & Leite, M. F. (2014). Decoding calcium signaling across the nucleus. *Physiology (Bethesda, Md.), 29*(5), 361–368. https://doi.org/10.1152/PHYSIOL.00056.2013.

Oliver, E. I., Jabloñski, M., Buffone, M. G., & Darszon, A. (2023). Two-pore channel 1 and Ca2+ release-activated Ca2+ channels contribute to the acrosomal pH-dependent intracellular Ca2+ increase in mouse sperm. *The Journal of Physiology, 601*(14), 2935–2958. https://doi.org/10.1113/JP284247.

Olson, S. D., Fauci, L. J., & Suarez, S. S. (2011). Mathematical modeling of calcium signaling during sperm hyperactivation. *Molecular Human Reproduction, 17*(8), 500–510. https://doi.org/10.1093/molehr/gar040.

Orta, G., Vega-Beltran, J. L., Hidalgo, D., Santi, C. M., Visconti, P., & Darszon, A. (2018). CatSper channels are regulated by protein kinase a. RA117.001566 *Journal of Biological Chemistry, jbc.* https://doi.org/10.1074/jbc.RA117.001566.

O'Toole, C. M. B., Arnoult, C., Darszon, A., Steinhardt, R. A., & Florman, H. M. (2000). Ca2+ entry through store-operated channels in mouse sperm is initiated by egg ZP3 and drives the acrosome reaction. *Molecular Biology of the Cell, 11*(5), 1571–1584. https://doi.org/10.1091/MBC.11.5.1571.

Oz, M., El Nebrisi, E. G., Yang, K. H. S., Howarth, F. C., & Al Kury, L. T. (2017). Cellular and molecular targets of menthol actions. *Frontiers in Pharmacology, 8*(JUL), https://doi.org/10.3389/FPHAR.2017.00472.

Palty, R., Silverman, W. F., Hershfinkel, M., Caporale, T., Sensi, S. L., Parnis, J., ... Sekler, I. (2010). NCLX is an essential component of mitochondrial Na+/Ca2+ exchange. *Proceedings of the National Academy of Sciences of the United States of America, 107*(1), 436–441. https://doi.org/10.1073/PNAS.0908099107.

Pang, Z. P., & Südhof, T. C. (2010). Cell biology of Ca2+-triggered exocytosis. *Current Opinion in Cell Biology, 22*(4), 496–505. https://doi.org/10.1016/J.CEB.2010.05.001.

Paredes, R. M., Etzler, J. C., Watts, L. T., Zheng, W., & Lechleiter, J. D. (2008). Chemical calcium indicators. *Methods, 46*(3), 143–151. https://doi.org/10.1016/J.YMETH.2008.09.025.

Pathak, T., & Trebak, M. (2018). Mitochondrial Ca2+ signaling. *Pharmacology & Therapeutics, 192*, 112–123. https://doi.org/10.1016/J.PHARMTHERA.2018.07.001.

Periasamy, M., & Kalyanasundaram, A. (2007). SERCA pump isoforms: Their role in calcium transport and disease. *Muscle & Nerve, 35*(4), 430–442. https://doi.org/10.1002/MUS.20745.

Prasad, V., Okunade, G., Liu, L., Paul, R. J., & Shull, G. E. (2007). Distinct phenotypes among plasma membrane Ca2+-ATPase knockout mice. *Annals of the New York Academy of Sciences, 1099*, 276–286. https://doi.org/10.1196/ANNALS.1387.029.

Puga Molina, L. C., Luque, G. M., Balestrini, P. A., Marín-Briggiler, C. I., Romarowski, A., & Buffone, M. G. (2018). Molecular basis of human sperm capacitation. *Frontiers in Cell and Developmental Biology, 6*, 72. https://doi.org/10.3389/fcell.2018.00072.

Putney, J. W. (1986). A model for receptor-regulated calcium entry. *Cell Calcium, 7*(1), 1–12. https://doi.org/10.1016/0143-4160(86)90026-6.

Ren, D., Navarro, B., Perez, G., Jackson, A. C., Hsu, S., Shi, Q., ... Clapham, D. E. (2001). A sperm ion channel required for sperm motility and male fertility. *Nature, 413*(6856), 603–609. https://doi.org/10.1038/35098027.

Riederer, E., Cang, C., & Ren, D. (2023). Lysosomal ion channels: What are they good for and are they druggable targets. *Annual Review of Pharmacology and Toxicology, 63*, 19–41. https://doi.org/10.1146/ANNUREV-PHARMTOX-051921-013755.

Rizzuto, R., De Stefani, D., Raffaello, A., & Mammucari, C. (2012). Mitochondria as sensors and regulators of calcium signalling. *Nature Reviews. Molecular Cell Biology, 13*(9), 566–578. https://doi.org/10.1038/NRM3412.

Rodrigues, T., Piccirillo, S., Magi, S., Preziuso, A., dos Santos Ramos, V., Serfilippi, T., ... Lariccia, V. (2022). Control of Ca2+ and metabolic homeostasis by the Na+/Ca2+ exchangers (NCXs) in health and disease. *Biochemical Pharmacology, 203*. https://doi.org/10.1016/J.BCP.2022.115163.

Rodriguez, E. A., Campbell, R. E., Lin, J. Y., Lin, M. Z., Miyawaki, A., Palmer, A. E., ... Tsien, R. Y. (2017). The growing and glowing toolbox of fluorescent and photoactive proteins. *Trends in Biochemical Sciences, 42*(2), 111–129. https://doi.org/10.1016/J.TIBS.2016.09.010.

Rodriguez, P. C., Satorre, M. M., & Beconi, M. T. (2012). Effect of two intracellular calcium modulators on sperm motility and heparin-induced capacitation in cryopreserved bovine spermatozoa. *Animal Reproduction Science, 131*(3–4), 135–142. https://doi.org/10.1016/J.ANIREPROSCI.2012.03.015.

Ruete, M. C., Lucchesi, O., Bustos, M. A., & Tomes, C. N. (2014). Epac, rap and Rab3 act in concert to mobilize calcium from sperm's acrosome during exocytosis. *Cell Communication and Signaling: CCS, 12*(1), https://doi.org/10.1186/S12964-014-0043-0.

Sacchetto, R., Bertipaglia, I., Giannetti, S., Cendron, L., Mascarello, F., Damiani, E., ... Zanotti, G. (2012). Crystal structure of sarcoplasmic reticulum Ca2+-ATPase (SERCA) from bovine muscle. *Journal of Structural Biology, 178*(1), 38–44. https://doi.org/10.1016/J.JSB.2012.02.008.

Saleem, H., Tovey, S. C., Rahman, T., Riley, A. M., Potter, B. V. L., & Taylor, C. W. (2013). Stimulation of inositol 1,4,5-trisphosphate (IP3) receptor subtypes by analogues of IP3. *PLoS One, 8*(1), https://doi.org/10.1371/JOURNAL.PONE.0054877.

Sánchez-Cárdenas, C., Romarowski, A., Orta, G., De la Vega-Beltrán, J. L., Martín-Hidalgo, D., Hernández-Cruz, A., ... Darszon, A. (2021). Starvation induces an increase in intracellular calcium and potentiates the progesterone-induced mouse sperm acrosome reaction. *The FASEB Journal, 35*(4), https://doi.org/10.1096/fj.202100122R.

Sánchez-Carranza, O., Torres-Rodríguez, P., Darszon, A., Treviño, C. L., & López-González, I. (2015). Pharmacology of hSlo3 channels and their contribution in the capacitation-associated hyperpolarization of human sperm. *Biochemical and Biophysical Research Communications, 466*(3), 554–559. https://doi.org/10.1016/j.bbrc.2015.09.073.

Sánchez-Tusie, A. A., Vasudevan, S. R., Churchill, G. C., Nishigaki, T., & Treviño, C. L. (2014). Characterization of NAADP-mediated calcium signaling in human spermatozoa. *Biochemical and Biophysical Research Communications, 443*(2), 531–536. https://doi.org/10.1016/J.BBRC.2013.12.011.

Santi, C. M., Martínez-López, P., Vega-Beltrán, J. L., De, Butler, A., Darszon, A., ... Salkoff, L. (2010). The SLO3 sperm-specific potassium channel plays a vital role in male fertility. *FEBS Letters, 584*(5), 1041–1046. https://doi.org/10.1016/j.febslet.2010.02.005.

Santonastasi, M., & Wehrens, X. H. T. (2007). Ryanodine receptors as pharmacological targets for heart disease. *Acta Pharmacologica Sinica, 28*(7), 937–944. https://doi.org/10.1111/J.1745-7254.2007.00582.X.

Santulli, G., Nakashima, R., Yuan, Q., & Marks, A. R. (2017). Intracellular calcium release channels: An update. *The Journal of Physiology, 595*(10), 3041–3051. https://doi.org/10.1113/JP272781.

Sargsyan, Y., Bickmeyer, U., Gibhardt, C. S., Streckfuss-Bömeke, K., Bogeski, I., & Thoms, S. (2021). Peroxisomes contribute to intracellular calcium dynamics in cardiomyocytes and non-excitable cells. *Life Science Alliance, 4*(9), https://doi.org/10.26508/LSA.202000987.

Sassone-Corsi, P. (1997). Transcriptional checkpoints determining the fate of male germ cells. *Cell, 88*(2), 163–166. https://doi.org/10.1016/S0092-8674(00)81834-6.

Shah, K. R., Guan, X., & Yan, J. (2022). Diversity of two-pore channels and the accessory NAADP receptors in intracellular Ca2+ signaling. *Cell Calcium, 104*. https://doi.org/10.1016/J.CECA.2022.102594.

Shao, J., Fu, Z., Ji, Y., Guan, X., Guo, S., Ding, Z., ... Shen, Y. (2016). Leucine zipper-EF-hand containing transmembrane protein 1 (LETM1) forms a Ca2+/H+ antiporter. *Scientific Reports, 6*. https://doi.org/10.1038/SREP34174.

Simms, B. A., & Zamponi, G. W. (2014). Neuronal voltage-gated calcium channels: Structure, function, and dysfunction. *Neuron, 82*(1), 24–45. https://doi.org/10.1016/J.NEURON.2014.03.016.

Skerrett-Byrne, D. A., Anderson, A. L., Bromfield, E. G., Bernstein, I. R., Mulhall, J. E., Schjenken, J. E., ... Nixon, B. (2022). Global profiling of the proteomic changes associated with the post-testicular maturation of mouse spermatozoa. *Cell Reports, 41*(7), https://doi.org/10.1016/J.CELREP.2022.111655.

Skerrett-Byrne, D. A., Teperino, R., & Nixon, B. (2024). ShinySperm: navigating the sperm proteome landscape. *Reproduction, Fertility, and Development, 36*(13), https://doi.org/10.1071/RD24079.

Song, M., Chen, D., & Yu, S. P. (2014). The TRPC channel blocker SKF 96365 inhibits glioblastoma cell growth by enhancing reverse mode of the Na+/Ca2+ exchanger and increasing intracellular Ca2+. *British Journal of Pharmacology, 171*(14), 3432–3447. https://doi.org/10.1111/BPH.12691.

Sutton, K. A., Jungnickel, M. K., & Florman, H. M. (2003). Of fertility, cystic fibrosis and the bicarbonate ion. *Nature Cell Biology, 5*(10), 857–859. https://doi.org/10.1038/NCB1003-857.

Sutton, K. A., Jungnickel, M. K., Wang, Y., Cullen, K., Lambert, S., & Florman, H. M. (2004). Enkurin is a novel calmodulin and TRPC channel binding protein in sperm. *Developmental Biology, 274*(2), 426–435. https://doi.org/10.1016/J.YDBIO.2004.07.031.

Suurväli, J., Boudinot, P., Kanellopoulos, J., & Rüütel Boudinot, S. (2017). P2X4: A fast and sensitive purinergic receptor. *Biomedical Journal, 40*(5), 245–256. https://doi.org/10.1016/J.BJ.2017.06.010.

Szabadkai, G., & Duchen, M. R. (2008). Mitochondria: The hub of cellular Ca2+ signaling. *Physiology (Bethesda, Md.), 23*(2), 84–94. https://doi.org/10.1152/PHYSIOL.00046.2007.

Takei, G. L. (2024). Molecular mechanisms of mammalian sperm capacitation, and its regulation by sodium-dependent secondary active transporters. *Reproductive Medicine and Biology, 23*(1), https://doi.org/10.1002/RMB2.12614.

Tateno, H., Krapf, D., Hino, T., Sánchez-Cárdenas, C., Darszon, A., Yanagimachi, R., & Visconti, P. E. (2013). Ca2+ ionophore A23187 can make mouse spermatozoa capable of fertilizing in vitro without activation of cAMP-dependent phosphorylation pathways. *Proceedings of the National Academy of Sciences of the United States of America, 110*(46), 18543–18548. https://doi.org/10.1073/pnas.1317113110.

Teves, M. E., Roldan, E. R. S., Krapf, D., Strauss, J. F., Bhagat, V., & Sapao, P. (2020). Sperm differentiation: The role of trafficking of proteins. *International Journal of Molecular Sciences, 21*(10), https://doi.org/10.3390/IJMS21103702.

Tiffner, A., & Derler, I. (2021). Isoform-specific properties of orai homologues in activation, downstream signaling, physiology and pathophysiology. *International Journal of Molecular Sciences, 22*(15), https://doi.org/10.3390/IJMS22158020.

Treiman, M., Caspersen, C., & Christensen, S. B. (1998). A tool coming of age: Thapsigargin as an inhibitor of sarco-endoplasmic reticulum Ca2+-ATPases. *Trends in Pharmacological Sciences, 19*(4), 131–135. https://doi.org/10.1016/S0165-6147(98)01184-5.

Treviño, C. L., Santi, C. M., Beltrán, C., Hernández-Cruz, A., Darszon, A., & Lomeli, H. (1998). Localisation of inositol trisphosphate and ryanodine receptors during mouse spermatogenesis: Possible functional implications. *Zygote (Cambridge, England), 6*(2), 159–172. http://www.ncbi.nlm.nih.gov/pubmed/9770782.

Trötschel, C., Hamzeh, H., Alvarez, L., Pascal, R., Lavryk, F., Bönigk, W., ... Kaupp, U. B. (2020). Absolute proteomic quantification reveals design principles of sperm flagellar chemosensation. *The EMBO Journal, 39*(4), https://doi.org/10.15252/EMBJ.2019102723.

Vangheluwe, P., Sepúlveda, M. R., Missiaen, L., Raeymaekers, L., Wuytack, F., & Vanoevelen, J. (2009). Intracellular Ca2+- and Mn2+-transport ATPases. *Chemical Reviews, 109*(10), 4733–4759. https://doi.org/10.1021/CR900013M.

Visconti, P. E. (2009). Understanding the molecular basis of sperm capacitation through kinase design. *Proceedings of the National Academy of Sciences, 106*(3), 667–668. https://doi.org/10.1073/pnas.0811895106.

Walensky, L. D., & Snyder, S. H. (1995). Inositol 1,4,5-trisphosphate receptors selectively localized to the acrosomes of mammalian sperm. *The Journal of Cell Biology, 130*(4), 857–869. https://doi.org/10.1083/JCB.130.4.857.

Wang, G., Guo, Y., Zhou, T., Shi, X., Yu, J., Yang, Y., ... Sha, J. (2013). In-depth proteomic analysis of the human sperm reveals complex protein compositions. *Journal of Proteomics, 79*, 114–122. https://doi.org/10.1016/J.JPROT.2012.12.008.

Wennemuth, G., Babcock, D. F., & Hille, B. (2003). Calcium clearance mechanisms of mouse sperm. *The Journal of General Physiology, 122*(1), 115–128. https://doi.org/10.1085/jgp.200308839.

Wennemuth, G., Westenbroek, R. E., Xu, T., Hille, B., & Babcock, D. F. (2000). CaV2.2 and CaV2.3 (N- and R-type) Ca2+ channels in depolarization-evoked entry of Ca2+ into mouse sperm. *The Journal of Biological Chemistry, 275*(28), 21210–21217. https://doi.org/10.1074/jbc.M002068200.

West, A. E., Chen, W. G., Dalva, M. B., Dolmetsch, R. E., Kornhauser, J. M., Shaywitz, A. J., ... Greenberg, M. E. (2001). Calcium regulation of neuronal gene expression. *Proceedings of the National Academy of Sciences of the United States of America, 98*(20), 11024–11031. https://doi.org/10.1073/PNAS.191352298.

Weyand, I., Godde, M., Frings, S., Weiner, J., Müller, F., Altenhofen, W., ... Kaupp, U. B. (1994). Cloning and functional expression of a cyclic-nucleotide-gated channel from mammalian sperm. *Nature, 368*(6474), 859–863. https://doi.org/10.1038/368859A0.

Weyer, A. D., & Lehto, S. G. (2017). Development of TRPM8 antagonists to treat chronic pain and migraine. *Pharmaceuticals (Basel, Switzerland), 10*(2), https://doi.org/10.3390/PH10020037.

Willegems, K., & Efremov, R. G. (2017). Structural details of the ryanodine receptor calcium release channel and its gating mechanism. *Advances in Experimental Medicine and Biology, 981*, 179–204. https://doi.org/10.1007/978-3-319-55858-5_8.

Wootton, L. L., Argent, C. C. H., Wheatley, M., & Michelangeli, F. (2004). The expression, activity and localisation of the secretory pathway Ca2+-ATPase (SPCA1) in different mammalian tissues. *Biochimica et Biophysica Acta – Biomembranes, 1664*(2), 189–197. https://doi.org/10.1016/j.bbamem.2004.05.009.

Wu, M., Wu, C., Song, T., Pan, K., Wang, Y., & Liu, Z. (2023). Structure and transport mechanism of the human calcium pump SPCA1. *Cell Research, 33*(7), 533–545. https://doi.org/10.1038/S41422-023-00827-X.

Xu, H., & Van Remmen, H. (2021). The sarcoendoplasmic reticulum calcium ATPase (SERCA) pump: A potential target for intervention in aging and skeletal muscle pathologies. *Skeletal Muscle, 11*(1), https://doi.org/10.1186/S13395-021-00280-7.

Yanagimachi, R. (1994). Mammalian fertilization. In J. D. K. E. Neill (Ed.). *The physiology of reproduction*. Raven Press.

Yeste, M., Ahmad, A., Viñolas, E., Recuero, S., Bonet, S., & Pinart, E. (2024). Inhibition of forward and reverse transport of Ca2+ via Na+/Ca2+ exchangers (NCX) prevents sperm capacitation. *Biological Research, 57*(1), https://doi.org/10.1186/S40659-024-00535-9.

Yoo, J. (2022). Structural basis of Ca2+ uptake by mitochondrial calcium uniporter in mitochondria: A brief review. *BMB Reports, 55*(11), 528–534. https://doi.org/10.5483/BMBREP.2022.55.11.134.

Zajac, M., Modi, S., & Krishnan, Y. (2022). The evolution of organellar calcium mapping technologies. *Cell Calcium, 108*. https://doi.org/10.1016/J.CECA.2022.102658.

Zeng, X.-H., Yang, C., Kim, S. T., Lingle, C. J., & Xia, X.-M. X.-M. (2011). Deletion of the Slo3 gene abolishes alkalization-activated K+ current in mouse spermatozoa. *Proceedings of the National Academy of Sciences, 108*(14), 5879–5884. https://doi.org/10.1073/pnas.1100240108.

Zhang, J., Guan, X., Shah, K., & Yan, J. (2021). Lsm12 is an NAADP receptor and a two-pore channel regulatory protein required for calcium mobilization from acidic organelles. *Nature Communications, 12*(1), https://doi.org/10.1038/S41467-021-24735-Z.

Zhou, H., Iwasaki, H., Nakamura, T., Nakamura, K., Maruyama, T., Hamano, S. I., ... Mikoshiba, K. (2007). 2-Aminoethyl diphenylborinate analogues: Selective inhibition for store-operated Ca2+ entry. *Biochemical and Biophysical Research Communications, 352*(2), 277–282. https://doi.org/10.1016/J.BBRC.2006.10.174.

Zong, P., & Yue, L. (2023). Regulation of presynaptic calcium channels. *Advances in Neurobiology, 33*, 171–202. https://doi.org/10.1007/978-3-031-34229-5_7.

CHAPTER TWELVE

Paternal contributions to mammalian zygote – Beyond sperm-oocyte fusion

Peter Sutovsky[a,b,*], Michal Zigo[a], Filip Tirpak[a], and Richard Oko[c]

[a]Division of Animal Sciences, University of Missouri, Columbia, MO, United States

[b]Department of Obstetrics, Gynecology and Women's Health, University of Missouri, Columbia, MO, United States

[c]Department of Biomedical and Molecular Sciences, Queen's University, Kingston, ON, Canada

*Corresponding author. e-mail address: sutovskyp@missouri.edu

Contents

Abstract

Contrary to a common misconception that the fertilizing spermatozoon acts solely as a vehicle for paternal genome delivery to the zygote, this chapter aims to illustrate how the male gamete makes other essential contributions , including the sperm

Current Topics in Developmental Biology, Volume 162
ISSN 0070-2153, https://doi.org/10.1016/bs.ctdb.2025.02.002

borne-oocyte activation factors, centrosome components, and components of the sperm proteome and transcriptome that help to lay the foundation for pregnancy establishment and maintenance to term, and the newborn and adult health. Our inquiry starts immediately after sperm plasma membrane fusion with its oocyte counterpart, the oolemma. Parallel to and following sperm incorporation in the egg cytoplasm, some of the sperm structures (perinuclear theca) are dissolved and spent to induce development, others (nucleus, centriole) are transformed into zygotic structures enabling it, and yet others (mitochondrial and fibrous sheath, axonemal microtubules and outer dense fibers) are recycled as to not stand in its way. Noteworthy advances in this research include the identification of several sperm-borne oocyte activating factor candidates, the role of autophagy in the post-fertilization sperm mitochondrion degradation, new insight into zygotic centrosome origins and function, and the contributions of sperm-delivered RNA cargos to early embryo development. In concluding remarks, the unresolved issues, and clinical and biotechnological implications of sperm-vectored paternal inheritance are discussed.

1. Introduction – fertilization beyond sperm-oolemma fusion

Detailed theoretical analysis, using mathematical modeling, reveals that sex is a very inefficient way of reproducing. The inefficiency lies in making male offspring. In the majority of sexual species only the female contributes energy and resources to the young. In contrast the males rarely contribute more than the minimum – a tiny sperm carrying genes, but devoid of other resources.

Quote from *The Evolution of Life* (Gamlin & Vines, 1987).

It is a matter of common public misperception that the fertilizing spermatozoon acts solely as a vehicle for paternal genome delivery to the zygote (Gamlin & Vines, 1987). This chapter takes a somewhat andro-centric approach to illustrate that the male gamete makes other essential contributions to the zygote, as well as it helps to lay the foundation for pregnancy establishment and maintenance to term, and the newborn and adult health. Our inquiry starts immediately after sperm plasma membrane fusion with its oocyte counterpart, the oolemma. During the ensuing process of zygotic development, some of the sperm components are dissolved and spent to mediate the activation of development, others are transformed into zygotic structures enabling it, and yet others are recycled as to not stand in its way (Fig. 1). Major progress has been achieved in the quarter of century that expired since our previous comprehensive update on paternal contributions (Sutovsky & Schatten, 2000), including the identification of multiple sperm-borne oocyte activating factor candidates,

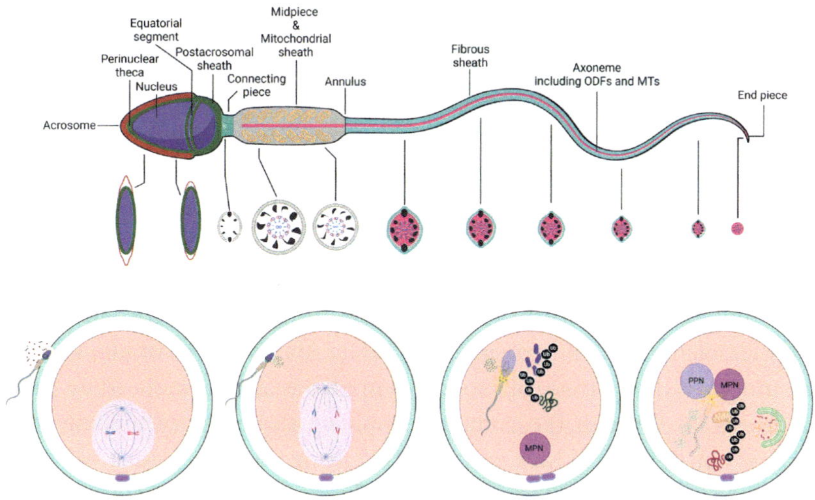

Fig. 1 Fate of the accessory sperm structures and organelles after fertilization. Sperm nucleus becomes the paternal pronucelus (PPN) while the sperm centriole seeds the zygotic centrosome and sperm aster necessary for PPN apposition with the maternal pronucelus (MPN). Sperm head perinuclear theca and sperm tail fibrous sheath disperse in the ooplasm to release oocyte activating factors (from perinucelar theca) and possibly other protein and RNA cargos (from perinculear theca and fibrous sheath) to regulate zygotic develoment. Sperm mitochondria, outer dense fibers (ODF) and axonemal microtubules (MTs) will be recycled when the zygote enters mitosis, with the help of proteasomal proteolysis and autophagy, both of which are dependent on protein ubiquitination (Ub). *Image created and licensed in BioRender (Tirpak, 2024).*

the implication of autophagy in the post-fertilization sperm mitochondrion degradation, new insight into zygotic centrosome origins and function, and the contributions of sperm-delivered RNA cargos to early embryo development.

Famously summed up by the Missouri-born physiologist Carl Richard Moore, fertilization is a series of irreversible, precisely orchestrated events (Moore, 1916), enabled by synergistic gamete transport mechanisms to reach the site of fertilization near the oviductal isthmus. Here, the spermatozoa settle down transiently, binding to the apical surface of the oviductal epithelia by their acrosomes, thus establishing the oviductal sperm reservoir. Upon sensing the ovulatory signal, the oviduct-bound spermatozoa undergo the orchestrated, terminal process of sperm capacitation and hyperactivation (Austin, 1952; Chang, 1951), which facilitates their detachment from reservoir and prepares them for fertilization. Subsequently, sperm-oocyte recognition takes place, mediated by the oocyte and/or cumulus cell issued chemoattractants [e.g., progesterone (Guidobaldi, Teves, Unates, Anastasia, & Giojalas, 2008)], sperm rheotaxis, oviductal peristaltic and temperature gradients [reviewed by (Suarez, 2016)]. Next, the sperm binding to zona pellucida (ZP), acrosomal exocytosis and sperm-zona penetration involve the surfaces and extracellular matrix of the sperm head acrosomal ridge and the oocyte zona pellucida, respectively (Florman & Wassarman, 1985). The zona induced acrosomal exocytosis, aided by the hyperactivated sperm tail movement and zona lysins creates the fertilization slit guiding the sperm head, and eventually, sperm tail to the perivitelline space between the zona and the oolemma (Zimmerman et al., 2011). Another observation, at least in the mouse, is that acrosomal exocytosis in vivo is already induced by passage of the sperm through the cumulus, implying that primary binding to the zona is not essential (Hino et al., 2016). Sperm-oocyte plasma membrane adhesion and fusion, is mediated by the acrosomal-exocytosis induced exposure of the IZUMO1 protein on the plasmalemma overlying the equatorial segment (Inoue, Ikawa, Isotani, & Okabe, 2005) and its binding to the oolemma's sperm receptor, JUNO (Bianchi, Doe, Goulding, & Wright, 2014). Gamete fusion thus sets the stage for the initiation and completion of zygotic development, including sperm incorporation in the ooplasm, assisted by the microfilament in the oocyte cortex, sperm perinuclear theca solubilization, oocyte activation, triggering of anti-polyspermy defense, zygotic centrosome reconstitution and formation of the sperm aster, pronuclear development and apposition, and recycling of the redundant accessory sperm structures [reviewed by (Sutovsky, 2018; Tumova, Zigo, Sutovsky, Sedmikova, & Postlerova, 2021)].

2. Perinuclear theca cargo

2.1 Perinuclear theca and its fate after fertilization

2.1.1 Perinuclear theca assembly and dispersion

Upon sperm-oolemma fusion, the sperm head perinuclear theca (PT) becomes exposed to the ooplasm, initially in the equatorial region and then distal of the post-acrosomal sheath (PAS) of the sperm head (Sutovsky, Oko, Hewitson, & Schatten, 1997). The PT is a protective capsule, an amalgamate of somatic and germ cell type cytosolic proteins, and specialized PT-specific proteins translated during the haploid phase of spermatogenesis, in the round spermatid. These proteins can be shuttled from the spermatid cytoplasmic lobe along the microtubule tracks of caudal manchette (Wu et al., 2007b) or diffused freely to be laid down as PT (Oko & Sutovsky, 2009). At fertilization, the PT must be removed to facilitate the decondensation of the sperm nucleus; it peels off the sperm nucleus in distinct layers, as shown by electron microcopy of bovine fertilization (Fig. 2A). While the relatively more rigid sub-acrosomal layer of PT may persist throughout zygotic development, the PT of PAS starts to dissolve in the oocyte cytoplasm even before sperm incorporation is completed (Fig. 2B and C; Wu et al., 2007a), releasing the PT-borne molecules that could contribute to fertilization and early development.

2.1.2 Varied perinuclear theca cargos

Included in the rather wide category of PT cargos are structural, cysteine-rich PT proteins [e.g., CRISP2 (Zhang, Bromfield, Helms, & Gadella, 2022a)], signaling molecules such as protein kinases and the sperm-borne oocyte-activating factors [SOAF; (Kang et al., 2020; Swann, Larman, Saunders, & Lai, 2004; Wu et al., 2007a)], transcription factors [e.g., STAT4 (Herrada & Wolgemuth, 1997; Lachance & Leclerc, 2011)], somatic cell type histones and testis-specific histone variants that could contribute to pronuclear development (Hamilton et al., 2021; Tovich, Sutovsky, & Oko, 2004; Aul & Oko, 2002; Tran et al., 2012), redox homeostasis proteins [e.g., GSTO2 (Hamilton et al., 2019)], and various species of mRNAs, non-coding RNAs and RNA-binding proteins (Yan et al., 2008). Recent comprehensive proteomic analysis of boar PT confirmed the presence of aforementioned proteins and identified additional classes of structural and cargo proteins that contribute to the regulation of oocyte activation and zygotic development. Particularly abundant are proteins of the endoplasmic reticulum (ER) and the ER-stress response elements, as well as proteasomal subunits, spliceosome components, and the Ras-pathway, nucleotide metabolism and Huntington disease related proteins (Zhang et al., 2022b).

2.2 Participation of perinuclear theca cargos in oocyte activation

2.2.1 Mechanism of oocyte activation

Oocyte activation is triggered by the protein release from the equatorial or postacrosomal PT at the site of sperm-oolemma fusion. Eventually, most of the PT is solubilized and dispersed around the incorporated sperm nucleus, with some PT proteins, such as WBP2NL and core histones also entering the sperm nucleus as it begins to decondense (Hamilton et al., 2019; Hamilton

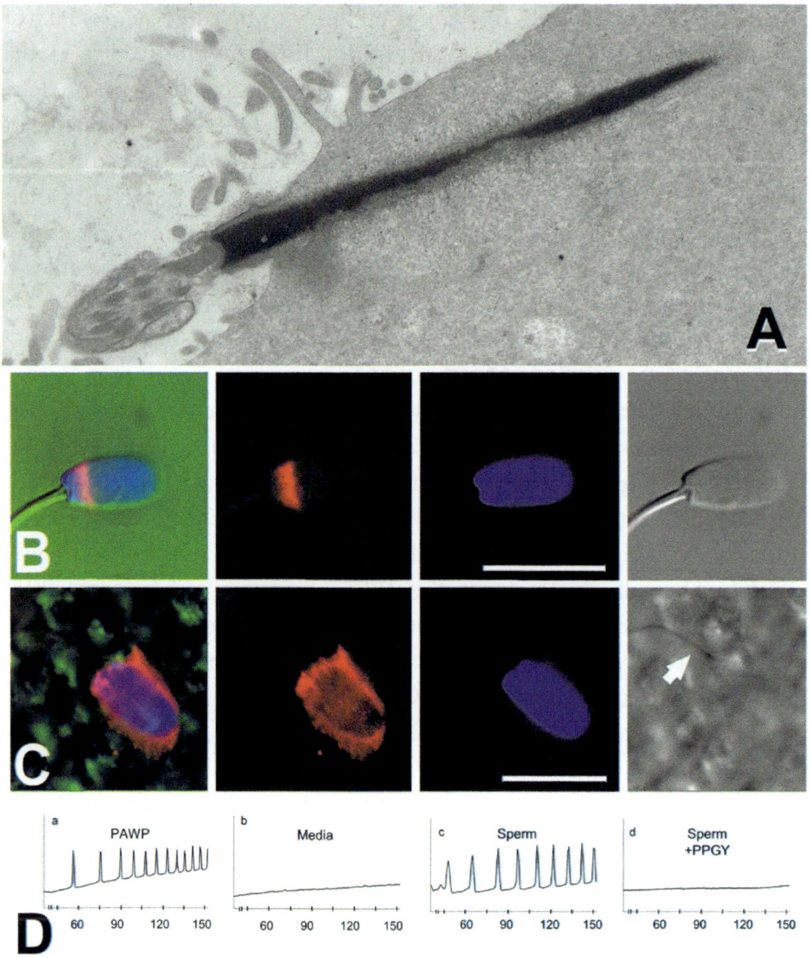

Fig. 2 Sperm perinuclear theca participation in mammalian oocyte activation. (A) Transmission electron micrograph of the sperm head incorporation into a bovine oocyte cytoplasm. Perinuclear theca is motly dispersed while the new nuclear envelope is starting to form around the decondensing sperm nucleus. Sperm tail is still ooutside of the oocyte, in the pervitelline space. (B, C) Immunolabeling of WBP2NL/PAWP (red) before porcine fertilization (B) and shortly after sperm head incorporation (C), documenting the dispersion of perinuclear theca in the ooplasm. (D) Oscillatory calcium release induced by the microinjection of WBP2NL/PAWP cRNA or human spermatozoa into human metaphase II oocytes. Controls included sperm-immobilization culture medium injection (Media) and sperm coinjection with an antagonist peptide designed to inhibit WBP2NL (sperm+PPGY). *Reprinted with permission from Wu et al. (2007a). (D) Reprinted with permission from Aarabi et al. (2014a).*

et al., 2021; Wu et al., 2007a). The general model of oocyte activation by the sperm PT-dispersion assumes that SOAF activates a soluble species of ooplasmic phospholipase C (PLC) enzyme that cleaves phosphatidyl inositol diphosphate (PIP2) into biologically active inositol-3-phosphate (IP3). The IP3 binds to its receptor (IP3R) on the calcium storing endoplasmic reticulum (ER), which releases Ca^{2+} ions, the master-messenger molecules of oocyte activation (Igarashi, Knott, Schultz, & Williams, 2007). Oocyte calcium release at fertilization has an oscillatory pattern in mammals, ascidians and mollusks, while being a solitary event (single spike) in other invertebrates (e.g., echinoderms) and lower vertebrates such as fishes and amphibians. Except for mammalian eggs being fertilized at metaphase II, the oscillatory eggs appear to be fertilized at metaphase I in most other taxa (Sugita, Takarabe, Kageyama, Kawata, & Ito, 2024). Repetitive oscillatory Ca^{2+} release from ER sequentially activates calcium-calmodulin kinase II (CaMKII), and protein kinases PKA and PKC, causing the opening of plasma membrane associated transient receptor potential channels (TRP-C), which refill ER with Ca^{2+}. The Ca^{2+} sensor on the ER membrane is the single transmembrane-spanning Ca^{2+}-binding protein, STIM1 (Koh, Lee, Wang, Cabot, & Machaty, 2009; Wang & Machaty, 2013). The calcium oscillation pattern and its duration influence the developmental competence of the fertilized egg (Ozil et al., 2005).

2.2.2 Properties of the sperm-borne oocyte activating factors

A series of studies using sperm structure fractionation by sonication, sperm freeze-thawing and serial detergent extractions, followed by comparative microinjections of said fractions and respective extracted spermatozoa into metaphase II oocytes, revealed the basic properties that a candidate SOAF components should meet (Kimura et al., 1998; Parrington, Jones, Lai, & Swann, 1999; Perry, Wakayama, & Yanagimachi, 1999; Perry, Wakayama, Cooke, & Yanagimachi, 2000; Swann, 1990). These conditions stipulate that SOAF should: (i) be a signaling molecule directly or indirectly causing ooplasmic PLC activation; (ii) induce oocyte Ca^{2+} oscillations or single release; (iii) trigger secondary anti-polyspermy defense via cortical granule (CG) exocytosis, as well as completion of oocyte meiosis, pronuclear development and embryo cleavage; (iv) be a male germ cell/testis specific gene product as the extracts of somatic cells from testis and other tissues are not known to induce calcium release upon oocyte injection; (v) be restricted to perinuclear theca of the equatorial segment and/or post-acrosomal sheath of the sperm head as the acrosomal fractions or sperm tails alone do not trigger oocyte activation via microinjection; (vi) be released by the fertilizing spermatozoon upon sperm-oolemma fusion; (vii) be soluble, present in the freeze-thaw sperm extracts and in the acrosome-reacted sperm heads; (viii) be absent from round spermatids requiring artificial activation of eggs injected with them via round spermatid microinjection (ROSI) and infertile round-headed spermatozoa lacking the PT and thus also lacking SOAF activity when injected in eggs by intracytoplasmic sperm injection (ICSI); and (ix) be evolutionarily conserved yet somewhat inter-specific promiscuous based on heterospecific sperm extract and SOAF component injections activating eggs.

2.2.3 Early SOAF candidates

The identity of the SOAF became something of a holy grail of fertilization biology in the wake of sperm extract microinjection experiments. The early candidate SOAF protein, a glucosamine-6-phosphate deaminase, named oscillin at the time to imply its involvement in the triggering of ooplasmic calcium oscillations (Parrington, Swann, Shevchenko, Sesay, & Lai, 1996) was disproved by further studies. Another candidate, the truncated c-Kit tyrosine kinase [tr-KIT; (Sette, Bevilacqua, Geremia, & Rossi, 1998)] showed multiple SOAF characteristics, including the ability to activate the ooplasmic phospholipase PLCG1 (Sette et al., 2002), and may still be considered a SOAF component and possible contributor to oocyte activation process, as well as a sperm quality biomarker (Tavalaee, Kiani-Esfahani, & Nasr-Esfahani, 2017). Beside soluble sperm PT proteins, nitric oxide (NO) donors (Jeseta et al., 2012; Thaler & Epel, 2003) and the sperm surface proteins with the integrin-binding RGD domains (Fenichel & Durand-Clement, 1998; White, Passipieri, Bunch, Campbell, & Pate, 2007) have been proposed to participate in oocyte activation. While NO donors could be delivered via PT solubilization, the RGD-dependent signaling pathways would be triggered by sperm-oolemma adhesion rather than by dissipation of the "factor" into ooplasm after sperm-oolemma fusion. Studies showing incomplete activation of anti-polyspermy defense when sperm-oolemma adhesion and fusion is bypassed by ICSI lend some credence to such hypothesis (Bianchi et al., 2014; Wortzman-Show, Kurokawa, Fissore, & Evans, 2007). As will be further discussed, the SOAF may be a multi-component, sperm-PT delivered factor acting in synergy with signaling pathways connecting egg surface to oocyte cortex and cytoplasm. Of note is the observation of tyrosine kinase recruitment to sperm-head surrounding fertilization cone early during sperm incorporation in the murine oocyte

(Wang, Hong, & Kinsey, 2021). Some of the SOAF components may be involved in the sequestration of SOAF into PT during spermatid elongation, others in timing and facilitation of SOAF release at fertilization, and yet others in regulating its activity or solubility. Three current leading candidates for SOAF components will be discussed next.

2.2.4 Phospholipase PLCZ1

Current favorite SOAF candidate, phospholipase C-zeta (PLCZ1) was identified by gene screen seeking potential male-germline-specific soluble phospholipase species (Saunders et al., 2002). It is a soluble PLC species found mainly in spermatozoa, of which the mRNA injection into oocytes triggers calcium oscillations similar to those induced by the spermatozoon. After oocyte activation, the overexpressed PLCZ appears to sequester in the pronuclei in some species (Ito et al., 2008). Some *Plcz1* gene mutations have been implicated in human male infertility/oocyte activation failure although studies have also detected expression outside of testis, which is not consistent with male germ-line specificity of SOAF. The *Plcz1* null male mice display spermatogenesis and sperm transport (within female reproductive tract) defects resulting in severe subfertility as most spermatozoa probably fail to reach oocytes after mating (Hachem et al., 2017; Nozawa, Satouh, Fujimoto, Oji, & Ikawa, 2018). This observation, along with recent studies of PLCZ1 translocation during sperm capacitation (Young, Grasa, Coward, Davis, & Parrington, 2009) suggest that it may be involved in the regulation of sperm motility and sperm capacitation. When probed by ICSI, the spermatozoa of null males failed to trigger calcium oscillations within the expected (based on wild type sperm ICSI) timeframe. However, some fertility and live born offspring was observed after null male mating with wild type females.

2.2.5 WBP2NL/PAWP

The WW-domain-binding protein 2 N-terminal-like (WBP2NL; alias PAWP) is a product of a gene expressed solely in the male germline, which resides exclusively in the post-acrosomal sheath of the sperm head PT. The recombinant mammalian WBP2NL protein injection into *Xenopus* eggs induced their activation including calcium release [a single calcium spark in frogs (Aarabi, Qin, Xu, Mewburn, & Oko, 2010)]. Aspects of oocyte activation process, including calcium oscillations and pronuclear development were observed in human eggs injected with recombinant protein or cRNA (Fig. 2D; Aarabi et al., 2014a). Competitive peptides and antibodies significantly hindered oocyte activation when coinjected with spermatozoa, both in frogs and mammals. Transfer of WBP2NL from sperm PAS to ooplasm, and even some integration into nascent paternal pronucleus have been documented (Wu et al., 2007a).

2.2.6 Extra-mitochondrial citrate synthase

The extra-mitochondrial citrate synthase (eCS) was first identified as a candidate SOAF in the newt (Harada et al., 2007). It was later proposed to initiate calcium oscillation and suppresses the age-dependent sperm dysfunction in the mouse. Mouse KO males have reduced fertility, exacerbated by age and while the first calcium spike is suppressed after ICSI of null spermatozoa, oscillations proceed normally thereafter (Kang et al., 2020). It was proposed that the reduced fertility of eCS null males may be due to increased polyspermy caused by delay of calcium oscillations after fertilization (Fukuoka et al., 2023). So far, little information has been given on spermatogenic dynamics and sperm localization patterns of eCS in mammals.

3. Sperm contributions to early zygotic development

3.1 Triggering of anti-polyspermy defense by the fertilizing spermatozoon

3.1.1 Oocyte plasma membrane depolarization

The endpoints of oocyte activation include triggering of anti-polyspermy defense, release from metaphase II arrest, completion of oocyte's second meiotic division (including the extrusion of second polar body), and development of the paternal and maternal pronuclei. There are at least two distinct mechanisms contributing to anti-polyspermy defense, including the primary (fast) block caused by the electric potential change of oocyte plasma membrane (OPM/oolemma depolarization) in the external fertilizers [amphibians, invertebrates (Komondor et al., 2023)]; and secondary (slow) block elicited by the cortical granule (CG) exocytosis resulting in zona hardening via proteolytic cleavage of ZPA/ZP2 component of zona pellucida (Burkart, Xiong, Baibakov, Jimenez-Movilla, & Dean, 2012). The primary block causes rapid OPM depolarization and the fusion of sperm plasma membrane with OPM also contributes paternal, sperm plasma membrane (SPM) proteins that diffuse into oolemma (Longo, 1989). Alteration of oocyte-cortical microfilament networks and microvilli are also observed. In mammals, or at least in mice, the polyspermy block at the oolemma level may also be contributed by the shedding of oolemma sperm receptor protein JUNO (Bianchi & Wright, 2014). Interestingly, this event may require sperm-oolemma fusion as it is not observed following the intracytoplasmic sperm injection (ICSI), which may explain why oocyte activation patterns after ICSI deviate from those observed after conventional in vitro fertilization (Wortzman-Show et al., 2007).

3.1.2 Zona hardening and other contributing mechanisms

The slow polyspermy block following CG exocytosis is caused by the proteolytic cleavage of ZPA/ZP2 protein by the action of CG exudate rich in ovastacin (oocyte actacin/ASTL), a metalloendopeptidase that makes the zona refractory to sperm binding. However, as the ASTL KO mouse shows, the ZP hardening alone is not a sufficient mechanism of polyspermy block (Burkart et al., 2012). Besides JUNO shedding, the rapid efflux of ooplasmic zinc ions after sperm-oolemma fusion, termed the zinc spark, could be a contributing factor. Similar to calcium ions, the zinc cation acts as a signaling molecule. Oocyte content of Zn^{2+} increases during maturation, particularly in the cortical zinc vesicles. Consequently, mammalian oocytes acquire the ability to issue the zinc spark during meiotic maturation. Zinc release and consequent decrease in Zn^{2+} ion centration following the zinc spark may be cell cycle related, and necessary for the completion of second meiosis (Kim et al., 2011). Zinc ions trapped in the zona after the spark could play multiple roles in anti-polyspermy defense, via direct alteration of ZP glycoprotein or by exerting an effect on the capacitated spermatozoa near the zona surface. Direct sperm alteration by the reuptake of Zn^{2+} lost during the capacitation induced efflux is a possible mechanism, supported by the demonstration of partial capacitation reversal (decapacitation) by Zn-reloading in vitro (Kerns, Zigo, Drobnis, Sutovsky, & Sutovsky, 2018). Active sperm repulsion by oocyte zinc and/or loss of oocyte chemoattractant activity may be occurring (Kerns et al., 2020). Further to these mechanisms, the anti-polyspermy defense in vivo could be co-regulated by the zona binding oviductal fluid proteins [e.g., osteopontin, UCHL1 (Hao et al., 2006; Yi et al., 2007b)], as well as by the timing of sperm release from oviductal sperm reservoir, oocyte cytoplasmic maturity, ZP integrity and oocyte quality changes related to female reproductive aging (Wang, Day, & Wu, 2003).

3.2 Pronuclear development

3.2.1 Transformation of gamete chromatin into zygotic pronuclei

The major events to zygotic development triggered by oocyte activation include the decondensation of the sperm nucleus and oocyte chromosomes; reconstitution of nuclear envelope (NE) bearing the nuclear pore complexes (NPC) around the nascent maternal pronucleus (MPN) and paternal pronucleus (PPN); replacement of sperm nuclear protamines by histones in the PPN; and the reconstitution of nucleolus precursor bodies (NPB) in preparation for the resumption of ribosomal RNA (rRNA) synthesis. Maternal and paternal DNA replication and centromere remodeling is also observed as an overture to first mitotic division (Fulka, Rychtarova, & Loi, 2020). Minor activation of transcriptional activity at this stage of development appears to be predominantly associated with PPN, perhaps as a quality control mechanism assuring the integrity of paternal DNA after protamine replacement. Such a complex remodeling of sperm nucleus and oocyte chromosomes requires multiple ooplasmic factors including histones (Gao et al., 2004), nucleosomal SWI/SNF complex ATPases (Ho & Crabtree, 2010), nucleoporins (Sutovsky, Simerly, Hewitson, & Schatten, 1998), ribonucleoproteins (Mao et al., 2018), proteasomes [to degrade protamines after removal from the sperm nucleus (Chen, Sun, Zhang, Davie, & Meistrich, 1998; Sutovsky et al., 2004)]; and nucleoplasmin [a chromatin remodeling, histone chaperone protein (McLay & Clarke, 2003)]. Glutathione (GSH, gamma-glutamyl-cysteinyl-glycine), an antioxidant and a disulfide-bond (S-S) reducing agent, supports sperm nucleus decondensation by destabilizing the resident protamines (Perreault, Barbee, & Slott, 1988). The bulk of ooplasmic glutathione is synthesized during oocyte maturation and enters the glutathione cycle in which the reduced GSH oxidizes itself (GS-SG) to reduce substrate proteins S-S bonds. The oxidized glutathione is then regenerated by a GSH-reductase, to re-enter the cycle (Perreault, Wolff, & Zirkin, 1984; Perreault et al., 1988; Sutovsky & Schatten, 1997). A synergistic mechanism may loosen the sperm chromatin even before fertilization: The zinc ion release during capacitation that notably starts in the sperm head (Kerns et al., 2018), probably encompasses the loss of chromatin stabilizing zinc bridges (Bjorndahl & Kvist, 2011), thus loosening sperm nuclear structure shortly before sperm-egg binding.

3.2.2 Nuclear envelope reformation/reconstitution

Sperm NE is stripped of NPC and most of the nuclear lamina components during spermiogenesis, at the time of spermatid elongation (Sutovsky et al., 1999b); it is completely removed from the sperm nucleus following sperm-oolemma fusion at fertilization and reformed as soon as the decondensation of the sperm nucleus commences (Fig. 3). At that time, the nuclear lamina is also rebuilt, and the NPCs repopulate the nascent NE, to re-establish the nucleo-cytoplasmic transport (Sutovsky et al., 1998). In oocytes of some species, including ungulates and humans, stacks of flat, NPC perforated membranes, termed annulate lamellae are abundant across the ooplasm, perhaps to assure rapid NE assembly following fertilization (Rawe, Olmedo, Nodar, Ponzio, & Sutovsky, 2003).

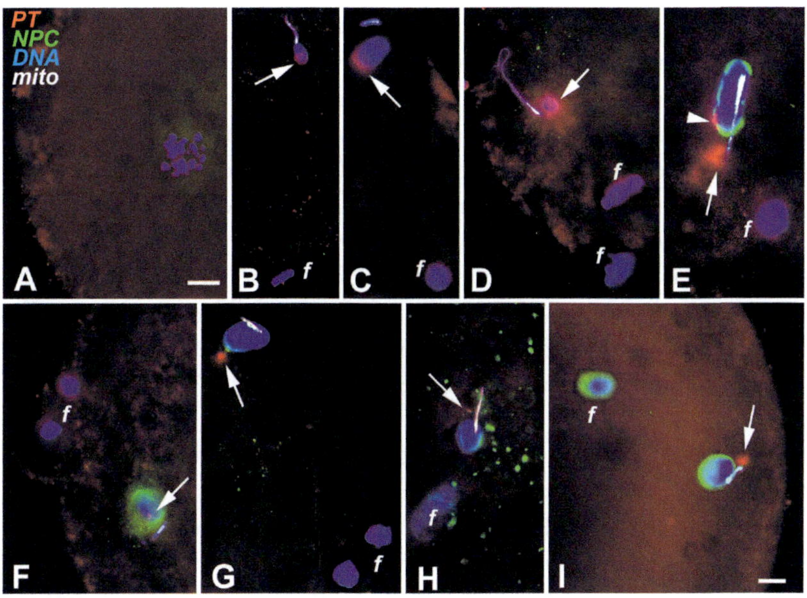

Fig. 3 Reformation of the nuclear envelope (NE) and nuclear pore complexes (NPC; green) during bovine fertilization, shown in relation to the dispersion of sperm perinuclear theca (PT; red). (A) A metaphase-II-arrested oocyte. (B–D) The removal of sperm PT (arrows) from the surface of sperm nucleus (blue) during sperm incorporation into oocyte cytoplasm, as the female chromosomes *(f)* complete the second meiosis. (E) A nascent paternal pronucleus with a ring of NPC on the newly reconstituted NE, with some remnants of PT nearby (arrow) and others still attached to it. The nascent female/maternal pronucleus *(f)*, still devoid of NPC. (F, G) Labeling of NPC on the surface of paternal pronucleus, with a clump of PT-derived material (arrow) and two female polar bodies *(f)* nearby. (H, I) Progressive dispersion of the PT (arrows) concomitant with the formation of female/maternal pronuclei *(f)*, now surrounded by an NE with NPC. Scale bars = 10 μm. *Reprinted with permission from Sutovsky et al. (2003a).*

3.2.3 Protamine-histone reversal

Protamines, the DNA hypercondensing sperm-specific nuclear proteins (Balhorn, 1982) replace nuclear histones during spermiogenesis, the haploid phase of spermatogenesis, more specifically, at the onset of round spermatid elongation (Chen et al., 1998) This developmental event shuts down transcription of spermatid DNA, at the same time endowing them with added DNA protection and streamlined sperm head shape conducive to sperm motility in viscous environment of the female reproductive tract. Then, the reversal of the sperm nuclear protamine-histone swap occurs at the early stage of sperm nucleus decondensation, after sperm head incorporation in the ooplasm, to promote PPN development, its DNA replication and reactivation of transcriptional activity of paternal DNA. Besides somatic type histones, the oocyte and embryo specific histones are involved in the reversal process. Protamines in the sperm nucleus are rapidly replaced by the oocyte-derived histones (e.g., linker H1FOO) within 60 min after sperm incorporation in the mouse egg (Gao et al., 2004), preceding the minor zygotic genome activation (ZGA). Paternally contributed, sperm PT-borne histones could be incorporated to help with early stages of PPN development (Hamilton et al., 2021; Tovich et al., 2004). In the two- to four-cell stage mouse embryos, the oocyte derived histones are replaced by the embryo-derived ones, an event coinciding with the activation of embryo transcription (major ZGA); (Gao et al., 2004; Teranishi et al., 2004).

3.2.4 Nucleologenesis and zygotic genome activation

Nucleolus is the cellular organelle responsible for the synthesis, processing, storage and export of ribosomal RNA (rRNA), thus considered the site of ribosome biogenesis. It is assembled around the transcription sites harboring ribosomal genes, the nucleolus organizer regions [NORs; (Sutovsky, Jelinkova, Antalikova, & Motlik, 1993)]. Transcriptionally active nucleolus has a lacelike, reticulate structure composed of the fibrillar center (FC), the site of rDNA transcription, dense fibrillar component (DFC) where nascent rRNA is processed and granular component (GC), responsible for rRNA packaging for the transport from nucleolus to cytoplasm. The nucleolus enters dormancy at the end of oocyte growth as the rDNA transcription vanes, turning into a single, compact nucleolus precursor body (NPB) inside the nucleus (germinal vesicle; GV) of a fully grown mammalian oocyte (Sutovsky et al., 1993). Similarly, the round spermatid nucleolus becomes inactive and is eventually dismantled prior to sperm nucleus hypercondensation as the spermatid starts to elongate. Nucleolus precursor bodies reassemble shortly after the onset of pronuclear development after fertilization, both in the PPN and MPN. The reassembly of NPBs appears to occur mainly from the ooplasmic pool of ribonucleoproteins (RNPs) and is essential for zygotic development. It is followed by the transition back from compact to reticulate structure, reinstating the FC, DFC and GC (Flechon & Kopecny, 1998), a hallmark of the onset of rDNA transcription and rRNA processing heralding the major ZGA (MZGA)(Tomanek, Kopecny, & Kanka, 1989) (Fig. 4). Removal of NPB from germinal vesicle stage oocytes prior to resumption of meiosis (enucleolation) causes failure of embryo development when such micromanipulated oocytes are matured and fertilized in vitro (Fulka & Langerova, 2014). Sperm head and/or tail structures could contain RNPs which could contribute to this process after fertilization (Kaneda et al., 2024), but may not be sufficiently abundant to rescue the paucity of ooplasmic

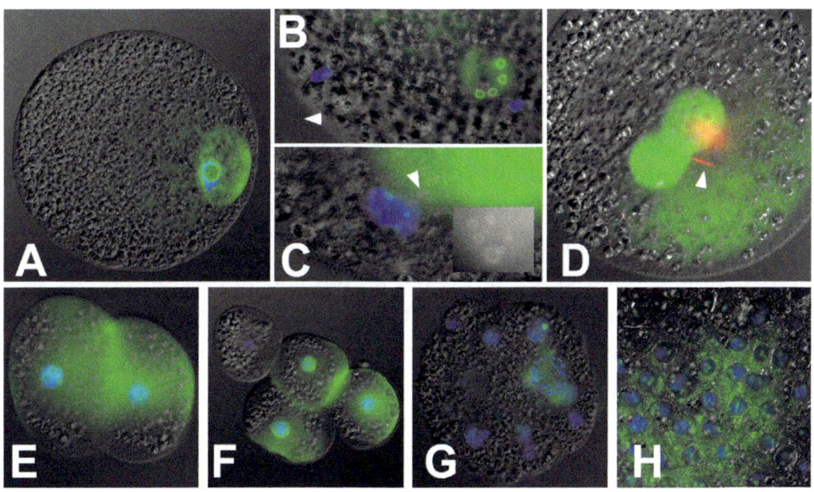

Fig. 4 Nuclear (blue) and nucleolar (green) transformation visualized in porcine oocytes and embryos by using a polyclonal antibody recognizing the ubiquitin-modified proteins and ubiquitin-tail ribonucleoproteins (green). (A) Single large nucleolus precursor body (NPB), inside the nucleus of an immature, germinal vesicle (meiotic prophase I) stage oocyte. (B–D) Multiple NPBs form inside the maternal pronucleus (B; MPN with four NPBs; sperm head is already incorporated while the tail still outside) and paternal pronucleus (C) shortly after fertilization, becoming more prominent at the time of pronuclear apposition (D). Inset in C shows gray scale detail of the nascent paternal pronucleus. Arrowheads in B-D denote sperm tail midpiece. Sperm mitochondrial sheath in D is labeled red with MitoTracker probe. (E) Two cell embryo. (F) Four cell embryo. (G) Morula stage embryo with more irregularly shaped, transcriptionally activated nucleoli. (H) Detail of a blastocyst showing somatic cell like, fully activated nucleoli.

RNPs after NPB removal from the GV. On the other hand, mRNA knockdown of oocyte RNP UBA52, which is also present in the sperm head (Sutovsky, unpublished data) impairs embryo development after IVF (Mao et al., 2018) and the obliteration of ubiquitin ligase HECW2 (syn. NEDL2) increases the number of NPB-like nuclear inclusions in the porcine embryos (Mao, Zigo, Zuidema, Sutovsky, & Sutovsky, 2021), linking ubiquitin-proteasome system with zygotic nucleogenesis.

3.2.5 *Paternal genome remodeling*

While mammalian embryos are generally transcriptionally silent until MZGA after one or more cellular divisions, depending on species, an early, minor outburst of transcription takes place in the pronuclei, primarily driven by the PPN (Aoki, Worrad, & Schultz, 1997). This could be a development-informing transcription and/or a proofreading mechanism acting in conjunction with zygotic DNA surveillance and repair mechanisms based on cell cycle checkpoints (Ladstatter & Tachibana-Konwalski, 2016), which appears to be independent of DNA replication in preparation for first embryo cleavage but likely affected by zygotic epigenome remodeling. Both paternal and maternal epigenomes contributed to the zygote are established during gametogenesis and partially erased and reestablished early after fertilization (Reik & Walter, 2001). Proper epigenome remodeling including paternal genome and histone code reprogramming (Guo et al., 2014) has a profound impact on zygotic development and pregnancy establishment and maintenance (Ross & Canovas, 2016). This is possibly co-regulated by sperm-contributed small regulatory RNAs, as will be discussed below. In addition to epigenome remodeling, one copy of X chromosome becomes inactivated by the Xist, a non-coding RNA transcribed from one X chromosome in the female embryos, to ensure proper gene expression dosage (Brown et al., 1991).

3.3 Centrosomal inheritance

3.3.1 *Centriole vs centrosome*

Centrosome is the cellular microtubule organizing center (MTOC) that guides cell division during meiosis and mitosis and centers the interphase microtubule (MT) network (Schatten, 2022). Conventional centrosome is composed of two centrioles and a hallo of peri-centriolar material (PCM). Centrioles are the symmetrical barrel-shaped organelles with a unique arrangement of nine microtubule triplets, with a pinwheel-like arrangement on cross-section. The zygotic centrosome is unique in its biparental inheritance and roles in microtubule nucleation into a star-shaped sperm aster that guides MPN and PPN apposition, organelle redistribution, organization of mitotic spindle poles, and eventually first embryo cleavage (Manandhar, Schatten, & Sutovsky, 2005).

3.3.2 Centrosome reduction during gametogenesis

Spermatid centriole is somatic cell-like early on, and its remodeling starts during spermatid elongation. The proximal centriole docks to the distal pole of the spermatid nucleus, the site of future implantation fossa, while the distal centriole generates the doublets of future axoneme, effectively acting as a cilium-like basal body (Sutovsky & Manandhar, 2006, 2007). It is not certain if PCM proteins are lost at this stage or transformed into a dense mass of capitulum in which the sperm proximal centriole is embedded. Progressively, the triplets of distal centrioles are dismantled, leaving an empty distal centriolar fossa in fully elongated spermatids. Complementary centrosome reduction occurs during female gametogenesis. Female germ cells lose centrioles but retain abundant pericentriolar material in their cytoplasm. Consequently, the meiotic oocyte spindle poles have PCM but no centrioles, a centrosomal reduction pattern that is complementary to mature spermatozoa having one centriole and no PCM (Manandhar et al., 2005). Formation of the zygotic centrosome and sperm aster after fertilization thus requires paternally contributed sperm proximal centriole as well as maternal, ooplasmic PCM.

3.3.3 Zygotic centrosome reconstitution

The assembly of zygotic centrosome starts concomitantly with the incorporation of sperm tail midpiece in the ooplasm, perhaps even before the principal piece of tail is fully engulfed, as suggested by images of nascent sperm aster in bovine zygotes [see Fig. 2 in (Sutovsky, 2018)]. Liberation of the sperm proximal centriole requires the disassembly of the sperm tail connecting piece, observable by electron microscopy as the breakdown of its striated columns and the dispersion of capitulum (Sutovsky, Navara, & Schatten, 1996b). In its mature state, the zygotic centrosome is a complete one, composed of PCM and two centrioles, one paternal, sperm contributed centriole, and one daughter centriole, formed *de novo* by the duplication of the paternal one. This biparentally-contributed zygotic centrosome gives rise to the microtubule array of the sperm aster, guiding pronuclear apposition in the anticipation of syngamy, though in mammals, the MPN and PPN never fuse. At this time, the whole zygotic centrosome duplicates in a manner similar to somatic cell, mitotic centrosome duplication that follows the cell cycle, to raise two poles of the future mitotic spindle ahead of first embryo cleavage. Non-rodent mammalian centrosomal inheritance is summarized in Fig. 5 (Sutovsky, 2010).

3.3.4 Rules and exceptions

Paternally contributed zygotic centrosome seems to be a rule in mammals, from marsupials to primates, humans included. One notable exception from the centrosomal rule is the uniparental, maternal inheritance of the centrosome in rodents. Complete dismantling of both centrioles occurs

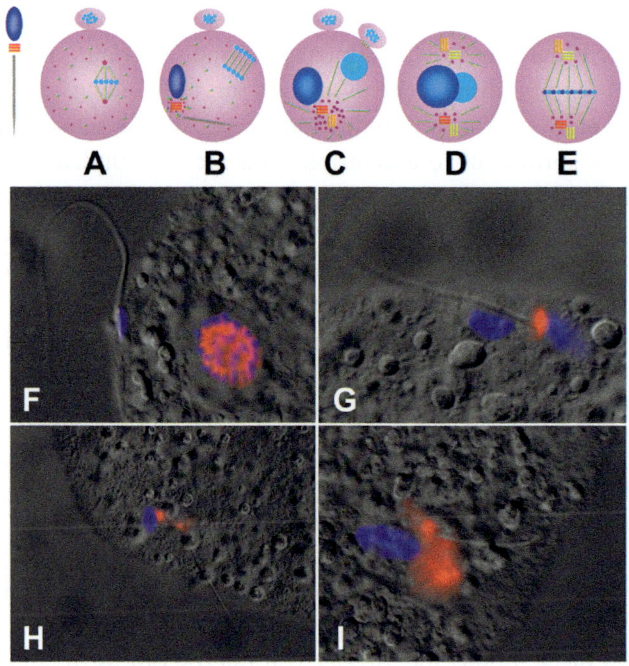

Fig. 5 Centrosomal inheritance is biparental in all mammals except rodents. (A–E) At fertilization (A, B), the fertilizing spermatozoon contributes the proximal centriole (red) while the oocyte provides pericentriolar material (PCM; pink) and tubulins (green). Sperm nucleus (dark blue) becomes the paternal pronucleus while the oocyte chromosomes decondense into maternal pronucleus after the completion of second meiosis and extrusion of the second polar body. Sperm centriole inside the nascent zygotic centrosome forms the microtubule sperm aster required for pronuclear apposition (C). Subsequently, the zygotic centrosome duplicates (D) to generate the microtubule organizing centers of the future mitotic spindle poles (E). (F–I) Labeling of oocyte spindle (F) and sperm aster (G–I) microtubules (red) during bovine in vitro fertilization. *Reprinted with permission from Sutovsky (2010).*

during spermiogenesis in the mouse (Manandhar, Sutovsky, Joshi, Stearns, & Schatten, 1998). Consequently, murine oocyte microtubules form multiple small cytoplasm asters in absence of the paternal centriole (Schatten, Simerly, & Schatten, 1985) and the first mitotic spindle poles are acentriolar, composed solely of PCM (Londono-Vasquez, Rodriguez-Lukey, Behura, & Balboula, 2022). The cytoplasmic asters then push the pronuclei together from the periphery towards the center of the zygote to achieve pronuclear apposition. It is not clear how or why such an alternative mode of centrosomal inheritance evolved. Rodents are considered evolutionarily more primitive but younger in evolutionary ages in comparison to other eutherian orders, which could result in loss of reproductive specializations and gain of redundant features. Rapid evolution in rodents is due to short pregnancy, large litters, and short time to puberty, together resulting in a short generational interval (Upham & Landis, 2023).

4. Paternal mitochondria and post-fertilization sperm mitophagy

4.1 The tale of mitochondrial Eve and Steve

4.1.1 Why mitochondria matter

Mitochondria are semi-autonomous organelles with their own genome, and a transcriptional and protein synthesizing machinery relying on mitochondrial transcription factors and resident mitochondrial ribosomes. They evolved by the integration of endosymbiotic bacteria with proto-eukaryotes, thus belonging to a class of organelles also including marine algae nitroplasts and plant chloroplasts (Massana, 2024). Mammalian mitochondrial genome contains up to 37 genes, including 13 protein coding genes including respiratory chain enzyme genes, mitochondrial tRNA coding genes, and two genes coding 12S and 16S rRNAs. Most other structural and regulatory mitochondrial proteins are thus encoded by the nuclear genome. Besides ATP production, cellular functions of mitochondria include cellular homeostasis and apoptosis. Consequently, mitochondrial health and mitochondrial DNA (mtDNA) inheritance patterns exert great influence on individuals' health (somatic and reproductive), fitness and lifespan (Song, Ballard, Yi, & Sutovsky, 2014). Within the focus of reproductive function and fertility, all of gamete quality, early development, and pregnancy establishment and maintenance are influenced by mitochondrial health. Mitochondrial genome is considered fragile, and prone to mutations and deletions as it lacks protection equivalent to histones shielding nuclear DNA. It has a simple DNA repair mechanism considered low-efficient in comparison to the sophisticated nuclear DNA repair machinery ensuring the integrity of complex mammalian genomes comprising tens of thousands of coding sequences. Consequently, the rate of mtDNA mutation is high and exacerbated by a mitochondrial genome bottleneck in the germ cells, which possess only a few mitochondria as well as by exceptionally rare recombination events.

4.1.2 *Ex ovo omnia*

Everything alive comes from an egg, proclaimed the frontispiece of Harvey's *Exercitationes de generatione animalium (1651),* a bold ovist manifesto that certainly holds true when it comes to the origin of mitochondria in our somatic cells and gametes. Given the fragility of mtDNA, it is assumed that a developmental advantage would be bestowed on a mammalian zygote if the paternal mitochondria and mtDNA, potentially compromised during spermatogenesis, epididymal sperm maturation, and within the female reproductive tract would better be eliminated, as it appears to be the case in all mammals studied thus far (Zuidema & Sutovsky, 2019). So, come Mother's Day, one would be obliged to give thanks to their mom for the gift of healthy cellular energy generators. Such a uniparental, maternal mode of mitochondrial organelle and genome inheritance, referred to as the Mitochondrial Eve paradigm (Sutovsky & Song, 2017) is the guiding principle and a powerful tool behind evolutionary lineage tracing. It was brought back to public attention by 2020 Nobel Prize for Medicine, awarded to Svante Pääbo for his pioneering discoveries concerning the genomes of extinct hominins and human evolution, enabled by his methods to extract and analyze ancient mtDNA. While inheriting competent mitochondria is considered a mother's gift, inheritance of corrupted oocyte-derived mitochondria is indeed a mother's curse, that could manifest itself in somatic mitochondrial disease of organs that rely most heavily on oxidative phosphorylation, such as heart and muscle (Kramer & Bressan, 2019).

4.2 Post-fertilization sperm mitophagy

4.2.1 *Seek and destroy*

How is the elimination of potentially corrupted sperm-borne mtDNA achieved during early development? As early as 1940, an article published in *Nature* pondered this issue, proposing a dispersion hypothesis (Gresson, 1940).

With less than one hundred sperm borne (i.e., paternal) mitochondria, the paternal contribution to zygotic mitochondrial gene pool would be diluted by thousands of ooplasmic mitochondria, each carrying multiple copies of complete maternal mitochondrial genome. This, however, would not guarantee a complete elimination of sperm-derived mtDNA, easily detectable and quantifiable by polymerase chain reaction (PCR) based techniques in the zygotes and early embryos (St John et al., 2000; Wolff, Nafisinia, Sutovsky, & Ballard, 2013a; Wolff, Sutovsky, & Ballard, 2013b). Another seemingly logical solution, jettisoning of the sperm tail once the sperm head enters the oocyte was disproved by electron microscopy of mammalian zygotes and fluorescent tracing of the oocyte-incorporated sperm mitochondrial sheaths following in vitro fertilization (Ankel-Simons & Cummins, 1996; Sutovsky et al., 1996b). The observation of accelerated sperm mitochondrion degradation concomitant with first embryo mitosis in the bovine zygote then led to the discovery of sperm mitochondrion ubiquitination inside the fertilized bovine and rhesus monkey eggs (Sutovsky et al., 1999a; Sutovsky et al., 2000). It was followed by the demonstration of proteasomal inhibitors preventing paternal mitochondrion degradation in the zygotes of domestic pigs, a species in which this seek and destroy mission is accomplished even before the first embryo cleavage (Sutovsky, McCauley, Sutovsky, & Day, 2003b). Ubiquitin, a small covalent protein modifier, is the central molecule of substrate-specific protein recycling via ubiquitin-proteasome system (UPS), an essential regulator of cell cycle progression and mitophagy, among many other cellular mechanisms.

4.2.2 The million-dollar question – proteasome-autophagy interplay

The substrate proteins targeted by UPS are tagged with multi-ubiquitin chains that make them visible to the 26S proteasome, a multi-subunit holoenzyme with three distinct proteolytic activities degrading proteins one molecule at a time. So, how can a pathway that recycles one protein molecule at a time degrade an entire organelle? Early studies of sperm mitophagy revealed ultrastructural alteration of sperm mitochondria, such as swelling and association with multivesicular bodies and lysosomes, consistent with the activity of autophagic mechanisms (Fig. 6A–F; Sutovsky et al., 2000). As the toolbox for autophagy research grew larger, studies in nematode *C. elegans* built on these early observations, showing that not only the individual mitochondrial proteins but the whole sperm-contributed mitochondrial organelle is degraded in bulk via autophagic/mitophagic machinery in which the ubiquitinated organelles become engulfed in the autophagophore which subsequently fuses with a lysosome

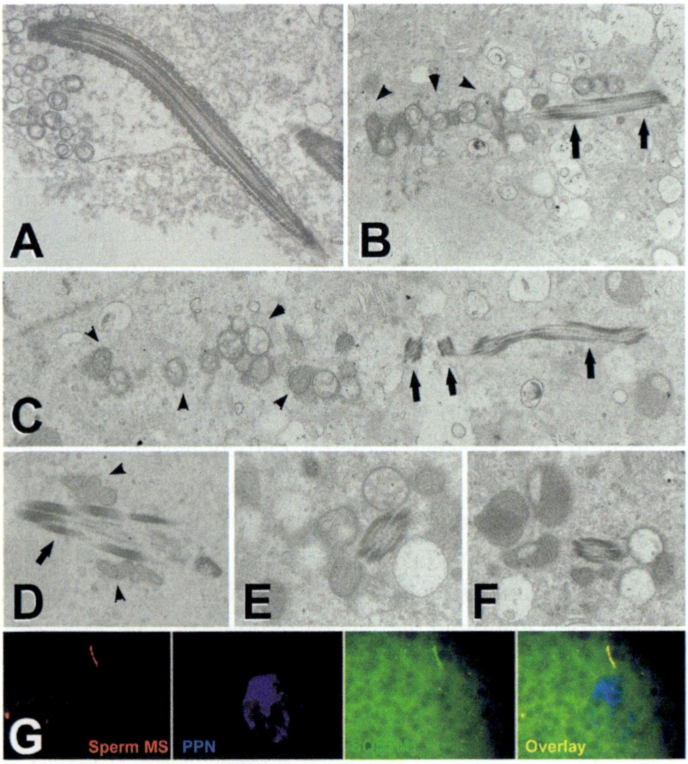

Fig. 6 Post-fertilization sperm mitophagy in the mammalina zygotes. (A–F) Mitophagy during bovine in vitro fertilization, documented by transmisison electron microscopy. Panel (A) shows an intact bovine sperm tail mitochondrial sheath with the proximal section of the principal piece visible in the lower left corner. Panels B-F document the detachment and degradation of sperm mitochondria (arrowheads) from the sperm axoneme denoted by outer dense fibers and microtubule doublets (arrows). Note three intact oocyte mitochondria in the upper left quarter of panel (F). (G) Pro-autophagic ubiquitin receptor protein SQSTM1 associates with the sperm mitochondrial sheath inside a porcine zygote within 8 h after in vitro fertilization. Panels (A–F) are previousluy unused images of serial sections from data sets described in published studies (Sutovsky et al., 2000; Sutovsky, 2003). *Panel (G) was adapted with permission from Song et al. (2014).*

(Al Rawi et al., 2011; Sato & Sato, 2011). Our own follow up studies uncovered coordination between UPS and autophagy during porcine sperm mitophagy (Song, Yi, Sutovsky, Meyers, & Sutovsky, 2016). An early event is the binding of pro-autophagic ubiquitin receptor, sequestosome 1 (SQSTM1; Fig. 6G) to sperm mitochondrial sheath within the first hours after sperm incorporation in the ooplasm. This universal autophagy signal may be aided by the destabilization of sperm outer mitochondrial membrane by the action of the valosin containing protein (VCP), a ubiquitinated protein translocase tasked with their extraction from organelle membranes and homing to the 26S proteasome. More recently, specific types of multi-ubiquitin chains and autophagy proteins such as LC3, GABARAP and FNDC1 were implicated in the nematode and mammalian sperm mitochondrion degradation (Molina, Lim, & Boyd, 2019; Song et al., 2016; Zuidema, Jones, Song, Zigo, & Sutovsky, 2023).

4.2.3 Candidate determinants and substrates of post-fertilization sperm mitophagy

A two-pronged UPS-autophagy approach to post-fertilization sperm mitophagy may be promoted by ubiquitination of sperm mitochondrial protein such as membrane protein prohibitin (Thompson, Ramalho-Santos, & Sutovsky, 2003) and mitochondrial transcription factor TFAM (Antelman et al., 2008) during spermatogenesis. Like Shakespeare's Hamlet characters delivering their own sentence to the King of Denmark, sperm mitochondria thus carry their ubiquitin tag to the egg at fertilization, predestining them for a quick demise. In this model (Fig. 7; Zuidema & Sutovsky, 2019), sperm mitochondria become ubiquitinated during spermatogenesis. This tag is masked by mitochondrial capsule cross-linking with disulfide bonds (S-S) during epididymal passage, to be relieved by ooplasmic reducing agents (e.g., glutathione) following fertilization. To make sperm mitochondria even less likely to propagate in the zygote, key mitochondrial transcription and sustenance factors could be degraded and/or relocated from sperm mitochondria to other sperm structures before fertilization, as seems to be the case of TFAM (Antelman et al., 2008). Our recent study in search of sperm mitophagy determinants employed a unique mammalian cell free system composed of demembranated porcine spermatozoa and oocytes extracts prepared from metaphase II-arrested, developmentally competent porcine oocytes (Song & Sutovsky, 2018). Spermatozoa were thus used as a bait for ooplasmic autophagy proteins, identified after coincubation and thorough washing via label-free quantitative proteomics of the extract-exposedspermatozoa (Zuidema et al., 2023). Three classes of proteins were identified: (i) Oocyte-specific proteins detected on spermatozoa after coincubation with oocyte extract, including autophagic ubiquitin receptors, mitochondrial membrane proteins, peroxiredoxins and proteins associated with pronuclear development (nucleoplasmin); (ii) Shared sperm and oocyte proteins, increased in

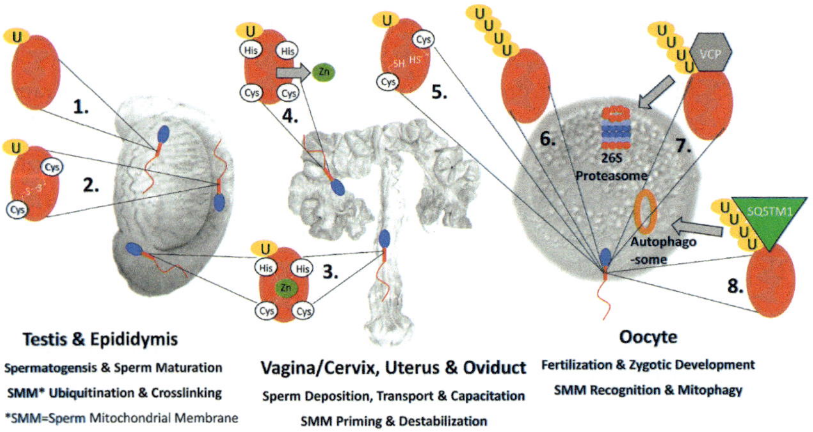

Fig. 7 Proposed mechanism of porcine post-fertilization sperm mitophagy. Ubiquitination of sperm mitochondrial proteins occurs during spermiogenesis in testis **(1)** and becomes cross-linked by disulfide bonds **(2)** and zinc bridges **(3)** during epididymal sperm maturation, concealing the ubiquitin tag until the sperm tail is incorporated in the oocyte cytoplasm. Zinc bridge stabilization is removed by the sperm capacitation-induced zinc ion efflux **(4)**, and the mitochondrial sheath destabilization is furthered by the removal of disulfide bonds (S-S) after sperm tail incorporation in the oocyte cytoplasm **(5)**. This event exposes ubiquitin tags added during spermiogenesis and makes the sperm mitochondria prone to further ubiquitination **(6)**. Ubiquitinated sperm mitochondrial membrane proteins are at first recognized and extracted by ooplasmic protein dislocase VCP which serves them to 26S proteasome for degradation **(7)**, making the mitochondrial sheath recognizable to ubiquitin-binding autophagy adaptors such as SQSTM1, resulting in the engulfment and digestion by autophagosome **(8)**. *Reprinted with permission from Zuidema and Sutovsky (2019).*

spermatozoa after coincubation, including autophagy regulators/activators, ubiquitin-proteasome system proteins, inner mitochondrial membrane, and mitochondrial electron transport chain proteins; and (iii) Shared sperm & oocyte proteins, decreased in spermatozoa after extract coincubation, including mitochondrial proteins (membrane & matrix), mitochondrial electron transport chain proteins, and autophagy/mitophagy regulators, both autophagy promoting and inhibitory ones. Simplified interpretation of these findings is that sperm mitochondria exposed to oocyte cytoplasm become infiltrated by autophagic/mitophagic proteins targeting them for degradation, while mitophagy limiting proteins are removed from them, thus enabling degradation of structural mitochondrial membrane proteins and respiratory chain enzymes.

5. Accessory sperm structures

5.1 Acrosome

5.1.1 Acrosomal structure

The mammalian sperm acrosome can be defined as a specialized exocytotic organelle, unique to the male-gamete. This is based on its attributes such as the origin from Golgi complex, high content of lysosomal and specialized proteolytic enzymes, and membrane properties and receptors akin to other types of exocytotic vesicles, including synaptic vesicles. The acrosome is composed of the inner and outer acrosomal membrane (IAM and OAM, respectively) between which the dense, homogeneous acrosomal matrix is layered. It sits on the subacrosomal perinuclear theca (sPT) and its distal portion is overlaid by the outer leaf of equatorial segment theca (esPT), also containing PT component. The apical acrosome is overlaid by a modified plasma membrane in which various receptors for oviductal epithelial cells, seminal plasma proteins and zona pellucida binding proteins are sequestered into membrane rafts. Similarly, the plasma membrane over the equatorial segment is enriched with ligands and receptors involved in sperm-oolemma adhesion.

5.1.2 Acrosome during and after fertilization

Upon semen deposition in the female reproductive tract, the acrosome binds seminal plasma proteins, some of which persist after sperm transport through uterus and oviduct, to mediate sperm binding to oviductal sperm reservoir (Tumova et al., 2021). Subsequent sperm capacitation event primes the acrosome for exocytosis by removing cholesterol from overlying plasma membrane, and assembling the egg receptors and signaling molecules involved in acrosomal exocytosis into membrane rafts. Primed spermatozoa undergo acrosomal exocytosis process during which the OAM and plasma membrane form hybrid vesicles of the acrosomal shroud, exposing the acrosomal matrix and IAM (Stival et al., 2016). This shroud remains on the zona surface as the spermatozoon progressively digests fertilization slit and burrows deeper in the zona (Yi, Manandhar, Oko, Breed, & Sutovsky, 2007a). Upon reaching the perivitelline space between the zona and the oolemma, only the IAM and sPT remain intact. These structures appear to disperse shortly after sperm incorporation in the ooplasm to enable the decondensation of the sperm nucleus (Sutovsky, Manandhar, Wu, & Oko, 2003a). To the contrary, the acrosome-sPT complex may persist in the ooplasm of primate eggs fertilized by ICSI, which may interfere with the development of PPN and result in a mitotic block, as will be discussed in the concluding section.

5.2 Flagellum

5.2.1 Sperm tail structure

Mammalian sperm flagellum contains motile apparatus built around a typical, cilium-like 9 + 2 arrangement of microtubules (9 interconnected doublets + central pair) supported on the outside by sturdy but flexible outer dense fibers (ODF) emanating from the striated columns of the sperm tail connecting piece. While the sperm tail midpiece is wrapped in a helical mitochondrial sheath, the principal piece is covered by a transversally ribbed fibrous sheath built around a pair of opposing longitudinal columns (Barth, Perry, Hamilton, Sutovsky, & Oko, 2025; Sutovsky & Manandhar, 2007). Relatively little attention has been paid to the post-fertilization fate of these paternally contributed structures.

5.2.2 Fate of flagellar accessory structures

While the post-fertilization recycling of the mitochondrial sheath has been studied extensively, and addressed in the mitochondrial inheritance section, less is known about the post-fertilization processing of ODF, FS and axonemal microtubule. Early studies of bovine fertilization revealed that the FS dissolves within a few hours of sperm tail incorporation (Sutovsky et al., 1996b). This early dispersion pattern has recently been confirmed by imaging of human zygotes (Kai, Iseki, Matsui, & Yamashita, 2024). In a similar fashion, the capitulum and striated columns of the connecting piece disassemble and dissolve shortly after fertilization to facilitate the formation of zygotic centrosome, as described in centrosomal inheritance section. To the contrary, the flagellar ODF and microtubules appear to be highly stabilized and resistant to rapid degradation as much as they are resistant to detergent extraction when sperm proteins are probed biochemically. Therefore, they remain detectable until, or even after first embryo cleavage (Asch, Simerly, Ord, Ord, & Schatten, 1995; Kai et al., 2024; Sutovsky et al., 1996b). The size, volume and solubility of sperm head perinuclear theca and sperm tail fibrous sheath predestines these structures for a vehicle role in the delivery of sperm-borne proteins and RNAs, discussed next.

6. Ribonucleic acids

6.1 Sperm transcriptome

6.1.1 Types of sperm borne RNAs

Spermatozoan transcriptome is composed of a suite of unique RNAs that are delivered to an oocyte upon fertilization. Human sperm RNA was identified for the first time in 1999 (Miller et al., 1999). The first global transcriptomic study of human spermatozoa identified around 3000 different protein coding transcripts (Ostermeier, Dix, Miller, Khatri, & Krawetz, 2002). A later study showed an inter-sample variation when only 564 coding transcripts out of 2712 unique ones detected in 10 pooled human sperm samples overlapped (Zhao et al., 2006). A substantial portion of coding spermatozoan RNAs are fragmented though (Sendler et al., 2013). Next-generation sequencing techniques allowed the identification of more than 22,000 transcripts, including both the coding and non-coding RNAs in human spermatozoa, among them messenger RNAs (mRNAs; both fragmented and non-fragmented), ribosomal RNAs (rRNAs; most of which is fragmented), mitochondrial RNAs (mitoRNAs), micro RNAs (miRNA), Piwi-interacting RNAs (piRNAs), long non-coding RNAs (lncRNA), small interfering RNAs (siRNAs), and transfer (t)RNA-derived small RNAs (tsRNAs) (Krawetz et al., 2011; Sendler et al., 2013), and other RNAs summarized in Table 1. It is thus evident that the sperm transcriptome carries important information about male fertility as summarized and reviewed in (Hernández-Silva, Caballero-Campo, & Chirinos, 2022).

6.1.2 Origin of sperm RNAs

A large portion (~80 %) of sperm RNA originates from the testes (Wang et al., 2004) and this could include the carryover of spermatogenic RNAs that are no longer intact, as well as specific coding and regulatory RNAs expressed late in spermatogenesis and inserted in the sperm structures for the purpose of influencing post-fertilization events. There is an ongoing discussion about whether the sperm RNAs are merely the relics of spermatogenesis or are being selectively retained since a large subset of spermatid cytoplasmic RNAs are absent from mature spermatozoa (Santiago, Silva, Howl, Santos, & Fardilha, 2021). Epididymis is most likely another source of sperm RNA, as spermatozoa were shown to contain transcripts of epididymal origin, e.g. Semg1 and Semg2 transcripts (Jodar, Sendler, & Krawetz, 2016; Lalancette et al., 2009). Along with other cargos, RNAs are transferred to spermatozoa via epididymosomes, microvesicles of heterogeneous size (40–100 nm) secreted by the epididymal epithelium (da Silveira et al., 2018). The third origin, which is still an ongoing controversial topic, is the spermatozoon itself, more specifically, transcripts that are derived from genomic regions packed by histones and enriched in H3K4me3 (Sendler et al., 2013). Genes with higher levels of H3K4me3 are usually coding developmental

Table 1 Spermatozoan RNAs.

Type of RNA	Length	Expression	Total content of RNA in a spermatozoon	Localization in spermatozoa	Postulated function in reproduction				References
					Spermatogenesis	Post testicular maturation	Postfertilization events	Other	
Long RNAs	>200 nucleotides by the definition		~50–90 fg of long RNA						
Coding RNA	variable	Expressed in all tissues		Head, midpiece, and tail	Availability and preservation of all elements required for sperm generation.	Sperm capacitation-related translation of proteins within mitochondria	Might assist in the formation of a functional centrosome as requisite for zygotic division, oocyte activation	Possible intergenerational non-Mendelian inheritance	Corral-Vazquez et al. (2021); Goto, O'Brien, and Eddy (2010); Gur and Breitbart (2006), Jodar et al. (2016); Manandhar et al. (2005); Sendler et al. (2013)
messenger (m) RNA									
Long non-coding (lnc)RNA	>200–10,000 nt	Expressed in all tissues		Outer membrane/ perinuclear theca	Spermatogenesis regulation and packaging of the paternal genome	Might play a role in fertilization events	Might be involved in the embryogenesis gene regulation		Corral-Vazquez et al. (2021); Jiang et al. (2016), Jodar et al. (2013); Johnson, Mackie, Jodar, Moskovtsev, and Krawetz (2015); Liang et al. (2014); Ostermeier et al. (2005); Qiu, Ren, and Yan (2016); Sahlu et al. (2020)

Short non-coding (snc)RNA	<200 nt by the definition		~0.3 fg of sncRNA					Joshi, Sethi, Mehta, Kumari, and Rajender (2023); Krawetz et al. (2011)
micro (mi) RNA	20–24 nt	Expressed in most tissues		Nuclear or perinuclear compartment, andt also in tail	Transcriptional gene silencing in sperm maturation	Early embryonic histone replacement; early embryogenesis; epigenetic modification	Possible intergenerational non-Mendelian inheritance	Gòdia et al. (2018); Johnson et al. (2011a); Salas-Huetos et al. (2015); Salas-Huetos et al. (2014); Santiago et al. (2021); Yuan et al. (2016)
PIWI-interacting (pi)RNA	24–32 nt	Mammalian male germline, spermatocytes, spermatids, spermatozoa		Tail	Protection of male germline from transition	Genome recognition and consolidation-confrontation; epigenetic modification		Bourc'his and Voinnet (2010); Cui, Fang, Shi, Qiu, and Ye (2018); Giebler, Greither, Müller, Mösinger, and Behre (2018); Krawetz et al. (2011); Pantano et al. (2015)
tRNA-derived small (ts) RNA	18–40 nt	Spermatozoa		Head		Epigenetic modification and early embryo development		Chen et al. (2016); Chen et al. (2020); Hua et al. (2019); Peng et al. (2012); Schuster et al. (2016)
rRNA-derived small (rs) RNA	15–50 nt	Spermatozoa		Head				Jodar et al. (2013); Johnson et al. (2011b); Zheng et al. (2021)
endogenious-small interfering (si)RNA	~21 nt	Expressed in most tissues		Head		Post-transcriptional regulation of gene expression		Gòdia et al. (2018); Han et al. (2009); Yuan et al. (2016)

(continued)

Table 1 Spermatozoan RNAs. (*cont'd*)

Type of RNA	Length	Expression	Total content of RNA in a spermatozoon	Localization in spermatozoa	Postulated function in reproduction				References
					Spermatogenesis	Post testicular maturation	Postfertilization events	Other	
small nuclear (sn)RNA	~150 nt	Expressed in most tissues		Head	Splicing regulation in spermatogonial stem cells				Concha et al. (1993); Wang et al. (2020); Zhou et al. (2022)
small nucleolar (sno)RNA	60–170 nt	Expressed in most tissues		Head	Ribosomal and rRNA biogenesis				Lui and Lowe (2013)
repetitive/ transposable elements		Expressed in most tissues							Joshi et al. (2023); Krawetz et al. (2011); Santiago et al. (2021)
transcription start sites (TTS)/ promoter-associated RNA a.k.a. quiescent (q) RNA		Expressed in most tissues							Joshi et al. (2023); Krawetz et al. (2011); Santiago et al. (2021)
small YRNA fragments (YsRNA)	85–115 nt	Expressed in most tissues		Head			The first stages of embryo development to initiate DNA damage repair		Elzer et al. (2023); Krawetz et al. (2011)

factors, which suggests that these sites might be marked for post-fertilization transcription in the paternal pronucleus of pre-cleavage stage zygotes (Hammoud et al., 2009).

6.1.3 mRNAs

A single spermatozoon contains ~50–90 fg of long RNA [>200 nucleotides (Goodrich, Anton, & Krawetz, 2013)] that comprise both coding and non-coding RNAs, which is around 200 × less than a somatic cell. A single study identified a total of 12,853 mRNA transcripts from 12 healthy donors, from which 5100 mRNAs were classified as highly expressed and 780 mRNAs were classified as ubiquitously expressed (Corral-Vazquez et al., 2021). Gene Ontology (GO) analysis of highly expressed spermatozoan mRNA identified the following enriched biological processes: *Chromosome Organization, Spermatogenesis, Microtubule-based Process, Protein Modification by Small Protein Conjugation or Removal, Reproduction, Ribonucleoprotein Complex Localization,* and *Cellular Response to Stress* (Corral-Vazquez et al., 2021). The top ten most abundant sperm mRNAs were Prm1, Prm2, Tnp1, CemP1, Golga6l2, Smcp, Lelp1, Cirbp, Tssk6, and Dus3l. Although the potential involvement of these mRNAs in early embryonic development needs to be explored, reports of the altered spermatozoal mRNA profiles in asthenozoospermic (Bansal, Gupta, Sankhwar, & Rajender, 2015; Caballero-Campo et al., 2020; Jodar, Kalko, Castillo, Ballescà, & Oliva, 2012), teratozoospermic (Platts et al., 2007), oligozoospermic (Montjean et al., 2012), idiopathic infertile (Bansal et al., 2015; Burl, Clough, Sendler, Estill, & Krawetz, 2018; García-Herrero et al., 2010; Garrido et al., 2009), and cryptorchid (Nguyen, Delaney, & Kolon, 2009) males suggest sperm RNA participation in the events following fertilization.

6.2 Non-Coding RNAs

6.2.1 Long non-coding RNAs

A single study reported that as many as 7521 long non-coding (lnc)RNA sequences, 45.78 % were found to be long intergenic non-coding RNA, followed by antisense lncRNA (36.96 %), sense lncRNA – both intronic and gene-overlapping that accounted for 4.93 % and < 1 % of other lncRNA (Corral-Vazquez et al., 2021). Several classes of lncRNAs were reported in human (Jodar et al., 2015; Sendler et al., 2013) and mouse (Zhang et al., 2017) spermatozoa. One of the lncRNA class members includes the RNA U small nuclear family, a component of sperm spliceosome, suggesting a role in the early fertilization events (Sendler et al., 2013). Another class of lncRNA includes natural antisense transcripts (NATs), which in spermatozoa overlap with genes associated with the first stages of embryogenesis – suggesting their function (Li et al., 2002; Ostermeier, Goodrich, Moldenhauer, Diamond, & Krawetz, 2005; Sendler et al., 2013). Additional classes of lcnRNAs, with yet-to-be-established functions in embryogenesis, are chromatin-associated RNA (CAR), small-nuclear ILF3/NF30-associated RNA (snaR), and circular RNAs (circRNAs) (Gòdia, Swanson, & Krawetz, 2018; Jodar, Selvaraju, Sendler, Diamond, & Krawetz, 2013). Intronic-retained elements (RE) including more than 200 distinct REs have also been reported in human spermatozoa although their function in embryonic developmen is not known (Sendler et al., 2013).

6.2.2 Micro RNAs

A single spermatozoon contains ~0.3 fg of small non-coding (snc)RNA of fewer than 200 nucleotides by definition, but most commonly of 18–30 nucleotides (Goodrich et al., 2013). These can be categorized into the following groups; (i) repeat-associated small RNAs (~65 % of all sncRNA), (ii) piRNAs (~16.9 % of all sncRNA), (iii) transcription start sites (TSS) or promoter-associated RNAs (~10.8 % of all sncRNAs), (iv) miRNAs (~6.9 % of all sncRNAs); and other less studied RNAs such as small nuclear (sn)RNAs, small nucleolar (sno)RNAs, tRNA-derived small (ts)RNAs, and YRNA, an emerging class of small ncRNAs (Krawetz et al., 2011). The best characterized sncRNAs are miRNAs that inhibit or activate translation or target mRNAs for degradation, thus regulating gene expression. The miRNAs were reported for the first time in human spermatozoa almost two decades ago (Ostermeier et al., 2005), and the discoverers of this snRNA class (Lee, Feinbaum, & Ambros, 1993; Wightman, Ha, & Ruvkun, 1993) were awarded a Nobel prize in 2024. Commonly seen miRNA association with TSS/promoters, histones, and repetitive elements implies involvement in the regulation of early embryonic histone replacement, early embryonic gene expression, and epigenetic modification of embryonic genome during development (Johnson et al., 2011a; Yuan et al., 2016). This proposition was supported by the observation that spermatozoa collected from fertile donors were enriched in miRNAs that regulate genes involved in cell differentiation, development, morphogenesis, and embryogenesis (Salas-Huetos et al., 2014). Furthermore, 736 miRNAs were differentially expressed between normozoospermic infertile patients and normozoospermic fertile individuals (Salas-Huetos et al., 2016). The sperm miRNA profile of men with several forms of infertility highlighted specific miRNAs associated with sperm motility, concentration, and age (Salas-Huetos et al., 2015). Lifestyle and environment-associated alterations in semen parameters linked with cigarette smoking (Marczylo, Amoako, Konje, Gant, & Marczylo, 2012; Metzler-Guillemain et al., 2015), living at environmentally polluted sites (Li, Li, Liu, Song, & Liu, 2012) or increased nut consumption (Salas-Huetos et al., 2018) were associated with differential miRNA expression when compared to the respective control groups. It appears probable that the effects of miRNA and other sncRNAs may be passed across generations epigenetically, affecting the progeny's phenotypes independently of Mendelian inheritance. Apart from miRNAs, mature human spermatozoa contain intact miRNA precursors (pri-miRNA, 100–150 nt) that might be involved in early embryo development (Sendler et al., 2013; Vassena et al., 2011).

6.2.3 piRNAs

Another major class of sncRNA are the PIWI-interacting RNAs (piRNAs). These can be distinguished from miRNAs by their larger size, RNase-III-independent mechanism of synthesis, and formation of the effector complexes with PIWI proteins as opposed to AGO family proteins that form effector complexes with miRNAs and siRNAs (Lin, 2007; Siomi, Sato, Pezic, & Aravin, 2011). Suggested functions of piRNAs include: (i) involvement in genome recognition, confrontation and consolidation upon fertilization (Bourc'his & Voinnet, 2010; Krawetz et al., 2011), (ii) transposable element silencing at specific stages of embryogenesis (Reuter et al., 2011), and (iii) epigenetic programming and intergenerational/transgenerational inheritance (Donkin et al., 2016; Peng & Lin, 2013; Tyebji, Hannan, & Tonkin, 2020).

6.2.4 Other regulatory RNAs

A relatively novel class of regulatory sncRNA, the tRNA-derived small (ts) RNA, also referred to as tRNA-derived fragments (tRFs), was reported to be highly enriched in the sperm head (Peng et al., 2012; Schuster et al., 2016). Multiple studies suggest an involvement of tsRNA in transgenerational, epigenetic inheritance (Chen et al., 2016; Guo et al., 2021; Hernandez et al., 2024; Hua et al., 2019; Ma et al., 2024; Sharma et al., 2016; Tomar et al., 2024; Yan & Zhai, 2016). New data also points to the role of tsRNA in early cleavage events during porcine preimplantation embryo development (Chen et al., 2020). Like tsRNA, the rRNA-derived small (rs)RNAs have been getting more attention recently. The rsRNAs are the most abundant RNA class in sperm RNA profile (Jodar et al., 2013; Johnson et al., 2011b). Similarly to miRNAs and tsRNAs, the rsRNAs have been implicated in contributing to transgenerational paternal inheritance (Zheng et al., 2021). Yet other sncRNA species in spermatozoa and their implied functions include small interfering (si)RNA, implicated in post-transcriptional events upon fertilization and early embryonic development in mice (Yuan et al., 2016) and zygotic gene expression in *C. elegans* (Han et al., 2009), small nuclear (sn)RNA, small nucleolar (sno) RNA, repetitive/ transposable elements, quiescent (q)RNA – all with unknown functions in post-fertilization events; and small YRNA fragments (YsRNA) that might serve in DNA damage repair mechanism (Krawetz et al., 2011) or enhance RNA chaperoning (Elzer, Bremser, & Zischler, 2023) during embryonic development.

7. Conclusions

7.1 Unresolved issues

7.1.1 SOAF identity

As aforementioned, the PLCZ1 is currently favored as the SOAF candidate, while WBP2NL warrants attention as well. However, the localization of the PLCZ1 across most of the sperm head and tail in mammalian species does not support an exclusive SOAF role (Aarabi et al., 2012; Gonzalez-Castro, Amoroso-Sanches, Stokes, Graham, & Carnevale, 2019; Sugita et al., 2024). Male *Plcz1*-null mice are subfertile but not sterile (Hachem et al., 2017; Nozawa et al., 2018), and thus far, the direct transfer of PLCZ1 from sperm head to the ooplasm during fertilization has not been documented. Of note, the subfertile *Plcz1*-null males show a statistically significant upregulation of *Wbp2nl* gene expression in the testis (Hachem et al., 2017), and their spermatozoa are sterile in conventional in vitro fertilization (IVF), necessitating ICSI. It is thus possible that the deletion of *Plcz1* gene slows down sperm transport within female reproductive tract and hinders sperm capacitation and sperm-zona interactions. As the sperm capacitation involves calcium signaling mediated by phospholipase family enzymes (Arrais, Burla Dias, de Souza, Curcio, & Mello, 2024; Ickowicz, Finkelstein, & Breitbart, 2012), the *Plcz1*-null males could be subfertile in vivo due to reduced ability to detach from the oviductal sperm reservoir or incomplete priming of the sperm acrosome and perinuclear theca, including failed processing of other SOAF components during capacitation. Similar to PLCZ1, some of the data argue against the exclusive role of WBP2NL as the SOAF factor. Development into blastocyst was rarely observed in the recombinant protein-injected mammalian oocytes, though solubility issues with recombinant protein and/or lack of proper processing of pre-protein translated after mRNA/cRNA injection into oocytes could be a culprit. The *Wbp2nl*$^{-/-}$ male mice are reportedly fertile (Satouh, Nozawa, & Ikawa, 2015), but it is notable that its somatic homolog *(Wbp2)* is highly expressed in the murine testes (Hamilton et al., 2018). Both WBP2NL and WBP2 proteins end up exclusively packaged in the mouse sperm postacrosomal PT, the agreed-upon exclusive localization of SOAF, and have similar SOAF properties [i.e., trigger pronuclear development and cortical granule exocytosis in the microinjected oocytes (Hamilton et al., 2018)], making cross-compensation plausible.

Neither PLCZ1 nor WBP2NL may be the sole SOAF components, with other candidates still under consideration. In this context, it is of interest to revisit the pre-SOAF "contact hypothesis" of oocyte activation, involving plasma membrane ligand-receptor binding triggering the signaling cascade of oocyte activation. The proposed interaction involved the binding of sperm head plasma membrane disintegrins of the ADAM metalloproteinase family, to their oolemma counterpart, the integrin family extracellular matrix receptors (White et al., 2007). Lending support to such pathway's contribution to oocyte activation, mouse ova after ICSI, which bypasses sperm-oolemma adhesion, do not establish efficient anti-polyspermy defense (Wortzman-Show et al., 2007) and no shedding of JUNO, observed after IVF is seen after ICSI (Bianchi et al., 2014). Furthermore, synthetic disintegrin RGD domain peptides bind oolemma and trigger CG exocytosis in mouse eggs (Tatone & Carbone, 2006), and cause calcium release, but no sustained oscillations in bovine eggs (Campbell, Reed, & White, 2000). Integrins have been linked to ooplasmic PLC/IP3 signaling pathway involved in calcium oscillations during fertilization.

7.1.2 Histones

Addressing unresolved issues in fertilization biology in one of his final papers (Yanagimachi, 2022), the late, great Ryuzo Yanagimachi suggested that the PT stored histones could be involved in oocyte activation. Although not elaborated, one could consider the onset of PPN development, though the mechanistic link to the pathway activating ooplasmic phospholipases is not obvious. Whether the single sperm PT contains enough histones for them to be essential for the first phase of PPN development, remains to be determined. Our recent studies allude to such a possibility (Hamilton et al., 2021).

7.1.3 Centrosomal inheritance

As explained previously, the distal centriole of mammalian spermatozoa is dismantled during spermiogenesis, while the resultant centriolar vault may still contain residues of proteolytic enzymes involved in this process, and centriolar or centrosomal proteins, and perhaps even occasional remnants of centriolar microtubule triplets, particularly in immature, testicular spermatozoa as the degradation of distal centriole appears to continue during epididymal sperm maturation (Simerly et al., 2016). Such partially degraded distal centriole of mammalian spermatozoa (Fishman et al., 2018) has recently been likened to atypical centrioles seen in the insects (Fishman et al., 2017). These atypical insect centrioles are somewhat reminiscent of a massive distal centriole seen in ratite birds' spermatozoa (Soley, 2016). While the basal body function of distal centriole and its structures in the biogenesis of sperm tail is not disputed, one should be cautious about attributing a role in centrosomal inheritance to such a vestigial structure, even if it carries residual microtubule nucleating proteins (Fouquet, Kann, Combeau, & Melki, 1998). Occasionally, a two-piece sperm aster can be observed in the zygotes [e.g., Fig. 2E in (Sutovsky, 2018)], which could arise from two distinct sperm-contributed centrioles, or be produced by mechanical damage to zygote during processing for sperm aster imaging. It is thus unclear whether the partially degraded or structurally unconventional distal centriole could contribute to zygotic centrosome formation and duplication in mammals. Also unexplained is why do rabbits, the lagomorphs evolutionarily related to rodents with centriole-free spermatozoa, have a blended mode of centrosomal inheritance wherein spermatozoa have a proximal centriole similar to higher mammals but form both a sperm aster and presumably acentriolar cytoplasmic asters in the zygotes just like the mouse (Terada, Simerly, Hewitson, & Schatten, 2000).

7.1.4 Mitochondrial inheritance

Many questions remain unanswered with regard to how the species specificity of maternal mitochondrial inheritance is assured and how are sperm mitochondria disfavored from mtDNA replication and fission. Early studies of domestic mouse crosses with *Mus spretus* found that F1 crosses were heteroplasmic while F2 pups reverted back to maternal mtDNA inheritance (Kaneda et al., 1995; Shitara, Hayashi, Takahama, Kaneda, & Yonekawa, 1998), likely because the recognition and elimination of paternal mitochondria after fertilization was dependent on the mitochondrial outer membrane proteins (or mitochondrial capsule) proteins encoded by nuclear DNA. Similarly, sperm mitochondrial sheaths are detectable past their natural lifespan inside hybrid embryos created by IVF of domestic cow oocyte with gaur *(Bos gaurus)* spermatozoa (Sutovsky et al., 2000). So far, the chaperone and regulatory proteins and enzymes associated with ubiquitin-dependent mitophagy do not appear to be highly divergent in their gene or protein sequences, as the UPS and ubiquitin-dependent autophagy are highly conserved. Pathways regulating the import of protein and nucleic acid degradation enzymes across outer mitochondrial membrane and into mitochondrial interior may provide further clues (Zuidema et al., 2023). The segregation of mitochondrial transcription and sustenance factors from sperm mitochondria during spermiogenesis also deserves further attention. The TFAM, mentioned above was found to undergo ubiquitination and segregate from sperm mitochondria into sperm head PAS and sperm tail principal piece in domestic pig model (Antelman et al., 2008). The failure of such segregation, potentially helping sperm mitochondria survive and multiply in the ooplasm, was mentioned as a possible contributor to transgenerational paternal mtDNA derived heteroplasmy in humans (Luo et al., 2018). Same pattern of TFAM segregation was recently "discovered" in human spermatozoa (Lee et al., 2023), while incorrectly hailed as a new, previously unknown mechanism. An early study of abnormal human blastocysts developed in vitro detected paternal mtDNA in some of them (St John et al., 2000), though it would be difficult to rule out that this could come from accessory spermatozoa on the exterior of such embryos.

7.1.5 Accessory sperm structures, sperm transcriptome and transgenerational inheritance

Besides the perinuclear theca, the fibrous sheath appears to be the fastest dissolving accessory sperm structure after fertilization. While it is known to harbor the A-kinase sequestering AKAP proteins to regulate motility (Carr & Newell, 2007), as well as components of glycolytic pathway to fuel it (Krisfalusi, Miki, Magyar, & O'Brien, 2006), little is known about other FS protein cargos that could exert paternal influence on early development.

These could include proteins involved in sequestration of small RNAs conveying paternal epigenetic effects and transgenerational paternal influences on offspring, resulting in paramutation (Chandler, 2007). Such a non-Mendelian inheritance could be enforced by the sperm delivered miRNAs targeting the ooplasmic mRNA encoding the enzymes involved with epigenome and histone code regulation in the zygote (Rodgers, Morgan, Leu, & Bale, 2015). Alternatively, the epigenome-regulating proteins could be sequestered in soluble sperm structures and activated in the zygote. One example of such paternally contributed protein is ten-eleven translocation 3 (TET3) protein, a methylcytosine dioxygenase present in fully differentiated spermatozoa but lacking from round spermatids grown in situ and also from spermatid-like haploid cells developed from induced pluripotent stem cells (iPSC) in an in vitro spermatogenesis culture system (Khampang et al., 2021). Such a mechanism could explain how substance abuse and other reprotoxic exposures of fathers could be vectored by the fertilizing spermatozoon to their sons' or even grandsons' epigenomes. Another unresolved issue is whether the sperm mRNAs could still be translated after ejaculation. This concerns specifically the mitochondrial genome as the mitochondrial matrix with its resident ribosomes could indeed rejuvenate the pool of mtDNA-encoded proteins during sperm transport in the female reproductive tract (Gur & Breitbart, 2006). Although the presence of various ribosomal subunit proteins in the spermatozoa has been established (Bisconti et al., 2022), there is yet to be conclusive evidence of protein synthesis in the sperm mitochondria offered, for example by the incorporation of precursor amino acids in said mitochondrial proteins in isolated, cultured sperm isolates free of somatic cell contamination.

7.2 Clinical aspects and biotechnological considerations

7.2.1 Putting Humpty-Dumpty back together

Aneuploidy and oocyte quality loss due to advancing reproductive age is perhaps the most recognized and researched contributing cause of human infertility (Biswas et al., 2021). Consequently, there have been concerted efforts to improve the developmental potential of eggs from patients of advanced reproductive age. Relevant to mitophagy and mitochondrial inheritance, this includes mitochondrial rejuvenation by transfer of healthy cytoplasm and mitochondria from eggs of prime reproductive age donors (Morimoto et al., 2023). Also, efforts are underway to boost autophagy, as a general way of promoting proteostasis, DNA repair and damaged organelle recycling (Sun et al., 2024), and perhaps in relation to post-fertilization sperm mitophagy as a way to ward off paternal heteroplasmy.

7.2.2 When ICSI goes wrong

Oocyte activation failure, associated with the ICSI procedure, could arise from several underlying causes. Among them, the persistence of the acrosome and subacrosomal PT after ICSI can interfere with the early stages of sperm nuclear decondensation and PPN development, as the subacrosomal PT has distinct protein composition and somewhat lesser solubility than equatorial and postacrosomal PT (Yu, Xu, Yi, Sutovsky, & Oko, 2006). The failure of acrosomal structure processing after ICSI as a potential contributor to sex chromosome abnormalities in ICSI children was recognized early after ICSI introduction into clinics, mainly because of studies conducted in rhesus monkey model for the study of ICSI mechanism (Luetjens, Payne, & Schatten, 1999; Sutovsky et al., 1996a). These failures can be mitigated by the use of piezo-driven ICSI injector which efficiently disrupts the acrosomal structure, while also stimulating the release of SOAF from the PAS and thusly inducing oocyte calcium oscillations (Yanagida et al., 2001). Piezo-driven ICSI can also overcome oocyte activation and sperm head decondensation failures in large animal species that are notoriously refractory to oocyte activation after conventional ICSI (Katayama et al., 2005). Also not clear is whether acrosomal exocytosis in domestic mice and other species truly starts in the oviduct before sperm contact with ZP, soon after acrosomal surface remodeling at sperm capacitation. Sperm capacitation is accompanied by the relief of zinc bridges in the sperm structures, and fluidization of plasma membrane over the acrosome, equatorial segment and postacrosomal sheath, which could promote SOAF release and acrosome processing after ICSI (Kerns et al., 2020; Zigo et al., 2022). In that case, routine in vitro capacitation (IVC) could help human ICSI outcomes. Molecular mechanism of sperm capacitation regulating the priming of SOAF for fertilization could also be exploited for the contraceptive purposes. Recent proof of concept study identifies the soluble adenylyl cyclase (sAC), a central enzyme of sperm capacitation signaling, as a plausible target of an on demand male oral contraceptive (Balbach et al., 2021).

7.2.3 SOAF dysfunction

Structural defects of the sperm nucleus and PT (nuclear vacuoles and craters/diadems) and molecular level defects such as insufficient amounts of

SOAF components or the SOAF-encoding genes being affected by deleterious genetic mutations [reviewed in (Sutovsky et al., 2024; Tirpak, Hamilton, Schnabel, & Sutovsky, 2024)] could cause activation failure by failing to release SOAF from PT, or by releasing inactive, damaged SOAF molecules in the ooplasm, respectively. Thus, recombinant SOAF proteins or their mRNAs could be coinjected with spermatozoa during ICSI to treat recurrent activation failure. Assisted oocyte activation by the sperm PLCZ1 coinjection has been proposed as a possible therapeutic measure to mitigate oocyte activation failure following ICSI, which could be done by the coinjection of mRNA, cRNA or recombinant protein with the spermatozoon (Parrella, Medrano, Aizpurua, & Gomez-Torres, 2024). On the other hand, simpler, micromanipulation avoiding chemical treatments, such as calcium ionophores appear to be reasonably safe and at least somewhat efficient (Cardona Barberan et al., 2023). Further to their therapeutic potential, the SOAF components should not be overlooked as biomarkers of sperm quality, fertility and pregnancy maintenance. For example, the WBP2NL/PAWP protein is present in a well-defined location (postacrosomal sheath) of morphologically normal spermatozoa while defective spermatozoa, particularly those with microcephalic or macrocephalic heads may have abnormally high (or low) levels of it, sometimes expressed ectopically outside of the PAS. Thus, sperm levels of WBP2NL correlate with fertility in bulls (Kennedy et al., 2014) and men (Aarabi et al., 2014b). The *Wbp2nl* gene is also expressed in the stem-cell derived male germ cells (Nourashrafeddin et al., 2014) and in certain types of breast cancers [e.g., most recent study by (Sanchez-Spitman et al., 2024)], which afford it a designation of cancer-testis antigen and breast cancer biomarker (Sutovsky et al., 2024).

7.2.4 Sperm-vectored influences on pregnancy establishment, fetal development and postnatal fitness.

Beyond direct contribution of organelles, proteins and RNAs, paternal inheritance is concerned with the intricacies of the paternal genomic and epigenomic influences on early embryo development, pregnancy establishment and pregnancy maintenance. Early pregnancy is the most sensitive period during which most of the pregnancy losses occur, affecting reproductive performance in the agriculturally important animals, and human reproductive health. This is in part attributable to dysregulation of imprinted genes expressed from paternal copy, and dosage and integrity of genes with biallelic expression during early development (Fallon et al.,

2023; Tirpak et al., 2024). This area of inquiry remains underappreciated, with misconceptions abounding about the parental origin of infertility, putting pressure on female partners in couples struggling with unexplained infertility. The health and integrity of paternally inherited, sperm-contributed elements can be reflected by sperm quality, as is the case of sperm SOAF content, DNA integrity, proteostasis and mitochondrial health (Sutovsky et al., 2024; Tirpak et al., 2024). Compromised, developmentally immature paternal genomes, epigenomes, organelles (centrosome, perinuclear theca, mitochondria), transcripts and proteins are also likely factors contributing to low success rate of round spermatid injection (ROSI), the method used to fertilize oocytes with haploid testicular cells biopsied from azoospermic patients (Niederberger, 2016), though an initial follow up survey of ICSI babies did not find major deviations from normal postnatal development within the first two years of age (Tanaka et al., 2018). As discussed above, the epigenetic history of paternal health (in utero and postnatal), fitness, reprotoxic exposures and even recreational drug abuse could be transmitted to next generation at fertilization, through sperm genome and epigenome (Lo et al., 2024). This is also the case of potential leakage and replication of paternal mitochondrial genome in the zygote, resulting in offspring heteroplasmy, which is commonly seen in cloned animals but may also occasionally occur in naturally conceived animals and humans (Song et al., 2014). Human paternal heteroplasmy, first described in 2002 (Schwartz & Vissing, 2002) then, after a prolonged hiatus confirmed and associated with a multi-generational heritable mitochondrial disease (Luo et al., 2018), could potentially contribute to pregnancy failure or postnatal reproductive health issues. Altogether, a better understanding of sperm transcriptome, epigenome and mitochondrial genome, as well as research on their post-fertilization processing will be important for detecting, managing and preventing aberrant paternal inheritance patterns in assisted reproductive therapy (ART).

7.2.5 Development blocks due to failed sperm structure recycling

Incomplete remodeling of paternal genome, epigenome and histone code, reflected by aberrant protamination and histone modification of the paternal DNA (Dogan et al., 2015; Stiavnicka et al., 2020) could trip zygotic cell cycle checkpoints, resulting in developmental roadblocks. Although permissive to oocyte activation and MPN development, low ooplasmic glutathione levels hinder the relief of disulfide bonds in the sperm nucleus after fertilization, causing the failure of PPN development

(Sutovsky & Schatten, 1997). Incorrect protamination, aberrant histone code and DNA fragmentation may not always prevent early development but eventually abort before or around the time of implantation, a major milestone associated with pregnancy failure. Such a pattern of successful fertilization coupled with failed repair of fragmented paternal DNA offers a possible explanation of paternally contributed recurrent pregnancy loss in human ART couples (Kennedy, Ahlering, Rodriguez, Levy, & Sutovsky, 2011; Naglot, Thapliyal, Tomar, & Yadav, 2023). On the other hand, altered spermatid proteostasis reflected in the retention of spermatid proteins in the sperm nuclei containing nuclear vacuoles and other DNA/chromatin defects may serve as a prognostic marker of successful embryo implantation after embryo transfer, enabling clinical decision making on single vs. multiple embryo transfer in human ART, to prevent unwanted multiple births (Ahlering et al., 2020).

Beyond post-fertilization sperm mitophagy, little is known about the incidence and pathology of failed sperm tail processing in the mammalian zygotes. We found evidence of sperm tail processing failures in the rat model of endometriosis, a painful, fertility reducing condition arising in humans from the ectopic intraperitoneal implantation of the shed uterine epithelia. In this transgenerational study, embryos produced by mating of the female rats with surgically induced endometriosis developed embryo cleavage delays and blocks at 6–8 cell stage. Some of the arrested embryos displayed sperm tails spanning two adjacent blastocysts that remained connected by microtubular midbody, assembled around the persisting axoneme (Fig. 8) (Stilley, Woods-Marshall, Sutovsky, Sutovsky, & Sharpe-Timms, 2009). Anomalies in this model also included meiotic/mitotic spindle and chromosome defects in oocytes and embryos, both paramount to the success of early embryo development. Persistence of the sperm axoneme was also shown in human embryos arrested at the zygote/1 cell stage (Asch et al., 1995) and 1 to 4-cell stage (Kai et al., 2024), though it is not clear if the persistence of sperm axonemes was a cause or a consequence of embryo cleavage failure. Low quality of in vitro matured or endometriosis affected oocytes, including insufficient GSH levels and elevated reactive oxygen species (ROS) could hinder timely degradation of sperm flagellum.

7.2.6 Centrosomal failure

Another form of human developmental arrest has been attributed to paternal centrosomal failure (Asch et al., 1995), while the subsequent attempts to restore sperm centriole function after ICSI failed. There was also a concern

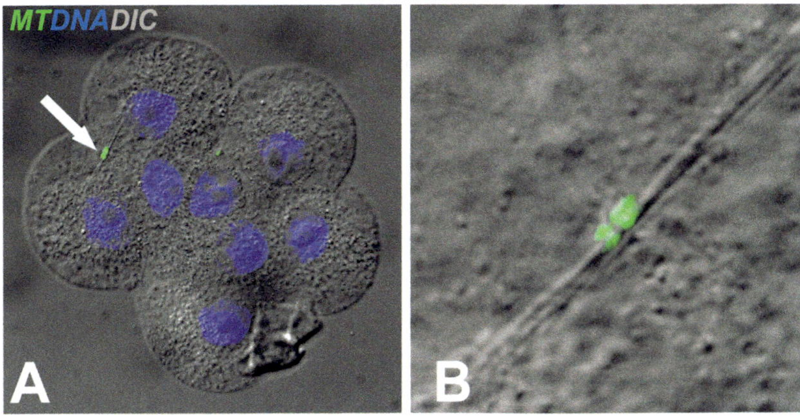

Fig. 8 Cleavage block in a 7-cell embryo flushed from the oviduct of a mated female rat with surgically induced endometriosis at a time point after mating corresponding to the blastocyst stage in control rats. (A) The sperm tail spans two adjacent blastomeres that remain connected by the microtubular midbody, assembled around the persisting sperm axoneme. (B) Detail of panel (A), with midbody microtubules (MT) labeled with anti-tubulin antibody and merged with a parfocal differential interference contrast (DIC) image. *Adapted with permission from Stilley et al. (2009).*

regarding ROSI success in overcoming the immaturity of spermatid centrosome, until the first validated births of ROSI babies were reported (Tanaka et al., 2015). Same concern is valid about spermatogenesis in vitro, the next frontier in the treatment of male infertility in which round spermatids can be generated but seldom advanced into elongation. Similar to biopsied spermatids of patients, those derived in vitro from patient-specific induced pluripotent stem cells (iPSC) will probably have a somatic-cell like centrosome. This spermatid-like cell centrosome may not function properly to promote pronuclear apposition after ROSI, thus depressing the rate of development to blastocyst, as studied in the rhesus monkey iPSC spermatid model (Khampang et al., 2021). Further progress in spermatid elongation in vitro is needed to increase the rate of blastocyst formation for embryo transfer. Meanwhile, there is some potential for the diagnostic value of the sperm centriole competence to form a sperm aster, applied to the tests based on human sperm exposure to frog egg extracts (Simerly et al., 1999) or interspecific human sperm-rabbit oocyte ICSI (Terada et al., 2000).

7.3 A male's worth

As shown in Fig. 1, the fertilizing spermatozoon contributes important organelles and molecules to the zygote. Besides delivering a haploid set of paternal chromosomes, the sperm-released SOAF provides the cue for oocyte activation, and the sperm-contributed proximal centriole serves as the scaffold for the formation of the zygotic centrosome. Meanwhile, the mitochondrial sheath and sperm tail accessory structures are degraded by the ooplasmic protein recycling systems, to avoid aberrant mitochondrial genome inheritance and mechanical interference with embryo cleavage. Thus, some sperm structures are transformed into zygotic components, while others are recycled to facilitate zygotic development. Less obvious are the contributions of structural (histones) and signaling proteins (kinases, transcription factors), and messenger and regulatory RNAs harbored by the sperm head perinuclear theca and sperm tail fibrous sheath. In a complementary manner, the minuscule sperm cell thus is the key to unlock the developmental potential of a much larger female gamete, the oocyte (Fig. 9) (Wolf, 1928), pre-stocked with organelles and molecules required

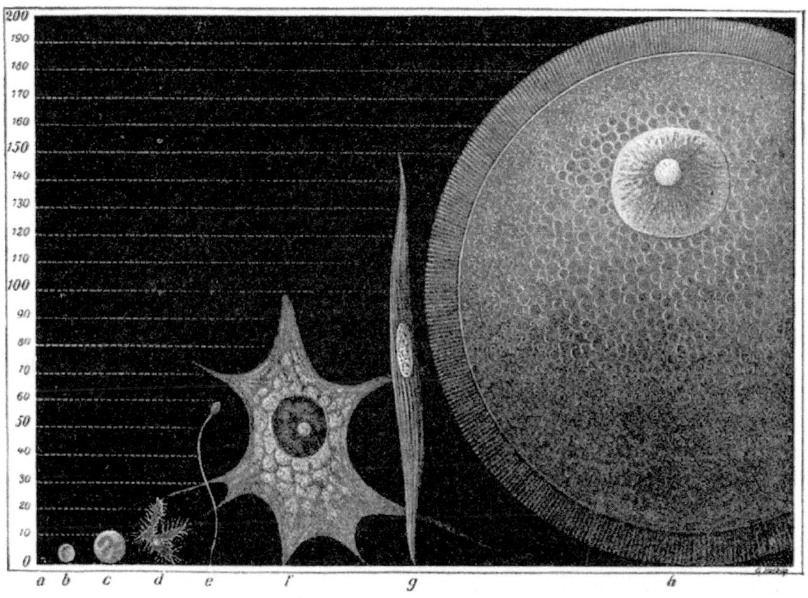

Fig. 9 Illustration from an early 20th century book comparing the size of a spermatozoon with that of an oocyte and assorted somatic cells. Gametogenesis results in the production of highly specialized haploid germ cells with unique features not found in any other cell type. The purpose of selective elimination of certain gametic organelles and of the remodeling of others during gametogenesis is to form male and female gametes with complementary endowments which give rise to a developmentally competent zygote at fertilization. *From Wolf (1928).*

for the onset and sustenance of embryo development up until the point of major zygotic genome activation. Our evolving understanding of paternal contributions to early embryo impacts the development and safeguarding of human ART, reproductive management in agriculturally important livestock species, and also the management of early pregnancy loss in both human and veterinary medicine. Long term impacts on fetal and postnatal health, and transgenerational effects imparted by sperm RNAs should also be considered.

Acknowledgments

We would like to acknowledge the support from the past and current members of our laboratories and countless external collaborators who contributed to studies reviewed in this article. Our research on paternal organelle inheritance is currently funded by the USDA NIFA Animal Reproduction Program grant number 2020–67015-31017 (PS). Seed funding for the Sutovsky laboratory is provided by the College of Agriculture, Food and Natural Resources, University of Missouri. Work on porcine models was made possible by the support received from the staff of the National Swine Research and Resource Center, University of Missouri, jointly funded by the National Institute of Allergy and Infectious Disease, the National Institute of Heart, Lung and Blood, and the Office of the Director (U42OD011140).

References

Aarabi, M., Balakier, H., Bashar, S., Moskovtsev, S. I., Sutovsky, P., Librach, C. L., & Oko, R. (2014a). Sperm-derived WW domain-binding protein, PAWP, elicits calcium oscillations and oocyte activation in humans and mice. *The FASEB Journal, 28*, 4434–4440.

Aarabi, M., Balakier, H., Bashar, S., Moskovtsev, S. I., Sutovsky, P., Librach, C. L., & Oko, R. (2014b). Sperm content of postacrosomal WW binding protein is related to fertilization outcomes in patients undergoing assisted reproductive technology. *Fertility and Sterility, 102*, 440–447.

Aarabi, M., Qin, Z., Xu, W., Mewburn, J., & Oko, R. (2010). Sperm-borne protein, PAWP, initiates zygotic development in Xenopus laevis by eliciting intracellular calcium release. *Molecular Reproduction and Development, 77*, 249–256.

Aarabi, M., Yu, Y., Xu, W., Tse, M. Y., Pang, S. C., Yi, Y. J., ... Oko, R. (2012). The testicular and epididymal expression profile of PLCzeta in mouse and human does not support its role as a sperm-borne oocyte activating factor. *PLoS One, 7*, e33496.

Ahlering, P., Sutovsky, M., Gliedt, D., Branson, K., Miranda Vizuete, A., & Sutovsky, P. (2020). Sperm content of TXNDC8 reflects sperm chromatin structure, pregnancy establishment, and incidence of multiple births after ART. *Systems Biology in Reproductive Medicine, 66*, 311–321.

Al Rawi, S., Louvet-Vallee, S., Djeddi, A., Sachse, M., Culetto, E., Hajjar, C., ... Galy, V. (2011). Postfertilization autophagy of sperm organelles prevents paternal mitochondrial DNA transmission. *Science (New York, N. Y.), 334*, 1144–1147.

Ankel-Simons, F., & Cummins, J. M. (1996). Misconceptions about mitochondria and mammalian fertilization: Implications for theories on human evolution. *Proceedings of the National Academy of Sciences of the United States of America, 93*, 13859–13863.

Antelman, J., Manandhar, G., Yi, Y. J., Li, R., Whitworth, K. M., Sutovsky, M., ... Sutovsky, P. (2008). Expression of mitochondrial transcription factor A (TFAM) during porcine gametogenesis and preimplantation embryo development. *Journal of Cellular Physiology, 217*, 529–543.

Aoki, F., Worrad, D. M., & Schultz, R. M. (1997). Regulation of transcriptional activity during the first and second cell cycles in the preimplantation mouse embryo. *Developmental Biology, 181*, 296–307.

Arrais, A. M., Burla Dias, A. J., de Souza, C. L. M., Curcio, A. G., & Mello, M. R. B. (2024). Inhibition of phospholipase C reduces the capacitation of cryopreserved ovine sperm. *Theriogenology, 213*, 19–23.

Asch, R., Simerly, C., Ord, T., Ord, V. A., & Schatten, G. (1995). The stages at which human fertilization arrests: Microtubule and chromosome configurations in inseminated oocytes which failed to complete fertilization and development in humans. *Human Reproduction (Oxford, England), 10*, 1897–1906.

Aul, R. B., & Oko, R. J. (2002). The major subacrosomal occupant of bull spermatozoa is a novel histone H2B variant associated with the forming acrosome during spermiogenesis. *Developmental Biology, 242*, 376–387.

Austin, C. R. (1952). The capacitation of the mammalian sperm. *Nature, 170*, 326.

Balbach, M., Ghanem, L., Rossetti, T., Kaur, N., Ritagliati, C., Ferreira, J., ... Levin, L. R. (2021). Soluble adenylyl cyclase inhibition prevents human sperm functions essential for fertilization. *Molecular Human Reproduction, 27*.

Balhorn, R. (1982). A model for the structure of chromatin in mammalian sperm. *The Journal of Cell Biology, 93*, 298–305.

Bansal, S. K., Gupta, N., Sankhwar, S. N., & Rajender, S. (2015). Differential genes expression between fertile and infertile spermatozoa revealed by transcriptome analysis. *PLoS One, 10*, e0127007.

Barth, A., Perry, V. E. A., Hamilton, L. E., Sutovsky, P., & Oko, R. (2025). The ultrastructure and composition of bovine spermatozoa. *Advances in Anatomy, Embryology, and Cell Biology, 240*, 1–63.

Bianchi, E., Doe, B., Goulding, D., & Wright, G. J. (2014). Juno is the egg Izumo receptor and is essential for mammalian fertilization. *Nature, 508*, 483–487.

Bianchi, E., & Wright, G. J. (2014). Izumo meets Juno: Preventing polyspermy in fertilization. *Cell Cycle (Georgetown, Tex.), 13*, 2019–2020.

Bisconti, M., Leroy, B., Gallagher, M. T., Senet, C., Martinet, B., Arcolia, V., ... Hennebert, E. (2022). The ribosome inhibitor chloramphenicol induces motility deficits in human spermatozoa: A proteomic approach identifies potentially involved proteins. *Frontiers in Cell and Developmental Biology, 10*, 965076.

Biswas, L., Tyc, K., El Yakoubi, W., Morgan, K., Xing, J., & Schindler, K. (2021). Meiosis interrupted: The genetics of female infertility via meiotic failure. *Reproduction (Cambridge, England), 161*, R13–R35.

Bjorndahl, L., & Kvist, U. (2011). A model for the importance of zinc in the dynamics of human sperm chromatin stabilization after ejaculation in relation to sperm DNA vulnerability. *Systems Biology in Reproductive Medicine, 57*, 86–92.

Bourc'his, D., & Voinnet, O. (2010). A small-RNA perspective on gametogenesis, fertilization, and early zygotic development. *Science (New York, N. Y.), 330*, 617–622.

Brown, C. J., Ballabio, A., Rupert, J. L., Lafreniere, R. G., Grompe, M., Tonlorenzi, R., & Willard, H. F. (1991). A gene from the region of the human X inactivation centre is expressed exclusively from the inactive X chromosome. *Nature, 349*, 38–44.

Burkart, A. D., Xiong, B., Baibakov, B., Jimenez-Movilla, M., & Dean, J. (2012). Ovastacin, a cortical granule protease, cleaves ZP2 in the zona pellucida to prevent polyspermy. *The Journal of Cell Biology, 197*, 37–44.

Burl, R. B., Clough, S., Sendler, E., Estill, M., & Krawetz, S. A. (2018). Sperm RNA elements as markers of health. *Systems Biology in Reproductive Medicine, 64*, 25–38.
Caballero-Campo, P., Lira-Albarrán, S., Barrera, D., Borja-Cacho, E., Godoy-Morales, H. S., Rangel-Escareño, C., ... Chirinos, M. (2020). Gene transcription profiling of astheno- and normo-zoospermic sperm subpopulations. *Asian Journal of Andrology, 22*, 608–615.
Campbell, K. D., Reed, W. A., & White, K. L. (2000). Ability of integrins to mediate fertilization, intracellular calcium release, and parthenogenetic development in bovine oocytes. *Biology of Reproduction, 62*, 1702–1709.
Cardona Barberan, A., Bonte, D., Boel, A., Thys, V., Paredis, R., Machtelinckx, F., ... Heindryckx, B. (2023). Assisted oocyte activation does not overcome recurrent embryo developmental problems. *Human Reproduction (Oxford, England), 38*, 872–885.
Carr, D. W., & Newell, A. E. (2007). The role of A-kinase anchoring proteins (AKaps) in regulating sperm function. *Society of Reproduction and Fertility Supplement, 63*, 135–141.
Chandler, V. L. (2007). Paramutation: From maize to mice. *Cell, 128*, 641–645.
Chang, M. C. (1951). Fertilizing capacity of spermatozoa deposited into the fallopian tubes. *Nature, 168*, 697–698.
Chen, H. Y., Sun, J. M., Zhang, Y., Davie, J. R., & Meistrich, M. L. (1998). Ubiquitination of histone H3 in elongating spermatids of rat testes. *The Journal of Biological Chemistry, 273*, 13165–13169.
Chen, Q., Yan, M., Cao, Z., Li, X., Zhang, Y., Shi, J., ... Zhou, Q. (2016). Sperm tsRNAs contribute to intergenerational inheritance of an acquired metabolic disorder. *Science (New York, N. Y.), 351*, 397–400.
Chen, X., Zheng, Y., Lei, A., Zhang, H., Niu, H., Li, X., ... Zeng, W. (2020). Early cleavage of preimplantation embryos is regulated by tRNA(Gln-TTG)-derived small RNAs present in mature spermatozoa. *The Journal of Biological Chemistry, 295*, 10885–10900.
Concha, I. I., Urzua, U., Yañez, A., Schroeder, R., Pessot, C., & Burzio, L. O. (1993). U1 and U2 snRNA are localized in the sperm nucleus. *Experimental Cell Research, 204*, 378–381.
Corral-Vazquez, C., Blanco, J., Aiese Cigliano, R., Sarrate, Z., Rivera-Egea, R., Vidal, F., ... Anton, E. (2021). The RNA content of human sperm reflects prior events in spermatogenesis and potential post-fertilization effects. *Molecular Human Reproduction, 27*.
Cui, L., Fang, L., Shi, B., Qiu, S., & Ye, Y. (2018). Spermatozoa expression of piR-31704, piR-39888, and piR-40349 and their correlation to sperm concentration and fertilization rate after ICSI. *Reproductive Sciences (Thousand Oaks, Calif.), 25*, 733–739.
da Silveira, J. C., de Ávila, A., Garrett, H. L., Bruemmer, J. E., Winger, Q. A., & Bouma, G. J. (2018). Cell-secreted vesicles containing microRNAs as regulators of gamete maturation. *The Journal of Endocrinology, 236*, R15–R27.
Dogan, S., Vargovic, P., Oliveira, R., Belser, L. E., Kaya, A., Moura, A., ... Memili, E. (2015). Sperm protamine-status correlates to the fertility of breeding bulls. *Biology of Reproduction, 92*, 92.
Donkin, I., Versteyhe, S., Ingerslev, L. R., Qian, K., Mechta, M., Nordkap, L., ... Barrès, R. (2016). Obesity and bariatric surgery drive epigenetic variation of spermatozoa in humans. *Cell Metabolism, 23*, 369–378.
Elzer, D., Bremser, M., & Zischler, H. (2023). Human sperm heads harbor modified YsRNA as transgenerationally inherited non-coding RNAs. *Frontiers in Genetics, 14*, 1294389.
Fallon, L., Diaz-Miranda, E., Hamilton, L., Sutovsky, P., Zigo, M., Spencer, T. E., & Ortega, M. S. (2023). The development of new biomarkers of spermatozoa quality in cattle. *Frontiers in Veterinary Science, 10*, 1258295.

Fenichel, P., & Durand-Clement, M. (1998). Role of integrins during fertilization in mammals. *Human Reproduction (Oxford, England), 13*(4), 31–46.

Fishman, E. L., Jo, K., Ha, A., Royfman, R., Zinn, A., Krishnamurthy, M., & Avidor-Reiss, T. (2017). Atypical centrioles are present in Tribolium sperm. *Open Biology, 7*.

Fishman, E. L., Jo, K., Nguyen, Q. P. H., Kong, D., Royfman, R., Cekic, A. R., ... Avidor-Reiss, T. (2018). A novel atypical sperm centriole is functional during human fertilization. *Nature Communications, 9*, 2210.

Flechon, J. E., & Kopecny, V. (1998). The nature of the 'nucleolus precursor body' in early preimplantation embryos: A review of fine-structure cytochemical, immunocytochemical and autoradiographic data related to nucleolar function. *Zygote (Cambridge, England), 6*, 183–191.

Florman, H. M., & Wassarman, P. M. (1985). O-linked oligosaccharides of mouse egg ZP3 account for its sperm receptor activity. *Cell, 41*, 313–324.

Fouquet, J. P., Kann, M. L., Combeau, C., & Melki, R. (1998). Gamma-tubulin during the differentiation of spermatozoa in various mammals and man. *Molecular Human Reproduction, 4*, 1122–1129.

Fukuoka, M., Kang, W., Katano, D., Horiike, S., Miyado, M., Tanaka, M., ... Yamada, M. (2023). Overdue calcium oscillation causes polyspermy but possibly permits normal development in mouse eggs. *International Journal of Molecular Sciences, 25*.

Fulka, H., & Langerova, A. (2014). The maternal nucleolus plays a key role in centromere satellite maintenance during the oocyte to embryo transition. *Development (Cambridge, England), 141*, 1694–1704.

Fulka, H., Rychtarova, J., & Loi, P. (2020). The nucleolus-like and precursor bodies of mammalian oocytes and embryos and their possible role in post-fertilization centromere remodelling. *Biochemical Society Transactions, 48*, 581–593.

Gamlin, L., & Vines, G. (1987). *The evolution of life*. New York: Oxford University Press.

Gao, S., Chung, Y. G., Parseghian, M. H., King, G. J., Adashi, E. Y., & Latham, K. E. (2004). Rapid H1 linker histone transitions following fertilization or somatic cell nuclear transfer: Evidence for a uniform developmental program in mice. *Developmental Biology, 266*, 62–75.

García-Herrero, S., Garrido, N., Martínez-Conejero, J. A., Remohí, J., Pellicer, A., & Meseguer, M. (2010). Ontological evaluation of transcriptional differences between sperm of infertile males and fertile donors using microarray analysis. *Journal of Assisted Reproduction and Genetics, 27*, 111–120.

Garrido, N., Martínez-Conejero, J. A., Jauregui, J., Horcajadas, J. A., Simón, C., Remohí, J., & Meseguer, M. (2009). Microarray analysis in sperm from fertile and infertile men without basic sperm analysis abnormalities reveals a significantly different transcriptome. *Fertility and Sterility, 91*, 1307–1310.

Giebler, M., Greither, T., Müller, L., Mösinger, C., & Behre, H. M. (2018). Altered PIWI-LIKE 1 and PIWI-LIKE 2 mRNA expression in ejaculated spermatozoa of men with impaired sperm characteristics. *Asian Journal of Andrology, 20*, 260–264.

Gòdia, M., Swanson, G., & Krawetz, S. A. (2018). A history of why fathers' RNA matters. *Biology of Reproduction, 99*, 147–159.

Gonzalez-Castro, R. A., Amoroso-Sanches, F., Stokes, J. E., Graham, J. K., & Carnevale, E. M. (2019). Localisation of phospholipase Czeta1 (PLCZ1) and postacrosomal WW-binding protein (WBP2 N-terminal like) on equine spermatozoa and flow cytometry quantification of PLCZ1 and association with cleavage in vitro. *Reproduction, Fertility, and Development, 31*, 1778–1792.

Goodrich, R. J., Anton, E., & Krawetz, S. A. (2013). Isolating mRNA and small noncoding RNAs from human sperm. *Methods in Molecular Biology, 927*, 385–396.

Goto, M., O'Brien, D. A., & Eddy, E. M. (2010). Speriolin is a novel human and mouse sperm centrosome protein. *Human Reproduction (Oxford, England), 25*, 1884–1894.

Gresson, R. A. R. (1940). Presence of the sperm middle-piece in the fertilized egg of the mouse (Mus musculus). *Nature*, 425.

Guidobaldi, H. A., Teves, M. E., Unates, D. R., Anastasia, A., & Giojalas, L. C. (2008). Progesterone from the cumulus cells is the sperm chemoattractant secreted by the rabbit oocyte cumulus complex. *PLoS One, 3*, e3040.

Guo, Y., Bai, D., Liu, W., Liu, Y., Zhang, Y., Kou, X., ... Gao, S. (2021). Altered sperm tsRNAs in aged male contribute to anxiety-like behavior in offspring. *Aging Cell, 20*, e13466.

Guo, F., Li, X., Liang, D., Li, T., Zhu, P., Guo, H., ... Xu, G. L. (2014). Active and passive demethylation of male and female pronuclear DNA in the mammalian zygote. *Cell Stem Cell, 15*, 447–459.

Gur, Y., & Breitbart, H. (2006). Mammalian sperm translate nuclear-encoded proteins by mitochondrial-type ribosomes. *Genes & Development, 20*, 411–416.

Hachem, A., Godwin, J., Ruas, M., Lee, H. C., Ferrer Buitrago, M., Ardestani, G., ... Parrington, J. (2017). PLCzeta is the physiological trigger of the Ca(2+) oscillations that induce embryogenesis in mammals but conception can occur in its absence. *Development (Cambridge, England), 144*, 2914–2924.

Hamilton, L. E., Lion, M., Aguila, L., Suzuki, J., Acteau, G., Protopapas, N., ... Oko, R. (2021). Core histones are constituents of the perinuclear theca of murid spermatozoa: An assessment of their synthesis and assembly during spermiogenesis and function after gametic fusion. *International Journal of Molecular Sciences, 22*.

Hamilton, L. E., Suzuki, J., Acteau, G., Shi, M., Xu, W., Meinsohn, M. C., ... Oko, R. (2018). WBP2 shares a common location in mouse spermatozoa with WBP2NL/PAWP and like its descendent is a candidate mouse oocyte-activating factor. *Biology of Reproduction, 99*, 1171–1183.

Hamilton, L. E., Zigo, M., Mao, J., Xu, W., Sutovsky, P., O'Flaherty, C., & Oko, R. (2019). GSTO2 isoforms participate in the oxidative regulation of the plasmalemma in eutherian spermatozoa during capacitation. *Antioxidants (Basel), 8*.

Hammoud, S. S., Nix, D. A., Zhang, H., Purwar, J., Carrell, D. T., & Cairns, B. R. (2009). Distinctive chromatin in human sperm packages genes for embryo development. *Nature, 460*, 473–478.

Han, T., Manoharan, A. P., Harkins, T. T., Bouffard, P., Fitzpatrick, C., Chu, D. S., ... Kim, J. K. (2009). 26G endo-siRNAs regulate spermatogenic and zygotic gene expression in Caenorhabditis elegans. *Proceedings of the National Academy of Sciences of the United States of America, 106*, 18674–18679.

Hao, Y., Mathialagan, N., Walters, E., Mao, J., Lai, L., Becker, D., ... Prather, R. S. (2006). Osteopontin reduces polyspermy during in vitro fertilization of porcine oocytes. *Biology of Reproduction, 75*, 726–733.

Harada, Y., Matsumoto, T., Hirahara, S., Nakashima, A., Ueno, S., Oda, S., ... Iwao, Y. (2007). Characterization of a sperm factor for egg activation at fertilization of the newt Cynops pyrrhogaster. *Developmental Biology, 306*, 797–808.

Hernandez, R., Li, X., Shi, J., Dave, T. R., Zhou, T., Chen, Q., & Zhou, C. (2024). Paternal hypercholesterolemia elicits sex-specific exacerbation of atherosclerosis in offspring. *JCI Insight, 9*.

Hernández-Silva, G., Caballero-Campo, P., & Chirinos, M. (2022). Sperm mRNAs as potential markers of male fertility. *Reproductive Biology, 22*, 100636.

Herrada, G., & Wolgemuth, D. J. (1997). The mouse transcription factor Stat4 is expressed in haploid male germ cells and is present in the perinuclear theca of spermatozoa. *Journal of Cell Science, 110*(Pt 14), 1543–1553.

Hino, T., Muro, Y., Tamura-Nakano, M., Okabe, M., Tateno, H., & Yanagimachi, R. (2016). The behavior and acrosomal status of mouse spermatozoa in vitro, and within the oviduct during fertilization after natural mating. *Biology of Reproduction, 95*, 50.

Ho, L., & Crabtree, G. R. (2010). Chromatin remodelling during development. *Nature, 463*, 474–484.

Hua, M., Liu, W., Chen, Y., Zhang, F., Xu, B., Liu, S., ... Wu, L. (2019). Identification of small non-coding RNAs as sperm quality biomarkers for in vitro fertilization. *Cell Discovery, 5*, 20.

Ickowicz, D., Finkelstein, M., & Breitbart, H. (2012). Mechanism of sperm capacitation and the acrosome reaction: Role of protein kinases. *Asian Journal of Andrology, 14*, 816–821.

Igarashi, H., Knott, J. G., Schultz, R. M., & Williams, C. J. (2007). Alterations of PLCbeta1 in mouse eggs change calcium oscillatory behavior following fertilization. *Developmental Biology, 312*, 321–330.

Inoue, N., Ikawa, M., Isotani, A., & Okabe, M. (2005). The immunoglobulin superfamily protein Izumo is required for sperm to fuse with eggs. *Nature, 434*, 234–238.

Ito, M., Shikano, T., Oda, S., Horiguchi, T., Tanimoto, S., Awaji, T., ... Miyazaki, S. (2008). Difference in Ca2+ oscillation-inducing activity and nuclear translocation ability of PLCZ1, an egg-activating sperm factor candidate, between mouse, rat, human, and medaka fish. *Biology of Reproduction, 78*, 1081–1090.

Jeseta, M., Marin, M., Tichovska, H., Melicharova, P., Cailliau-Maggio, K., Martoriati, A., ... Bodart, J. F. (2012). Nitric oxide-donor SNAP induces Xenopus eggs activation. *PLoS One, 7*, e41509.

Jiang, G. J., Zhang, T., An, T., Zhao, D. D., Yang, X. Y., Zhang, D. W., ... Gao, S. H. (2016). Differential expression of long noncoding RNAs between sperm samples from diabetic and non-diabetic mice. *PLoS One, 11*, e0154028.

Jodar, M., Kalko, S., Castillo, J., Ballescà, J. L., & Oliva, R. (2012). Differential RNAs in the sperm cells of asthenozoospermic patients. *Human Reproduction (Oxford, England), 27*, 1431–1438.

Jodar, M., Selvaraju, S., Sendler, E., Diamond, M. P., & Krawetz, S. A. (2013). The presence, role and clinical use of spermatozoal RNAs. *Human Reproduction Update, 19*, 604–624.

Jodar, M., Sendler, E., & Krawetz, S. A. (2016). The protein and transcript profiles of human semen. *Cell and Tissue Research, 363*, 85–96.

Jodar, M., Sendler, E., Moskovtsev, S. I., Librach, C. L., Goodrich, R., Swanson, S., ... Krawetz, S. A. (2015). Absence of sperm RNA elements correlates with idiopathic male infertility. *Science Translational Medicine, 7* 295re296.

Johnson, G. D., Lalancette, C., Linnemann, A. K., Leduc, F., Boissonneault, G., & Krawetz, S. A. (2011a). The sperm nucleus: Chromatin, RNA, and the nuclear matrix. *Reproduction (Cambridge, England), 141*, 21–36.

Johnson, G. D., Mackie, P., Jodar, M., Moskovtsev, S., & Krawetz, S. A. (2015). Chromatin and extracellular vesicle associated sperm RNAs. *Nucleic Acids Research, 43*, 6847–6859.

Johnson, G. D., Sendler, E., Lalancette, C., Hauser, R., Diamond, M. P., & Krawetz, S. A. (2011b). Cleavage of rRNA ensures translational cessation in sperm at fertilization. *Molecular Human Reproduction, 17*, 721–726.

Joshi, M., Sethi, S., Mehta, P., Kumari, A., & Rajender, S. (2023). Small RNAs, spermatogenesis, and male infertility: A decade of retrospect. *Reproductive Biology and Endocrinology: RB&E, 21*, 106.

Kai, Y., Iseki, H., Matsui, T., & Yamashita, N. (2024). Morphological analysis of the human sperm tail during the early post-fertilization stage. *Reproduction (Cambridge, England), 167*.

Kaneda, H., Hayashi, J., Takahama, S., Taya, C., Lindahl, K. F., & Yonekawa, H. (1995). Elimination of paternal mitochondrial DNA in intraspecific crosses during early mouse embryogenesis. *Proceedings of the National Academy of Sciences of the United States of America, 92*, 4542–4546.

Kaneda, Y., Miyata, H., Xu, Z., Shimada, K., Kamoshita, M., Nakagawa, T., ... Ikawa, M. (2024). FBXO24 deletion causes abnormal accumulation of membraneless electron-dense granules in sperm flagella and male infertility. *Elife, 13.*

Kang, W., Harada, Y., Yamatoya, K., Kawano, N., Kanai, S., Miyamoto, Y., ... Miyado, K. (2020). Extra-mitochondrial citrate synthase initiates calcium oscillation and suppresses age-dependent sperm dysfunction. *Laboratory Investigation; A Journal of Technical Methods and Pathology, 100*, 583–595.

Katayama, M., Sutovsky, P., Yang, B. S., Cantley, T., Rieke, A., Farwell, R., ... Day, B. N. (2005). Increased disruption of sperm plasma membrane at sperm immobilization promotes dissociation of perinuclear theca from sperm chromatin after intracytoplasmic sperm injection in pigs. *Reproduction (Cambridge, England), 130*, 907–916.

Kennedy, C., Ahlering, P., Rodriguez, H., Levy, S., & Sutovsky, P. (2011). Sperm chromatin structure correlates with spontaneous abortion and multiple pregnancy rates in assisted reproduction. *Reproductive Biomedicine Online, 22*, 272–276.

Kennedy, C. E., Krieger, K. B., Sutovsky, M., Xu, W., Vargovic, P., Didion, B. A., ... Sutovsky, P. (2014). Protein expression pattern of PAWP in bull spermatozoa is associated with sperm quality and fertility following artificial insemination. *Molecular Reproduction and Development, 81*, 436–449.

Kerns, K., Sharif, M., Zigo, M., Xu, W., Hamilton, L. E., Sutovsky, M., ... Sutovsky, P. (2020). Sperm cohort-specific zinc signature acquisition and capacitation-induced zinc flux regulate sperm-oviduct and sperm-zona pellucida interactions. *International Journal of Molecular Sciences, 21.*

Kerns, K., Zigo, M., Drobnis, E. Z., Sutovsky, M., & Sutovsky, P. (2018). Zinc ion flux during mammalian sperm capacitation. *Nature Communications, 9*, 2061.

Khampang, S., Cho, I. K., Punyawai, K., Gill, B., Langmo, J. N., Nath, S., ... Easley, C. A. T. (2021). Blastocyst development after fertilization with in vitro spermatids derived from nonhuman primate embryonic stem cells. *F&S Science, 2*, 365–375.

Kim, A. M., Bernhardt, M. L., Kong, B. Y., Ahn, R. W., Vogt, S., Woodruff, T. K., & O'Halloran, T. V. (2011). Zinc sparks are triggered by fertilization and facilitate cell cycle resumption in mammalian eggs. *ACS Chemical Biology, 6*, 716–723.

Kimura, Y., Yanagimachi, R., Kuretake, S., Bortkiewicz, H., Perry, A. C., & Yanagimachi, H. (1998). Analysis of mouse oocyte activation suggests the involvement of sperm perinuclear material. *Biology of Reproduction, 58*, 1407–1415.

Koh, S., Lee, K., Wang, C., Cabot, R. A., & Machaty, Z. (2009). STIM1 regulates store-operated Ca2+ entry in oocytes. *Developmental Biology, 330*, 368–376.

Komondor, K. M., Bainbridge, R. E., Sharp, K. G., Iyer, A. R., Rosenbaum, J. C., & Carlson, A. E. (2023). TMEM16A activation for the fast block to polyspermy in the African clawed frog does not require conventional activation of egg PLCs. *The Journal of General Physiology, 155.*

Kramer, P., & Bressan, P. (2019). Mitochondria inspire a lifestyle. *Advances in Anatomy, Embryology, and Cell Biology, 231*, 105–126.

Krawetz, S. A., Kruger, A., Lalancette, C., Tagett, R., Anton, E., Draghici, S., & Diamond, M. P. (2011). A survey of small RNAs in human sperm. *Human Reproduction (Oxford, England), 26*, 3401–3412.

Krisfalusi, M., Miki, K., Magyar, P. L., & O'Brien, D. A. (2006). Multiple glycolytic enzymes are tightly bound to the fibrous sheath of mouse spermatozoa. *Biology of Reproduction, 75*, 270–278.

Lachance, C., & Leclerc, P. (2011). Mediators of the Jak/STAT signaling pathway in human spermatozoa. *Biology of Reproduction, 85*, 1222–1231.

Ladstatter, S., & Tachibana-Konwalski, K. (2016). A surveillance mechanism ensures repair of DNA lesions during zygotic reprogramming. *Cell, 167*, 1774–1787.e1713.

Lalancette, C., Platts, A. E., Johnson, G. D., Emery, B. R., Carrell, D. T., & Krawetz, S. A. (2009). Identification of human sperm transcripts as candidate markers of male fertility. *Journal of Molecular Medicine (Berl), 87*, 735–748.

Lee, R. C., Feinbaum, R. L., & Ambros, V. (1993). The C. elegans heterochronic gene lin-4 encodes small RNAs with antisense complementarity to lin-14. *Cell, 75*, 843–854.

Lee, W., Zamudio-Ochoa, A., Buchel, G., Podlesniy, P., Marti Gutierrez, N., Puigros, M., ... Temiakov, D. (2023). Molecular basis for maternal inheritance of human mitochondrial DNA. *Nature Genetics, 55*, 1632–1639.

Li, Y., Li, M., Liu, Y., Song, G., & Liu, N. (2012). A microarray for microRNA profiling in spermatozoa from adult men living in an environmentally polluted site. *Bulletin of Environmental Contamination and Toxicology, 89*, 1111–1114.

Li, T., Vu, T. H., Lee, K. O., Yang, Y., Nguyen, C. V., Bui, H. Q., ... Hoffman, A. R. (2002). An imprinted PEG1/MEST antisense expressed predominantly in human testis and in mature spermatozoa. *The Journal of Biological Chemistry, 277*, 13518–13527.

Liang, M., Li, W., Tian, H., Hu, T., Wang, L., Lin, Y., ... Sun, F. (2014). Sequential expression of long noncoding RNA as mRNA gene expression in specific stages of mouse spermatogenesis. *Scientific Reports, 4*, 5966.

Lin, H. (2007). piRNAs in the germ line. *Science (New York, N. Y.), 316*, 397.

Lo, J. O., Hedges, J. C., Chou, W. H., Tager, K. R., Bachli, I. D., Hagen, O. L., ... Easley, C. A. T. (2024). Influence of substance use on male reproductive health and offspring outcomes. *Nature Reviews Urology, 21*, 534–564.

Londono-Vasquez, D., Rodriguez-Lukey, K., Behura, S. K., & Balboula, A. Z. (2022). Microtubule organizing centers regulate spindle positioning in mouse oocytes. *Developmental Cell, 57*, 197–211 e193.

Longo, F. J. (1989). Incorporation and dispersal of sperm surface antigens in plasma membranes of inseminated sea urchin (Arbacia punctulata) eggs and oocytes. *Developmental Biology, 131*, 37–43.

Luetjens, C. M., Payne, C., & Schatten, G. (1999). Non-random chromosome positioning in human sperm and sex chromosome anomalies following intracytoplasmic sperm injection. *Lancet, 353*, 1240.

Lui, L., & Lowe, T. (2013). Small nucleolar RNAs and RNA-guided post-transcriptional modification. *Essays in Biochemistry, 54*, 53–77.

Luo, S., Valencia, C. A., Zhang, J., Lee, N. C., Slone, J., Gui, B., ... Huang, T. (2018). Biparental inheritance of mitochondrial DNA in humans. *Proceedings of the National Academy of Sciences of the United States of America, 115*, 13039–13044.

Ma, Z., Tang, N., Zhang, R., Deng, H., Chen, K., Liu, Y., & Ding, Z. (2024). Ribonuclease inhibitor 1 (RNH1) regulates sperm tsRNA generation for paternal inheritance through interacting with angiogenin in the caput epididymis. *Antioxidants (Basel), 13*.

Manandhar, G., Schatten, H., & Sutovsky, P. (2005). Centrosome reduction during gametogenesis and its significance. *Biology of Reproduction, 72*, 2–13.

Manandhar, G., Sutovsky, P., Joshi, H. C., Stearns, T., & Schatten, G. (1998). Centrosome reduction during mouse spermiogenesis. *Developmental Biology, 203*, 424–434.

Mao, J., O'Gorman, C., Sutovsky, M., Zigo, M., Wells, K. D., & Sutovsky, P. (2018). Ubiquitin A-52 residue ribosomal protein fusion product 1 (Uba52) is essential for preimplantation embryo development. *Biology Open, 7*.

Mao, J., Zigo, M., Zuidema, D., Sutovsky, M., & Sutovsky, P. (2021). NEDD4-like ubiquitin ligase 2 protein (NEDL2) in porcine spermatozoa, oocytes, and preimplantation embryos and its role in oocyte fertilizationdagger. *Biology of Reproduction, 104*, 117–129.

Marczylo, E. L., Amoako, A. A., Konje, J. C., Gant, T. W., & Marczylo, T. H. (2012). Smoking induces differential miRNA expression in human spermatozoa: A potential transgenerational epigenetic concern? *Epigenetics: Official Journal of the DNA Methylation Society, 7*, 432–439.

Massana, R. (2024). The nitroplast: A nitrogen-fixing organelle. *Science (New York, N. Y.), 384*, 160–161.
McLay, D. W., & Clarke, H. J. (2003). Remodelling the paternal chromatin at fertilization in mammals. *Reproduction (Cambridge, England), 125*, 625–633.
Metzler-Guillemain, C., Victorero, G., Lepoivre, C., Bergon, A., Yammine, M., Perrin, J., ... Nguyen, C. (2015). Sperm mRNAs and microRNAs as candidate markers for the impact of toxicants on human spermatogenesis: An application to tobacco smoking. *Systems Biology in Reproductive Medicine, 61*, 139–149.
Miller, D., Briggs, D., Snowden, H., Hamlington, J., Rollinson, S., Lilford, R., & Krawetz, S. A. (1999). A complex population of RNAs exists in human ejaculate spermatozoa: Implications for understanding molecular aspects of spermiogenesis. *Gene, 237*, 385–392.
Molina, P., Lim, Y., & Boyd, L. (2019). Ubiquitination is required for the initial removal of paternal organelles in C. elegans. *Developmental Biology, 453*, 168–179.
Montjean, D., De La Grange, P., Gentien, D., Rapinat, A., Belloc, S., Cohen-Bacrie, P., ... Benkhalifa, M. (2012). Sperm transcriptome profiling in oligozoospermia. *Journal of Assisted Reproduction and Genetics, 29*, 3–10.
Moore, C. R. (1916). On the superposition of fertilization in parthenogenesis. *Biological Bulletin, 31*, 137–180.
Morimoto, Y., Gamage, U. S. K., Yamochi, T., Saeki, N., Morimoto, N., Yamanaka, M., ... Yanagimachi, R. (2023). Mitochondrial transfer into human oocytes improved embryo quality and clinical outcomes in recurrent pregnancy failure cases. *International Journal of Molecular Sciences, 24*.
Naglot, S., Thapliyal, A., Tomar, A. K., & Yadav, S. (2023). Male contributory factors in recurrent pregnancy loss. *Reproductive Sciences (Thousand Oaks, Calif.), 30*, 2107–2121.
Nguyen, M. T., Delaney, D. P., & Kolon, T. F. (2009). Gene expression alterations in cryptorchid males using spermatozoal microarray analysis. *Fertility and Sterility, 92*, 182–187.
Niederberger, C. (2016). Re: Fourteen babies born after round spermatid injection into human oocytes. *The Journal of Urology, 195*, 1566.
Nourashrafeddin, S., Aarabi, M., Miryounesi, M., Ebrahimzadeh-Vesal, R., Zarghami, N., Modarressi, M. H., & Nouri, M. (2014). Expression analysis of PAWP during mouse embryonic stem cell-based spermatogenesis in vitro. *In Vitro Cellular & Developmental Biology. Animal, 50*, 475–481.
Nozawa, K., Satouh, Y., Fujimoto, T., Oji, A., & Ikawa, M. (2018). Sperm-borne phospholipase C zeta-1 ensures monospermic fertilization in mice. *Scientific Reports, 8*, 1315.
Oko, R., & Sutovsky, P. (2009). Biogenesis of sperm perinuclear theca and its role in sperm functional competence and fertilization. *Journal of Reproductive Immunology, 83*, 2–7.
Ostermeier, G. C., Dix, D. J., Miller, D., Khatri, P., & Krawetz, S. A. (2002). Spermatozoal RNA profiles of normal fertile men. *Lancet, 360*, 772–777.
Ostermeier, G. C., Goodrich, R. J., Moldenhauer, J. S., Diamond, M. P., & Krawetz, S. A. (2005). A suite of novel human spermatozoal RNAs. *Journal of Andrology, 26*, 70–74.
Ozil, J. P., Markoulaki, S., Toth, S., Matson, S., Banrezes, B., Knott, J. G., ... Ducibella, T. (2005). Egg activation events are regulated by the duration of a sustained [Ca2+]cyt signal in the mouse. *Developmental Biology, 282*, 39–54.
Pantano, L., Jodar, M., Bak, M., Ballescà, J. L., Tommerup, N., Oliva, R., & Vavouri, T. (2015). The small RNA content of human sperm reveals pseudogene-derived piRNAs complementary to protein-coding genes. *RNA (New York, N. Y.), 21*, 1085–1095.
Parrella, A., Medrano, L., Aizpurua, J., & Gomez-Torres, M. J. (2024). Phospholipase C zeta in human spermatozoa: A systematic review on current development and clinical application. *International Journal of Molecular Sciences, 25*.

Parrington, J., Jones, K. T., Lai, A., & Swann, K. (1999). The soluble sperm factor that causes Ca2+ release from sea-urchin (Lytechinus pictus) egg homogenates also triggers Ca2+ oscillations after injection into mouse eggs. *The Biochemical Journal, 341*(Pt 1), 1–4.

Parrington, J., Swann, K., Shevchenko, V. I., Sesay, A. K., & Lai, F. A. (1996). Calcium oscillations in mammalian eggs triggered by a soluble sperm protein. *Nature, 379*, 364–368.

Peng, J. C., & Lin, H. (2013). Beyond transposons: The epigenetic and somatic functions of the Piwi-piRNA mechanism. *Current Opinion in Cell Biology, 25*, 190–194.

Peng, H., Shi, J., Zhang, Y., Zhang, H., Liao, S., Li, W., ... Duan, E. (2012). A novel class of tRNA-derived small RNAs extremely enriched in mature mouse sperm. *Cell Research, 22*, 1609–1612.

Perreault, S. D., Barbee, R. R., & Slott, V. L. (1988). Importance of glutathione in the acquisition and maintenance of sperm nuclear decondensing activity in maturing hamster oocytes. *Developmental Biology, 125*, 181–186.

Perreault, S. D., Wolff, R. A., & Zirkin, B. R. (1984). The role of disulfide bond reduction during mammalian sperm nuclear decondensation in vivo. *Developmental Biology, 101*, 160–167.

Perry, A. C., Wakayama, T., & Yanagimachi, R. (1999). A novel trans-complementation assay suggests full mammalian oocyte activation is coordinately initiated by multiple, submembrane sperm components. *Biology of Reproduction, 60*, 747–755.

Perry, A. C., Wakayama, T., Cooke, I. M., & Yanagimachi, R. (2000). Mammalian oocyte activation by the synergistic action of discrete sperm head components: Induction of calcium transients and involvement of proteolysis. *Developmental Biology, 217*, 386–393.

Platts, A. E., Dix, D. J., Chemes, H. E., Thompson, K. E., Goodrich, R., Rockett, J. C., ... Krawetz, S. A. (2007). Success and failure in human spermatogenesis as revealed by teratozoospermic RNAs. *Human Molecular Genetics, 16*, 763–773.

Qiu, J. J., Ren, Z. R., & Yan, J. B. (2016). Identification and functional analysis of long non-coding RNAs in human and mouse early embryos based on single-cell transcriptome data. *Oncotarget, 7*, 61215–61228.

Rawe, V. Y., Olmedo, S. B., Nodar, F. N., Ponzio, R., & Sutovsky, P. (2003). Abnormal assembly of annulate lamellae and nuclear pore complexes coincides with fertilization arrest at the pronuclear stage of human zygotic development. *Human Reproduction (Oxford, England), 18*, 576–582.

Reik, W., & Walter, J. (2001). Genomic imprinting: Parental influence on the genome. *Nature Reviews. Genetics, 2*, 21–32.

Reuter, M., Berninger, P., Chuma, S., Shah, H., Hosokawa, M., Funaya, C., ... Pillai, R. S. (2011). Miwi catalysis is required for piRNA amplification-independent LINE1 transposon silencing. *Nature, 480*, 264–267.

Rodgers, A. B., Morgan, C. P., Leu, N. A., & Bale, T. L. (2015). Transgenerational epigenetic programming via sperm microRNA recapitulates effects of paternal stress. *Proceedings of the National Academy of Sciences of the United States of America, 112*, 13699–13704.

Ross, P. J., & Canovas, S. (2016). Mechanisms of epigenetic remodelling during preimplantation development. *Reproduction, Fertility, and Development, 28*, 25–40.

Sahlu, B. W., Zhao, S., Wang, X., Umer, S., Zou, H., Huang, J., & Zhu, H. (2020). Long noncoding RNAs: New insights in modulating mammalian spermatogenesis. *Journal of Animal Science and Biotechnology, 11*, 16.

Salas-Huetos, A., Blanco, J., Vidal, F., Mercader, J. M., Garrido, N., & Anton, E. (2014). New insights into the expression profile and function of micro-ribonucleic acid in human spermatozoa. *Fertility and Sterility, 102*, 213–222.e214.

Salas-Huetos, A., Blanco, J., Vidal, F., Godo, A., Grossmann, M., Pons, M. C., ... Anton, E. (2015). Spermatozoa from patients with seminal alterations exhibit a differential micro-ribonucleic acid profile. *Fertility and Sterility, 104*, 591–601.

Salas-Huetos, A., Blanco, J., Vidal, F., Grossmann, M., Pons, M. C., Garrido, N., & Anton, E. (2016). Spermatozoa from normozoospermic fertile and infertile individuals convey a distinct miRNA cargo. *Andrology, 4*, 1028–1036.

Salas-Huetos, A., Moraleda, R., Giardina, S., Anton, E., Blanco, J., Salas-Salvadó, J., & Bulló, M. (2018). Effect of nut consumption on semen quality and functionality in healthy men consuming a Western-style diet: a randomized controlled trial. *The American Journal of Clinical Nutrition, 108*, 953–962.

Sanchez-Spitman, A. B., Bohringer, S., Dezentje, V. O., Gelderblom, H., Swen, J. J., & Guchelaar, H. J. (2024). A genome-wide association study of endoxifen serum concentrations and adjuvant tamoxifen efficacy in early-stage breast cancer patients. *Clinical Pharmacology and Therapeutics, 116*, 155–164.

Santiago, J., Silva, J. V., Howl, J., Santos, M. A. S., & Fardilha, M. (2021). All you need to know about sperm RNAs. *Human Reproduction Update, 28*, 67–91.

Sato, M., & Sato, K. (2011). Degradation of paternal mitochondria by fertilization-triggered autophagy in C. elegans embryos. *Science (New York, N. Y.), 334*, 1141–1144.

Satouh, Y., Nozawa, K., & Ikawa, M. (2015). Sperm postacrosomal WW domain-binding protein is not required for mouse egg activation. *Biology of Reproduction, 93*, 94.

Saunders, C. M., Larman, M. G., Parrington, J., Cox, L. J., Royse, J., Blayney, L. M., ... Lai, F. A. (2002). PLC zeta: A sperm-specific trigger of Ca(2+) oscillations in eggs and embryo development. *Development (Cambridge, England), 129*, 3533–3544.

Schatten, H. (2022). Centrosomes in reproduction. *Advances in Anatomy, Embryology, and Cell Biology, 235*, 55–73.

Schatten, G., Simerly, C., & Schatten, H. (1985). Microtubule configurations during fertilization, mitosis, and early development in the mouse and the requirement for egg microtubule-mediated motility during mammalian fertilization. *Proceedings of the National Academy of Sciences of the United States of America, 82*, 4152–4156.

Schuster, A., Tang, C., Xie, Y., Ortogero, N., Yuan, S., & Yan, W. (2016). SpermBase: A database for sperm-borne RNA contents. *Biology of Reproduction, 95*, 99.

Schwartz, M., & Vissing, J. (2002). Paternal inheritance of mitochondrial DNA. *The New England Journal of Medicine, 347*, 576–580.

Sendler, E., Johnson, G. D., Mao, S., Goodrich, R. J., Diamond, M. P., Hauser, R., & Krawetz, S. A. (2013). Stability, delivery and functions of human sperm RNAs at fertilization. *Nucleic Acids Research, 41*, 4104–4117.

Sette, C., Bevilacqua, A., Geremia, R., & Rossi, P. (1998). Involvement of phospholipase Cgamma1 in mouse egg activation induced by a truncated form of the C-kit tyrosine kinase present in spermatozoa. *The Journal of Cell Biology, 142*, 1063–1074.

Sette, C., Paronetto, M. P., Barchi, M., Bevilacqua, A., Geremia, R., & Rossi, P. (2002). Tr-kit-induced resumption of the cell cycle in mouse eggs requires activation of a Src-like kinase. *The EMBO Journal, 21*, 5386–5395.

Sharma, U., Conine, C. C., Shea, J. M., Boskovic, A., Derr, A. G., Bing, X. Y., ... Rando, O. J. (2016). Biogenesis and function of tRNA fragments during sperm maturation and fertilization in mammals. *Science (New York, N. Y.), 351*, 391–396.

Shitara, H., Hayashi, J. I., Takahama, S., Kaneda, H., & Yonekawa, H. (1998). Maternal inheritance of mouse mtDNA in interspecific hybrids: Segregation of the leaked paternal mtDNA followed by the prevention of subsequent paternal leakage. *Genetics, 148*, 851–857.

Simerly, C., Castro, C., Hartnett, C., Lin, C. C., Sukhwani, M., Orwig, K., & Schatten, G. (2016). Post-testicular sperm maturation: Centriole pairs, found in upper epididymis, are destroyed prior to sperm's release at ejaculation. *Scientific Reports, 6*, 31816.

Simerly, C., Zoran, S. S., Payne, C., Dominko, T., Sutovsky, P., Navara, C. S., ... Schatten, G. (1999). Biparental inheritance of gamma-tubulin during human fertilization: Molecular reconstitution of functional zygotic centrosomes in inseminated human oocytes and in cell-free extracts nucleated by human sperm. *Molecular Biology of the Cell, 10*, 2955–2969.

Siomi, M. C., Sato, K., Pezic, D., & Aravin, A. A. (2011). PIWI-interacting small RNAs: The vanguard of genome defence. *Nature Reviews. Molecular Cell Biology, 12*, 246–258.

Soley, J. T. (2016). A comparative overview of the sperm centriolar complex in mammals and birds: Variations on a theme. *Animal Reproduction Science, 169*, 14–23.

Song, W. H., Ballard, J. W., Yi, Y. J., & Sutovsky, P. (2014). Regulation of mitochondrial genome inheritance by autophagy and ubiquitin-proteasome system: Implications for health, fitness, and fertility. *BioMed Research International, 2014*, 981867.

Song, W. H., & Sutovsky, P. (2018). Porcine cell-free system to study mammalian sperm mitophagy. *Methods in Molecular Biology.*

Song, W. H., Yi, Y. J., Sutovsky, M., Meyers, S., & Sutovsky, P. (2016). Autophagy and ubiquitin-proteasome system contribute to sperm mitophagy after mammalian fertilization. *Proceedings of the National Academy of Sciences of the United States of America, 113*, E5261–E5270.

St John, J., Sakkas, D., Dimitriadi, K., Barnes, A., Maclin, V., Ramey, J., ... De Jonge, C. (2000). Failure of elimination of paternal mitochondrial DNA in abnormal embryos. *Lancet, 355*, 200.

Stiavnicka, M., Garcia-Alvarez, O., Ulcova-Gallova, Z., Sutovsky, P., Abril-Parreno, L., Dolejsova, M., ... Nevoral, J. (2020). H3K4me2 accompanies chromatin immaturity in human spermatozoa: An epigenetic marker for sperm quality assessment. *Systems Biology in Reproductive Medicine, 66*, 3–11.

Stilley, J. A., Woods-Marshall, R., Sutovsky, M., Sutovsky, P., & Sharpe-Timms, K. L. (2009). Reduced fecundity in female rats with surgically induced endometriosis and in their daughters: A potential role for tissue inhibitors of metalloproteinase 1. *Biology of Reproduction, 80*, 649–656.

Stival, C., Puga Molina Ldel, C., Paudel, B., Buffone, M. G., Visconti, P. E., & Krapf, D. (2016). Sperm capacitation and acrosome reaction in mammalian sperm. *Advances in Anatomy, Embryology, and Cell Biology, 220*, 93–106.

Suarez, S. S. (2016). Mammalian sperm interactions with the female reproductive tract. *Cell and Tissue Research, 363*, 185–194.

Sugita, H., Takarabe, S., Kageyama, A., Kawata, Y., & Ito, J. (2024). Molecular mechanism of oocyte activation in mammals: Past, present, and future directions. *Biomolecules, 14.*

Sun, F., Ali, N. N., Londono-Vasquez, D., Simintiras, C. A., Qiao, H., Ortega, M. S., ... Balboula, A. Z. (2024). Increased DNA damage in full-grown oocytes is correlated with diminished autophagy activation. *Nature Communications, 15*, 9463.

Sutovsky, P. (2003). Ubiquitin-dependent proteolysis in mammalian spermatogenesis, fertilization, and sperm quality control: Killing three birds with one stone. *Microscopy Research and Technique, 61*, 88–102.

Sutovsky, P. (2010). Sperm capacitation, the acrosome reaction, and fertilization. In D. T. C. A. C. M. Peterson (Ed.). *Reproductive endocrinology and infertility: Integration modern clinical and laboratory practice* (pp. 389–421)LLC, New York: Springer Science + Business Media.

Sutovsky, P. (2018). Review: Sperm-oocyte interactions and their implications for bull fertility, with emphasis on the ubiquitin-proteasome system. *Animal, 12*, s121–s132.

Sutovsky, P., Hamilton, L. E., Zigo, M., Ortiz D'Avila Assumpcao, M. E., Jones, A., Tirpak, F., ... Sutovsky, M. (2024). Biomarker-based human and animal sperm phenotyping: The good, the bad and the uglydagger. *Biology of Reproduction, 110*, 1135–1156.

Sutovsky, P., Hewitson, L., Simerly, C. R., Tengowski, M. W., Navara, C. S., Haavisto, A., & Schatten, G. (1996a). Intracytoplasmic sperm injection for Rhesus monkey fertilization results in unusual chromatin, cytoskeletal, and membrane events, but eventually leads to pronuclear development and sperm aster assembly. *Human Reproduction (Oxford, England), 11*, 1703–1712.

Sutovsky, P., Jelinkova, L., Antalikova, L., & Motlik, J. (1993). Ultrastructural cytochemistry of the nucleus and nucleolus in growing rabbit oocytes. *Biology of the Cell/ Under the Auspices of the European Cell Biology Organization, 77*, 173–180.

Sutovsky, P., Manandhar, G., McCauley, T. C., Caamano, J. N., Sutovsky, M., Thompson, W. E., & Day, B. N. (2004). Proteasomal interference prevents zona pellucida penetration and fertilization in mammals. *Biology of Reproduction, 71*, 1625–1637.

Sutovsky, P., & Manandhar, G. (2006). Mammalian spermatogenesis and sperm structure: Anatomical and compartmental analysis. In C. De Jonge, & C. Barratt (Eds.). *The sperm cell* (pp. 1–30)Cambridge UK: Cambridge University Press.

Sutovsky, P., & Manandhar, G. (2007). Comparative histology and subcellular structure of mammalian spermatogenesis and spermatozoa. In H. Schatten, & G. Constantinescu (Eds.). *Histology, cellular and molecular biology of reproductive organs* (pp. 81–98). Ames IA: Iowa State Press, A Blackwell Publishing Company.

Sutovsky, P., Manandhar, G., Wu, A., & Oko, R. (2003a). Interactions of sperm perinuclear theca with the oocyte: Implications for oocyte activation, anti-polyspermy defense, and assisted reproduction. *Microscopy Research and Technique, 61*, 362–378.

Sutovsky, P., McCauley, T. C., Sutovsky, M., & Day, B. N. (2003b). Early degradation of paternal mitochondria in domestic pig (Sus scrofa) is prevented by selective proteasomal inhibitors lactacystin and MG132. *Biology of Reproduction, 68*, 1793–1800.

Sutovsky, P., Moreno, R. D., Ramalho-Santos, J., Dominko, T., Simerly, C., & Schatten, G. (1999a). Ubiquitin tag for sperm mitochondria. *Nature, 402*, 371–372.

Sutovsky, P., Moreno, R. D., Ramalho-Santos, J., Dominko, T., Simerly, C., & Schatten, G. (2000). Ubiquitinated sperm mitochondria, selective proteolysis, and the regulation of mitochondrial inheritance in mammalian embryos. *Biology of Reproduction, 63*, 582–590.

Sutovsky, P., Navara, C. S., & Schatten, G. (1996b). Fate of the sperm mitochondria, and the incorporation, conversion, and disassembly of the sperm tail structures during bovine fertilization. *Biology of Reproduction, 55*, 1195–1205.

Sutovsky, P., Oko, R., Hewitson, L., & Schatten, G. (1997). The removal of the sperm perinuclear theca and its association with the bovine oocyte surface during fertilization. *Developmental Biology, 188*, 75–84.

Sutovsky, P., Ramalho-Santos, J., Moreno, R. D., Oko, R., Hewitson, L., & Schatten, G. (1999b). On-stage selection of single round spermatids using a vital, mitochondrion-specific fluorescent probe MitoTracker(TM) and high resolution differential interference contrast microscopy. *Human Reproduction (Oxford, England), 14*, 2301–2312.

Sutovsky, P., & Schatten, G. (1997). Depletion of glutathione during bovine oocyte maturation reversibly blocks the decondensation of the male pronucleus and pronuclear apposition during fertilization. *Biology of Reproduction, 56*, 1503–1512.

Sutovsky, P., & Schatten, G. (2000). Paternal contributions to the mammalian zygote: Fertilization after sperm-egg fusion. *International Review of Cytology, 195*, 1–65.

Sutovsky, P., Simerly, C., Hewitson, L., & Schatten, G. (1998). Assembly of nuclear pore complexes and annulate lamellae promotes normal pronuclear development in fertilized mammalian oocytes. *Journal of Cell Science, 111*(Pt 19), 2841–2854.

Sutovsky, P., & Song, W. H. (2017). Post-fertilisation sperm mitophagy: The tale of Mitochondrial Eve and Steve. *Reproduction, Fertility, and Development, 30*, 56–63.

Swann, K. (1990). A cytosolic sperm factor stimulates repetitive calcium increases and mimics fertilization in hamster eggs. *Development (Cambridge, England), 110*, 1295–1302.

Swann, K., Larman, M. G., Saunders, C. M., & Lai, F. A. (2004). The cytosolic sperm factor that triggers Ca2+ oscillations and egg activation in mammals is a novel phospholipase C: PLCzeta. *Reproduction (Cambridge, England), 127*, 431–439.

Tanaka, A., Nagayoshi, M., Takemoto, Y., Tanaka, I., Kusunoki, H., Watanabe, S., ... Yanagimachi, R. (2015). Fourteen babies born after round spermatid injection into human oocytes. *Proceedings of the National Academy of Sciences of the United States of America, 112*, 14629–14634.

Tanaka, A., Suzuki, K., Nagayoshi, M., Tanaka, A., Takemoto, Y., Watanabe, S., ... Yanagimachi, R. (2018). Ninety babies born after round spermatid injection into oocytes: Survey of their development from fertilization to 2 years of age. *Fertility and Sterility, 110*, 443–451.

Tatone, C., & Carbone, M. C. (2006). Possible involvement of integrin-mediated signalling in oocyte activation: Evidence that a cyclic RGD-containing peptide can stimulate protein kinase C and cortical granule exocytosis in mouse oocytes. *Reproductive Biology and Endocrinology: RB&E, 4*, 48.

Tavalaee, M., Kiani-Esfahani, A., & Nasr-Esfahani, M. H. (2017). Relationship between potential sperm factors involved in oocyte activation and sperm DNA fragmentation with intra-cytoplasmic sperm injection clinical outcomes. *Cell Journal, 18*, 588–596.

Terada, Y., Simerly, C. R., Hewitson, L., & Schatten, G. (2000). Sperm aster formation and pronuclear decondensation during rabbit fertilization and development of a functional assay for human sperm. *Biology of Reproduction, 62*, 557–563.

Teranishi, T., Tanaka, M., Kimoto, S., Ono, Y., Miyakoshi, K., Kono, T., & Yoshimura, Y. (2004). Rapid replacement of somatic linker histones with the oocyte-specific linker histone H1foo in nuclear transfer. *Developmental Biology, 266*, 76–86.

Thaler, C. D., & Epel, D. (2003). Nitric oxide in oocyte maturation, ovulation, fertilization, cleavage and implantation: A little dab'll do ya. *Current Pharmaceutical Design, 9*, 399–409.

Thompson, W. E., Ramalho-Santos, J., & Sutovsky, P. (2003). Ubiquitination of prohibitin in mammalian sperm mitochondria: Possible roles in the regulation of mitochondrial inheritance and sperm quality control. *Biology of Reproduction, 69*, 254–260.

Tirpak, F. (2024). Fate of the accessory sperm structures and organelles after fertilization.

Tirpak, F., Hamilton, L. E., Schnabel, R. D., & Sutovsky, P. (2024). Biomarker-based high-throughput sperm phenotyping: Andrology in the age of precision medicine and agriculture. *Animal Reproduction Science, 271*, 107636.

Tomanek, M., Kopecny, V., & Kanka, J. (1989). Genome reactivation in developing early pig embryos: An ultrastructural and autoradiographic analysis. *Anatomy and Embryology, 180*, 309–316.

Tomar, A., Gomez-Velazquez, M., Gerlini, R., Comas-Armangué, G., Makharadze, L., Kolbe, T., ... Teperino, R. (2024). Epigenetic inheritance of diet-induced and sperm-borne mitochondrial RNAs. *Nature, 630*, 720–727.

Tovich, P. R., Sutovsky, P., & Oko, R. J. (2004). Novel aspect of perinuclear theca assembly revealed by immunolocalization of non-nuclear somatic histones during bovine spermiogenesis. *Biology of Reproduction, 71*, 1182–1194.

Tran, M. H., Aul, R. B., Xu, W., van der Hoorn, F. A., & Oko, R. (2012). Involvement of classical bipartite/karyopherin nuclear import pathway components in acrosomal trafficking and assembly during bovine and murid spermiogenesis. *Biology of Reproduction, 86*, 84. https://doi.org/10.1095/biolreprod.111.096842.

Tumova, L., Zigo, M., Sutovsky, P., Sedmikova, M., & Postlerova, P. (2021). Ligands and receptors involved in the sperm-zona pellucida interactions in mammals. *Cells, 10*.

Tyebji, S., Hannan, A. J., & Tonkin, C. J. (2020). Pathogenic infection in male mice changes sperm small RNA profiles and transgenerationally alters offspring behavior. *Cell Reports, 31*, 107573.

Upham, N. S., & Landis, M. J. (2023). Genomics expands the mammalverse. *Science (New York, N. Y.), 380*, 358–359.

Vassena, R., Boué, S., González-Roca, E., Aran, B., Auer, H., Veiga, A., & Izpisua Belmonte, J. C. (2011). Waves of early transcriptional activation and pluripotency program initiation during human preimplantation development. *Development (Cambridge, England), 138*, 3699–3709.

Wang, W. H., Day, B. N., & Wu, G. M. (2003). How does polyspermy happen in mammalian oocytes? *Microscopy Research and Technique, 61*, 335–341.

Wang, H., Hong, X., & Kinsey, W. H. (2021). Sperm-oocyte signaling: The role of IZUMO1R and CD9 in PTK2B activation and actin remodeling at the sperm binding sitedagger. *Biology of Reproduction, 104*, 1292–1301.

Wang, X., Li, Z. T., Yan, Y., Lin, P., Tang, W., Hasler, D., ... Liu, M. F. (2020). LARP7-mediated U6 snRNA modification ensures splicing fidelity and spermatogenesis in mice. *Molecular Cell, 77*, 999–1013 e1016.

Wang, C., & Machaty, Z. (2013). Calcium influx in mammalian eggs. *Reproduction (Cambridge, England), 145*, R97–R105.

Wang, H., Zhou, Z., Xu, M., Li, J., Xiao, J., Xu, Z. Y., & Sha, J. (2004). A spermatogenesis-related gene expression profile in human spermatozoa and its potential clinical applications. *Journal of Molecular Medicine (Berl), 82*, 317–324.

White, K. L., Passipieri, M., Bunch, T. D., Campbell, K. D., & Pate, B. (2007). Effects of arginine-glycine-aspartic acid (RGD) containing snake venom peptides on parthenogenetic development and in vitro fertilization of bovine oocytes. *Molecular Reproduction and Development, 74*, 88–96.

Wightman, B., Ha, I., & Ruvkun, G. (1993). Posttranscriptional regulation of the heterochronic gene lin-14 by lin-4 mediates temporal pattern formation in C. elegans. *Cell, 75*, 855–862.

Wolf, F. (1928). *Die Natur as Artz und Helfer. Das Neue Naturärtzliche Hausbuch*. Berlin & Leipzig: Deutsche Verlags Stuttgart.

Wolff, J. N., Nafisinia, M., Sutovsky, P., & Ballard, J. W. (2013a). Paternal transmission of mitochondrial DNA as an integral part of mitochondrial inheritance in metapopulations of Drosophila simulans. *Heredity (Edinb), 110*, 57–62.

Wolff, J. N., Sutovsky, P., & Ballard, J. W. (2013b). Mitochondrial DNA content of mature spermatozoa and oocytes in the genetic model Drosophila. *Cell and Tissue Research, 353*, 195–200.

Wortzman-Show, G. B., Kurokawa, M., Fissore, R. A., & Evans, J. P. (2007). Calcium and sperm components in the establishment of the membrane block to polyspermy: Studies of ICSI and activation with sperm factor. *Molecular Human Reproduction, 13*, 557–565.

Wu, A. T., Sutovsky, P., Manandhar, G., Xu, W., Katayama, M., Day, B. N., ... Oko, R. (2007a). PAWP, a sperm-specific WW domain-binding protein, promotes meiotic resumption and pronuclear development during fertilization. *The Journal of Biological Chemistry, 282*, 12164–12175.

Wu, A. T., Sutovsky, P., Xu, W., van der Spoel, A. C., Platt, F. M., & Oko, R. (2007b). The postacrosomal assembly of sperm head protein, PAWP, is independent of acrosome formation and dependent on microtubular manchette transport. *Developmental Biology, 312*, 471–483.

Yan, W., Morozumi, K., Zhang, J., Ro, S., Park, C., & Yanagimachi, R. (2008). Birth of mice after intracytoplasmic injection of single purified sperm nuclei and detection of messenger RNAs and MicroRNAs in the sperm nuclei. *Biology of Reproduction, 78*, 896–902.

Yan, M., & Zhai, Q. (2016). Sperm tsRNAs and acquired metabolic disorders. *The Journal of Endocrinology, 230*, F13–F18.

Yanagida, K., Katayose, H., Hirata, S., Yazawa, H., Hayashi, S., & Sato, A. (2001). Influence of sperm immobilization on onset of Ca(2+) oscillations after ICSI. *Human Reproduction (Oxford, England), 16*, 148–152.

Yanagimachi, R. (2022). Mysteries and unsolved problems of mammalian fertilization and related topics. *Biology of Reproduction, 106*, 644–675.

Yi, Y. J., Manandhar, G., Oko, R. J., Breed, W. G., & Sutovsky, P. (2007a). Mechanism of sperm-zona pellucida penetration during mammalian fertilization: 26S proteasome as a candidate egg coat lysin. *Society of Reproduction and Fertility Supplement, 63*, 385–408.

Yi, Y. J., Manandhar, G., Sutovsky, M., Li, R., Jonakova, V., Oko, R., ... Sutovsky, P. (2007b). Ubiquitin C-terminal hydrolase-activity is involved in sperm acrosomal function and anti-polyspermy defense during porcine fertilization. *Biology of Reproduction, 77*, 780–793.

Young, C., Grasa, P., Coward, K., Davis, L. C., & Parrington, J. (2009). Phospholipase C zeta undergoes dynamic changes in its pattern of localization in sperm during capacitation and the acrosome reaction. *Fertility and Sterility, 91*, 2230–2242.

Yu, Y., Xu, W., Yi, Y. J., Sutovsky, P., & Oko, R. (2006). The extracellular protein coat of the inner acrosomal membrane is involved in zona pellucida binding and penetration during fertilization: Characterization of its most prominent polypeptide (IAM38). *Developmental Biology, 290*, 32–43.

Yuan, S., Schuster, A., Tang, C., Yu, T., Ortogero, N., Bao, J., ... Yan, W. (2016). Sperm-borne miRNAs and endo-siRNAs are important for fertilization and preimplantation embryonic development. *Development, 143*, 635–647.

Zhang, M., Bromfield, E. G., Helms, J. B., & Gadella, B. M. (2022a). The fate of porcine sperm CRISP2 from the perinuclear theca before and after in vitro fertilizationdagger. *Biology of Reproduction, 107*, 1242–1253.

Zhang, M., Chiozzi, R. Z., Skerrett-Byrne, D. A., Veenendaal, T., Klumperman, J., Heck, A. J. R., ... Bromfield, E. G. (2022b). High resolution proteomic analysis of subcellular fractionated boar spermatozoa provides comprehensive insights into perinuclear theca-residing proteins. *Frontiers in Cell and Developmental Biology, 10*, 836208.

Zhang, X., Gao, F., Fu, J., Zhang, P., Wang, Y., & Zeng, X. (2017). Systematic identification and characterization of long non-coding RNAs in mouse mature sperm. *PLoS One, 12*, e0173402.

Zhao, Y., Li, Q., Yao, C., Wang, Z., Zhou, Y., Wang, Y., ... Qiao, Z. (2006). Characterization and quantification of mRNA transcripts in ejaculated spermatozoa of fertile men by serial analysis of gene expression. *Human Reproduction (Oxford, England), 21*, 1583–1590.

Zheng, X., Li, Z., Wang, G., Wang, H., Zhou, Y., Zhao, X., ... Sun, F. (2021). Sperm epigenetic alterations contribute to inter- and transgenerational effects of paternal exposure to long-term psychological stress via evading offspring embryonic reprogramming. *Cell Discovery, 7*, 101.

Zhou, S., Dong, J., Xiong, M., Gan, S., Wen, Y., Zhang, J., ... Gui, Y. (2022). UHRF1 interacts with snRNAs and regulates alternative splicing in mouse spermatogonial stem cells. *Stem Cell Reports, 17*, 1859–1873.

Zigo, M., Kerns, K., Sen, S., Essien, C., Oko, R., Xu, D., & Sutovsky, P. (2022). Zinc is a master-regulator of sperm function associated with binding, motility, and metabolic modulation during porcine sperm capacitation. *Communications Biology, 5*, 538.

Zimmerman, S. W., Manandhar, G., Yi, Y. J., Gupta, S. K., Sutovsky, M., Odhiambo, J. F., ... Sutovsky, P. (2011). Sperm proteasomes degrade sperm receptor on the egg zona pellucida during mammalian fertilization. *PLoS One, 6*, e17256.

Zuidema, D., Jones, A., Song, W. H., Zigo, M., & Sutovsky, P. (2023). Identification of candidate mitochondrial inheritance determinants using the mammalian cell-free system. *Elife, 12*.

Zuidema, D., & Sutovsky, P. (2019). The domestic pig as a model for the study of mitochondrial inheritance. *Cell and Tissue Research, 380*, 237–262.